The Discovery of the Native Flora of Britain & Ireland

A compilation of the first records for 1670 species and aggregates, covering Great Britain, Ireland, The Channel Isles and the Isle of Man

David Pearman

'The work proceeded sideways as much as forwards, and in fits and starts, like the progress of a drunken crab. At the end of five years I was back at the beginning, knowing better how to begin, covering the same ground in different ways'.
Peter Levi The Hill of Kronos. 1981.

'The footnotes engulfed and swallowed the text. They were ugly and ungainly, but necessary, Blackadder thought, as they sprang up like the heads of the Hydra, two to solve in the place of one solved'.
A.S. Byatt. Possession. 1990.

'Rowe's Rule: the odds are five to six that the light at the end of the tunnel is the headlight of an oncoming train'.
Paul Dickson 1939 – in Washingtonian, November 1978.

Published by the Botanical Society of Britain and Ireland, Bristol.
Copyright © Text D. Pearman 2017
ISBN 978-0-901158-52-9

Designed by Paul Westley, Norwich
Printed by Henry Ling Limited, Dorchester

The Botanical Society of Britain and Ireland (BSBI), www.bsbi.org, is a learned society of professional and amateur botanists dedicated to the study of, and interest in, the British and Irish vascular plant and charophyte flora.

Front cover: *Astragalus alpinus*, Smith, J. and Sowerby, J. *English Botany*
Title page: *Carex rariflora*, Smith, J. and Sowerby, J. *English Botany*
Rear cover: *Allium sphaerocephalum*, Smith, J. and Sowerby, J. *English Botany*

Contents

Foreword

Introduction

1.	The purpose of this compilation	7
2.	Criteria used in this compilation	9
3.	Previous Catalogues of First Dates	13
4.	Developments since Clarke and Druce	15
5.	Progress in the discovery of the British and Irish flora	16
6.	Printed Sources – Books and Journals	17
	Histories of Botany and Biographies, including county floras	19
7.	Other sources – Herbaria	19
	Correspondence	?
8.	Country histories – Ireland, Scotland, Wales	19
9.	The botanists	20
10.	Conventions, abbreviations etc.	20
11.	Explanation of layout	22
12.	Acknowledgements	23

Appendices

I	A. Significant Herbals and Floras, 1538 to date	25
	B. Census Catalogues	36
	C. Plant Lists and Field Guides	38
	D. Journals and Periodicals	38
II	Histories of Botany and Biographies (including county floras)	40
III	Herbaria and Collections of Correspondence	44
IV	Progress of discovery in Ireland, Scotland and Wales	47
V	The Botanists – the discoverers of our flora	50

Vice-counties of the British Isles 81

Species Accounts 83

Bibliography 429

Index of English names 469

Foreword

Who first described our native plants? This book attempts to answer that question, starting from almost the dawn of printing, with William Turner's *Libellus* of 1538. The herbals and folk lore of the five centuries or more before Turner must be the subject of another work, and, indeed, for much of the century after that early work of Turner's, our compilers were still trying to fit our plants into those described from the Mediterranean by Dioscorides and his followers. So, printed sources are the cornerstone of this work, supplemented by information from manuscripts and herbaria, and covering about 1670 species or aggregates of all the indisputably natives and archaeophytes, but including 40 or so species that some have argued as native in the last half-century. Once one omits the critical species and those restricted to the more remote parts of Britain and Ireland, it is a notable fact that well over 80% of this flora was discovered and 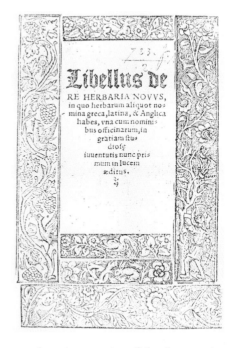 described by just after 1700. The only omissions are the microspecies of the three major genera of *Hieracium* (Hawkweeds), *Rubus* (Brambles) and *Taraxacum* (Dandelions), and an account of the discovery of those must be one for another work.

There will be users of this work who are already aware of much or all of the background, and of the manifold sources that contribute to the whole. For me, though I counted myself as a relatively well-informed field botanist, it has been a voyage of discovery, constantly becoming aware (or being made aware by others) of sources of which I was ignorant. For that reason I have not only provided very copious notes to many of the species, especially in instances where there cannot be any real certainty in the attribution to one species over another, (although it might be born in mind that many of these identifications are based on synonymies in Ray, Smith and others and usually follow those of my predecessors) but I have also supplied additional background in the form of appendices on the key herbals and floras, on the relevant journals, the important works on the history of botany, some of the national herbaria and a major section on the botanists who actually discovered the plants. Despite the caveat above there will inevitably still be differences of opinion (and, I hope, not too many errors). I'd be pleased to receive notices of any of these.

Truro, Cornwall, August 2017

Introduction

1. The purpose of this compilation

This work tries to set out how every species of plant and fern that we consider native and archaeophyte in Britain and Ireland was first recorded and discovered. This must primarily be of interest for the detailed information shown for each individual species, where, in addition to giving the describer and the date, it has frequently been possible to show extra information, including name of the discoverer and the location (assigning the latter to a vice-county), together with notes to supplement these details or to suggest alternatives. But in addition the work will show the progress of discovery, illustrating that the vast bulk of the English and Welsh flora was noticed and described by the time of the death of Ray and his contemporaries. Raven (1947) points out that Turner, Johnson and particularly Ray, laid such a firm foundation in blending science and fieldwork that most of the later findings were either in upland areas, in Scotland and Ireland particularly, or were critical species whose boundaries were elucidated in the 19th and 20th centuries. Some of the other plants first noticed after 1720 were of species whose claim to native status has often been disputed. (see sect. 5 for further details). For the date of discovery of any species can also be one of the criteria for assessing nativeness – thus, for *Fritillaria meleagris*, the fact that it was not discovered until well on into the 18th century, whereas all of its associated plants were discovered nearly 200 years earlier, is a factor in the likelihood of it being an introduction rather than part of our native flora.

For each species there are details given of when it was first recorded in print, adding, where appropriate, earlier records where there is evidence from manuscripts, herbaria, or in the published texts themselves. Thus Ray (1686) recorded *Herniaria glabra* from the Lizard and from Lankaster (1846) (and in Derham (1760: 281n)) it is possible to see that Ray actually saw it on his travels in 1667. This species had been described first by Gerard (1597), presumably from a plant in the London area. Babington (1836), however, showed that Ray's Lizard record was in fact another species, *H. ciliolata*, from the only place in the British Isles where it occurs. So Babington's becomes 'the first literature record' and Ray's 'first date of observation or evidence'.

I have tried to show, particularly for the more critical species, the often tortuous path to our current understanding, and for some of these (such as *Agrostis canina, Callitriche*) I have shown one date for the aggregate and then later dates for the current segregates. In instances where the segregate does not occur where the aggregate was described from, for instance *Erodium lebelii*, I have assumed that there is no need for a separate entry.

This is intended to be a compilation of data and their sources, and a re-appraisal of earlier interpretations. Of necessity I have given one view prominence over the others, yet I have attempted to give other interpretations and I have tried to show conflicting views, incorporating, to the best of my knowledge, all that has been published in the century since Clarke (1900) and Druce (1932), the only other authors to publish comprehensive details of first records. If specimens exist, or if the written description is such as to brook no alternative interpretation, then this is relatively easy, but if these criteria are not met then there will always be much uncertainty and room for reappraisal, as shown by recent work by Oswald & Preston (2011) and by Edgington (2013), and indeed in the new editions of Turner's and Johnson's works that have appeared since 1960 and are described in Appendix I. It is possible that I have been too deferential to Clarke, especially in instances where he was citing as his first record a member of a critical group that was not really clearly elucidated until later.

But how do we know what plant Turner, Lyte, Gerard and all the others up to at least Ray were actually describing? There are very few surviving herbarium specimens up to around 1680, and all too few of those are documented with more than a bare species name. It is true that some of the early printed works have illustrations, but these were almost invariably culled from European works and were often rather stylised woodcuts. Clarke (1900) in his introduction gives a good example of why he has accepted a pretty vague record. Taking *Anemone nemorosa*, he writes 'we find the plant pretty clearly referred to by Turner in his Herbal as a kind of *Ranunculus*; there is no evidence in his description that he knew it as a British plant, but there can be no reason to doubt it. The Herbal contains a good figure of the plant, and Turner says it grows "in woddes and shaddish places in April". This is enough'.

However, our species concepts are often much narrower than they were in the 16th and 17th centuries, and narrower too than even in the nineteenth century. It is possible that botanists might have recorded as one two or more closely allied species. By creating separate entries for many as *sensu lato* and *sensu stricto*, I hope I have minimized such possibilities. For an illuminating discussion on the pitfalls here see Oswald & Preston (2011: 116).

There is another complication in some of these early sources. Lyte and Gerard were translating European Herbals, and we do not know if their generalized locations refer to Europe or to the British Isles. Pulteney (see below, sect. 3) seems to have accepted most without question, but I have been more circumspect, largely following Clarke but occasionally accepting an earlier record, where, for instance, Turner or Lyte cite an English name. It seems unlikely that people would be referring to an imported plant in that age.

Most of the early works were obsessed with vain efforts to make the plants of the British Isles fit those described from the Mediterranean by Dioscorides (see, for instance, Oswald & Preston, 2011: 73) and others such as Pliny and Theophrastus, all of whose works were first printed in the West in the late fifteenth century. However, where there are localities cited, many of the plants described can be confirmed and in particular Ray's Cambridge Flora (1660) is an enormous help since he was dealing with a small relatively homogeneous area that has been both expertly botanised and expertly analysed since his day (see Babington 1860 & Oswald & Preston 2011: 115).

As is noted in the Foreword, when researching this work I found it necessary to consult so many works that I felt that some sort of summary would assist others who wished to read or research further, and these are set out in the five Appendices after the species accounts. Although the early history of botany in Britain and Ireland has been well-covered through the works of Raven (1947, 1950) and others, I am not aware of any recent *summary* account of all the published floras, particularly of those since 1800.

Therefore an account of all the key works from Turner's *Libellus* of 1538 up to date, together with the primary sources from journals and periodicals, is given in Appendix I. The histories of the progress of botany and important biographies appear in Appendix II, the major herbaria rich in historical specimens that have some accessible listing of their contents are in Appendix III, and the progress of discovery in Ireland, Scotland and Wales in Appendix IV. Finally brief details of all the botanists who appear in this book and the plants they discovered, together with references to further details of their lives and achievements, are set out in Appendix V.

The information contained in the species accounts that form the main part of this book is also available in spreadsheet format, for analysis purposes.

2. Criteria used in this compilation

Species and area covered

This compilation covers all the species accepted as natives or 'native or alien' in Preston *et al.* (2002), together with any additions in Stace (2010) and the very few since. It also includes the archaeophytes defined in Preston *et al.* (2004). For convenience only, I also include the few species that were treated as alien both in Preston *et al.* (2002) and subsequently, but had been treated as native in many National floras since Clapham *et al.* (1952). The first record is that of a *native population*; thus for a species that is local as a native, but widespread as a casual, such as *Marrubium vulgare*, I have attempted to provide dates for both the acceptance of the species and its first record from a native site.

Species concepts and nomenclature follow Stace (2010). Thus, as he does, all apomictic species are included, excepting those of *Hieracium*, *Rubus* and *Taraxacum*. I have included no subspecies or varieties, but it should be noted that Stace (2010) includes, as species, several taxa that were treated as subspecies in earlier floras. In some cases I have included synonymies back to earlier records, but this is an endless task, and users are directed to Smith (1824-8) for the best compilation of early synonyms including polynomials and paginated references to the works of Bauhin, and to Druce (1908 & 1928) and Clapham *et al.* (1987). Authorities are omitted, and can be found in Stace (2010).

The area covered is the Britain, that is England, Scotland and Wales, and Ireland, together with the Channel Islands and Isle of Man. For the biogeographical purists, where a Channel Islands record predates a record elsewhere in Britain and Ireland then I have tried to include both.

Earliest sources

The compilation commences with the records of William Turner (1538), for the reason that his was the first printed work by an Englishman writing with knowledge of the English countryside to mention in print more than a very few species, detailing with certainty not less than 50 species as occurring in the wild in the British Isles and possibly quite a few more. Perhaps more to the point, Turner, and his immediate successors Lyte and Gerard, were writers of a completely different genre to the earlier herbalists. The latter were concerned with 'useful' plants, whereas Turner and his successors aimed to describe all plants, whether useful or not.

The grete herball

whiche geueth parfyt knowlege and vnder
standyng of all maner of herbes & there gracyous vertues whiche god hath
ordeyned for our prosperous welfare and helth, for they hele & cure all maner
of dyseases and sekenesses that fall or mysfortune to all maner of creatoures
of god created/practysed by many expert and wyse maysters/as Auicenna &
other. &c. Also it geueth full parfyte vnderstandynge of the booke lately pryn
tyd by me(Peter treueris)named the noble experiens of the vertuous hand
warke of surgery.

Grete Herbal 1526

I have spent only a little time on researching vernacular names before Turner – and thus the awareness of a plant in the wild - but possibly such an exercise would give an earlier date of first evidence for quite a few common plants. Of course some native plants were described in medieval literature (such as in Chaucer and in the herbals), and this is especially true of trees and plants brought into gardens. In most of these sources it is not possible to separate plants in the garden from those collected in the wild, but Friar Henry Daniel (Harvey 1981, 1987) did record a large number of plants in his manuscript herbal, which probably dates from the 1370s, carefully noting in what sort of habitat, including wild habitats, they occurred, and even giving a few detailed locations (as noted in my species accounts). This is by far the most interesting and valuable source pre-Turner. Harvey (and Amherst (1895) before him) discuss a slightly earlier list, dating from c.1350, by Jon Gardener (Harvey 1985), which includes garden plants that must have been taken from the wild. In his work on medieval gardens, Harvey (1981) gives, in an appendix, more than 240 species grown in gardens in the period 800 to 1500AD, over half of which we would recognise as wild plants. It is slightly strange that between the writings of Amherst and Harvey this corpus of literature was hardly mentioned, even by Raven (1947). Other useful sources would include an excellent book on Anglo-Saxon trees (Hooke 2010), Allen (2001), also writing on Henry Daniel's records and Fitzherbert (1523), especially where he is describing cornfield weeds (see the edition by Skeat, 1882).

There are also two more recent reissues or compilations from important pre-Turner Herbals which should be noted. The first, *'Banckes's' Herbal* of c1525, has been re-issued by Larkey & Pyles (1941). This is an unillustrated work, which appears to be a compilation from various sources, and in particular from older English manuscript herbals (Rhode, 1922). Of the 207 entries roughly 60 can be identified with native plants, of which many, perhaps not surprisingly, are archaeophytes.

The second recent work on historical, and by that is meant pre-Elizabethan, English plant-names is Rydén (1984), based on the anonymous *Grete Herbal* of 1526. That work is based on a French work '*Le grant herbier*', which is in turn an expanded version of a 12th century Latin manuscript, but is further expanded and adapted for use in England and does have illustrations. Ryden lists nearly 500 English names covering c.300 species or genera (though many of the plants are imported herbalists' plants rather than natives), and analyses them between those that have their roots in Old English (up to c.1100 A.D.), in Middle English (from c.1100 to c.1500 A.D.) and those coined by translation from the French (and originally from the Latin) in the *Grete Herbal*. It makes interesting reading and he notes that c.175 are included in one way or another in Clapham *et al.* (1962). Arber (1912), which is perhaps the best introduction to the world of herbals, considers *Banckes's Herbal* the more interesting book, noting that it gives less space to the virtues of plants and, on the whole, more botanical information. Other compilations of English names, such as Britten & Holland (1886) and Grigson (1955), are, I think, almost entirely concerned with names in use post-Turner, though do contain the occasional earlier reference and the latter would be worth consulting for his excellent introduction and bibliography. But Grigson's later work (1974) contains at least 130 pre-Turner vernacular names, including, not surprisingly, almost all our native trees. And, of course, it should not be forgotten that Turner, Lyte and Gerard coined many vernacular names.

Mention too might be made of Earle (1880), who gives ten lists of vernacular names from the tenth to the fifteenth centuries, and indexes of Latin species names together with those in Saxon and English, with an excellent introduction. A final source must be Allen & Hatfield (2004), which chronicles the traditional uses of plant remedies, and thus their names, in Britain and Ireland.

In truth the whole question of the knowledge of early English vernacular names (that is pre-Turner, or, at the latest, Gerard) is, it seems to me, an under-explored topic, and the systematic interpreting and assigning of these might well be the subject of further investigation. It might be noted that a reviewer of Rydén (1984), though expressing reservations on his approach, described his work as helping to supply 'a shocking gap in English-language scholarship'. For all these reasons English and other vernacular names are also omitted in this compilation.

But it was Turner who, noting the chaotic state of nomenclature that existed in his time, commenced attempts to classify our flora and recorded, within that framework, the majority of our native plants known in his day. The neatness of having a substantial baseline outweighs any other starting date, and though I could have added many earlier references from the species accounts in Grigson (1955) alone this seems outside the bounds of this project. Thus Grigson's is a marvellous compendium of plant histories and uses, right back to Anglo-Saxon times and is quite the best in its field. It is well complimented by Richard Mabey's more recent *Flora Britannica* (1996). From the time of Turner onwards his successors often, but by no means always, made reference to plants described by their predecessors. Johnson, Ray, Hudson, Smith and many others all cited earlier works, and where post-Linnean Floras cite pre-Linnean polynomials this is especially valuable.

Dates of first record

The '**first literature date**' is principally that of the first published literature reference from Britain and Ireland, or, if in a foreign publication, only if specifically referring to Britain and Ireland.

For many species, our current taxonomic concept has evolved over a long period, and for some species this is still an ongoing process. For these, largely in the same spirit as Clarke (1900, 1909) and, more cautiously, Druce (1932), the first literature reference given has been that which seems to refer relatively unambiguously to what we now accept. There may well be inconsistencies here, and later clarifications are shown where relevant. Thus Withering (1804) seems to have set out quite satisfactorily a split of *Agrostis canina s.l.* into what we now call *A. canina s.s.* and *A. vineale*, even though our current species concept did not finally evolve until shown in Clapham *et al.* (1987), whereas the annual species of *Salicornia*, despite descriptions and some clarifications in Wood (1851), were a muddle until Ball & Tutin set out parameters in 1959. Some might well say that they are still a muddle. Other problems include those where there is a species which was later split into two, and there I have tried to provide dates for both.

I have retained Clarke's convention of occasionally citing further, later, records. As he writes in his Introduction 'It must not, however, be assumed that in all cases where more than one record is extracted the first is considered a doubtful one, as sometimes the object is merely to illustrate the gradual knowledge of the plant, or to give a localised record'.

For many species (c740, or 45% of the whole) it has been possible to find a record earlier than the date of the first published literature reference, the '**first date of evidence or observation**', or '**first record from the wild**'. This might be a reference contained within the actual literature record (*Alchemilla alpina* - 'We found it this year [1671] on a mountain in Westmerland' from Ray (1677: 11)), or from a manuscript source that was only published later (such as Ray's Journals or Gosselin's manuscript (*Carex punctata* - 'Guernsey. Gosselin, J., c.1790. See McClintock (1982: 170)) or a herbarium specimen (including such as the glosses by Thomas Penny on Gesner's herbarium sheets (*Hypochaeris glabra* - 'in Anglia' - annotations made by T. Penny on C. Gesner's plant illustrations dated as by 1565 in Foley (2006b)). I have not conducted any original manuscript work myself, other than to re-locate and work through the Pulteney ms. (sect 3, below).

3. Previous Catalogues of First Dates

William Clarke was the first to achieve (if not the first to attempt) a listing of the first records of British and Irish plants. His *First records of British plants* was published in parts in the *Journal of Botany*, commencing in 1892, and then printed in book form in 1897. The second (and final) edition of this was issued in 1900, with greatly extended citations and corrections and it is this that has been used in the present work. A short supplement appeared in Clarke (1909). The 1900 edition was reprinted, with the supplement, by Trollius Publications in 2004.

The basis of Clarke's work is the first published record, the first literature date, which may include an earlier date of discovery. This first record is frequently supplemented by a later date, as described above. I have retained almost all of these later records. He covered only native species, with only a handful of exceptions, and excluded ferns and their allies, and also any records from the Channel Islands.

Clarke's work is first-rate and thorough, and by and large, though I have been happy to follow him in most instances (probably still too many), I have erred on the side of caution, especially in critical taxa, and where I still have doubts I have tried to add later, more certain records. Where I have not improved on his first date, I have prefaced the citation with an asterisk (*) to acknowledge his work and defend myself, as there must be an element of conjecture in many of these early attributions, although it must be born in mind that Clarke was usually following the synonyms given in the works of Ray, Smith and others. But I have corrected his citations where necessary, without comment, and I have expanded the citations from the original source where he, for reasons of space, only quoted the first of several locations, and omitted other descriptive text. His Introduction and his Summary (an afterword, as it were) are well worth reading, although I have attempted to distil much of that into this introduction. He gives no hint of how he assembled his information, though we know that he had available to him not only the fine libraries in Oxford to where he retired in 1892, but some later editions of the works of the early botanists. We also know that he was a fine classical scholar (Britten 1911).

Druce gave much abbreviated details of first dates in his *Comital Flora* (1932), covering more species and often adding earlier dates from herbaria, using the rich resources of the Oxford herbarium. This was produced right at the end of his life, and whilst it is definitely valuable, particularly in extending Clarke's work to include the ferns and their allies, as well as records from the Channel Islands, it is marred by many errors. But it does complement Clarke's work, in taking as its criterion of (almost always) the first recorded date of the presence in the wild of the species in question. His references for the ferns, are, it must be admitted, woeful, with many omissions of early works and frequent citation errors. Like Clarke, Druce primarily covered native species, including only a few records of alien plants.

His citings are much briefer than Clarke's, if only because 'first records' are only part his much larger project. Druce acknowledges Clarke's book as an 'excellent little work', but he is much more optimistic, or even rash, in his attributions. Where his citings are still relevant I have expanded them where at all possible, but have included no distinguishing acknowledgement that Druce was the intermediary source, other than to point out some of the errors that researchers might perpetuate by using Druce.

There are a few queries in using his work that I have not been able to resolve, among these an occasional citing of an earlier date <u>after</u> his first date. Some of these relate to herbarium records, and I assume others are where the second citing is a tentative identification.

Since Druce there has been no national compilation, although Marren (1999) includes an extensive appendix on the first records of rare plants. Indeed, as long ago as 1947 Canon Raven was saying that it was high time that Clarke's work was revised and brought up to date (Raven 1947). However for the ferns and their allies, omitted in Clarke and so poorly dealt with in Druce, Edgington (2013) has provided a marvellous account of impeccable scholarship that I have been happy to follow, with only a couple of minor exceptions.

However, I have discovered in the Natural History Museum library a volume of two manuscripts bound together that have been attributed to Richard Pulteney (1730 – 1801), and, indeed, in his handwriting, which must date from around 1790.

The first, with a front sheet entitled '*Flora Anglica abbreviata M.S. (Dr Pulteney)*' in a different hand to the main text, consists of 164 pages, written one-side only, giving against the names of each species literature references and citations, covering the whole period from Turner to Hudson. It has been added to (also in Pulteney's hand-writing) as though it was a working document. It is based on the second edition of Hudson's *Flora Anglica* (1778), and the references cover the entire period from Turner up to Hudson. These are very valuable, since all other authors who give other authorities only go back as far as Johnson (1633) at least as far as the English sources are concerned. Unfortunately there are major gaps in the manuscript, which must date to before the copy arriving in the museum, since the pencil pagination runs consecutively.

The second has a title page '*Catalogue of English Plants with the names of first describers, or discoverers annexed. M.S. (Dr Pulteney)*', similarly in a different hand to the text. It appears to be a clean copy of the other, with just one source for each species, but it is complete, covering all the species in Hudson. This covers 76 sides (on 38 double pages), but, unlike the first includes lower plants.

Pulteney himself (1790: xii) refers to his plans for a '*Pinax*, in which it was my design to have distinguished, as far as I was able, the first discover of each species …', and later gives his reason for omitting that research from his book. This manuscript must have been the working document for that projected project.

This manuscript has been curiously ignored by those who have written about the progress of botany in Britain and Ireland. Green, Druce and Raven do not mention it; Desmond (1994) and Simpson (1960) list its existence, but only Clarke has even referred to it, but then only a short paragraph in the first instalment (1892) of his initial notes in the *Journal of Botany*. He notes 'that in many cases the references [Pulteney's] are not to actual first records and in other respects the information afforded is not always satisfactory'. I cannot agree.

I have been through the second manuscript, and compared every entry in it with my master. Pulteney, who was living at Blandford Forum in Dorset, was handicapped by not having access to other libraries. He was evidently not aware of Turner's *Libellus*, and although he knows of his *Names* had evidently not seen a copy. Similarly whilst he mentions Johnson's accounts of his trips to Kent and Hampstead of 1629 and 1632, he states that he has not been able to see a copy of them. In fact, of Johnson's works, he only had copies of the revision of Gerard and the first part of *Mercurius* (for full details see Kew & Powell (1932: 138 *et seq.*)). He gives, to my way of thinking, too much weight to the entries in L'Obel's works when there is little evidence that he (L'Obel) was writing from knowledge of occurrence in England. Furthermore he frequently cites Lyte's *Herbal* (1578), and here too is a quandary; not a few of the records there seem perfectly acceptable as first records, but when Lyte claims the plant 'for this country', is this really true, or is he just translating the Dutch of Dodoens? Raven (1947: 201) commented that many, but not all, did refer to Flanders. Those points aside, Pulteney's work is an extremely comprehensive summary, and he demonstrates the use of the information in his catalogue of the Dorset plants (1799). The manuscript certainly deserves better than Clarke's curt dismissal; it predates Clarke by a century, and definitely forms a part in the history of the discovery of our flora.

4. Developments since Clarke and Druce

These two works, with only minor caveats, do seem to have incorporated facts from all the published sources that were available to them. Those sources are covered in sects. 6 & 7 below, together with their appendices. But the last century has seen a wealth of new material, which can be summarised as follows:

- **New editions** of original texts, principally those of Turner, Johnson & Ray. This is occasionally a mixed benefit, as later editors have sometimes re-interpreted earlier attributions in an area of scholarship where there can be no certainty. I have tried to show all the differences. Special mention should be made of Oswald & Preston's (2011) edition of Ray's 1660 *Catalogue*, containing also a mass of background information on Ray's plants, his contemporaries and sources, which I have found easily the most reliable and interesting.
- **Conferences on the discovery of our Flora**. There have been at least two organised by the BSBI, one published with the talks in full (Noltie 1986) and a later 2010 meeting in Birmingham where the report is only a short summary of the papers delivered (Allen 2011).
- **Biographies and Histories of Botany**, the latter often containing much biographical detail. Here the works of John Raven (1947, 1950) are pre-eminent,

but Gunther (1922 and other works) was also a major contributor. David Allen and John Edgington too, in a range of books and journal articles, have added much background. There have been very many booklets and papers on individual botanists, particularly from Scotland.

- **Manuscripts**. These include McClintock's editing of Gosselin's Guernsey discoveries, but works by Foley, Henderson & Dickson, Horsman and others are all important sources. The compilation by Bridson *et al.* (1980) is of immense importance in identifying the existence of manuscript and their location.
- **Journals**. In addition to national and local journals, specialist publications such as *Archives of Natural History* and *Garden History* have been launched, with the latter containing much research from the garden historian John Harvey.
- **Herbaria.** Post-Clarke, Druce published works on the Oxford Herbaria (Druce & Vines 1907, Vines & Druce 1914, 1919). Since then Dandy (1958a) and Clokie (1964) have summarised the resources in the Sloane and Oxford Herbariums respectively, and perhaps these herbarium holdings are the biggest resource left untapped, with the historic holdings in the Natural History Museum the priority.
- **Databases.** I have searched the BSBI's Distribution Database (DDb) which includes all the plant records collected by the Biological Records Centre (BRC) and incorporated any records additional to my researches that I can verify.
- Many of the older works are now available **'on-line'**, either to view or download as pdfs, and some as print-on-demand **facsimiles.** I have tried to indicate any of these in Appendix I, but this is a fast-moving and seemingly ever-changing field.

5. The progress in the discovery of the British and Irish flora.

Clarke (1900) included a 'summary', at the end of his work, in which he traced the discovery of the flora, author by author. His total, by Ray's death, amounted to about 970 species. In the current project the picture is broadly similar, in that of the species and aggregates covered in the book 64% (1067 out of 1671) had been described (first literature date) by 1724.

But it is possible to identify a number of factors that might influence discovery and thus to refine these crude figures:

- Use – that is whether the plant had any medicinal use (real or imagined). That is the approach used by Bebber *et al.* (2007). Horticultural use too might be a small factor here.
- Remoteness. Apart from Johnson's trip to N. Wales, Heaton's stay in Ireland, the excursions of Willisel and Ray and Lhwyd's travels, there was very little exploration away from lowland areas before 1724, and almost none in Scotland.
- Rarity. In theory it should be more difficult to discover a rare plant. Therefore with increasing number of botanists and increasing travel, more rarities should be described.

- Critical species. These are defined here as those that have subtle differences which have only been elucidated gradually by later botanists; this too would include most of the aopomictic species. Of course the definition of a critical species is relatively subjective, but most of those so categorised are included in that key to difficult taxa, the *Plant Crib* (Rich & Jermy, 1998). The overwhelming majority were described after 1724 (322 out of 340).
- Status. It is interesting – and, of course, one of the possible pointers to alien status – just how many of the species whose status is argued about, were in fact found late (26 out of 38 so marked, and there are probably another dozen who might fall into this category).

Amending the totals by excluding the relevant categories gives the following:

Total number of species covered (excluding the microspecies of the three major critical groups)	1671	
Other apomicts and critical species	(334)	
Non-English species	(96)	
Probable alien species	(44)	
Net total	1197	
Of this net total, described by 1724	1037	87%
By 1800	1105	92%

I accept that there are degrees of subjectivity about this, but I would contend that anything contentious is so minor as to not significantly alter the overall picture.

6. Printed Sources

See Appendices I and II. Appendix I is arranged in four parts:

Herbals and Floras

The appendix contains details of all the significant works up to 1724, since by that date the vast majority of the non-critical native lowland flora had been discovered (see sect. 5 of the Introduction). From 1724 I have concentrated almost entirely on systematic floras, since it is through an examination of those that one is able to trace the developments of today's species concepts.

This section is not intended to be a complete summary of the botanical or taxonomic literature in the last five centuries, but more a selection of the **key** works that made a significant difference to the history of the knowledge of our flora, either by the records they included, or the advances they made to differentiating between one plant and another, whether by what we would now call taxonomic approaches or not. Thus I have included notes on Lyte, How and Merrett, however unsatisfactory are their works in one way or another, but little or nothing on, for instance, Bobart, Morison, Plukenet, Plot or even Petiver, Martyn and Blackstone. For these reference to Pulteney (1790, available online or as print on demand) would be the easiest, though notes on many are scattered elsewhere, perhaps best in Trimen & Dyer (1869) and Raven (1950). Similarly, some of the nineteenth century national floras, such as those by Baxter, Woodville and many others, are only of very minor interest to my thread.

Census catalogues

Because these have encouraged a much more comprehensive collection of records for each county, this has enabled early records to be better traced and catalogued.

Plant Lists and field guides

Those listed have been useful in tracing the progression of the acceptance of new taxa.

Journals and Periodicals

The rise of journals and the frequency of new species being first described in those was, largely, a post 1800 phenomenon. To put this into perspective, of the 180-odd non-critical English species described since then, almost half have first appeared in journals rather than floras. Of course many other works are referred to in this introduction and throughout the section on the individual species.

Prior to about 1790 only a few articles in the *Philosophical Transactions of the Royal Society* (which commenced in 1665) had included any significant new plant records. But from then (the commencement of the first from the Linnean Society) journals have been a very rich source of plant records. Indeed Allen (1986) considers that the run of comments on specimens circulated through botanical exchange clubs, circulated to their members, to have been the 'single most fruitful source of new information on the taxonomy of British vascular plants to appear on a regular basis'.

The major point of entry to this resource would often be Simpson (1960), who summarises many of the journals, lists the principal references to families, genera and species, and gives an extremely comprehensive (and, it must be said, sometimes relatively opaque) listing of sources, vice-county by vice-county.

This extraordinary proliferation of journals from around 1790, many short-lived, is ably summarised in Allen's paper (1996), from which many of the details given in this Appendix have been taken. Cope (2009) is very useful for the details on the Botanical Exchange Club (BEC) and its companions at the end of the nineteenth century. Particularly in the nineteenth century reports of many societies are often listed in others. Thus the Linnean Society *Transactions* have many notes from the Edinburgh journals, the *Journal of Botany* from the BEC and Watsonian BEC and *vice versa*.

Often the greatest problem for many botanists away from the libraries and museums in London, Edinburgh, Cardiff, Oxford and Cambridge, is locating these journals. Many of the well-known discoveries were already highlighted in Clarke and Druce, at least up to 1930, but not only are there frequently earlier hints in the literature, but also many of the journals listed below were, until recently, difficult to access. The Internet has profoundly changed this, with many, at least of the major sources, now available to all either as free access or with password.

Wherever possible I have tried to go back to the originating source, but I am sure there are some inconsistencies here. Another source of confusion might be that a paper was read in say, 1834, but the issue not published till 1835. I have tried to use the latter date as the first literature record and the former date as the earlier actual record, where relevant.

Histories of Botany & Biographies

These are summarized in Appendix II.

This is probably the area in which the most progress has been made in the century since Clarke. The collating and interpretation of the Herbaria at Oxford by Druce was largely published after Clarke's work, though of course some of the results were incorporated in Druce's *Comital Flora* (1932) albeit with maddening imprecision. Gunther's great work, *Early British Botanists*, was published in 1922, but curiously little used by Druce, and much more inexplicably, relatively sparingly referred to by Raven (1947), and even then often with only faint praise. And of course, Raven's magisterial works have helped enormously in the preparation of this compilation.

There is another point that seems worth making. Other than Green's (1904) work, below, which takes a somewhat different approach to the other authors cited here, almost all the historical and biographical work available is on botanists active up to and just beyond 1724, the year of the last edition of Ray's *Synopsis*. Pulteney (1790), as others have done, especially Allen (1993), points out the longeur on British botanical investigation after 1724, but then devotes very little space to Hudson and Withering, let alone many of the minor players of the eighteenth century. Perhaps an exception should be made for works on Scottish botanists (see Appendix IV), and for those botanists active in a particular county, where local floras often do include a worthwhile account of the discovery of their flora. However, by and large, the history of the botanical discoveries and taxonomic advances in British and Ireland from 1760 to date, and particularly for the twentieth century, remains to be written. Appendix I is an attempt to at least provide an outline.

7. Other sources

These are summarised in Appendix III, and comprise notes on Herbaria and published Correspondence. For Herbaria relatively little has been published and even less catalogued and made available, though several institutions are now digitising their collections and placing them online.

8. Developments in Ireland, Scotland and Wales

All the early published works, say pre Hudson's *Flora Anglica* (1762), consist almost entirely of records from England, with only a few from Wales. Some account then of early discoveries and botanical history for Ireland, Scotland and Wales might be of value, and this is included in Appendix IV.

9. The botanists – the discoverers of the flora

Appendix V lists around 500 individuals mentioned as the discoverers of species covered in the main body of this work. This gives their dates of birth and death (where known) and details of further sources for information on each, and includes for most, the plants they discovered.

Much is known about the personal travels and sources of the early authors (Turner, Gerard, Johnson and of course Ray), either from their own works or through the investigations of Raven (1947 & 1950), especially, but by many others too. Yet looking through my draft I felt that there is still more to do on some of the sources of authors such as How and Merrett, on clarification of some of Dillenius' sources for his revision of Ray's *Synopsis*, and even for many of Hudson's additions.

10. Conventions used in this compilation

Authors.

All are cited as on the title page, with date of publication, with the exceptions of:

Britten, Jackson & Stearn (1965) – cited as Stearn (1965).

Clarke (1900) – cited as Clarke (any references to his other works are given in full)

Curtis (1777-1798). Stevenson (1961) sets out the probable dating of the fascicules, which had been a source of confusion throughout the 150 years since publication. I have used his dates.

Druce (1932) – cited as Druce (any references to his other works are given in full).

L'Ecluse, not as Clusius.

Pena & L'Obel (all works) - all cited as L'Obel (and not Lobel)

Pulteney manuscripts from NHM – the manuscript is cited as just 'Pulteney' (all references to his other works are given in full).

Smith & Sowerby, *English Botany* and supplements – cited as E.B. & E.B.S. Note that each plate was dated with the month of issue, and these are the dates used here.

Presentation of the citations

I have tried to check every citation cited. Where that is in Latin, I have not, by and large, provided a translation, simply because that would have been a huge job, and one that is probably beyond my capabilities. I appreciate that this might well be a drawback for many users, and anyone able to recall Latin from their schooldays should remember that the mediaeval Latin is often very different from 'classical' Latin. Nevertheless, it is usually possible to follow the gist, and I would be happy to help in case of any difficulty.

I have been inconsistent in not including many of the accents, and in italicising names of people and places that might have been so presented in the original. I have been inconsistent too in expanding the 'tildes' in the original; that is the diacritical marks indicating an elided letter.

The use of the letters 'u' instead of 'v' has been modernised, as has 'i' for 'j' other than in citations of pagination, where the originals are so marked.

All quotations are indicated by '', but herbarium specimens are not enclosed in quote marks, unless they are from a cited publication.

Other comments:

[…] denote my explanations, such as modern names of localities.

Herbaria are shown in Bold.

Species names are only italicised for post-Linnean names. Pre-Linnean polynomials are not italicised, but the authorities cited in these, are, as in the original texts. Many of the authors up to and including Ray also italicised locations and discoverers. I have not followed this.

For polynomials, by and large, following Clarke and Druce, only the first of those cited is included.

Vice-county numbers have been added for all records with locations; this has done away with any need to note the county names in the citations, since the county concerned will be obvious from the number. A list and a map of these shown on p81.

Dating of undated records

In those instances I have used the date of death of the recorder. It would have been more accurate to have expressed these as 'ante', but this would make it impracticable to sort electronically. Others have used the period of assumed activity, or guessed. That problem is mainly caused by Druce - see notes below in Appendix 3 on Druce (1932) and against the various species.

Pagination in Turner and Johnson.

For Turner (1538, 1548 & 1551) I have adopted A, Ai etc., for the right-hand page, and A, Ai. verso for the left-hand page. For Turner (1562 & 1568) the pages are consecutively numbered.

Johnson (1629 & 1632) (his journeys into Kent and Hampstead). The former is numbered A, A1 etc., from the start of the Itinerary (and in Ralph's reprint), but Clarke and others have cited these by numerical sequence. The latter is numbered from the very start in Ralph and from the commencement of the Itinerary in Gilmour, and later compilers have chosen one or the other.

For both works I have commenced the pagination from the start of the actual Itinerary, using a numerical sequence, as in Gilmour (1972), whereas Druce numbered from the start of the preface. For the 1932 work Gilmour has the original page numbers on the facsimile.

n.b. Johnson's journey is set out like a diary. I have cited his text in quote marks, but added the locations, outside the quotes, derived from a reading of his text.

Herbaria

(this list includes only those cited in the text)

BATHG	Geology Museum, Bath	K	Kew
BFT	Belfast	LINN	Linnean Society
BIRM	Univ. of Birmingham	LIV	Liverpool
BM	NHM London	MANCH	Manchester
CGE	Cambridge	NMW	National Museum of Wales
CYN	Croydon N.H. Society	OXF	Oxford
DBN	Dublin, National Botanic Gdn	RNG	Reading
E	Edinburgh	UCSA	Swansea Univ College
G	Geneva	WAR	Warwick
GLAM	Glasgow Museum	YRK	Yorkshire Museum
HAMU	Newcastle Hancock Museum		

11. Explanation of layout

1st paragraph is the first literature reference

2nd paragraph is earlier evidence, if available

3rd paragraph contains any notes

Vice-county of the record and the date of publication or other evidence

Aconitum napellus

* 'Found by Rev. Edw. Whitehead [in 1819] in a truly wild state on the bank of a brook, and on the river Teme in Herefordshire'. - Purton (1821: 47n.).

VC36   **1821**

'by the side of the brook a few yards above Gossart Bridge between Ludlow and Burford, in abundance, 1799'. - Rev. E. Williams, m/s Flora of Shropshire (in Shropshire County archives).

VC40   1799

A very doubtful native even in Herefordshire and Shropshire. There are earlier records as a certain alien; for instance Druce cites 'Coln, Gloster; Iver, Bucks, Lightfoot, prior to 1786. *Sibthorp ms.*'.

12. Acknowledgements

I owe a principal debt to Chris Preston, not only for constant advice, comment and encouragement throughout the thirteen years that this project has taken, especially during his research, with Philip Oswald, into John Ray's *Cambridge Catalogue*, but also for reading through the entire final draft and saving me from very many pitfalls. Those that are left are entirely my responsibility.

The other major helpers and advisors, in alphabetical order, have been; David Allen, for initiating the project, for letting me have a copy of his collected notes and amendments, and for much comment and advice throughout the last 13 years; John Edgington, for access to all his notes and investigations, particularly into the (to me at least) baffling history of the discovery of our ferns, but also for his papers on discoveries in the London area and other topics; Philip Oswald, for sharing his thoughts and in particular for his translations of the medieval Latin; and Keith Spurgin, particularly at the start of the project for his introducing me to the mysteries of texts available online during his parallel work on rare English plants, but also for many illuminating discussions thereafter.

At the Natural History Museum (BM) both Fred Rumsey and Mark Spencer have been of great assistance in tracing specimens and with other comments and for arranging for access to the Sloane Herbarium; and Andrea Hart, the Librarian there, has produced books and the Pulteney manuscripts. At the Linnean Society Gina Douglas, Lynda Brooks, Elaine Charwat and Ben Sherwood have helped both in tracing references and letting me borrow texts. Elsewhere, Tom Cope (K) has helped both with advice and letting me see specimens. Donna Young and John Edmondson (LIV) have also found specimens for me, as has Christine Bartram (CGE) and Serena Marner (OXF). Clive Lovatt has given me the benefit of his comprehensive knowledge of the Avon Gorge and its environs and Mike Foley has amplified on his valuable work on early botanists. Peter Marren's work on *Britain's Rare Flowers* was an inspiration and he has expanded on this with further explanations when necessary. Henry Noltie offered me further help on his important publications on historical research; he also made very helpful comments on earlier drafts of this introduction. The wonderful website on Cambridge botany compiled by Gigi Crompton has been a joy to consult.

Other helpers have included: Ken Adams, Ian Bennallick, Michael Braithwaite, Arthur Chater, Clive Chatters, Graham Coles, John Crossley, Helena Crouch, David Earl, Gwynn Ellis, Stephen Evans, Holga Funk, Alistair Godfrey, Eric Greenwood, Paul Hackney, Geoffrey Halliday, Angus Hannah, Frank Horsman, George Hutchinson, Matthew Jebb, Colin Kelleher, Alan Leslie, Wendy McCarthy, Jim McIntosh, Charles Nelson, Mike Porter, Tim Rich, John Richards, James Robertson, Martin Sanford, Walter Scott, John Simpson (Accrington Libraries), Paul Smith, Phil Smith, Roger Smith, Clive Stace, Roy Vickery, Kevin Walker, Dan Watson & Geoffrey Wilmore.

I am sure that there have been others and I apologise for any omissions.

I would like to thank Paul Westley for the design and for constant help and suggestions, and Gwynn Ellis for constructing an English index at very short notice.

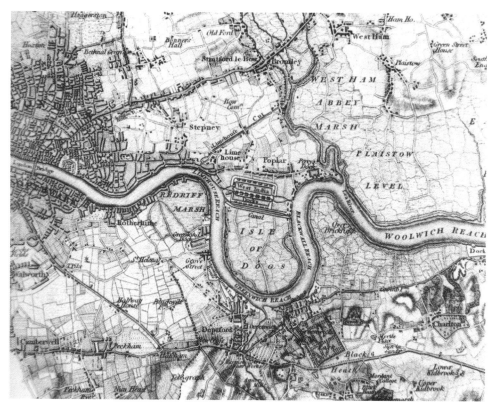

A view of the R. Thames in London, covering the stretch from London Bridge to Greenwich and beyond, and showing the sites of Rotherhithe and Redriff Marsh, mentioned in Gerard (1597) as the locations for *Berula erecta*, *Dianthus deltoides*, *Oenanthe aquatica s.l.* and other species. The map is from 'An Entirely New & Accurate Survey Of The County Of Kent, With Part Of The County Of Essex', by William Mudge, 1801.

Appendices

Appendix I

A. Herbals and Floras

Turner (1538). *Libellus de re herbaria novus.* Latin. Alphabetical.

Clarke was using the edition annotated by Jackson (1877). Another translation was edited by Stearn (1965), which closely followed Jackson's attributions, and a new translation (Rydén *et al.*) was published in 1999.

Turner (1548). *The Names of Herbes.* English. Alphabetical, but with a second list at the end under the heading 'newe founde Herbes whereof is no mention in any olde aunceint wryter'.

Neither of these two early works had illustrations and the identification of some of the plants with our native flora must be tenuous. Furthermore Turner had spent years in Europe between 1540 and 1547, and whilst many of the references in the *Names* are specifically from there, often too his descriptions of habitats might apply just as easily to Europe as to England. Turner coined or introduced many English vernacular names, and I wonder whether a distinction might be drawn between those where he was evidently coining an English name for something known by a German or a Dutch name or whatever and with those he reported (as 'anglis vocatur', 'called by our countrymen' or similar), which could well imply nativeness, assuming they were not from plants cultivated in gardens. In the latter category are many for which Clarke gives only a later first date, such as Alexanders (*Smyrnium olusatrum*), Hasyle tree (*Corylus avellana*), Nygteshade (*Solanum nigrum*), Service tree (*Sorbus torminalis*) and quite a few (at least fifteen) more. I have included comments for many of these, often citing Nelson (1959), but though the conundrum might be solvable by looking back to earlier works on husbandry, the result matters little in the overall scheme of things. All the editions cited above contain tables of Turner's scientific and vernacular names and their suggested modern scientific names, and the Stearn and Ryden editions have been checked with Clarke both for their synonymy and for any suggested changes. For notes on the citing of the pagination, see sect. 10 of the Introduction.

Turner came from Northumberland and worked for the Duke of Somerset at Syon Park, Middlesex, and several of his localised records are from those two areas.

Clarke was using the edition annotated by Britten (1887). Subsequently Stearn (1965) updated the nomenclature with the help of J.E. Dandy, making a very few changes.

Turner (1551, 1562, 1568). *Herbal.* English.

His major botanical work, published in three parts. The third part (1568) includes a new edition of the first part and a reissue of the second (without the dedication).

Both Raven (1947) and Oswald & Preston (2011) have pointed out that his works were not well known to later botanists; in fact Raven concludes that Johnson only knew the first part, Parkinson, none, and even Ray apparently only the first part. Goodyer had the complete issue of 1568 (Gunther 1922: 228). Clarke (in his 1892 paper) notes that he was using the 1568 edition, with his citations suitably annotated if they had appeared in the original part 1.

Nelson (1959) (cited in Rydén *et al.* (1999), but not in Stearn (1965)) covered all the species in Turner's works that appeared to be the first British record. These have been checked against other interpretations; on balance he was less conservative than Clarke, including many instances, particularly from the *Libellus*, that give no indication that they are actually in the wild in Britain.

There have been recent editions of the 1551 part by Chapman & Tweddle (1989, reissued 1995) and Parts 2 and 3 were covered in Chapman *et al.* (1995), but neither volume covers the amendments to Part 1 that were incorporated in Turner's 1568 re-issue. The 1989 edition of the 1551 part had no indexes, and thus is more difficult to use. None of the parts of *The Herbal* appear to be available online.

Turner, W. 1568. *The first and seconde partes of Vuillium Turner's Herball* (the re-issue)

L'Obel (1571, 1576, 1581, 1605). *Herbals.* Latin. Systematic.

Although of Flemish extraction, L'Obel made such a contribution to British botany that he should be included in this summary. He lived in England from late 1567 to 1571, with his colleague Pierre Pena, and then on his own or with his family for much of the period from 1590 to his death in 1616. He was a careful and critical observer, and corrected many of the errors in previous herbals. Turner had made no attempt at a systematic arrangement, but L'Obel at least tried. He travelled fairly widely in the south of England, with his works containing many references to British plants, with nearly 90 first records. Louis, in his exhaustive analysis (1980) of L'Obel's life, sources, influences and travels, analyses his locatable records: from his 1571 work over half of those (48) are from London and its environs, but also 15 from Bristol and area where he lived for a while before 1571.

I have altered the date of his first work to 1571 with reluctance. Although the title page has 1570, the colophon (the inscription at the end of the book) has 1571, and both Louis (1980) and Oswald & Preston (2011) follow this.

The contribution of his colleague Pena has rather been over-shadowed by history, but a useful counterview is expressed in a book review by Jackson (1899). There he notes that it is possible that Pena contributed most of the text for the 1571 work, but nothing thereafter. However Louis (1980) disputes much of this, with some vehemence.

Ray (1686) commented about him 'in his descriptions inexact, and in localities often mistaken through trusting to his memory', but this is too harsh. In a period when English botany was not particularly strong he made a great contribution.

At his death he was working on a much larger work, the *Illustrationes Plantarum*, but this was unfinished. See also the entry under L'Obel (1655) below.

Available online.

Lyte (1578). *Herbal.* English, Systematic.

A translation of L'Ecluse's translation of Dodoen's Herbal. He gives localities, such as 'in this coutrie', but it is not at all clear whether he is referring to England, or merely translating Dodoen's references to Flanders. Raven (1947) assumed most related to Flanders, but Pulteney adds most of these entries to his list of first dates. I have temporised, including most of these as a note against the relevant species.

Available online.

Gerard (1597). *Herbal.* English. Systematic.

Much has been written in criticism of both Gerard's botany and his plagiarism. Raven (1947: 208) even starts off his chapter on him 'Gerard was a rogue'. Jeffers (1967) offers a very different view, providing reasons for the incorporation of the works of others, and also goes some way towards reconstructing his extensive journeys throughout England, and describing his contacts and friends. He advances reasons for suggesting that a draft of at least Gerard's Book 1 (pps 1-176) was in existence by 1574. In that Book 1 there are 27 species that provide my first record, but I have not attempted to cite other than the 1597 date. Jeffers also (1969) produced a later, very useful, set of biographical details of all the people mentioned in or relevant to the *Herbal*. Henrey too (1975: 45) wonders if the criticisms of Gerard have been overdone. Of more relevance to this compilation is that Gerard's work is in English, is accessible, at least in the Johnson (1633) edition, and that it adds around 180 new species, and both for these and for other species discovered earlier, adds, for the first time, many detailed locations. The majority of these are from the areas around London (the 'Home Counties'), which seems to be the area where Gerard did most of his botany: indeed Raven (*op. cit.*) finds his records from further afield 'unreliable if not demonstrably false'. However I have found his records a major source of assistance. He includes good indexes to both the Latin polynomial and English names used in the work. See also notes under Johnson (1633) below).

Available online.

Johnson (1629). *Iter Plantarum Investigationis . . . in Agrum Cantianum; and Ericetum Hamstedianum.* [*Iter*]. Description of itinerary. Latin.

Johnson (1632). *Descriptio Itineris Plantarum Investigationis . . . in Agrum Cantianum A.D. 1632; et Enumeratio Plantarum in Ericeto Hampstediano locisq. vicinis crescentium.* [*Descriptio*]. Description of itinerary. Latin.

Both of these, referred to below as his Itineraries, were accounts of simpling trips to Kent and to Hampstead Heath in London, and are the first accounts of the flora of any area in Britain and Ireland. There is an excellent, beautifully produced, recent edition with a very full translation and notes, with maps of the itineraries, edited by John Gilmour (1972). The plant names given there are interpreted by Francis Rose, with a considerable number showing differences compared with those used by

Clarke who was using the reprint of Johnson's *Opuscula* in Ralph (1847), and seemingly relied on Hanbury & Marshall's *Flora of Kent* (1899), which has many speculative identifications. I have largely followed Rose in my compilation, but noted all the others. The citing of the pagination has been confused – see note under Conventions (Introduction, Sect. 10).

Johnson (1633). *Herbal.* [Ger. emac.] English. Systematic.

This is a much-enlarged, and much-corrected edition of Gerard (1597), and has been frequently referred to as '*Gerard emac.*', but I consider it sufficiently different to give Johnson the credit. He added a new introduction, describing in detail how his edition is to be used. In this is explained his alterations and additions, both for text and illustrations, noting † for much altered sections and ‡ for completely new accounts. He also corrected or changed spellings, but I have tried to cite the spellings used in the 1597 edition where relevant. Very many of the additions came for John Goodyer, who sent him full descriptions of over 100 species of which he used around 60. The 1633 introduction also includes a long history of botany and an uncomplimentary critique of Gerard's procedures and accomplishments. Almost all later works cite the 1633 work in their synonymies, but because Johnson retained Gerard's division into three books, with any additions being inserted into their proper place, it is relatively easy to trace back to the location and chapter in the original, although no modern index to that exists.

Johnson, T. 1632. *Descriptio Itineris Plantarum Investigationis*

It was reprinted in 1636 'line for line the same' (Jackson 1881: 27), but see, for instance, *Cardamine impatiens*. In that reprint Johnson mentioned (verso of last leaf) that he intended to make figures for the new plants that he had described. Indeed, in 1641 he expressed his determination to publish the histories of the plants in a joint work with Goodyer, and to add figures of as many as he was able (Kew & Powell 1932: 140). Kew & Powell (1932: 126ff) discuss the possible fate of these figures.

There is a 1975 Dover reprint of the 1633 work and a modern synonymy (Harvey 1982). Harvey made no claims to be a critical botanist, though he was assisted in this by Richard Gorer who at least was a distinguished horticulturalist, but nevertheless I find many of the attributions ambitious, and, by and large, have not included them. Interestingly he also includes a reverse index giving a binomial name for all of Johnson's polynomial illustrations, which is something that has always intrigued me: how many of the species described in pre-Linnean works, especially those claimed for the British Isles, have never been satisfactorily identified?

Johnson (1634) (2nd part 1641). *Mercurius Botanicus.* Latin. Alphabetical.

The work is an alphabetical listing, derived from his second edition of Gerard (1633) and of his earlier works, together with names and a few localities derived from his 1634 trip to Bristol, via Marlborough and Bath, and their return via Salisbury, Southampton and Guildford. A second part, a supplement, was issued in 1641, incorporating material from his 1639 trip to North Wales and from Parkinson (1640) and others, including Stonehouse, Bowles and Goodyer as well as L'Obel (Kew & Powell, 1932).

This is only available in the exact reprint by Ralph (1847), along with the earlier works on Kent and Hampstead of 1629 and 1632, published as Johnson's *Opuscula*. This is the source used by Clarke. No translation or interpretation exists, but the outlines are very well-given in Raven (1947).

Kew & Powell (1932) feel that this is Johnson's most important work 'the first British Flora; the first work in which the known British plants were enumerated; separated from those of the Herbals; and dealt with alone'. They add interesting background to all of Johnson's works, and include a map of the vice-counties that Johnson visited – 33 in total. Theirs is a first-class biography in every respect. Altogether, Johnson added over 170 plants to the British flora – the two parts of the *Mercurius* list a total of 898 species.

Parkinson (1629). *Paradisi in Sole; Paradisus Terrestris.* English. Systematic.
(1640). *Theatrum Botanicum . . . an Herball.* English. Systematic.

The first is primarily a gardening book, the first in England, covering nearly a 1000 plants, but has only a very few new records of wild plants. But the second is perhaps the major British and Irish herbal. It is his life's work, but was overshadowed by Johnson's edition of Gerard, which was rushed out knowing that Parkinson's own work was nearly finished. Parkinson also had access to L'Obel's unpublished *Stirpium Illustrationes* (see the note on that under L'Obel (1655) below). Quite a few of Parkinson's localities are noted with great detail, presumably from his wide circle of contacts, since he himself rarely left London. Raven (1947: 255 *et seq.*) devotes a long, fascinating and extremely detailed chapter on the *Theatrum*, good on both the records and those who contributed records.

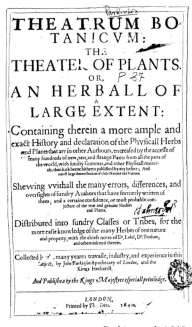

Parkinson, J. 1640.
Theatrum Botanicum . . . an Herball

No later translation or interpretation known, but copies of each are online.

How (1650). *Phytologia Britannica natales exhibens lndigenarum Stirpium sponte emergentium.* Latin. Alphabetical. (published anonymously, but the authorship revealed by Merrett).

A bare list of plants, with very little information other than the names, mainly derived from Johnson's *Mercurius*, re-arranged in alphabetical order. It was little regarded by his contemporaries. However this work added records of a few species, mainly from Ireland (from Reverend Heaton), together with many records from W.T. Stonehouse and one or two others, including the enigmatic Mr. Sare from Newmarket. Druce

(1897: cxi) mentions that John Watlington (d.1659) was probably the discoverer of many of the Berkshire plants listed in How from 'Redding' or nearby, and this is confirmed by the many references in the Ashmole MS in the Bodleian to him (Ashmole) and Watlington botanising together (John Edgington *pers.comm.*).

How's own interleaved copy, with ms. additions, is in Goodyer's library at Magdalen College, received by him on 30 Apr. 1659.

No later translation or interpretation known. No online digitised copy found. EEBO print on demand.

L'Obel (1655). *Stirpium Illustrationes … Accurante Gul. How.* Latin. Systematic.

A selection by William How from the manuscripts that L'Obel left at his death in 1616. How extracted details of 223 of L'Obel's descriptions (out of a total of 835), but with no illustrations, despite the title. Parkinson (1640) had evidently seen these, and had incorporated some of the new species. There has been lasting controversy regarding How's accusations of Parkinson's alleged plagiarism, but Gunther (1922: 251-253) gives the full details and Raven (1947: 267-269) largely demolishes How's case. L'Obel's complete manuscript is with Goodyer's papers at Oxford.

How's selection contains 78 references to British plants (*pace* Louis 1980), of which those with localities (42) were all from London or W. Kent, possibly from L'Obel's later sojourn in England. However L'Obel did make an expedition to Chester and Wales in 1609 or 1610, mentioning 'a rare blue *Pinguicula* and a small yellow *Pulsatilla*' (see also Raven (1947: 239), but neither of those is in his 1655 work.

An important point in the context of this compilation is the dating of the records. Gunther (1922: 252) relates that he had 'the original manuscript from which the book was printed before us [i.e. at Magdalen], with the excerpts from L'Obel, with How's additions pasted or pinned thereto, and the leaves of the MS. exactly as they were marked for the compositor, and returned by the printers to the editor'. In the printed book 'How's notes, which appear in small type near the margin …'. This means that all the species new to the British Flora in this work can be dated to before L'Obel's death in 1616.

No later translation or interpretation known, but available online.

Ray (1660, 1663, 1685). *Catalogus Plantarum circa Cantabrigiam nascentium*, with supplements in 1663 and 1685. Latin. Alphabetical.

Long described as the first county flora, with the benefit of two recent translations (Ewen & Prime 1975; Oswald & Preston 2011). But in this context it is much more than just a county flora; not only does it introduce first records of over 50 species, usually with locations, but he gives detailed synonymies and corrects errors of previous authors. He also has an excellent bibliography. The Oswald & Preston edition gives a wealth of background information and the notes that accompany the text are endlessly informative.

Merrett (1666). *Pinax Rerum Naturalium Britannicarum.* Latin, no illustrations. Alphabetical.

Lists of plants, almost all with English names, but also containing lists of animals, birds, fishes and fossils. The plant names are only occasionally referenced to earlier writers, but have many locations.

His work was almost immediately superseded by Ray's *Catalogus Plantarum Angliae* of 1670, but contains first records for over 40 species (Clarke reckoned 46). He had access to Goodyer's manuscripts and he employed Thomas Willisel to search for plants (Raven, 1947 and others). Druce (1897) notes that William Browne is named for some records and probably supplied many more. Merrett must have had other correspondents of whom we know no details. He cut out much of the dross in How and asterisks new species, but, on the whole, his work is a disappointment (cf. Raven (1947: 320) and Ray describes it as 'Dr Merrett's bungling *Pinax*'.

No later translation or interpretation known, but his synonyms are cited in Ray and others, including Smith. Available as print on demand & online.

Ray (1670, 1677, 1688). *Catalogus Plantarum Angliae et Insularum adjacentium. . .* Latin. Alphabetical.

A major step forward, with a list of 1050 native plants, practically all of which he had seen himself, other than a few from L'Obel and Johnson (marked with an asterisk). He made special efforts to omit errors, most of which arose 'because of the brevity and insufficiency of the descriptions of the older botanists and the consequent and reckless multiplying of species' (Raven 1950). Thus this could fairly be called the first English flora, stripped of the foreign plants of the herbalists and the dross in How and Merrett, and was the basis of all to follow, though further improved in his *Synopsis* (see below). For the second edition in 1677 Ray includes a note at the front of an additional 30 or so new plants, together with a further nine observed by Dr Dent around Cambridge. Plants recorded from Cambridgeshire are marked with a 'C'. In 1688, after the issue of the second volume of the *Historia*, there was issued a brief (20pp) *Fasciculus* of additional discoveries, some of which had already appeared in the *Historia*, but with some new, including a list of plants received from Thomas Lawson. Keynes (1951) reproduces a page of the 1677 edition marked up by Ray for the 1688 *Fasciculus*.

Ray, J. 1670. *Catalogus Plantarum Angliae et Insularum adjacentium*

Not translated, but all are available online and the first two as EEBO reprints.

Morison (1672). *Plantarum Umbelliferanum distributio nova.*

Morison (1680, 1699) *Plantarum Historiae Universalis Oxoniensis* parts 2 & 3.

A few new records, but with relatively little advance in knowledge. Despite the title the species dealt with cover the known world. Morison died in 1683 and work was completed, in an abridged form, by Jacob Bobart, jnr. But the 'Morisonian' herbarium at Oxford, which was organised by Bobart, is of value since it is a collection of plants made to illustrate his 1680 and 1699 works. See Clokie (1964) and Vines & Druce (1914).

Available online.

Plot. (1677). *The Natural History of Oxfordshire.*
See Raven (1950: 249-250). Ray not impressed by him and sceptical about his records.

Available online.

Ray (1686, 1688, 1704). *Historia plantarum generalis.* Vol 1 1686; vol 2 1688; vol 3 1704. Vol 2 also contains a preface of plants from Lawson, and addenda and emendanda (pp 1849 – 1921). Latin. Systematic.

Ray's *magnum opus*, well-covered in Raven (1950) but adding relatively little to this compilation over and above his other works cited here. Ray adds the letter 'A' to those that 'occur spontaneously in England' with a supplementary list of those that were accidentally omitted, just before the index. This work had no illustrations, but Petiver (1713, 1715b) issued an extensive catalogue of plates.

No translation or reprint exists. but it is available on line and from EEBO (print on demand).

This was the major source of information on plants of the known world for the next century – indeed such a venture has never been attempted again, other than in outline in the *Index Kewensis*. Sir Hans Sloane's own copy, annotated by him to show where specimens exist in the volumes in his herbarium, is in the Sloane Herbarium; this is of great assistance in tracing specimens there.

Ray (1690, 1696, 1724). *Synopsis methodica stirpium britannicarum.* Ed 2 1696; ed 3 (by Dillenius, and often cited as such) 1724. Latin. Systematically arranged, as are all subsequent entries in this section.

This was intended to be a development of the earlier *Catalogus*, but due to a dispute with his publishers, appeared as a totally new work, but based on the *Historia*. With the systematic arrangement, the extra species, the greater accuracy of the synonyms, the relegation to an appendix of the uses and medicinal qualities, it was a huge advance on the *Catalogus*. It includes cryptograms, fungi and seaweeds.

The 1690 appendix includes observations by Bobart, Sherard (including many from Jersey), Plukenet and Doody, together with emendations and a supplement by Tancred Robinson. The 1696 edition has appendices of plants from Petiver and Doody

The third edition, compiled by Dillenius after Ray's death, contains many alterations, with records from Richardson, Sherard, Lhwyd, Rand, Dale and others. These are marked there with an asterisk. It also has a chapter on the '*Plantarum dubiarum*' of How and Merrett – plants listed by them that have never been identified – as well as a further list of plants from L'Obel (1655) that Dillenius was unsure about either their identification or whether they occurred in Britain.

The *Synopsis* was the standard British flora for the next 70 years, until Hudson's *Flora Anglica* in 1762. Stearn (1973) relates how some of the surviving copies have been annotated by some of the great botanists of the 18th century, and he compares this work with the floras of Bentham & Hooker and Clapham *et al.*, pointing out that it fulfils the same function as these of field guide and source of reference.

No translation exists, but there is a reprint of the 3rd edition by the Ray Society (Stearn 1973), with an introduction by Stearn. Editions 1 and 2 are available on-line and as print on demand.

Keynes (1951, reissued and expanded, with corrections, 1976) gives an extremely comprehensive and interesting bibliography of all of Ray's works.

Petiver (1713, 1715b). *Herbarii Britannici … clariss. … D. Raii Catalogus*. Tab. 1-50 & 51-72.

Petiver (1715a). *Botanicum Hortense IV. Giving an Account of Divers Rare Plants, Observed the Last Summer A. D. 1714. in Several Curious Gardens about London, and Particularly the Society of Apothecaries Physick-Garden at Chelsea*. [and other articles along the same lines].

Petiver (1716). *Graminum, muscorum, fungorum, submarinorum, &c. Britannicorum concordia*, etc.]

See Raven (1950: 233). Petiver, although a very influential and active figure in his age, was more of a collector than a scientist and more of an entomologist than a botanist. The Ray volumes illustrate many of the plants in Ray's *Historia*, and are valuable for that, though the drawings are very small. However specimens of many of these are also in his herbarium in the Sloane collection.

Hudson (1762, 2nd ed 1778, 3rd ed. 1798). *Flora Anglica*. Latin, no illustrations.

An excellent compilation, the first to use Linnean binomials, yet citing the polynomials of earlier authors back to Johnson (1633). He did not refer to How or Morison and only occasionally to Merrett. He, and later writers, cited Johnson (1633) rather than Gerard, and Ray's 1724 edition rather than his earlier works.

His descriptions were concise and the localities good and clear, and the layout of the text and the separation of the species is much cleaner than his predecessors. This flora effectively supplanted Ray's *Synopsis*. Available online and as print on demand.

Withering (1776 *et seq.*). *A Botanical Arrangement of British plants*. English, as are all subsequent entries in this section. No illustrations, as in all subsequent entries in this section, unless so marked.

1776 2 vols, 2nd ed 1787 2 vols and third vol 1792, 3rd ed 1796 3 vols, 4th ed 1801 4 vols, 5th ed 1812 4 vols (there were two later editions and seven more abridged by W. Macgillivray). The first volume, in all editions other than the second, consisted of background, and arrangement of genera. Later volumes commence pagination with vol 2 and then are consecutive. References and concordance back to Johnson (1633), along with Parkinson, Plot, Petiver and Ray.

Most available online.

Lightfoot (1777). *Flora Scottica.*

Included here as the first to cover many of the Scottish species. Not only was he accompanied on his trip by Scottish botanists, but also John Hope made his own records available to him (see Appendix IV for fuller details).

Available online.

Curtis (1777-1798). *Flora Londinensis*, continued by **Graves & Hooker (1817-1828)**.

This is included here although the work as a British flora is incomplete, and is not arranged in any systematic order. Yet the descriptive text, the detail on the locations and the folio illustrations are so good that it is repeatedly cited. The original edition has 432 plates, and though the intention was to describe and illustrate plants from London, in the quest for sales the scope was widened after about 1792. The second edition contained a further 216 plates. A useful concordance of the plates and their dates of issue is given by Stevenson (1961).

General comment up to 1800.

Pulteney (1790) and Green (1904) are the only two authors who have covered the botanists and their works for the whole period with any comprehensiveness, but I have not found them really critical enough about the floras that they deal with. Raven (1947) certainly was (see Oswald & Preston 2011: 89), but he only covers up to Ray. An excellent, but perhaps harsh and not disinterested critique, is found in Smith (1798, and repeated in his 1824 work). Here he gives Ray unreserved praise, Dillenius less so. He states that Hudson was the model for flora writers of the later 18th century, praises his specific characters and descriptions, but does not exempt him, Withering's early editions, and, indeed, all the others post-Ray, from criticism for applying the names found in foreign authors (and here he includes Linnaeus too) to the names of Ray and for copying the synonyms that they have used, without going back to the original works. Lightfoot is dismissed utterly. Only Withering's 3rd ed. and Curtis' *Fl. Londinensis* are exempted. Smith ends by hoping that he has gone back to basics in his own works, and made the synonyms, which are such a feature of these, as perfect as possible.

Smith & Sowerby (1790-1814, supp. 1831-1866), 2nd ed. 1835-1846). *English Botany.* Illustrations

Smith (1800-1804). *Flora Britannica.*

Smith (1824-1828). *The English Flora*

Syme (1863-1886). *English Botany*, 3rd ed.

All English, with a complete set of illustrations in the *English Botany*, but there covering algae, bryophytes and lichens as well as flowering plants and ferns.

Of course there is no systematic order in the original edition, which appeared in parts, monthly, over almost 25 years, though the second and third editions were so arranged. The text, short in the early volumes, but fuller in the later issues and in the supplements, has much on the history and habitat, and that, coupled with the superb paintings, makes it attractive and useful to this day. Many species were included in the native flora on very flimsy evidence, but made good illustrations!

Smith's *English Flora* is summation of his life's work, with full descriptions and full synonymy, of especial use to the historical botanist. There is a small section on corrections and additions in the middle of Vol 4 (pps 262-274, missing from the 2nd edn, presumably incorporated into text). There are still no keys, and though there are indexes in each volume, there is no master index. However each index includes the historical polynomials as well as the Linnean binomials.

The plates were largely re-drawn for Syme's edition, which should probably be described as a work in in its own right, so thoroughly was it revised and re-written.

Hooker, W.J. (1821). *Flora Scottica.*

Hooker, W.J. (1830 *et seq.*, up to 1861). *The British Flora.*

Although the former brought Lightfoot's work up-to-date, and the latter incorporated advances since Smith's floras, neither contributes much to this compilation or to the advance of taxonomic knowledge.

Babington (1843 *et seq.*, 10 eds. up to 1922). *Manual of British botany.*

Perhaps the premier flora of the middle and late nineteenth century, and the first, other than the 5th ed of Hooker's *British flora*, to adopt the natural system, thus bringing British and Irish botany into line with that of France and Germany. He was a critical botanist (see Bentham, below), and responsible for elucidating many of the difficult genera, such as *Atriplex*, *Arctium*, *Fumaria*, *Cerastium*, the aquatic *Ranunculus* species, *Saxifraga*, *Potamogeton*. He also worked on two of the really critical groups, *Hieracium* and *Rubus*, possibly with less lasting success.

Bentham (1858 et seq., 4 eds. up to 1878), followed by

Bentham & Hooker. (1887 *et seq.*, 5 further eds. up to 1924). *Handbook of the British Flora.*

Butcher & Strudwick (1930). *Further illustrations of British plants.*

Butcher (1961). *A New Illustrated British flora.*

The antithesis to Babington (and to Watson, see Appendix II, Census catalogues), taking a much broader species view (treating over 300 less), describing instead a range of varieties. A major innovation was the adoption of dichotomous keys, which we now take for granted, but does not seem to have been incorporated in a full flora until Clapham *et al.* (1952). Bentham was probably the standard flora, at least for the relatively inexperienced, for the next century, but because of his wider species concept, and dismissal of the work of those who had investigated the flora more critically, was never really accepted by more experienced botanists. This position lasted right up to the publication of Clapham *et al.* (1952). Butcher & Strudwick was more than a further series of illustrations, as it included many critical species not in Bentham & Hooker, with descriptions and occasionally references back to the originator of the taxon concept.

Moss (1914-1920). *The Cambridge British Flora*, with distribution maps (based on geographical counties) and illustrations.

This was an ambitious attempt to produce a comprehensive critical British flora. Moss listed three aims: to register the present state of knowledge with respect to British plants, including classification, nomenclature, characteristics and distribution; to attempt to relate British plants to allied forms in foreign countries; and to stimulate further research, particularly in the areas of variation and distribution. Only two parts were ever published, but these are magisterial in their scope. See Bunting *et al.* (1995) for the background and history of the project.

Clapham, Tutin & Warburg [CTW] (1952, 1962, 1987). *Flora of the British Isles.*

The lack of a flora that was comprehensive, modern and taxonomically rigorous had long been a *cri-de-coeur* in British and Irish botany, and this work fulfilled that very long-standing need, and immediately supplanted all its predecessors. It covered more species, including many aliens, with proper dichotomous keys, distribution notes and a fair number of synonyms (certainly the best recent source for these). It did, though, replace the statuses on criteria pioneered by Watson (native, denizen, colonist and casual) with something much more woolly and uncritical and thus preserved many species that really have no place in the native flora of the British Isles. For a very interesting critique see Lousley (1953). A series of illustrations to complement this were issued by Roles (1957-1965).

Stace (1991, 1997, 2010). *New Flora of the British Isles.*

The successor to CTW, fully-keyed, with up-to-date taxonomy, many more alien species, but with shorter accounts and fewer synonyms.

Sell & Murrell (1996 →). *Flora of Great Britain and Ireland.*

A critical infraspecific flora, treating a considerable number of varieties. In effect this is the flora that has been looked for and promised for almost a century – through the incomplete efforts of Moss, the '*New Students British Flora*' of the 1930s (see Briggs & Gorringe 2002) and the abortive efforts of Prof. Valentine and his colleagues in the 1970s.

General note. Most, if not all of the national floras before Smith & Sowerby, contained no alien species at all, but many were covered in that publication as if they were natives, often on the most optimistic of grounds. All subsequent floras, with the exception of Bentham, have tried to distinguish between native and introduced status, though those from CTW onwards with much less care and with a broader brush than those of the late Victorian era. A pleasing exception to the above is in the preface to Ray (1660), where he lists plants that he suspects might not be native in Cambridgeshire (see Oswald & Preston (2011: 131)).

B. Census catalogues

Turner & Dillwyn (1805). *The Botanist's Guide through England and Wales.*
The first to try and summarise distribution by county, but it contained a mass of unreliable evidence, which was gradually sifted by Watson's works. Much of the information came directly from Camden's *Britannia* (see below).

Watson (1835-1837). *New Botanist's Guide.*
Updating Turner & Dillwyn.

Watson (1847-59, supp 1860). *Cybele Britannica.*

Watson (1868, 1869, supp. 1872). Compendium to *Cybele Britannica*.

Watson (1873, 1883 with later supplements). *Topographical botany.*
Watson's works, based on his conception of the vice-county system (1852), laid the ground for all subsequent distributional analysis, and greatly stimulated local botanists, their collection of records old and new and the formation of regional herbaria (see Allen 1986). All of these led to the publication of many new county floras and many articles and notes in national and local journals. He took a relatively narrow view of species.

Moore & More (1866). *Contributions towards a Cybele Hibernica.*

Colgan & Scully (1898). *Contributions towards a Cybele Hibernica.* Ed. 2.
The latter not only updates Moore & More, but contains, at the front, a comprehensive list of papers relating to the flora of Ireland.

Praeger (1901). *Irish Topographical Botany.*
Praeger mentions (1901: lxxxii) that when he started this exercise, only 11 of the 40 Irish VCs were known well botanically, with 22 hardly known at all.

Druce (1932). *Comital Flora.*
Not merely a compilation of distributions, but containing also first records and other comments, but with, as has been mentioned earlier, many errors.

Scannell & Synott (1972, 1987). *Census Catalogue of the flora of Ireland.*

Ellis (1983). *Flowering Plants of Wales.*

Stace *et al.* (2003). *Vice-county census catalogue of the vascular plants of Great Britain.*

Perhaps this is the place to mention the editions of Camden's *Britannia* by Gibson and Gough, which included plant lists.

Bishop Gibson revised Camden's original in 1695, with further editions in 1722, 1753 and 1772. In each of these he provided lists of the 'less rare' plants found in those counties. For the 1695 edition John Ray provided all the details for England, other than for Middlesex, where James Petiver's list was used. Edward Lhwyd provided a single list for the whole of Wales. No plants were included for Scotland or Ireland. Only minor alterations were provided in his later editions, with no indication of their provenance.

Gough produced a further revised edition in 1789, followed by a second in 1806. There are no details of who provided the additional plant details, including those for Scotland and Ireland. Those for England and Wales amount to only relatively modest additions to Gibson's records.

Carter (1955a & b, 1960, 1986), in his accounts of botanical explorations for each Welsh county, has no hesitation (and gives no reasons) for attributing the new records in Gough's edition to Lhwyd, even though they must have been made almost a century earlier. I have not yet been able to discover anything in Lhwyd's correspondence to support Carter's claims.

C. Plant Lists and field guides

Edinburgh Catalogue (1836, four editions, last 1865).

London Catalogue of British plants (1844, eleven eds. last 1925). Started by Watson, replaced the above, and used as a standard list until Druce's time.

Hooker, J.D. (1870, three eds. to 1884). *The Student's flora of the British Islands.* Not keyed.

Hayward (1872, nineteen eds. up to 1930). *The botanist's pocket book.* A field flora, with keys, arranged in the order of the *London Catalogue*.

Druce (1908, 1928a). *List of British plants.*

Clapham (1946). *Check-list of British vascular plants.*

Dandy (1958b). *List of British vascular plants.*

Kent (1992). *List of vascular plants of the British Is*les.

D. Journals and Periodicals
(*Free access, † with password)

Principal

Royal Society (Philosophical Transactions). *1665→.

Linnean Society (including all the other journals in this stable).†1791→.

Magazine of Natural History (Loudon). *(some) 1828–1842.

Botanical Miscellany. *1829-1833; then (Hooker's) Journal of Botany 1834-1842; London Journal of Botany 1842-1848; then Hooker's Journal of Botany and Kew Garden Miscellany 1849-1857.

Companion to the Botanical Magazine. *1835-1837.

Magazine of Zoology and Botany, 1836-1838; then Annals of Natural History 1838-1840; then Annals and Magazine of Natural History. *1841-1967.

Phytologist. *Series 1, 1841-1854, Series 2, 1855-1863.

Botanical Gazette. *1849-1851.

Journal of Botany. *1863-1942. 80 vols.

Botanical Locality Record Club 1873-1886.

Watsonian Botanical Exchange Club 1884-1934.

Botanical Exchange Club *(some) (Thirsk BEC 1858-1865; London BEC 1866-1868; BEC 1869 – 1878; BEC of British Isles 1879- 1948.

Botanical Society of the British Isles. *(most) Year book 1949-1953; Proceedings 1954-1969.

Watsonia. *1949-2010 (this, and its successor, below, would be the prime source of recent taxonomic developments in Britain & Ireland).

New Journal of Botany. †2011→.

BSBI News.*1972→

Others of import

Science Gossip, Hardwicke's Science Gossip. *1865-1893, New Series 1894-1902.

Annals of Botany 1887→.

New Phytologist *(Jstor)1902→.

Journal of Ecology *(Jstor)1913→.

Journal of the Society for the History (previously Bibliography) of Natural History. 1936→.

Scotland

Botanical Society of Edinburgh (under various titles) *(some) 1836→

Scottish Naturalist 1871-1891; then *Annals of Scottish NH 1892-1911; then Scottish Naturalist (Zoology) etc.

Irish journals

Royal Irish Academy Transactions *(Jstor) 1787-1907; Proceedings 1841-.

Scientific Transactions of the Royal Dublin Society

Irish Naturalist *(Jstor) 1892-1924, Irish Naturalists Journal *(Jstor) 1925→.

Local Journals

No attempt has been made to access this vast field of records, other than where I have been specifically directed to a particular reference. The three below are only listed because they cover a larger area than one county, and also because I consider that there has often been a tradition in the north of England to publish locally rather than in London-based journals.

Naturalist 1837→ (a complicated history).

North Western Naturalist 1926-1951, 1953-1955.

Vasculum *1915→.

Appendix II

Histories of Botany & Biographies

Pulteney (1790) Sketches of the Progress of Botany in England.

Vol. 1 covers from earliest times up to Willisel and Plot, but including all of Ray; Vol. 2 covers from Sibbald to Linnaeus, but with only a perfunctory mention of Hudson and none of Withering and Lightfoot. Much of what he wrote is included in Raven's books (see below), but I feel that what he wrote well repays study for a more detailed background for each botanist. Note, however, that some authorities feel that his work was uneven – see, for instance, Kew & Powell (1932) who felt that Pulteney did not place enough stress on Johnson's contribution to British botany, particularly because he gave credit to How as the first sketch of a general list of English plants, and thus failed to realise the character and content of the *Mercurius*.

He also gives notes (p.352ff) of the early county floras published in the late 17th and 18th centuries, as does Martyn (1763: viii).

There is a most interesting copy in the library of the Linnean Society, which belonged to B.D. Jackson, is extremely heavily annotated by him, and also includes notes from the Rev. W.W. Newbould. It was used too by R.H. Jeffers for his works on Gerard (1967, 1969).

Green (1904) A history of botany in the United Kingdom.

This covers progress from earliest times, but in view of the fact that Gunther and Raven only cover up to c.1700, and Pulteney up to the time of Linnaeus, it is particularly valuable for its coverage of the period following the adoption of the Linnaean system. Two-thirds of the book covers from Hudson and Withering through to the end of the Victorian era, with Smith, the Hookers, Babington, and Bentham particularly receiving extensive coverage. But he deals with only the main participants, and those not sequentially, but in different parts of the book. From the perspective of British and Irish field botany a coherent thread of discovery and elucidation is elusive.

Oliver (1913) Makers of British Botany.

A collection of essays, mainly by other botanists, on 17 figures prominent in the development of botany in Britain and Ireland. But the emphasis is more on the theoretical side, with only five or six of the subjects particularly relevant to this work.

Gunther (1922) Early British Botanists.

This must be essential reading for all plant historians, simply because it describes the remarkable collection of botanical books and manuscripts collected by the Hampshire botanist, John Goodyer, and bequeathed to Magdalen College, Oxford after his death in 1664. Goodyer described in detail some 350 species, of which about 150 are extant, and these comprise, in many cases, the earliest that are extant in the English language. They are included in full in Gunther's book. Goodyer's collection was virtually ignored for 250 years, until 'discovered' by Canon Vaughan in 1901 and then by Druce (Raven 1947: 291). But it was Gunther, the librarian at Magdalen, who provided the definitive account of Goodyer's life and discoveries together with a catalogue of his remarkable library, which consisted of just about every botanical book ever published up to his death, many annotated by their authors or subsequent owners and by Goodyer himself. It was Gunther too who collected, collated and bound all the manuscripts that came to Magdalen with Goodyer's library (see Gunther 1922: 229 *et seq.*) The last section of the book consists of a mass of lists from 16th and early 17th century botanists and gardeners, of which very little has, to my knowledge, been interpreted. However, included here are 16 pages of extra records accumulated by How before his death in 1656, of which very few have been interpreted but might well be valuable. There is also (p.372) a note of some of Goodyer's excursions – not only in his home area but to Essex, London, Surrey and Bath, among others.

This book has been curiously ignored, possibly because for many years it was virtually impossible to obtain, although there was a reprint as a hardback in 1971 and now as a paperback in 2010. As mentioned above Raven cited it sparingly, and was slightly sniffy about Gunther's over-egging of Goodyer's importance as a botanist (slightly ironic in view of his virtual hagiography of Ray). Whatever the merits of Goodyer as a botanist, his collection is of absolutely unparalleled importance and there must be scope for more interpretation of his manuscript sources.

Raven (1942, 1950) John Ray and (1947) English Naturalists from Neckham to Ray.

But the two most important insights in the last century into early botany, and quite possibly the best books available, are Canon Raven's two works. Of course both of these fully cover the published works, adding interpretation, insight and context, but they also deal with the known manuscript sources, including much of the raw material in Gunther (1922). There is little doubt that Raven was a better botanist than Gunther or his assistants, so Raven's interpretations are more reliable. Raven too incorporates Ray's *Itineraries* (see below), and, possibly of even more value, uses these together with detective work based on deep scholarship to interpret the records published in his *Catalogus* to weave an account of Ray's years of travel, and in the process providing many new dates for first observations. Ray's *Itineraries* were first extracted by and published by George Scott in 1760 (Stearn 1986), but were re-printed, together with extracts of his correspondence from a selection made after his death, in two volumes edited by Edwin Lankester and published by the Ray Society (Lankester 1846, 1848), and those two are the versions which I have used in this compilation. Raven is very dismissive indeed of the quality of Lankester's editing, though he exempts his botanical helper, C.C. Babington, from this criticism. Nevertheless, a reading of the 1846 volume in particular is really illuminating, and being able to use the actual dates of Ray's discoveries has added much to this work.

Raven is good too on the field workers, adding an element of criticism where appropriate.

Raven's works are essential reading for any student of the discovery of the British flora up to Ray's death. Adequately indexed, they lack only bibliographies, which is frustrating when dealing with early, unfamiliar works. However many of the references might be traced in Gunther's catalogue of Goodyear's library (Gunther (1922: 197-232).

Later works.

I have consulted with profit Kent's *Index to Botanical monographs* (1967) for assistance in tracing the often tortuous path or understanding of our critical genera, and wish that there was an up-to-date version available. It should be noted though that the family and genus accounts in Sell & Murrell (1996→) almost always contain a useful guide to the literature. Stearn's (1973) introductory chapters to the facsimile reprint of Ray (1724) and of Linnaeus' *Flora Anglica* (1754 & 1759) are very useful, with, *inter alia*, sections on contributors to that edition of the *Synopsis*, the abbreviations used and a summary of differences in nomenclature between the above works of Linnaeus and Dandy (1958b), with references back to Ray (1724). He too stresses the importance of Smith's *English Flora* (1824-28), with its comprehensive indexes of pre-Linnaean nomenclature.

One might note the works of the distinguished plant historian John Harvey, whose works have so illuminated the history of the introduction of garden plants to these islands. Many of these are in articles, especially in the periodical *Garden History*, but his work on *Medieval Gardens* (Harvey 1981) brings together much of this research, and is of particular relevance for the study of plants (particularly archaeophytes) introduced by man, and then escaping out into the wild. A very useful bibliography of his publications is in Harvey (1998).

There has been little substantial work on general British and Irish botanical history since Raven, although a conference in Edinburgh produced a very interesting report (Noltie 1986), covering many of the principal discoverers of our flora. However mention must be made of Oswald & Preston's (2011) edition of Ray's *Cambridge Catalogue* of 1660. The introductory chapters contain a great deal that is relevant to this work, with extensive biographical notes on the authors cited by Ray, the identification of Ray's plants and much more; it has been of considerable assistance.

Finally Blanche Henrey's (1975) tour-de-force should be noted, covering all botanical and horticultural books up to 1800 and often beyond. For this project the text adds little to what is available elsewhere, but the bibliographies are immensely comprehensive (though not all that easy to use) and have been of much assistance.

Biographies

Many of the discoverers of our flora have been the subject of biographies, articles or accounts in local floras. Where known these have been referenced in sect. 9 of the Introduction, the Botanists, and I would be glad to know of omissions. Other authors are referenced in the species accounts, but certainly the papers by Michael Foley, Frank Horsman (on the Teesdale flora and its discoverers), Charles Nelson (on Irish botanists and through the journal of the Society for the Bibliography of Natural History, for which he was the long-serving editor), Henry Noltie and his predecessors on Hope, and Jean Whittaker, on the seventeenth century botanists Lawson and Nicolson, have added much to our knowledge of little-known figures.

But perhaps special mention should be made of the editing of Joshua Gosselin's *Flora Sarniensis* (Guernsey) by David McClintock (McClintock 1982). Publication of this manuscript and examination of his herbarium, dating from the 1790s, has added about 16 actual dates of discovery to my list.

County Floras

Pulteney (1790) has notes on the eighteenth century works, although I have not found any of these a useful source of first records.

But commencing with Trimen & Dyer's *Middlesex* (1869) (which has excellent accounts of Petiver and especially Buddle, as well as much else), many of the subsequent county floras published have chapters on botanical exploration. Some are perfunctory, but of the older floras the best are Lees' *West Yorkshire* (1889), De Tabley's *Cheshire* (1899), Hanbury & Marshall's *Kent* (1899), where B.D. Jackson wrote the historical summary, Davey's *Cornwall* (1909) and Druce's Thames Valley series, particularly *Berkshire* (1897) and *Oxfordshire* (1886, 1927), although many others contain useful details even if only in the species accounts themselves, such as Wilson's *Westmorland* (1938). Of the more recent Philip Oswald's contributions to the *Floras of Shropshire* (Sinker *et al*. 1985) and *Montgomery* (Trueman *et al*. 1995) are models of their kind, as are David Allen's chapters in the *Floras of Hampshire* (Brewis *et al*. 1996), *Dorset* (Bowen 2000) and especially *Isle of Wight* (Pope *et al*. 2003). Coles (2011) devotes a whole book to South Yorkshire, as does Greenwood (2015) to North Lancashire. Perhaps the above selection is slightly invidious – other recent county floras, including those for Bedfordshire, Cumbria, Durham, Hertfordshire, Isle of Man, Jersey, Norfolk, Staffordshire, Suffolk, Warwickshire, Worcestershire and parts of Yorkshire all have good accounts of the discovery of their flora.

Carter's highly detailed series on plant explorations not only covered all the Welsh counties (see App. IV, Wales), but also Dorset, Shropshire (somewhat woefully) and the Isle of Man.

Some of those above, particularly that of Lees, are of even more value for their commentaries on species recorded by the early botanists, traditionally interpreted as one thing, but which he felt, based on his local knowledge, might be errors.

Simpson (1960) contains, amongst much else, details of all county floras up to that date. Druce (1933) contains a much more succinct summary for each county.

Appendix III

Herbaria and correspondence

Herbaria

The British and Irish specimens held in museums go back at least to the late seventeenth century and possibly before. It is very difficult to say how many new early records could be produced by a comprehensive trawl through, but my feeling is that there is a significant relatively untapped source in these. Certainly the herbaria in London, Oxford and Cambridge have been extensively used for their relevant county floras, but even there, particularly in the Sloane Herbarium at the **BM**, there seems to be scope to find new records. This is especially true, of course, for critical groups, where more recent taxonomic work could elicit fresh records from specimens formerly identified as something else, but I would not make the slightest claim to expertise in any of these, and have relied on research that has been published or otherwise brought to my attention. None of the major holdings are comprehensively catalogued, though moves are underway to at least partially achieve this in several. Some of the most important are covered at least in outline.

1. Druce and the Oxford Herbaria

In his *Comital Flora*, Druce (1932) briefly cites many instances of early specimens in the various Herbaria housed in the Department of Botany in the University of Oxford. He had worked on these for nearly 50 years, written two accounts of the whole (Vines & Druce 1897, 1919), one on the *Dillenium Herbaria* (Druce & Vines 1907) and another on the *Morisonian Herbarium* (Vines & Druce 1914).

However there is no mention of these works in the introduction to the *Comital Flora*. In his preface to the *Flora of Oxfordshire* (ed. 2, 1927) he does expand, stating (introduction p. cxxvii) that the Morison Herbarium was collected by J. Bobart about 1690, the Du Bois Herbarium about 1700, that from Sherard about 1720, and those from Dillenius about 1730. These are, more or less, the dates he used in the *Comital Flora*.

In her excellent account of the Oxford Herbaria Clokie (1964) gives much more background to each Herbaria and on the men who assembled them. She states that many, or indeed most of the sheets are undated, and it seems far safer to use, for 'first date' purposes, the date of the collector's death, or the date of death of any contributor when those are named. This advice I too have followed. To give an example: for *Puccinellia fasciculata*, Druce (1932) states (as *Glyceria Borreri*) that 'Mr Stonestreet has a specimen in Herb. Du Bois, c. 1700'. There is no evidence (*pace* Clokie) that Du Bois gave his herbarium to Oxford before his death in 1740, but William Stonestreet died in 1716, so it seems safe to use that date to the first actual record.

There are many other records in Druce (1932) for Du Bois which he has dated either 1700 or c.1700, and more again for Dillenius dated c.1726. All these have been ignored and replaced using the forgoing criteria. The underlying specimens have not been inspected. However it should be noted that Druce's approximated dates have been used by flora writers, notably Kent (1975) in his flora of Middlesex, and by later analysers, such as Bebber *et al.* (2007).

a. **The Dillenian Herbaria (Druce & Vines 1907).** Druce determined and gave modern names to most of the specimens in all these herbaria, which comprise the *Synopsis* together with an *Appendix*, the *Hortus Elthamensis* and the *Historia Muscorum*. That for the *Synopsis* is arranged in the order of edition 3, 1724, but as Druce makes notes on p.1, 'very many specimens are undated, and these may have been collected before the compilation of the *Synopsis*, but the greater number were evidently collected after its publication in 1724, with the intention of more fully illustrating and describing the species, probably for use in an Appendix or in another edition'. Clokie (1964) too makes it crystal clear that this herbarium mainly contains material collected <u>after</u> his revision of Ray's work (1724). Apart from the records that must come from the Welsh trip, below, there are hardly any dates on the specimens and all those that are dated are post-1724. Although Druce in his *Comital Flora* gave the conventional date of 1730, it does not seem possible other than to date these as at Dillenius's death (1747), unless the specimens originated from a collector who died before that date.
In the citations in the species accounts, Druce gives the text from the herbarium sheet together with his indications (in brackets) of whether they were first British or County records. Yet by no means are all of those 'first British records' reflected in the entries in his *Comital Flora*, and I cannot see any reason for this.

To my knowledge neither did he publish any revisions to that work in his lifetime nor has there been any other than minor re-determinations of species in the century since. I have, perforce, accepted them *in toto*.

The volume also has several useful introductory chapters. These include a Memoir of Dillenius, a Diary of his Journey into Wales (with many new records, all of which are dateable) and correspondence with Brewer, and from Richardson and others.

b. **The Sherardian Herbarium.** Clokie (1964) confirms that Dillenius used this collection for his third edition of Ray's *Synopsis*, as he was in process of re-arranging the Herbarium. She notes that Dillenius 'always gives more details about the locality and collector in the book than he does on the herbarium sheet'. Very few sheets are dated. No detailed account of the whole collection exists, but Clokie's summary is an excellent entrée.

c. **The Morisonian Herbarium. (Vines & Druce (1914).** This was formed by Jacob Bobart the Younger (1640 – 1719) for the purpose of illustrating Morison's *Historia* (see Clokie 1964). Clokie says that few are dated, and of those that are, none is earlier than 1680. Very few are localised. If accounts appear in the *Historia*, then they are so dated, but it seems that he continued to add to the herbarium till his death as this was the working herbarium at Oxford until the setting up of the Sherardian Herbarium.

d. **The Du Bois Herbarium (Druce 1928).** Druce's paper is a very detailed account of the British plants contained in this herbarium and it is an interesting list, containing about 350 entries, though very few dates of collection are given, and no explanation is provided for his choice of dates in the title '1690-1723'. Again, I have dated all undated specimens as of Du Bois's death in 1740, though many (?most) must have been collected a considerable time before that.

Druce states that the collection came to Oxford in the time of Humphrey Sibthorp (1712-1797), but Clokie can find no evidence of this.

As with the Dillenian Herbaria as far as I am aware there has been no critical re-examination of Druce's determinations and I have accepted all of them.

2. The Sloane Herbarium

In contrast with matters covering the Oxford herbaria, this vast and important early (pre-1750) resource remains largely uncatalogued and almost all of it is without modern nomenclature. However, a comprehensive summary of its broad contents has been made by Dandy (1958a). There is evidence of selective use of the contents, as in the Middlesex floras cited above, but there can be little doubt that a comprehensive review of the thousands of specimens might result in more first dates of species, though most of these are undated and unlocalised, so that at best the records can only be dated to the compiler of each herbarium.

3. Other herbaria

To my knowledge there exists no detailed account of the contents of any other of the major national herbaria. Kent & Allen (1984) list the known herbaria held at institutions and a few in private hands, together with the collectors whose specimens are found in those herbaria. It is possible that the next few years will see more of their holdings digitised and placed on the Web, through initiatives such as the BSBI sponsored 'herbaria@home' project and others. It is very likely that earlier records of many critical species will be discovered by judicious searching through these resources.

4. Correspondence

Several collections of letters have survived, and are often useful for finding the first reference to something that was later published. The most important are summarised here under principals. A search for other collections might commence with Bridson *et al.* (1980).

Babington. (1897).

Lhwyd. Gunther (1945).

Martyn. (Gorham 1830)

Ray. These were collected by Derham (1760), but unfortunately he retained some and epitomised the rest, seemingly destroying most of the originals of the latter (see Whittaker (1986) for a fuller critique). Lankester (1848) edited many of those that remained, with the botanical assistance of Babington. Gunther (1928) edited a further collection, these with the botanical assistance of Druce. This last has a useful concordance of the letters in his and Lankester's volumes. A final, fuller selection of some of these that have a particular bearing on Cambridge botany has been published by Thompson (1974) (and see also Oswald and Preston 2011).

Richardson. (Turner 1835, and in Sloane, below).

Sloane (in British Library). Not published, but comprehensively indexed, see Scott (1904, reprinted 1971 and available as a print-on-demand). 'An inexhaustible treasure trove' (John Edgington *in litt.*).

Smith. (Smith 1832).

Appendix IV

Progress of discovery in Ireland, Scotland and Wales

Irish Flora

Ireland has a considerable body of publications devoted to the discovery of its flora. The honour of 'the first Irish botanist' is accorded to **Richard Heaton**, an English Protestant priest, who was first appointed to an Irish benefice in 1633, and remained there until 1641, at the start of the Civil War. Although he returned at the Restoration in 1660, any botanical records that survive were published by 1650 and thus must have been made by 1641 and are dated as such in this compilation – see the biography by Walsh (1978) and also the earlier account in Raven (1947: 302 *et seq*.). Heaton's records appear in How (1650) – but see also his claim to the record of *Drosera anglica* in Parkinson (1640). These records include for certain *Dryas octopetala*, *Gentiana verna*, *Scilla verna* and possibly *Epipactis atrorubens*. Praeger (1909a) and Walsh (1978) considered that How's record of *Euphorbia hyberna* must have come from Heaton, and they also admit the possibility, as do Colgan & Scully (1898), that it was Heaton who first found *Saxifraga umbrosa* which is referred to in How, but with no details or locality.

Sherard lived in Ireland from 1690-1694, and **Edward Lhwyd** visited in 1699 and 1700, adding several plants. But by 1700 Ireland was still very little known botanically – indeed Mitchell (1975) states that the sum total was around 50 plants included in Ray's Synopsis. Mitchell also adds a lovely quote from Baily (1833) '… when England and France had their provincial Floras, the Botany of this island was as much known as that of an island in the Pacific'.

Threckeld (1726) produced the first Irish flora (see also Nelson 1979). To his own records he added more from Heaton, and Molyneux (who in turn had records from Sherard and Ray). After that date discoveries are scattered through the literature, with the strands drawn together through *Flora Hibernica* (Mackay 1836), *Cybele Hibernica* (Moore & More 1866, Colgan & Scully 1898), *Irish Topographical Botany* (Praeger 1901) and *The botanist in Ireland* (Praeger 1934). Colgan & Scully's work has a useful list of sources at the beginning, and is the most important and comprehensive of these works. For Praeger (1901) details and maps of his itineries are available at http://www.botanicgardens.ie/herb/books/itbitineraries.htm. Praeger (1949) has details of most Irish botanists, together with sources for obituaries or further reading (available online, with very few later additions). Both Simpson (1960), in detail, and Mitchell (2000), in outline, are good references, together with Mitchell's two earlier papers on 17th century Irish botany (1974, 1975). In addition the Glasnevin Botanic Garden website has a very extensive list of floras, papers etc that can be downloaded http://www.botanicgardens.ie/contents.htm. Webb (1986) gives an excellent and very enjoyable account of 'The heyday of Irish Botany, 1866-1916'.

Perhaps mention should be made of Webb's *Irish Flora* (1943 and many editions since up to 2012). This is a properly keyed flora but contains, so far as I am aware, no first notices of plants.

Exploration of the Scottish Flora

There are very few published records of Scottish plants before Lightfoot (1777). Ray only reached as far as Stirling, on his 1671 trip, and his works contain but a few records, from Edinburgh and north of Berwick. Sutherland (1683) published a list of the plants cultivated in the Physic Garden in Edinburgh, and marked with an 'S' those indigenous or cultivated in Scotland, adding only *Ligusticum scoticum*. There is little of relevance to this project in these but he does list c.550 Scottish native plants (Robertson, 2001). Sibbald (1684) has only a handful of localities, and even as late as Hudson national floras contain little of substance. It was Lightfoot, albeit using information accumulated by Hope together with the results of his trip with Pennant in 1772, who really commenced the published history of Scottish botany, adding over 25 new species to the British and Irish flora. After Lightfoot, records of Scottish plants appeared in all national floras, culminating in Hooker's *Flora Scotica* (1821) and thereafter.

In the last century there has been considerable research into the exploration of the Scottish flora, particularly that of the mountains, with many valuable publications which have enabled not a few dates of first observations to be added.

Professor **John Hope** (1725 – 1786) was the King's Botanist for Scotland from 1761, founder of the new Botanic Garden in Edinburgh, and very influential in eighteenth century Scottish botany. He probably planned a catalogue of Scottish plants, but made his records and herbarium available to Lightfoot whose work stole much of his thunder (Morton 1986, Noltie 2011). He organised excursions in Scotland by his students, one of whom, James Robertson, see below, made major discoveries. He had a large and important Scottish herbarium, which has not been traced, but the hand-written catalogue of the *hortus siccus*, dated 1768, survives. It was published anonymously but there is next to no doubt that the author was I.B. Balfour, the Regius Keeper at Edinburgh, and it is referenced as such (Balfour 1907). This is all the more certain as Balfour, under his own name, had published earlier notes on the same subject in the *Annals of Scottish N.H*. This catalogue, published too late for Clarke to incorporate, and seemingly overlooked by Druce, contains much original information (though not, of course, now backed by the actual specimens). It was added to, seemingly up to his death, but it is apparently possible to distinguish between the original and later annotations.

One of Hope's protégées was **James Robertson** and it is only recently that his contributions have become better known, though many of the records are contained in the catalogue of Hope's herbarium (see Balfour, above). Dickson (1986) published his records for Bute, but a much more substantial contribution came from Henderson & Dickson (1994). Their work involved the publication of several manuscripts that were either inaccessible or unknown, comprising Robertson's tours through north and west Scotland from 1767 to 1771. His records have been annotated by them and add many first field records, but I have not used the plant lists of the Islands given in their Chapter 7.

Two other botanists have also had light shed on their contributions to the discovery of Scotland's montane flora - **John Stuart** (Mitchell 1986 & 1992) and **John Walker** (Taylor 1959). For Stuart, the case is simply and persuasively set out, but for Walker, Taylor only sketches the background, leaving others to interpret the information that he draws attention to. This would be particularly important for willows, for Walker was very interested in these, and wrote his drafts, according to Taylor, between 1764 and 1774. Smith practically ignored any reference to Walker in his various works covering the period 1804 to 1828.

The common thread through the careers of the above has been the interaction with John **Lightfoot**. His *Flora Scotica* (1777) was compiled from his own records from his visit in 1772, but also from those of Hope, Robertson, Stuart and others, not always with due acknowledgement. It was not well received upon its publication, but this is not the place for a discussion on this, merely to point out that the works on the other botanists cited above, together with an excellent biography of Lightfoot and catalogue of his herbarium (Bowden 1989), have greatly clarified the dates of the actual discoveries of the plants published, particularly those by Lightfoot himself. See also Slack (1986) for an examination of Lightfoot's first records and of the species illustrated by him.

A slightly later botanist was **Robert Brown**, the subject of a biography by Mabberley (1985).

The key figure of **George Don** has been well-served by historians, but none have treated him more fully or fairly than Druce in a major paper (1904). Here, commencing with a long memoir, Druce deals fully with all of Don's reputed discoveries. He shows that almost all were probably real, and where he made errors then they were compounded by others. He also reproduced in full all the species of Don's *Herbarium Britannicum* (1804-6), reproducing the information on their labels, and followed this by notes on Don's private herbarium and the full text of the plants and animals of Forfar, reprinted from Headrick's '*General view of the Agriculture of the County of Angus, or Forfarshire*', 1813.

The whole, completed by letters from Don to Winch, runs to 200 pages, and to my mind Druce successfully deals with the innuendos and criticisms of Hooker, Arnott (in particular), Watson, Syme and many others that had obscured Don's major contribution to Scottish botany. Even today there is still a slight cloud over Don's name, and a reading of Druce's apologia should be compulsory for all doubters.

Smith and Sowerby's *English Botany* (1790-1814) was a great impetus to the recording of these discoveries, and by the conclusion of that, with its later supplements, the Scottish flora was almost completely known. For a relatively recent summary of the discovery of the Scottish flora, see Fletcher (1959) and papers in Noltie (1986). Among other topics, Lusby & Wright (1996) cover very well the discovery of 45 of the more iconic Scottish species, and Scott (2016) too gives similar details for some of the rarer montane species.

Welsh Flora

There has been no separate flora of Wales, and botanical exploration there has been detailed along with English works ever since Johnson's trip of 1638. Ray's *Itineries* cover trips there, as do papers from Lhwyd, Dillenius and others, as well as Riddelsdell's (1905) editing of the trip by Lightfoot and Banks. Of recent works Ellis (1974, 1983) briefly summarised the discovery of the Welsh flora and Jones (1996) covered that of Snowdonia. A really detailed account of the botanical exploration of every county has been compiled by P.W. Carter, over the period 1946 until his death in 1971. Some of these are cited here, and references to all are given in Ellis (1983) and in Carter (1986).

Appendix V

The Botanists – the discoverers of our flora

This appendix covers people who have discovered any plant in this compilation. For almost every entry there is a reference to a source where more might be ascertained about each. Works where there is major coverage are in **bold**. In addition many of the major compilers of herbals or national floras are also included.

The key source referred to is Desmond (1977 & 1994) (D), where the entries include, in addition to dates of birth and death, main activities and publications, and usually a list of sources of biographies (if any) and obituaries which would give further details of their lives and achievements. But although this work is relatively readily available in libraries, it seems, at prices of £275 and upwards, out of the reach of most readers, so these references are supplemented by those botanists covered by the *Dictionary of National Biography* (DNB), by individual biographies or fuller accounts in Gunther (1922) and Raven (1947, 1950), as well as selected county floras and references to obituaries in the mainstream botanical journals. The Wild Flower Society, for their centenary in 1986, included in one of their bulletins (No. 407: Autumn 1986), a useful little account of the significant members past and present. Where all these avenues have failed, often in relation to people who just happened to have sent plants to better known authors, I have had to limit details to a rough date for the period during which they seem to have been active.

For other sources (listed after DNB & D) see Botanical Exchange Club (BEC); Britten & Boulger (BB); Journal of Botany (JB); Kent & Allen (K); Praeger (1949) (P); BSBI Proceedings (Pr); Watsonia (W); BSBI Year Book (Y). See also Pulteney (1790), Green (1904), Allen (1986: app.2) and Oswald & Preston (2011: 79-96), though none of these sources are referenced if covered adequately by Gunther or Raven. Full details of all the sources are in the Bibliography.

Two other sources that frequently give further details for early botanists are Dandy (1958a) and Clokie (1964) in their accounts of the Sloane and the Oxford Herbaria.

There are biographies for not a few of the early botanists in the volumes of Rees's *Cyclopaedia*, but these have not been referenced here. Sir W. Smith contributed those of Petiver and Ray.

Alternative spellings of names are shown in brackets.

There are a few (<40) where I have been unable to trace a date of birth or death. Many of these might be casual finders who have brought a plant to the attention of someone more expert. These are indicated as '*floruit*' or '*fl*', and in addition to the plants that they are associated with I have added the county to assist future tracing of more details. Those who are assumed to be still alive are undated.

To give some context to the names of the botanists listed I have added the plants that they either discovered or were associated with, listing both the first literature records and an earlier evidence if applicable, up to a total of around 15 species. Those who are associated with more are left blank, as are compilers, identifying plants that others had seen but not identified, unless there is persuasive evidence that they themselves were the discoverers. The boundary between discoverer and describer is not always clear-cut, and in addition, a discoverer of a plant at the time that it entered in the literature might be 'upstaged' by an earlier herbarium specimen or the like. In those cases I have tried to give the credit to both, but any duplication will be readily apparent.

Abbot, R. c.1560-1618. DNB. D. Pulteney, Raven (1947).
Epipactis palustris.

Abbot, Rev. R. 1761?-1817. BB.
Epipactis purpurata.

Abbott, R.J.
Senecio eboracensis.

Airy Shaw, H.K. 1902-1985. D. Kew Bulletin.
Equisetum ramosissimum.

Alchorne, S.A. 1727-1800. D.
Briza minor. Hudson's successor as Demonstrator of Plants at the Chelsea Physic Garden.

Anderson, G. 1773-1817. D. van der Lande (2016).
Salix myrsinifolia.

Andrews, C.R.P. 1870-1951. D.
Milium vernale.

Andrews, J. fl.1710s-1760s. D.
Potamogeton gramineus, Potamogeton obtusifolius.

Arnett, J.E. c.1850-1940. D. K.
Centaurium scilloides.

Arnott., G.A.W. 1799-1868. DNB. D.
Potamogeton praelongus.

Arsene, L. 1875-1959. D.
Limonium auriculae-ursifolium.

Ashby, F. 1660-1743. Kent (1975).
Fritillaria meleagris.

Ashfield, C.J. c.1817-1877. D.
Pulmonaria obscura.

Attenborough, T.W. 1883-1973. D.
Limonium auriculae-ursifolium.

Atwood, Miss M.M. c.1810-1885. D. Allen (1986).
Sorbus bristoliensis.

Auquier, P. fl.1970s.
Festuca huonii in Jersey, 1973 (deceased by 1987).

Babington, C.C. 1808-1895. DNB. D. JB. Babington (1897), Allen (1999).
Surely the preminent critcal botanist of the nineteenth century. *Anacamptis laxiflora, Armeria alliacea, Epilobium obscurum, Fumaria capreolata, Hypericum linariifolium,*

Isoetes echinospora, Myriophyllum alterniflorum, Ornithopus pinnatus, Orobanche hederae, Pilosella peleteriana, Potamogeton coloratus, Ranunculus ophioglossifolius, Ruppia cirrhosa, Spiranthes aestivalis, Vicia parviflora. He was also responsible for elucidating a great number of critical species.

Babington, Rev. C. 1821-1889. DNB. D. JB.
Diphasiastrum complanatum. He was C.C. Babington's cousin.

Backhouse, J. jnr., [fils]. 1825-1890. DNB. D. JB.
Cystopteris alpina, Polygala amarella, Sagina filicaulis, Viola rupestris.

Backhouse, J. snr. 1794-1869. DNB. D.
Athyrium flexile, Carex appropinquata, Minuartia stricta, Polygala amarella, Viola rupestris.

Bacon, G. Miss 1874-1949 D. Pr. Y. (later Mrs G. Foggitt).
Carex microglochin.

Bailey, C. 1838-1924. D. BEC.
Erodium lebelii, Myosotis stolonifera.

Baker, E.G. 1864-1949. D.
Viola kitaibeliana.

Baker, J.G. 1834-1920. DNB. D. BEC. JB.
Erodium lebelii, Ranunculus penicillatus.

Baker, Rev. W.L. 1752-1830. D.
Cephalanthera rubra.

Balfour, Prof. J.H. 1808-1884. DNB. D.
Carex magellanica, C. norvegica, Gymnadenia densiflora, Phyllodoce caerulea, Sagina nivalis.

Balfour, Sir A. 1630-1694. D.
Discoverer of *Ligusticum scoticum* per Desmond.

Ball, P.W.
Salicornia specialist 1959.

Banister, J. 1654-1692. D.
Potamogeton polygonifolius. Desmond gives birth as 1650.

Barnard, A.M. Miss 1825-1911. D.
Bromus interruptus.

Bauhin, C. 1560-1624.
Probably no first records, but his immensely influential works cited by all who followed in the next century.

Bayley, W. 1529-1592. D. Gunther.
Pulicaria dysenterica.

Beattie, J. 1735-1810. D.
Carex binervis, C. laevigata, Linnaea borealis.

Beeby, W.H. 1849-1910. D.
Juncus capitatus, *J. pygmaeus*, *Potamogeton rutilus*.

Beeke, Rev. H. 1751-1837. D.
Euphorbia lathyris, *Lotus angustissimus*.

Bellers, F. 1687-1742. D.
Gymnocarpium robertianum.

Bennett, A. 1843-1929. D. BEC. JB.
Najas marina.

Bennett, W. 1804-1873. D.
Teucrium botrys, see Salmon (1931: 52).

Bentham, G. 1800-1884. DNB. D.
Rumex rupestris.

Bicheno, J.E. 1785-1851. DNB. D.
Juncus gerardii.

Binks, J. 1766-1817. D. Graham (1988).
Gentiana verna.

Blackstone, J. 1712-1753. DNB. D. Kent (1949).
Fritillaria meleagris, *Limonium bellidifolium*. His local flora of the plants around Harefield, Middlesex (Blackstone 1737) gives many first locations of species previously described, but with no such details.

Blomefield, L. 1800-1893. DNB. D. (Jenyns).
Salicornia fragilis.

Bobart, J. jnr. 1641-1719. DNB. D. Vines & Druce (1914).
Brachypodium pinnatum s.l., *Bromopsis erecta*, *Carex dioica*, *Elymus caninus*, *Leontodon saxatilis*, *Microthlaspi perfoliatum*, *Myriophyllum alterniflorum*, ?*Orchis simia*, *Persicaria lapathifolia*, ?*Poa angustifolia*, *Poa nemoralis*, *Polystichum setiferum*, *Quercus petraea*.

Bobart, Tilleman, fl.1650s-1720s. D. Vines & Druce (1914).
Carex nigra, *Poa nemoralis*.

Boel, W. (Boelius). <1616.
Carex hirta in Middlesex. A professional plant collector, he was sent to Spain by Parkinson in 1608.

Bolton, J. <1735-1799. DNB. D.
Gymnocarpium, *Woodsia*.

Borrer, W. 1781-1862. DNB. D. Smail (1974).
Perhaps, with Babington, the pre-eminent critical botanist of the 19th century, with many (>20) first records, a very high number considering how well explored the British flora was by that date. Most are of critical species, but in addition include *Carex vaginata*, *Euphorbia stricta*, *Leersia oryzoides*, *Lonicera xylosteum*, *Nuphar pumila*, *Polygonum maritimum*, *Trifolium bocconei*.

Bossey. fl.1810s.
Spiranthes aestivalis. A possible discoverer of this in the New Forest - see Chatters (2009).

Bowles, G. c.1604-1672. D. Raven (1947).
Cicuta virosa, Drosera anglica/intermedia, Eriophorum vaginatum, Helianthemum oelandicum, Himantoglossum hircinum, Impatiens noli-tangere, Lythrum hyssopifolia, Matthiola sinuata, Ophrys sphegodes, Radiola linoides, Teesdalia nudicaulis, Wahlenbergia hederacea.

Bowman, J.E. 1785-1841. DNB. D.
Elatine hydropiper, Fumaria muralis, Juncus foliosus.

Bradshaw, M.E.
Alchemilla.

Brand, W. 1807-1869. D.
Astragalus alpinus, Carex lachenalii.

Brebner, Rev. J. 1839-1933. D. K.
Schoenus ferrugineus.

Bree, Rev. W.T. 1787-1863. D.
Dryopteris aemula, D. submontana, Rosa agrestis.

Brewer, S. 1670-1743. DNB. D. Pulteney. Hyde (1931).
Dianthus gratianopolitanus, Glyceria notata, Isoetes echinospora, Luronium natans.

Brichan, Rev. J.B. fl.1842-1845. D. K.
Potamogeton praelongus in Nairnshire.

Briggs, T.R.A. 1836-1891. D.
Hypericum undulatum, Pyrus cordata.

Britten, J. 1846-1924. DNB. D. BEC.
Veronica catenata.

Britton, C.E. 1872-1944. D. BEC.
Centaurea debeauxii.

Broad, W. fl.1620s-1630s. D.
Baldellia ranunculoide, Nardus stricta, Stratiotes aloides. Friend of Johnson.

Brodie, J. 1744-1824. D.
Centaurium littorale, Moneses uniflora, Potamogeton filiformis, Potamogeton praelongus.

Bromfield, W.A. 1801-1851. DNB. D.
Clinopodium menthifolium, Orobanche picridis, Polygala serpyllifolia, Vulpia ciliata.

Brown(e), Rev. W. c1628-1678. D. Druce (1927), Gunther.
Chrysosplenium alternifolium, Myosotis discolor, Orchis militaris.

Brown, J. fl.1812.
James Brown Jnr., of Perth, a nurseryman. Probable discoverer of *Phyllodoce caerulea* and co-discoverer of *Sagina saginoides.*

Brown, R. 1773-1858. DNB. D. Mabberley (1985).
Alopecurus magellanicus, *Dryopteris oreades*, [*Phyllodoce caerulea*], *Sagina maritima*, *Trichophorum alpinum*.

Brown, R. c.1767–1845. D. Hardy (2011).
Robert Brown of Perth, a nurseryman, possibly the son of J. Brown. Probable discoverer of *Phyllodoce caerulea*.

Browne, Sir T. 1605-1682. DNB. D.
Sueada vera.

Bruce, A. c1725-1805. D.
Callitriche hermaphroditica, (*Eriophorum latifolium*), *Polygonatum verticillatum*.

Bruinsma, J.
Callitriche palustris.

Brunker, J.P. 1885-1970. D.
Dactylorhiza traunsteinerioides.

Bryant, Rev. H. 1721-1799. D.
Carex extensa, *Crassula tillaea*.

Bucknall, C. 1849-1921. D. BEC.
Stachys alpina.

Buckner, L. fl.1620s-1650s. D.
Baldellia ranunculoides, *Stachys germanica*. Friend of Johnson.

Buddle, Rev. A. c.1660-1715. DNB, D. Trimen & Dyer (1869), Green.
Dandy (1958a) calls him the outstanding taxonomist of his age. He left a MS flora, drafted by 1708 (in Sloane Mss 2970 - 2980), and his Herbarium (Sloane vols. 116 to 125, BM). *Arctium minus*, *Carex extensa*, *Chenopodium ficifolium*, *Equisetum ramosissimum*, *Festuca rubra*, *Galium uliginosum*, *Geranium rotundifolium*, *Juncus gerardii*, *J. subnodulosus*, *Potamogeton pusillus*, *P. trichoides*. His herbarium also contains the first record of a number of critical segregates.

Bull, M.M. 1820-1879. D. JB.
Ranunculus paludosus.

Burchell, W.J. 1781-1863. D. K.
Carex aquatilis.

Burgess, Rev. J. 1725-1795. D.
Ajuga pyramidalis.

Butcher, R.W. 1897-1971. D. W.
Crassula aquatica.

Campbell, M.S. 1903-1982. D. W.
Dactylorhiza ebudensis.

Campbell, W.H. 1814-1883. D.
Scrophularia umbrosa.

Cargill, J. c.1565-1616. D.
Trientalis europaea.

Cautley, Miss C. fl.1830s. K.
Viola persicifolia in S. Lincs. See Gibbons (1975).

Chapple, J.F.G. 1911-1990. W.
Alchemilla acutiloba.

Cheshire, W. fl.1850s. D.
Potamogeton nodosus in Warwicks.

Clark. fl.1730s-1740s. D. (Clerk).
Polycarpon tetraphyllum in Jersey.

Clark. fl.1830s.
Carex lachenalii in Angus. With Dickie 1836, see Sowerby E.B.S 2815.

Clarke, J. fl.1630s-1650s. D.
Rhynchspora alba in Berkshire. Accompanied Thomas Johnson.

Cobbing, P.
Serapias parviflora.

Coleman, Rev. W.H. c.1816-1863. DNB. D.
Bunium bulbocastanum, Oenanthe fluviatilis.

Coles, W. 1626-1662. D. (Cole).
Daphne laureola. The probable discoverer of *Parentucellia viscosa.*

Coombe, D.E. 1927-1999. W.
Trifolium occidentale.

Corbyn, S. ?-1673. D. Druce 1912b & Preston, in press.
The possible discoverer of *Melampyrum cristatum, Ranunculus lingua, Tephroseris palustris and Torilis arvenis.* But his letters were reporting their existence as wild plants, which might have been found by others.

Corder, T. 1812-1873. D.
Bupleurum falcatum.

Cotton, R. fl.1800s.
Rubus arcticus in E. Perth. Sent to Sowerby, see E.B. 1505 (1806).

Cox, Sir C.W.M. 1899-1982. DNB.
Artemesia norvegica.

Craven, H.E. born c.1863, fl.1900s. Coles (2011).
Orobanche reticulata in Yorkshire.

Croall, A. 1809-1885. D.
Vaccinium microcarpum. Joint author of *Nature Printed British Seaweeds.*

Crossfield, G. the elder. 1754-1820. D.
Ranunculus omiophyllus.

Crowe, J. 1750-1807. D.
Carex diandra, Festuca longifolia

Cullum, Rev. Sir J. 1733-1785. D.
Carex montana, Festuca longifolia, Genista pilosa, Muscari neglectum, Thymus serpyllum, Veronica verna.

Curnow, W. c.1809-1887. D. JB.
Lavatera cretica, Poa infirma.

Curtis, W. 1746-1799. DNB. D.
Carex depauperata, Epilobium roseum, Gymnocarpium robertianum, Leucojum aestivum, Puccinellia rupestris.

Dale, J. d. ?1662. D. Gunther.
Foeniculum vulgare. Manuscripts in Goodyer's collection include catalogues of British plants, probably by Dale, possibly amounting to as many 1075 species, a very high number for such an early date.

Dale, S. 1659-1739. DNB. D. Boulger (1883).
Neighbour and executor of Ray. *Atriplex laciniata, Avenula pratensis, Bromus racemosus s.l., Carex strigosa, Chenopodium hybridum, Filago gallica, Koeleria macrantha, Limonium humile, Orchis anthropophora, ?Rosa tomentosa, Schoenoplectus tabernaemontani, Silene noctiflora, Valerianella dentata, V. rimosa, Vulpia fasciculata.*

Dalton, Rev. J. 1764-1843. D.
Scheuchzeria palustris.

Dandy, J.E. 1903-1976. D. W.
Potamogeton.

Dare, G. fl.1680s-1690s. D. (Daire).
Hymenophyllum tunbrigense in W. Kent.

Davall, E. 1763-1798. DNB. D.
Carex davalliana.

Davey, F.H. 1868-1915. D. BEC. JB. Bates & Spurgin (1994).
Echium plantagineum, Euphrasia vigursii.

Davies, Rev. H. 1739-1821. DNB. D.
Juncus subnodulosus, Rosa sherardii, Tephroseris integrifolia.

Davy, Lady J.C. 1865-1955. D. Pr.
Carex microglochin. Friend of Druce.

Deakin, R. 1808-1873. D.
Epilobium obscurum.

Dent, P. 1628/9-1689. DNB. D.
Papaver dubium. Friend of Ray.

Dick, R. 1811-1866. DNB. D. See Smiles (1878).
Calamagrostis scotica, Hierochloe odorata.

Dickenson, Rev. S. 1730-1823. D.
Agrostis gigantea.

Dickie, G. c1812-1882. DNB. D. JB.
Carex lachenalii, C. rupestris, Cystopteris dickieana, Eleocharis uniglumis.

Dickinson, J. c.1805-1865. DNB. D.
Carex flava s.s.

Dickson, J. 1738-1822. DNB. D. Druce (1917a).
Carex bigelowii, C. depauperata, C. saxatilis, Cerastium cerastoides, C. diffusum, C. pumilum, Crepis mollis, Draba norvegica, Erigeron borealis, Eriophorum latifolium, Gentiana nivalis, Juncus castaneus, Kobresia simpliciuscula, Lotus angustissimus, Salix myrsinites, Saxifraga cernua, Veronica alpina, V. fruticans.

Dillenius, J.J. 1684-1747. DNB. D. Druce & Vines (1907).
More a compiler than a discoverer, but the latter include *Fumaria purpurea, Hypericum maculatum, Koeleria vallesiana, Salix aurita, Utricularia australis.*

Dodsworth, Rev. M. 1654-1697. D. Britten (1909).
Eleocharis acicularis, Lysimachia thyrsiflora, Ribes alpinum.

Don, D. 1799-1841. DNB. D.
Myosotis secunda, Polygala calcarea.

Don, G. 1764-1814. DNB. D. Green. See particularly Druce (1904).
Very many Scottish discoveries.

Dony, J.G. 1899-1991. W.
Carex cespitosa.

Doody, S. 1656-1706. DNB. D. Trimen & Dyer (1869).
A friend of Petiver and Plukenet, and colleague of Ray. *Bromus commutatus, Carex pilulifera, Chenopodium glaucum, Deschampsia flexuosa, Dryopteris affinis s.l., Equisetum fluviatile, Gastridium ventricosum, Koeleria macrantha, Polygonum rurivagum, Schedonorus arundinaceus, Schedonorus giganteus.*

Druce, F. 1873-1941. D. BEC. JB.
Alchemilla micans.

Druce, G.C. 1850-1932. DNB. D. JB. Allen (1986a).
Bromus interruptus, Dactylorhiza praetermissa, Gymnadenia borealis, Koeleria vallesiana, Poa humilis, Potamogeton nodosus. He also elucidated a number of critical groups.

Drummond, J. c.1784-1863. DNB. D.
Spiranthes romanzoffiana.

Drummond, T. c.1790-1835. DNB. D.
Juncus balticus, Salix lanata.

Du Bois, C. 1656-1740. DNB. D. See Druce (1928).
He appears to have been a collector of specimens, rather than a discoverer in his own right.

Dunlop, G. fl.1831-1853.
Scrophularia umbrosa in Berwicks.

Duthie, J.F.1845-1922. D. JB.
Polygala amarella var. *austriaca* in Kent.

Eales, L. fl.1660s-1690s. D.
Gentianella amarella in Herts.

Edmonston, T. 1825-1846. DNB. D. (Edmonstone).
Arenaria norvegica, Carex magellanica, Cerastium nigrescens.

Ellis, A.E. 1909-1986. W.
Limonium paradoxum.

Fabricius, F.W.P. 1742-1817.
Carex limosa. German naturalist, who received his medical doctorate from the University of Edinburgh in 1767.

Falconer, J. d.1560. DNB. D.
Pulsatilla vulgaris. Raven (1947) and others refer to Turner's reference to 'Maister Flakonner's boke' is the earliest English record of a herbarium. Note that he died in 1560, not 1547, as given in Desmond.

Farrar, D.R.
Trichomenes speciosum gametophyte.

Finlay, Capt. J. c.1760-1802. D.
Juncus capitatus, Romulea columnae.

Fisher, H.S. d.1881. D.
Erodium lebelii.

FitzRoberts, J. fl.1680s-1720s. D. (also known as Robinson, J). Foley (2008).
Gentianella campestris, Melica nutans in Westmoreland.

Foggitt, W. 1835-1917. D. BEC.
Ranunculus penicillatus.

Forster, E. 1765-1849. DNB. D.
Fumaria bastardii, Gentianella uliginosa, Luzula forsteri, Utricularia australis, Valerianella carinata, V. dentata, Viola arvensis.

Forster, T.F. 1761-1825. DNB. D.
Juncus capitatus, Lotus glaber, Myosotis laxa, Viola lactea.

Forster, T.I.M. 1789-1860. DNB. D.
Stellaria neglecta.

Foulkes, Rev. R. c.1702-1729. D. (Fowkes, Fowlkes).
Saxifraga hirculus. Assumed to be this person.

Fowler, Rev. W. 1835-1912. D. JB.
Selinum carvifolia.

Fox, Rev. H.E. 1841-1926. D.
Elytrigia campestris.

Francis, Rev. R.B. c.1768-1850. D.
Dryopteris cristata.

Fraser-Jenkins, C.R.
Dryopteris specialist.

Fryer A. 1826–1912. DNB. D. JB.
Potamogeton nodosus.

Gardiner, J.C. 1905-1989. D. W.
Carex caespitosa.

Garnsey, Rev. H.E.F. 1826-1903. D. JB.
Carex vulpina.

Garth, R. fl.1550s-1597. D. Gunther, Raven (1947).
Pinguicula lusitanica, *P. vulgaris*, *Primula farinosa.*

Gay C. fl.1830s.
Pilosella peleteriana. A Frenchman who made a two day visit to Guernsey in 1832. See McClintock (1975).

Gerard(e), J. 1545-1612. DNB. D. Jeffers (1967, 1969).
Around 170 first records.

Gibb, of Inverness. fl.1810s.
Primula scotica in Caithness. See Hooker (1819).

Gibson, G.S. 1818-1883. DNB. D. JB.
Filago pyramidata, *Minuartia stricta.*

Gibson, S. 1789-1894. D.
Spergularia rupicola.

Gillenius, A. <1620.
Eleogiton fluitans. See Britten (1909).

Glasspoole, H.G. 1825-1887. D. JB.
Carex trinervis.

Glück, C.M.H. 1868-1940. K.
Veronica catenata.

Glynn, T. fl.1630s. (Glyn). D. Raven (1947), Pulteney.
Geum rivale. A friend of Johnson.

Godfery, M.J. 1856-1945. D.
Epipactis leptochila.

Goldie, J. 1793-1886. D. JB.
Rumex longifolius.

Goodenough, Rev. S. 1743-1827. DNB. D.
Agrostis canina, *Carex* species, *Leersia oryzoides.*

Goodyer, J. 1592-1664. DNB. D. Gunther.
  A very large number of discoveries.

Goss, H. G. fl.1857-1907. K.
  Dactylorhiza purpurella in Cumberland.

Gosselin, J. 1739-1813. D. McClintock (1982).
  At least 16 first records, all from Guernsey.

Graham, R. 1786-1845. DNB. D.
  Astragalus alpinus, Ononis reclinata.

Grant, J. d.1930. D.
  Carex recta.

Graveson, A.W. 1893-1979.
  Teucrium chamaedrys. Hertfordshire botanist, who made important Dorset records, and was correspondent and companion of Lousley.

Green, H.E. 1886-1973. D. W.
  Senecio cambrensis.

Gregory, Mrs E.S. 1840-1932. D.
  Prunella laciniata.

Greville, R.K. 1794-1866. DNB. D.
  Astragalus alpinus, Carex norvegica, Dryopteris affinis, Equisetum pratense.

Griffith, J.W. 1763-1834. D.
  Poa glauca, Saxifraga cespitosa, Saxifraga rosacea.

Groult, Rev. P. fl.1800s. D.
  Carex davalliana in Bath, N. Somerset.

Groves, H. 1855-1912. D. JB.
  Elatine hydropiper, Ranunculus ophioglossifolius.

Guille, Miss M.E. d.1903. D.
  Ophioglossum lusitanicum.

Gunthorp, Dr. <1666.
  Viola odorata. From 'Cornwal', in Merrett (1666).

Hagen, H. fl.1850s.
  Anogramma leptophylla in Jersey, see Newman (1853a).

Halliday, G.S.
  Crepis praemorsa.

Hambrough, A. J. c.1820-1861. D.
  Arum italicum.

Hampton, M.
  Sorbus domestica.

Hankey, J.A. d.1882. D.
  Fallopia dumetorum.

Harriman, Rev. J. 1760-1831. DNB. D.
Gentiana verna, Kobresia simpliciuscula.

Hart, H.C. 1847-1908. D. JB. P.
Arabis alpina.

Haslam, S.H. d.1856. D.
Polygonum maritimum.

Haworth, A.H. 1768-1833. DNB. D.
Cyperus fuscus.

Hayward, I.M. 1872-1949. D. W.
Poa infirma.

Heaton, Rev. R. fl.1620s-1660s. D. Walsh (1978), Raven (1947).
Dryas octopetala, ?Epipactis atrorubens, ?Euphorbia hyberna, Gentiana verna, ?Saxifraga spathularis, Scilla verna in Ireland and *Melittis melissophyllum* in England.

Hemstead, Rev. J. 1746-1824. D.
Cuscuta europaea.

Henslow, H.D. d.1843. K.
Salicornia europaea.

Henslow, Rev. J.S. 1796-1861. DNB. D.
Carex spicata, Eleocharis uniglumis, Euphrasia pseudokerneri, Fumaria vaillantii, Limonium recurvum, Poa angustifolia.

Hesketh, T. 1561-1613. DNB. D.
Andromeda polifolia, Mertensia maritima, Saxifraga hypnoides.

Heslop Harrison, J.W. 1881-1967. D. W.
Euphrasia heslop-harrisonii , Potamogeton epihydrus, (*Gymnadenia densiflora*).

Hibbert, E. fl.1870s.
Carex ornithopoda in Derbyshire. Discoverer, with others, see Babington (1874a).

Hill, Waldron. fl.1780s.
Verbascum virgatum in Worcs. Friend of Stokes (in Withering 1787).

Holbech, Rev. C. 1782-1837. D.
Aster linosyris.

Holliday, W.H. 1834-1909. D. See Druce (1927: cxvii).
Apium repens.

Hood, Mrs E.G. fl.1870s.
Gentianella ciliata. A painting of this from Bucks., 1873, see Pope (2003).

Hooker, Sir J.D. 1817-1911. DNB. D.
Potamogeton praelongus.

Hooker, Sir W.J. 1785-1865. DNB. D.
Luzula arcuata.

Hope, J. 1725-1786. DNB. D. Noltie (2011), Morton (1986).
Carex lasiocarpa, Ranunculus reptans. These were listed by him, but we do not know if he was the discoverer.

Hore, Rev. W.S. 1807-1882. D. JB.
Trifolium incarnatum subsp. *molinerii*.

Hort, F.J.A. 1828-1892. DNB. D. JB.
Ranunculus peltatus.

Hosking, A.J. ?1874-1938. K.
Salicornia dolichostachya in W. Norfolk.

Houston, L.
Sorbus.

Houstoun, W. c1695-1733. DNB. D.
Carum verticillatum.

How, W. 1620-1656. DNB. D. Gunther, Raven (1947).
Artemisia campestris, Centunculus minimus, Coeloglossum viride, Dryas octopetala, Euphorbia hyberna, Gentiana verna, Linum perenne, Melittis melissophyllum, Ophrys sphegodes, Scilla verna, Silene otitis, Viola lutea, Zostera marina.

Hoy, J.B. d.1843. D.
Moneses uniflora.

Hubbard, C.E. 1900-1980. D. W.
Festuca huonii.

Hudson, W. 1734-1793. DNB. D.
He undoubtedly discovered speices, but gives no hint as to whether his additions were his records other than *Carex bigelowii*.

Hume, A.O. 1829-1912. DNB. D. JB.
Alchemilla monticola, Fumaria occidentalis.

Humphrey, W. d<1792. D.
Vicia lutea.

Hunnybun, E.W. 1848-1918. D. BEC. JB.
Luzula pallidula.

Hunter, A. fl.1836-1874. K.
Scrophularia umbrosa in W. Lothian.

Hunter, C. 1675-1757. DNB. D. Horsman (2011).
Possibly *Betula nana*. The DNB notes that he contributed a list of wild plants of co. Durham to Gibson's second edition of Camden's Britannia (1722).

Hunter, E. fl.1790-1824. D. See Kent (1975).
Maianthemum bifolium in Middlesex.

Huon, A.
Festuca armoricana in Jersey, 1970.

Hutcheon, K.
Carex salina.

Hutcheon, K. E.
Carex salina.

Hutcheson, J.C. fl.1869-1878. K.
Juncus alpinoarticulatus in Angus.

Hutchins, Miss E. 1785-1815. D.
Alchemilla monticola, Fumaria occidentalis.

Ingall, T. c.1799-1862. K.
Teucrium botrys.

Ingrouille, M.J.
Limonium specialist.

Irvine, A. 1793-1873. DNB. D.
Stellaria pallida.

Jalas, P. 1920-1999.
Thymus.

Jenyns, Rev. L. 1800-1893. D. (Blomefield).
Fumaria densiflora, Potamogeton coloratus.

Johns, Rev. C.A. 1811-1874. DNB. D. JB.
Trifolium strictum.

Johnson, R. d.1689. D.
Potentilla fruticosa.

Johnson, T. c.1597-1644. DNB. D. Kew & Powell (1932). Also Gunther, Raven (1947). Around 170 first records both from all his expeditions and in his revision of Gerard.

Johnston, H.H. 1856-1939. D.
Euphrasia scottica.

Jones, Rev. J.P. 1790-1857. DNB. D.
Allium schoenoprasum, Ulex gallii.

Jorden, G. 1783-1871. D.
Thymus pulegioides.

Kayse, R. fl.1680s. (Kaise).
Hornungia petraea. Bristol, see Ray (1690); probably also *Polypodium cambricum.*

Kemp, Miss F. fl.1870s.
Romulea columnae.

Kemp, R.F.O.
Gagea bohemica in Radnor, 1965. Mycologist.

Keogh, Rev. J. c.1681-1754. DNB. D. (K'Eogh).
Euphorbia hyberna.

Khan, M.A.
Brachypodium rupestre.

Kidston, R. 1852-1924. D. JB.
Dryopteris cambrensis.

Kingstone. fl.1720s.
Saxifraga hirculus in Cheshire. See Ray (1724).

Knight, W. 1786-1844. DNB. D.
Cystopteris dickieana.

Knowlton, T. 1691-1781. DNB. D. Henrey (1986).
Orobanche hederae, Romulea columnae.

L'Ecluse, C. 1526-1609. (Clusius). Oswald & Preston (2011).
Clinopodium acinos, Erica cinerea, Salvia verbanacea.

L'Obel, M. 1538-1616. DNB. D. (Lobel). See particularly Louis (1980), Gunther, Raven (1947).
At least 90 first records.

La Gasca y Segura, W. 1776-1839. D.
Persicaria mitis.

Lambert, A.B. 1761-1842. DNB. D.
Carex davalliana, Cirsium tuberosum.

Lawson, Rev. T. 1630-1691. DNB. D. Whittaker (1986) & Raven (1948).
Allium scorodoprasum, [Campanula rapunculus], Hymenophyllum wilsonii, Lysimachia thrysiflora, Phegopteris connectilis, Sagina maritima.

Lea, Miss. fl.1830s.
Diphasiastrum complanatum.

Lees, E. 1800-1887. DNB. D. JB.
Oenanthe pimpinelloides, Pyrus pyraster.

Leighton, W.A. 1805-1889. DNB. D. JB.
Epipactis leptochila, Scrophularia umbrosa, Sorbus anglica, Trichophorum cespitosum s.s.

Lemon, Sir, C. c.1784-1868. D.
Erica ciliaris.

Ley, Rev. A. 1842-1911. D. JB.
Sorbus leyana, S. minima, Sorbus stenophylla.

Lhywd, E. 1660-1709. DNB. D. Ray. Green. Gunther (1945).
The second Keeper of the old Ashmolean Museum in Oxford. According to Gunther (1945) also known as Llhuwyd, Llwyd, Lloyd, Floyd, Floid, ffloid, Lhuyd, Lhuid & Llhuyd. Gunther includes a paper on Lhwyd's life by R. Ellis. Some of his records are in Ray's works, others in numerous letters to his friends, many of which were collected by Gunther (1945) and/or published in the Philosophical Transactions of the Royal Society. See also Chater (1984). *Adiantum capillus-veneris, Arenaria ciliata, Asplenium viride, Cardamine hirsuta, Cerastium nigrescens, ??Cystopteris alpina, Dryopteris expansa,*

Gagea serotina, Isoetes lacustris, Juncus triglumis, Lycopodium annotinum, Phegopteris connectilis, Polystichum lonchitis, Saxifraga hirsuta, Thalictrum alpinum, Woodsia alpina, W. ilvensis.

Lightfoot, Rev. J. 1735-1788. DNB. D. See especially Bowden (1989).
Many Scottish records. It is not easy to separate those that he recorded on his 1772 tour from those reported to him by Hope, Stuart and others, but his discoveries, with Pennant, include *Arctostaphylos alpinus, Carex pauciflora, Corallorhiza trifida, Epilobium anagallidifolium, Goodyera repens, Juncus trifidus, Loiseleuria procumbens, Luzula spicata, Minuartia sedoides.*

Lingwood, R.M. d.1887. D.
Fumaria densiflora.

Linton, Rev. E.F. 1848-1928. D. BEC. JB.
Alchemilla glabra, A. xanthochlora, Limonium paradoxum.

Lister, M. 1638-1712. DNB. D. Raven (1947).
Carex distans.

Lloid, M. fl.1640s D. (Lloyd).
Oxyria digyna.

Lloyd, G. 1804-1889. D.
Fumaria capreolata. Almost certainly this man.

Long, J.W. 1864-1948. D.
Gaudinia fragilis.

Lorkin, R. fl.1620s-1630s. D. (Larkin).
Berula erecta in W. Kent. Companion of Thomas Johnson.

Lorraine, H. 1793-1883.
Alchemilla glabra, Dactylorhiza purpurella. She was a watercolour artist, whose family owned Kirkharle [20m NW. of Newcastle], the birthplace of 'Capability' Brown. The paintings are dated and located, and were named by Prof. A. J. Richards.

Lousley, J.E. 1907-1976. D. W.
Alisma gramineum.

Lowe, A.J.
Senecio eboracensis.

Lucas, J. fl.1790s.
Draba aizoides from Stout Hall, Swansea, see Wade *et al.* (1994).

Lucas, Rev. W.H. fl.1850s.
Gladiolus illyricus from S. Hants. See Babington (1857).

Luxford, G. 1807-1854. DNB. D.
Fallopia dumetorum.

Lynam, J. 1812-1885. D.
Sisyrinchium bermudiana.

Lyons, I. 1739-1775. D.
Phleum phleoides.

Lyte, H. c1529-1607. DNB. D.
It is not easy to know whether he was describing new plants from Britain or from the continent, but possibly *Aegopodium podagraria, Chenopodium vulvaria, Glechoma hederacea, Lapsana communis, Lepidium coronopus, Salsola kali.* Pulteney gives him credit for many more.

MacCalla, W. c.1814-1849. D. P. (M'Alla).
Erica mackaiana.

Mackay, J. 1772-1802. D. see Ingram & Noltie (1981).
Carex laevigata, Eleocharis multicaulis, Festuca altissima, Minuartia rubella, Myosotis alpestris, Poa flexuosa, Sagina saginoides, Sorbus pseudofennica,

Mackay, J.T. 1775-1862. DNB. D. P.
Arenaria ciliata, Asplenium onopteris, Erica erigena.

Mackechnie, R. 1902-1978. D. W. K.
Calamagrostis purpurea, Rumex aquaticus.

Macpherson, Sir J. ?1713-1765. ?DNB.
Eriocaulon aquaticum. Possibly John M., a native of Skye and a minister there, or his son, also John (1745-1821), the first baronet.

MacRitchie, W. 1754-1837. D.
Callitriche hermaphroditica, Potamogeton coloratus.

Manningham, Rev. T. 1684-1750. D.
Cerastium diffusum, Ceratophyllum submersum, Petrorhagia nanteuilii.

Mansel-Pleydell, J.C. 1817-1902. DNB. D. JB.
Valerianella eriocarpa.

Marett, Miss J. fl.1850s.
Anogramma leptophylla in Jersey 1853, see Le Sueur (1984).

Marquand, E.D. 1848-1918. D. BEC. JB.
Galium constrictum.

Marshall, Rev. E.S. 1858-1919. D. BEC. JB.
Alchemilla glomerulans, Carex chordorrhiza, Cochlearia micacea.

Martyn, J. 1699-1768. DNB. D.
?Ribes uva-crispa.

Matthews, A. ?d.1841. D.
Probably this man, the discoverer of *Ophrys fucifera* in E. Kent before 1828, see Smith (1829). Only in Desmond (1977).

Merrett, C. 1614-1695. DNB. D.
At least 44 first records.

Milford, J. fl.1830s.
Romulea columnae. Discoverer of this species in S. Devon and author of the account in Loudon's *Mag. of N.H.*

Milne-Redhead, E.W.B.H. 1906-1996. W.
Cerastium brachypetalum.

Mitten, W. 1819-1906. D.
Carex montana.

Moggeridge, M. 1803-1882. D.
Wolffia arrhiza.

Molyneux, Sir T. 1661-1733. DNB. D.
Saxifraga spathularis.

Montford, H.M. fl.1930s.
Koeningia islandica. Discoverer in Skye, 1934.

Moore, D. 1807-1879. DNB. D. JB.
Carex appropinquata, C. buxbaumii, Inula salicina.

Moore, J.J.
Minuartia recurva 1966.
Aged 87 in 2014, a missionary in Zambia.

Moore, T. 1821-1887. DNB. D. JB.
Dryopteris affinis s.s., Glyceria declinata, G. notata.

More, A.G. 1830-1895. DNB. D. JB. P.
Spergularia.

More, Miss F.M. d.1909. D.
Neotinea maculata.

Morison, R. 1620-1683. DNB. D.

Morley, C.L. c.1646-1702. DNB. D.
Tripleurospermum inodorum.

Moss, C.E. 1870-1930. D. BEC. JB.
Polygonum aviculare s.s., *arenastrum* & *rurivagum.*

Mott, F.T. 1825-1908. D.
Prunella laciniata.

Mount, Rev. W. 1545-1602. DNB. D. Gunther.
Gunther gives a list of plants recorded by Mount from around his home near Maidstone in Kent. These were written by Mount in his copy of L'Obel (1581), which, of course, was illustrated, and with the help of the Kew botanist, O. Stapf, Gunther gives a modern name to a selection, but it does seem as though many of these are tenuous, although Raven (1947) does not criticise these specifically. They are noted against the species concerned.

Moyle ?R. (or ?J.) fl.1570s. (Muyle).
Asplenium marinum and *Potentilla sterilis*, mentioned in L'Obel (1571). He,

and his descendants lived at Bake, nr. St Germans, East Cornwall. John Moyle, M.P., of that address lived c.1522-1586. (https://www.geni.com/people/John-Moyle/6000000006444616620).

Murray, Rev. R.P. 1842-1908. D.
Spartina anglica.

Naesmyth, Sir J. d.1779. K.
It is assumed that this is he who found *Betula nana* in Ross-shire, but this is not mentioned in the account in DNB.

Neck, Rev. A. 1769-1852. D.
Bupleurum baldense.

Neill, P. 1776-1851. DNB. D.
Phyllodoce caerulea.

Nelmes, E. 1895-1959. D.
Carex flava group.

Newbery, W. d.<1797. D.
Lobelia urens.

Newbould, Rev. W.W. 1819-1886. DNB. D. JB.
Agrimonia procera, Fumaria muralis, Rumex rupestris, Sagina apetala s.s., Stellaria pallida.

Newman, E. 1801-1876. DNB. D. JB.
Anogramma leptophylla, Ferns.

Newton, H. ?fl.1850s. D.
Carex ornithopoda. Discoverer in Derbyshire, with others, see Babington (1874a).

Newton, J. 1639-1718. D.
Friend of Ray.

Nicholson, J. d. c.1880. ?D. Gibbons (1975).
Viola persicifolia. Assumed to be this person. Desmond reports there are plants in herb. T. Baker at Whitby.

Nicolson, W. 1655-1727. D. Whittaker (1981).
His manuscript diary (very well annotated in Whittaker's work), contains no new records, but valuable information on many of his contempories.

O'mahony, Rev. T. 1823-1879. D. (O'Mahoney).
Simethis mattiazzii.

Oaks. fl.1760s. (Oakes).
Poa alpina. Sent specimen from ?Argyll to John Hope by 1767.

Oliver, D. 1830-1916. D. JB. (Britten 1917).
Euphrasia salisburgensis, Najas flexilis.

Ostenfeld, C.E.H. 1873-1931. JB. K.
Alchemilla wichurae.

Pamplin, W. 1806-1899. D. JB.
Fumaria muralis, Gentianella germanica.

Parker, Miss. fl.1830s.
Epipactis phyllanthes in Surrey.

Parkinson, J. 1567-1650. DNB. D. Parkinson (2007).
Around 35 first records.

Parsons, J. 1742-1785. D.
Ranunculus reptans.

Pearsall, W.H. 1891-1964. DNB. D. BEC.
Callitriche brutia.

Peete ,W. 1771-1848. D.
Polygala calcarea.

Pena, P. (c.1530->1600). D. See particularly Louis (1980), Gunther, Raven (1947). Jackson (1899).
The joint discoverer with L'Obel of many plants during the latter's first stay in England from 1567-1571.

Pennant, T. 1726-1798. DNB. D.
The companion of Lightfoot on their 1772 tour, and presumably, joint discoverer with him of many Scottish species.

Penny, Rev. T. c.1530-1589. DNB. D. Gunther, Raven (1947), Foley (2006).
Arnoseris minima, Baldellia ranunculoides, Cirsium heterophyllum, Rubus chamaemorus, Sedum rosea, Trollius europaeus. Our knowledge of Penny's contribution to the discovery of the British flora has been greatly advanced by the issue of the plant illustrations that lie behind Conrad Gesner's *Historia Plantarum*. (Zoller *et al.* 1972 – 1987). These illustrations were heavily annotated, by others as well as by Penny, but Foley has cogently argued that Penny must have made his notes between his arrival in Europe in 1565, and his return to England in 1569, and that they must have been based on English knowledge that he had before he left here in 1565. 26 species (not all native plants) are so annotated, mainly as 'in Anglia', but including some with actual locations; 13 of these are first evidences and have been included in this compilation. He sent details to L'Ecluse and Pulteney (1790) suggests that he communicated plants to Gerard and L'Obel too.

Petiver, J. c.1658-1718. DNB. D. Trimen & Dyer (1869).
Alisma lanceolatum, Anthemis arvensis, Callitriche brutia, Carex distans, Chenopodium urbicum, Melampyrum arvense, Oreopteris limbosperma, Polygonum arenastrum, Rorippa palustris, Veronica anagallis-aquatica s.s.

Philipson, P.
Gentianella ciliata in Bucks, 1982.

Philipson, W.R. fl.1930s.
Agrostis. Author of monograph (1937).

Phillips, Mrs. fl.1850s.
Gladiolus illyricus. From Isle of Wight, see More (1862). The wife of a local clergyman

Pickard, J.F. 1876-1943. D.
Orobanche reticulata.

Piquet, J. 1825-1912. D. JB.
Callitriche obtusangula, Poa infirma.

Pitchford, J. c.1737-1803. D.
Carex elata, Holosteum umbellatum, Orobanche purpurea.

Pitt, E. fl.1650s-1670s. D.
Sorbus domestica in Worcestershire.

Planchon, J.E. 1823-1888. D.
Ulex gallii.

Plot, R. 1640-1696. DNB. D. (Plott).
Arenaria serpyllifolia s.s., Eleocharis acicularis, Potamogeton berchtoldii, ?P. friesii, Potentilla anglica, Sagina apetala.

Plukenet, L. 1642-1706. DNB. D. (Plucknet) Trimen & Dyer (1869). Pulteney.
Atriplex pedunculata, Helianthemum apenninum, Potamogeton friesii, Ranunculus circinatus, Salvia pratensis.

Polwhele, Rev. R. 1760-1838. D.
Sorbus devoniensis.

Praeger, R.E. 1865-1933. DNB. D.
Asplenium onopteris.

Price, Rev. R. fl.1820s.
Phyteuma spicatum in W. Sussex. See Borrer in E.B.S. 2598.

Prior, (R.C.) A. 1809-1902. D. JB.
Carex vulpina.

Proctor, M.C.F.
Sorbus.

Pugsley, H.W. 1868-1947. D. W.
Dactylorhiza kerryensis, Gentianella uliginosa and also discovered or described many species in the *Euphrasia* and *Limonium* critical groups.

Purchas, Rev. W.H. 1823-1903. D. JB.
Sorbus eminentiformis.

Quekett, E.J. 1808-1847. DNB. D.
Vicia parviflora.

Ralfs, J. 1807-1890. D. JB.
Echium plantagineum, Viola kitaibeliana.

Rand, I. d.1743. DNB. D.
Chenopodium glaucum, Epilobium roseum, Gentianella anglica, Orobanche minor, Silene conica, Tripleurospermum inodorum, Vicia lathyroides.

Ray, J. 1627-1705. DNB. D. Raven (1950). Keynes (1951).
Over 210 first records, not including those in the third edition of his *Synopsis* by Dillenius. See also his Letters in Derham (1760) (and Whittaker (1986) for how much Derham lost!), Lancaster (1848), Gunther (1928, 1934) and Thompson (1974).

Rayer, J. 1735-1797. D.
Althaea hirsuta, Fumaria parviflora.

Rayner, J.F. 1854-1947. D. BEC.
Galium strictum.

Rea, C. 1861-1946. D.
Alisma gramineum.

Reader, H.P. 1850-1929. D.
Diphasiastrum complanatum.

Reeves ,W.W. 1819-1892. D. JB.
Ranunculus tripartitus.

Relham, Rev. R. 1754-1823. D.
Carex hostiana.

Rich, T.C.G.
Sorbus.

Richardson, R. 1663-1741. DNB. D. Memoir in Turner (1835).
Isoetes echinospora, Melica nutans, Ranunculus peltatus, Rhinanthus angustifolius, Stellaria nemorum, Trichomanes speciosum. Turner's work contains only a selection of his letters – possibly one-eighth. He mentions others in his preface, and there are a very large number in the Sloane collection.

Riddlesdell, Rev. H.J. 1866-1941. D. BEC.
Apium repens.

Roberts, J. 1792-1849 (D.)
Limonium procerum. Desmond gives 1828 as the date of death, but Jones (1996) shows this is an error. See also Dallman, A bygone Welsh botanist in *NW Naturalist* 1930: 1.

Robertson, J. fl.1760s-1770s. D. See especially Henderson & Dickson (1994).
Arctostaphylos alpinus, Carex maritima, Eriocaulon aquaticum, Goodyera repens, Loiseleuria procumbens, Minuartia sedoides, Orthilia secunda, Oxytropis halleri, Primula scotica, ?Saxifraga cespitosa, Sorbus rupicola.

Robertson, W. d.1847. D.
Rosa micrantha.

Robson, E. 1763-1813. DNB. D.
Ribes spicatum, Ribes uva-crispa.

Rogers, Rev. W.M. 1835-1920. D. BEC. JB.
Hypericum undulatum.

Rolph (Rolfe) R.A. 1855-1921. D. BEC. JB.
Dactylorhiza purpurella.

Rose, F.R. 1921-2006. W.
Gymnadenia borealis.

Saintloo, E. d.1578. D.
Cirsium eriophorum.

Salmon, C.E. 1872-1930. D. BEC. JB.
Alchemilla glaucescens, Erodium lebellii, Limonium normannicum, Myosotis stolonifera, Poa infirma.

Salmon, J.D. c.1802-1859. DNB. D.
Cyperus fuscus.

Salt, J. 1759-1810. D. Coles (2011).
Carex elongata, Carex ornithopoda.

Salusbury, Sir J. 1567-1612. DNB. D. Gunther.
Erigeron acer, Matthiola sinuata, Veronica spicata.

Sandwith, C.I. 1871-1961. D.
Scorzonera humilis.

Sandwith, N.Y. 1901-1965. D. Pr.
Eleocharis austriaca, Scorzonera humilis.

Sandys, Rev. G.W. ?1812-1848. D. K.
Presumably the finder of *Viola reichenbachiana.*

Sare. <1650.
Only in How (1650). From Newmarket, Cambs. Discover of *Artemisia campestris, Linum perenne, Silene otites* and others. It has not been possible to find anything about him, or in what way How obtained his records.

Saunders, W.W. 1809-1879. DNB. D. JB.
Potamogeton coloratus.

Scampton, J. fl.1690s-1710s. D.
Calamagrostis canescens in Leics.

Scarles. fl.1770s.
Orobanche purpurea in Norfolk. Discoverer of this in 1779, EB 423.

Scott, H. <1746.
Limonium bellidifolium in Norfolk. Discoverer of this, see Blackstone (1746).

Scott, R. 1757-1808. D. Nelson (1979).
Utricularia intermedia s.l.

Scott, W. E.
Pilosella flagellaris.

Seward, J. fl.1890s. D.
Hypericum maculatum in Worcs.

Sherard, J. 1666-1738. DNB. D.
Alopecurus bulbosus, Carex acutiformis, Galium uliginosum, Melampyrum arvense, Orchis purpurea, Silene conica, Vicia lathyroides.

Sherard, W. 1658-1728. DNB. D.
Agrostis stolonifera, Galium parisiense, Subularia aquatica and many from Jersey.

Shoolbred, W.A. 1852-1928. D. BEC. JB.
Alchemilla wichurae, Carex chordorrhiza, Utricularia stygia.

Shuttleworth, R. J. 1810-1874. D.
Rorippa islandica.

Sibbald, Sir R. 1641-1722. DNB. D.
Ligusticum scoticum, Sibbaldia procumbens.

Sibthorp, J. 1758-1796. DNB. D.
?*Apium repens, Oenanthe silaifolia, Ranunculus circinatus.*

Silliard, Z. fl.1640s. D.
Drosera anglica from Ireland.

Skrimshire, W. 1766-1830. D.
Orobanche purpurea.

Slack, A.A.P. 1913-1998. W.
Homogyne alpina.

Sledge, W.A. 1904-1991. W.
Carex flava.

Sloane, Sir H. 1660-1753. DNB. D.
Callitriche platycarpa, Sarcocornia perennis

Smith, A. fl.1790s.
*Cuscuta europaea.*Aberdeen, 1797 (sent specimen to Sowerby).

Smith, A.J.E. fl.1950s. Jour. of Bryology.
Epipactis sancta in N. Northumberland, 1958. Possibly the distinguished bryologist, 1935-2012, author of the Moss Flora (1978), the Liverwort Flora (1990) and instrumental in setting up the Society's Bryophyte Mapping Scheme in 1960.

Smith, C. c.1715-1762. DNB. D.
Pinguicula grandiflora.

Smith, Mrs W.A. fl.1840s. D.
Epipogium aphyllum from Herefords.

Smith, Mrs. fl.1790s.
Cephalanthera rubra from E. Gloucestershire, 1797, EB 437.

Smith, Rev. G.E. 1804-1881. DNB. D. JB.
Eleocharis parvula, Epipactis phyllanthes, Ophrys fuciflora.

Smith, Sir J.E. 1759-1828. DNB. D.
There are very many first records in his floras and in *English Botany*, but it is not known how many were his discoveries.

Smith, T. fl.1620s. D.
Rhynchspora alba in Berkshire. Accompanied Thomas Johnson.

Smith, W.E. & H. Smith fl.1880s.
Euphrasia foulaensis. Probably the Rev. W.E. Smith (and his wife?) who presented specimens to BM in 1884.

Snooke, W.D. 1787-1857. D.
Marrubium vulgare.

Sole, W. 1741-1802. DNB. D.
Carex depauperata, C. digitata, C. humilis.

Spence, M. 1853-1919. D. JB.
Rorippa islandica.

Stables, W.A. 1810-1890. D.
Potamogeton praelongus.

Stace, C.A.
Taxonomist and collaborator on *Brachypodium* and *Limonium*.

Stackhouse, J. 1742-1819. DNB. D.
Viola lactea.

Stephens, H.O. 1816-1881. D.
Allium sphaerocephalum.

Stephenson, Rev. T. 1855-1948. D.
Epipactis dunensis.

Stephenson, T.A. 1898-1961. D.
Epipactis dunensis.

Stevens, Rev. L. ?1654-1725. D. (Stephens)
Physospermum cornubiense, Scirpoides holoschoenus, ?Sorbus domestica.

Stewart, O. 1920-1998. W.
Calamagrostis purpurea.

Stillingfleet, B. 1702-1771. DNB. D.
Deschampsia setacea, Mibora minima.

Stokes, J. 1755-1831. D.
Cardamine hirsuta, Lemna gibba, Stellaria neglecta.

Stonehouse, Rev. W. 1597-1655. D. Gunther, Raven (1947).
Antennaria dioica, Orchis ustulata, Viola palustris. See also Gunther, *J. Bot* (1920) 58: 170-173.

Stonestreet, Rev. W. d.1716. D.
Eriophorum latifolium, Euphorbia portlandica, Euphrasia nemorosa, Polypodium

interjectum, Puccinellia fasciculata, Raphanus maritimus, Ruppia cirrhosa, Salicornia ramosissima, Stellaria palustris.

Stow, S.C. Miss. C.1870-1956. D.
Vulpia unilateralis in 1903, see Stace (1961).

Stribley, M.
Cystopteris diaphana.

Stuart Rev. J. 1743-1821. DNB. D. (Stewart). (Mitchell 1986 & 1992).
Carex bigelowii, C. capillaris, C.pauciflora, C. saxatilis, Erigeron borealis, Gnaphalium supinum, Juncus biglumis, J. castaneus, Luzula spicata, Melampyrum sylvaticum, Poa alpina, Salix lapponum, S. myrsinites, S. phylicfolia, S. reticulata.

Sutherland, J. ?1638- 1719. D.
Allium scorodoprasum.

Swan, G. 1917-2012. Y.
Trichophorum cespitosum s.s.

Syme, J.T.I. 1822-1888. DNB. D. JB. (Boswell, Boswell-Syme).
Ophioglossum azoricum.

Symmons, P. fl.1969.
Atriplex longipes from Dorset.

Symons, Rev. J. 1778-1851. D.
Epilobium roseum.

Talbot, Sir G. fl.1920-1930. K.
Teucrium chamaedrys in E. Sussex.

Taschereau, P.M.
Atriplex specialist.

Tate, R. 1840-1901. D. See Scott & Palmer (1987).
Polygonum boreale.

Tatham, J. jnr. 1793-1875. D.
Minuartia stricta.

Taylor, Sir G. 1904-1993. DNB. D.
Potamogeton.

Tebbutt, C.F. fl.1950s.
Diapensia lapponica. Archaeologist and ornithologist from Huntingdonshire.

Teesdale, R. c.1740-1804. D.
Carex filiformis, Potamogeton alpinus, P. gramineus.

Tellam, R.V. 1826-1908. D. JB.
Ranunculus tripartitus.

Templeton, J. 1766-1825. D. P.
Orobanche alba.

Templeton, R. 1802-1892. D. (K.) P.
It is possible that he was the Templeton who was the discoverer of *Carex rupestris* with Dickie, 1836. Son of J. Templeton. See also Nash & Ross (1980).

Thiselton-Dyer, W.T. 1843-1928. DNB. D.
Sorbus admonitor.

Thompson, J. c.1778-1866 D.
Carex magellanica.

Thompson, S. 1818-1881. D.
Minuartia stricta.

Thompson. F.W. fl.1930s.
Festuca huonii in Jersey.

Thomson, R. D. 1811-1864.
(*Potamogeton praelongus*). Dates supplied by M.E. Braithwaite.

Thorpe, J. 1682-1750. DNB. D.
Asplenium obovatum.

Tilden, R. fl.1700s. D.
Gentianella germanica in Herts. Dandy (1959: 219) cites a specimen from him in Petiver's 'Hortus Siccus'.

Todd, Miss E.M. 1859-1949. D. K.
Galium constrictum.

Tofield, T. 1730-1779. D. Skidmore *et al.* (n.d.).
Carex digitata, Peucedanum palustre, Vicia bithynica.

Townsend, F. 1822-1905. D. JB.
Centaurium tenuiflorum, Exaculum pusillum.

Townson, R. fl.1790s-1800s. D.
Saxifraga cernua, S. rivularis in Inverness and Perthshire.

Tozer, Rev. J.S. c.1790-1836. D.
Erica ciliaris.

Travis, W.G. 1877-1958. D. Pr.
Epipactis dunensis.

Trevelyan, Sir W.C. 1797-1879. DNB. D. JB.
Armeria arenaria, Carex ericetorum, Polygonum maritimum, Romulea columnae.

Trimen, H. 1843-1896. DNB. D. JB.
Viola kitaibeliana, Wolffia arrhiza.

Trimmer, Rev. K. 1804-1887. D.
Petrorhagia prolifera, Potamogeton trichoides.

Trist, P.J.O. 1908-1996. W.
Elytrigia campestris.

Trow, A.H. 1863-1939. D.
Limosella australis.

Tucker, R. 1832-1905. D. JB.
Gentianella anglica.

Tunstal, Mrs T. fl.1620s. D.
Cephalanthera longifolia, Cypripedium calceolus. She lived at Hornby, E of Lancaster, only a few miles from Ingleton, the sites of these plants.

Turner, Dawson. 1775-1858. D.
Carex punctata, Circaea alpina, Festuca arenaria.

Turner, Rev. W. c1508-1568. DNB. D.
Around 250 first records.

Turton, W. 1762-1835. D.
Draba aizoides.

Tutin, T.G. 1908-1987. D. W.
A collaborator on *Salicornia*.

Tyacke, N. 1812-1900. D.
Lamium confertum.

Uvedale, Rev. R. 1642-1722. DNB. D. Boulger (1891).
Dactylorhiza praetermissa. His superb herbarium is in Hb. Sloane, with specimens from all the leading botanists of the day.

Vice, W.A. 1852-1937. D. (1870-1936. K.).
Euphrasia rivularis.

Walker, Rev. J. 1731-1803. DNB. D. Taylor (1959).
Blysmus rufus, ?Oxytropis halleri, Salix arbuscula, S. lapponum and possibly other Willows if only his works were subject to modern scholarship (*pace* Taylor, *op. cit.*).

Walker, S. 1924-1985.
Dryopteris expansa. 1961. President B.P.S 1975-1979.

Wallace, E.C. 1909-1986. D. W.
Calamagrostis purpurea.

Walters, S.M. 1920-2005. W.
Alchemilla minima, Aphanes arvensis s.s,

Warburg, E.F. 1808-1966. D. Pr.
Sorbus species.

Watlington, J. d. 1659. D. Druce (1897).
Discoverer of plants around Reading, Berkshire, whose records are in How (1650). He was a friend and botanical advisor to Elias Ashmole, and also knew Thomas Johnson, who came to visit him in Reading.

Watson, H.C. 1804-1881. DNB. D. Allen (1986), Egerton (2010).
Athyrium distentifolium, Lotus subbiflorus, Viola reichenbachiana.

Watt, J. 1736-1819. DNB. D.
Centaurium pulchellum. Pioneer of steam engine!

Webb, F.M. 1841-1880. D. JB.
Cochlearia micacea.

Webster, G. 1851-1924. D. BEC. JB.
Dactylorhiza praetermissa, Orobanche reticulata.

Westcombe, T. d 1893. D. JB.
Athyrium flexile with Backhouse.

Wheldon, J.A. 1862-1924. D. JB.
Epipactis dunensis.

White, F.B.W. 1842-1894. DNB. D. JB.
Juncus alpinoarticulatus.

White, J.W. 1846-1932. D. BEC. JB.
Prunella laciniata, Sorbus spp.

White, Miss M. fl.1830s. D.
Ornithopus pinnatus in Isles of Scilly.

White, T. 1724-1797 D.
Orobanche hederae, 1781, vc45 in Curtis *Fl Lond*. Probably this person, known also as T. Holt-White and Gilbert White's younger brother.

Whitehead, J. 1833-1896. D. JB. (Britten 1897a).
Carex ornithopoda.

Whitehead, Rev. E. 1789-1827. D.
Aconitum napellus,

Whitfield, Rev. fl.1780s.
Carex depauperata in W. Kent. Named as the joint discoverer in Curtis (1791).

Wigg, L. 1749-1829. DNB. D.
Chenopodium chenopodioides.

Wilkins, Miss C. fl.1840s. (Wilson).
Simethis mattiazzii in Hampshire.

Wilkinson, M.J.
Festuca.

Williams, Miss. fl.1870s.
Gentianella ciliata in Bucks. Anon (1875).

Williams, Rev. E. 1762-1833. DNB. D.
Aconitum napellus, Elatine hexandra, Potamogeton alpinus.

Willisel, T. Baptised 1621, d. c.1675. DNB. D. Raven (1947 & 1950). (Willisell).
Collected for Merrett, Morison, Sherard and Ray, and a field companion of the last. *Alchemilla alpina, Asplenium septentrionale, Cephalanthera damasonium, Epilobium alsinifolium, Epipactis atrorubens, Equisetum hyemale, Lathyrus palustris, Lobelia*

dortmanna, Potentilla fruticosa, P. tabernaemontani, Sedum villosum, Silene nutans, S. viscaria, Sonchus palustris, Vaccinium uliginosum, Veronica triphyllos.

Willughby, F. 1635-1672. DNB. D. Preston (2015).
Lepidium heterophyllum. Colleague of Ray.

Wilmer, J. 1697-1769. D. K.
Alopecurus bulbosus.

Wilmott, A.J. 1888-1950. D. W.
Alchemilla acutiloba, Myosotis sicula, Sorbus spp.

Wilson, A. 1862-1949. D. W.
Myosotis stolonifera.

Wilson, W. 1799-1871. DNB. D.
Carex punctata, Cotoneaster cambrica, Cystopteris montana, Hymenophyllum wilsonii.

Winch, N.J. 1768-1838. DNB. D.
Centaurium littorale.

Witham, G. c.1610-c1680. D.
Pyrola media or *P. minor.*

Withering, W. 1741-1799. DNB. D.
Alopecurus aequalis, Potamogeton polygonifolius.

Wolley-Dod, A.H. 1861-1948. D. W.
Rosa caesia, R. obtusifolia, Rosa sherardii.

Wolsey, G. c.1814-1870. D.
Isoetes histrix, Ophioglossum lusitanicum.

Wood, D. fl.1730s.
Carex muricata from Cheshire, see David & Kelcey (1985).

Woodley, G. fl.1820.
Hedera hibernica in the Scilly Isles, see Woodley (1822).

Woodruffe-Peacock, Rev. A. 1858-1922. D. JB.
Vulpia unilateris.

Woods, J. 1776-1864. DNB. D.
Rosa agrestis, Salicornia pusilla, Salicornia ramosissima, Trifolium strictum.

Woodward, T.J. 1745-1820. DNB. D.
Carex depauperata, C. disticha, Potampgeton alpinus, (*Vicia parviflora*).

Wright, F.R.E. 1879-1966. D. Pr.
Coincya wrightii.

Yalden, T. 1750-1777. D.
Symphytum tuberosum.

Yeo, P.F. 1929-2010. W.
Euphrasia.

Vice-counties of the British Isles

Channel Isles
S (J) Jersey
S (G) Guernsey
S (A) Alderney
S (S) Sark

England I
1   W. Cornwall
2   E. Cornwall
3   S. Devon
4   N. Devon
5   S. Somerset
6   N. Somerset
7   N. Wilts
8   S. Wilts
9   Dorset
10 Wight
11 S. Hants
12 N. Hants
13 W. Sussex
14 E. Sussex
15 E. Kent
16 W. Kent
17 Surrey
18 S. Essex
19 N. Essex
20 Herts
21 Middlesex
22 Berks
23 Oxon
24 Bucks
25 E. Suffolk
26 W. Suffolk
27 E. Norfolk
28 W. Norfolk
29 Cambs
30 Beds
31 Hunts
32 Northants
33 E. Gloucs
34 W. Gloucs
35 see Wales
36 Herefs
37 Worcs
38 Warks
39 Staffs
40 Salop

Wales
35 Mons
41 Glam
42 Brecs
43 Rads
44 Carms
45 Pembs
46 Cards
47 Monts
48 Merioneth
49 Caerns
50 Denbs
51 Flints
52 Anglesey

England II
53 S. Lincs
54 N. Lincs
55 Leics
56 Notts
57 Derbys
58 Cheshire
59 S. Lancs
60 W. Lancs
61 S.E. Yorks
62 N.E. Yorks
63 S.W. Yorks
64 M.W. Yorks
65 N.W. Yorks
66 Co. Durham
67 S. Northumb
68 Cheviot
69 Westmorland
70 Cumberland

Isle of Man
71 Man

Scotland
72 Dumfriess
73 Kirkcudbrights
74 Wigtowns
75 Ayrs
76 Renfrews
77 Lanarks
78 Peebless
79 Selkirks
80 Roxburghs
81 Berwicks
82 E. Lothian
83 Midlothian
84 W. Lothian
85 Fife
86 Stirlings
87 W. Perth
88 M. Perth
89 E. Perth
90 Angus
91 Kincardines
92 S. Aberdeen
93 N. Aberdeen
94 Banffs
95 Moray
96 Easterness
97 Westerness
98 Argyll
99 Dunbarton
100 Clyde Is
101 Kintyre
102 S. Ebudes
103 M. Ebudes
104 N. Ebudes
105 W. Ross
106 E. Ross
107 E. Sutherland
108 W. Sutherland
109 Caithness
110 Outer Hebrides
111 Orkney
112 Shetland

Ireland
H1   S. Kerry
H2   N. Kerry
H3   W. Cork
H4   M. Cork
H5   E. Cork
H6   Co. Waterford
H7   S. Tipperary
H8   Co. Limerick
H9   Co. Clare
H10 N. Tipperary
H11 Co. Kilkenny
H12 Co. Wexford
H13 Co. Carlow
H14 Laois
H15 S.E. Galway
H16 W. Galway
H17 N.E. Galway
H18 Offaly
H19 Co. Kildare
H20 Co. Wicklow
H21 Co. Dublin
H22 Meath
H23 Westmeath
H24 Co. Longford
H25 Co. Roscommon
H26 E. Mayo
H27 W. Mayo
H28 Co. Sligo
H29 Co. Leitrim
H30 Co. Cavan
H31 Co. Louth
H32 Co. Monaghan
H33 Fermanagh
H34 E. Donegal
H35 W. Donegal
H36 Tyrone
H37 Co. Armagh
H38 Co. Down
H39 Co. Antrim
H40 Co. Londonderry

Species Accounts

Acer campestre

* 'Acutie foliorum cognitu facilis Aceris species illa, quam ... & nonnulli prope Oxoniam easdem oriri sponte nobis asserverint'. - L'Obel (1571: 443). 'Acer minus vulgare'. Between Nash & Queakes [in Thanet]. - Johnson (1632: 13).   vc22, 23   **1571**

Druce notes the carving of leaves of this species on St. Frideswide's shrine, Oxford, dated to 1289.

Achillea maritima

* 'Gnaphalium marinum ... At a place called Merezey [Mersea] sixe miles from Colchester neere unto the sea side'. - Gerard (1597: 516).   vc19   **1597**

Achillea millefolium

* 'Millefolium ... ab anglis ... Yarow'. - Turner (1538: B. iij).   **1538**

Achillea ptarmica

'called in Duche Wilder bertram, ... they ar also indented all about ye edges of the lefe ... groweth wheresoever I have sene it only about water sydes and in mersily meadows' - Turner (1562: 106). 'Ptarmica ... in the three great fieldes next adjoining to a village neere London called Kentish towne'. - Gerard (1597: 484).   **1562**

Turner is contrasting the plant he knows with Dioscorides' small twiggy bushling, 'not unlike Southernwood', see Chapman *et al.* (1995: 531). Clarke & Druce both cited Gerard as their first date.

Aconitum napellus s.l.

* 'Found by Rev. Edw. Whitehead [in 1819] in a truly wild state on the bank of a brook, and on the river Teme in Herefordshire'. - Purton (1821: 47n.).   vc36   **1821**

'By the side of the brook a few yards above Gossart Bridge between Ludlow and Burford, in abundance, 1799'. - Williams [c.1800].   vc40   1799

A very doubtful native even in Herefordshire and Shropshire. There are earlier records as a certain alien; for instance Druce cites 'Coln, Gloster; Iver, Bucks, Lightfoot, prior to 1786. *Sibthorp ms.*'.

Actaea spicata

* 'Christophoriana ... Herbe Christopher groweth in the north parts of England, neere unto the house of the right worshipfull Sir   vc66   **1597**

William Bowes'. - Gerard (1597: 829).

Clarke adds 'No doubt Sir W. B., of Barnard Castle, Durham'.

Adiantum capillus-veneris

'Capillus Veneris verus …found by Mr Lhwyd at Barry Head and Porth Kirig in Glamorganshire'. - Ray (1724: 123).   vc41   **1724**

'First recorded in the county [Glamorgan] by Edward Lhwyd in 1698 … at Barry Island and Parth Kirig [Porthkerry]'. - Wade et al. (1994). The reference comes from a letter from Lhwyd to Richardson, dated Sept 19th 1698 (Gunther (1945: 400)). But Edgington (2013) cites a specimen in **BM** 'Found by Mr Lhwyd on a sea rock called nine-acre-cliff in ye parish of Porthkerig in Glamorganshire, ex D. Bobart' and dates it to 1697. Edgington (*in litt.*) also draws my attention to a letter dated 23rd Feb 1698, Richardson to Sloane (British Library: Sloane ms. 4062, f. 276) confirming his suspicion of a 1697 date.   vc41   **1697**

There are possible earlier references, including Gerard (1597: 982) 'Capillus Veneris verus … it is a stranger to England: notwithstanding I haue heard … that it groweth in divers places of the West countrey of England' and in Parkinson (1640: 1049.1) 'Adianthum verum seu Capillus Veneris verum … some have reported it is found in Glocestershire'. Druce has 'A. fol. Coriandra verum', St Ives, Cornwall and Isle of Aran, Galway, Mr Lhwyd, 1700, in Herb. Du Bois. Webb & Scannell (1983) give the date of Lhwyd's Irish discovery as 1712.

Adonis annua

* 'Anthenus [Anthemis] … the thirde kinde … I have sene it in Englande but very rare … red mathes, alii, red mayde wed, alii, purple camomyle'. - Turner (1548: B. j).   **1548**

Adoxa moschatellina

* 'Minimus Ranunculus Septentrionalium herbido muscoso flore … In sylvosis et umbrosis frigidiusculis Angliae'. - L'Obel (1571: 300).   **1571**

Aegopodium podagraria

* 'A weede or an unprofitable plante … Herba Gerardi … In Englishe some call it Aishweed'. - Lyte (1578: 300). ' Herba Gerardi … At Wallingford-house and in Moorfields.' - Merrett (1666: 61)).   **1578**

Lyte adds words to the effect that once you have this taken root, it will remain willingly and yearly increase. Pulteney cites L'Obel (1571: 311), under Podagraria, but I can see nothing there to hint that it was known by him from England.
Kent (1975) claims Merrett's locations for Middlesex.

Aethusa cynapium

* 'Cicutaria tenuifolia. Dog's Parsley. Among stones, rubbish, by the wals of cities and townes almost everywhere'. - Gerard (1597: 905). **1597**

Pulteney cites L'Obel (1571: 327) 'Circutaria fatua, quae minus Foetida … ad margines & sepes paulo editiores et sicciores', but this has no reference to England.

Agrimonia eupatoria

* 'Agrimony ... groweth amonge bushes and hedges and in myddowes and woddes in all countries in great plentye'. - Turner (1551: P. vij). **1551**

Agrimonia procera

'Rev. W. W. Newbould, Beaumont, Island of Jersey, 15th July, 1842'. - Babington (1846). vc113 **1846**

Druce (1928: 472) cites a specimen in Hb. Du Bois, and dates it c. 1700. He adds 'although not definitely added to the British Flora till 1857, the plant was well known to the earlier botanists. Unfortunately no locality is given on these specimens'. 1740

Du Bois died in 1740. Druce (1928: 472) also notes 'A. odorata *Park*. and *Ray Historia p. 400, 1688*.

The first certain non-Channel Islands record is 'With [Rev. W. W. Newbould], Sept. 9th, 1852, on the rocky shore of Lough Neagh in the county of Antrim; and about the same time by Mr. Jos. Woods near to the Start Point in Devonshire, and near Gwithian in Cornwall'. - Babington (1853a: 363).

Agrostemma githago

* 'Githago ... est illa herba procera que in tritico flavescente existit, inde corollas apud morpetenses meos pueri in die divi baptistae, texunt, vulgus appellat CoccIe aut pople'. - Turner (1538: B. verso). **1538**

The latter part of the Turner entry might be translated 'among my people of Morpeth the children make garlands from it on the day of [St John] Baptist'. See Addyman (2016).

Agrostis canina s.l.

* as *A. vinealis* 'In pratis humidis frequens'. - Hudson (1762: 26). 'On some parts of Hounslow Heath, abundant, Dr. Goodenough'. - Withering (1796: ii. 127). **1762**

'Gramen miliaceum locustis minimis panicula fere arundinacea. Near Kerry block just by ye pole in Boggy ground. (The first record from Britain for *Agrostis canina* L.) The first record for Montgomeryshire of *A. tenuis, Sibth*.'. - Druce & Vines (1907: 152). vc47 1747

Kerry Pole and Block Wood are six miles SE of Newtown. The record from Druce & Vines is dated as 1747, the date of Dillenius' death.

Agrostis canina s.s.

as *A. canina var.* α 'in pratis, pascuis et ericetis humidiusculis' - Hudson (1778: 30) cited in Philipson (1937) under his *A. canina* var. *fascicularis*.

1937
1778

Agrostis capillaris

* 'Gramen pratense vulgare spica fere arundinacea *J.B.* ... in pascuis nihil vulgarius est hoc Gramine'. - Ray (1670: 154).

1670

Agrostis curtisii

* as *Agrostis canina var.* γ 'supra Hall Down prope Exeter et alibi in Devonia'. - Hudson (1778: 31). 'The first information I received of this grass was from my Gardener Robert Squibb, who sent me up some tufts of it from Piddletown Heath, Devonshire' [?Puddletown, Dorsetshire]. - Curtis (1798: vi. 12).

vc3   **1778**

Hall Down is Haldon Hill, just SW of Exeter.

Agrostis gigantea

'*A. nigra* (black squitch), in marley, clayey and other wet soil'. - Withering (1796: ii. 131).

vc39   **1796**

Withering's correspondant was Mr S. Dickenson, citing specimens from Blymhill, on the Staffs/Shropshire border. See Dickenson (1798). Withering equated his species to the 'gramen montanum mileaceum minus, radice repente. Dr Sherard gathered it on the Mountains of Mourn in Ireland', in Ray (1724: 402) [which in turn refers to Ray (1696: 256], but that surely is more likely to refer to *A. stolonifera*, see below. Curtis (1790: 35) includes *A. repens*, Couchy agrostis in his species of the genus which seems to represent this species, but provides no other details.

Agrostis stolonifera

'Gramen montanum miliaceum minus radice repente, Dr Sherard gathered it upon the Montains of Mourn in lreland'. - Ray (1696: 256). 'Gramen miliaceum majus panicula spadicea ... in dry hilly pastures'. - Petiver (1716: n. 118).

vcH38   **1696**

Nelson (1979b) cites the same polynomial from Ray (1724: 402) and notes that there is a specimen in the Morisonian Herbarium (see also Vines & Druce, 1914: 108). Sherard was in Ireland c1690 - 1694.

vcH38   1694

Clarke and Druce both cite the Petiver reference as their first date.

Agrostis vinealis

as *A. canina* var. β, Hudson (1778: 30) 'in ericetis aridiusculis' cited in Philipson (1937) under his *A. canina* var. *arida*. **1937**

Philipson was the first to articulate what had been suspected for 150 years, but this was not raised to species level until Clapham *et al.* (1987). **1778**

Aira caryophyllea

* 'Gramen nemorale Avenaceum alterum … Hoc idem Gramen aridis collibus theriotrophio Regiae Grynwich [Greenwich] Anglo-britanniae vicinis elapsis annis collegi'. - L'Obel (1605: 465). 'Gramen montanum panicula spadicea delicatiore'. - Ray (1660: 69). vc16 **1605**

Edgington (2011) writes 'probably c1567 on L'Obel's arrival in England from Flanders'.

Aira praecox

* 'Gramen parvum praecox spicâ laxâ canescente … In glareosis et sterilioribus plerumque nascitur'. - Ray (1670: 153). 'On Harefield Common, and elsewhere' [Middlesex]. - Blackstone (1737: 38). **1670**

Ajuga chamaepitys

* 'Chamepitys … Grounde Pyne … I here … that it is founde nowe in diverse places in england'. - Turner (1551: I. vj). 'In good plenty in Kent'. - Turner (1568: i. 131). **1551**

Ajuga pyramidalis

* 'I am assured by the Rev. Doctor Burgess of Kirkmichael that it is a native of Scotland, but I have not yet learned the particular place of its growth'. - Lightfoot (1777: 303). 'Supra montem Ben Nevis in Scotia D. Hope'. - Hudson (1778: 249). **1777**

'At Ben Nevis, Ft. William, plentifully in the burn of Killgower and Ord of Caithness'. - Hope, 1768, in Balfour (1907). vc97, 109 **1768**

The specimen in Hope from Caithness sounds as though it ought to be a record from J. Robertson, but there is no reference to it in his Journals (Henderson & Dickson 1994).

There is a record in How (1650: 17) of 'Bugula sive Consolida media, caerulea Alpina. *Bauhini*, Mountain Bugle or Siclewort: On Carnedh Lhewelyn in Wales', which Hudson and Pulteney take to be this species. But both E.B. 477 and Smith (1825) give this as *A. alpina*, which they show as a separate species; it was not until Babington (1851) that this was finally sunk into *A. reptans*.

Ajuga reptans

* 'Consolida media is called in english Bugle ... groweth in shaddowy places and moyst groundes'. - Turner (1548: H. ij).		**1548**

Alchemilla acutiloba

'This species was first noted as British by Wilmott, who in 1946 found a sheet in Herb. Mus. Brit. collected in Teesdale in 1933 by J.F.G. Chapple and previously labelled *A. pastoralis*'. - Walters (1949a).   vc66 **1949** / vc66 1933

Alchemilla alpina

* 'Alchimilla Alpina quinquefolia *C.B.* … We found it this year [1671] on a mountain in Westmerland beside a great pool or lake called Hulls-water [Ullswater], about 3 miles from Pereth in Cumberland, just over against Wethermellock [Watermillock]'. - Ray (1677: 11).   vc69, 70 **1677**

This was on Ray's trip to the north of England in 1671, with T. Willisel. See Raven (1950: 154).   vc69, 70   1671

Both Hudson (1778: 71) and Pulteney cite Gerard (1597: 837), where he has 'Pentaphyllum petrosum Heptaphyllum Clusii' from Savoy and 'if his memory fails him not, on the walls of Beeston Castle in Cheshire'. This seems very unlikely and should be disregarded. See *Bupleurum tenuissimum* for another unlikely record from Gerard from this locality.

Alchemilla filicaulis

* 'Alchimilla … Pes leonis ... our Ladies Mantel or syndow. It groweth in middowes like a Mallowe'. - Turner (1548: H. j).   **1548**

Clarke (1909) notes that his first edition erroneously gave 1848.

Alchemilla glabra

as *A. alpestris* Schmidt (*A. vulgaris* var. *glabra*), Linton (1895).   **1895**

Kirkharle [20m. NW. of Newcastle], vc67, from a painting by Henrietta Lorraine, 29th May 1824, identified as this species by John Richards (*pers. comm.*). Peeblesshire, 1834, C.C. Lacaita, det. Walters, **BM**.   vc67   1824

Alchemilla glaucescens

as *A. pubescens*, Salmon (1928).   vc64   **1928**

Ingleton, vc64, 1928, **BM**.

Salmon also lists records from Scarborough, 1914, but as probably a garden escape, and from near Edinburgh; the latter is cited by Druce (under *A. pubescens* Lam.) 'Yalden, Water of Leith, c.1780' as his first record. But in any event these are outside the native range.

Wilmott (1939) equates this species with that described by Hudson (1762: 59) as '*A. alpina pubescens minor* … in montibus Westmorelandicis'.

Alchemilla glomerulans

| 'From herb. E.S. Marshall (**CGE**), Ben Lawers, 1913'. - Wilmott (1922). | vc88 vc88 | **1922** 1913 |

Alchemilla micans

| '… collected as '*A. pastoralis*', (i.e. *A. monticola*) by Francis Druce at Langton Beck in Upper Teesdale on 12th August 1924'. - Swan & Walters (1988). | vc66 vc66 | **1988** 1924 |

Alchemilla minima

| 'Simon Fell, Ingleborough, M.W.Yorks. (vc64), S.M. Walters, 29-7-47, **CGE**'. - Walters (1949a). | vc64 vc64 | **1949** 1947 |

Alchemilla monticola

| as *A. pastoralis* 'Langdon Beck Inn, Upper Teesdale, Durham, 1903, A.O. Hume'. - Wilmott (1922). | vc66 vc66 | **1922** 1903 |

Alchemilla subcrenata

| 'In upper Teesdale … collected by Miss M.E. Bradshaw in 1951'. - Walters (1952). | vc66 vc66 | **1952** 1951 |

Alchemilla wichurae

| as *A. acutidens* var. *alpestriformis* 'Ben Lawers, 1911, C.H. Ostenfeld'. - Salmon (1914). | vc88 | **1914** |
| Meall Buidhe, vc88, W.A. Shoolbred, 1892, **BM**. | vc88 | 1892 |

See note in Walters (1949a).

Alchemilla xanthochlora

| as *A. vulgaris* L. *sensu restricto* (*A. pratensis* Schmidt), Linton (1895). | | **1895** |
| S. Lancs, 1809, G. Crossfield, det. Walters, **BM**. 'Crossfield … Fields in Arpley [nr. Warrington]'. - Salmon (1912) | vc59 | 1809 |

Alisma gramineum

| 'In several inches of water, muddy edge of Westwood Park Pool, Droitwich, Aug 15th, 1948. R.C.L. Burges'. - Lousley (1950). ' … first brought to general notice when Dr R.C.L. Burges distributed specimens from Westwood Park Pool near Droitwich through our Exchange Section in 1948'. - Lousley (1957). | vc37 | **1950** |
| 'In all probability it was the plant found by the Worcestershire Naturalists' club on their visits to the Pool in 1920 and 1930'. - | vc37 | 1920 |

Lousley (*op.cit.*).

Maskew (2014) gives the original discoverer as C. Rea.

Alisma lanceolatum

as var. β. *lanceolatum* (*lanceolata* With.), Babington (1843)		**1843**
'Prope Waltham Abbey in Essexia, 1715, det. Samuelsson 1932', ex Herb S. Dale, **BM**.	vc18	1715

This plant has a long and uncertain history. Hudson (1778: 158) refers to it, as *A. plantago-aquatica* var γ, *longifolia*, in Petiver (1713: xliii. 7.) who calls it Narrow Water Plantain, from London. Petiver separates it from *A. p.a.*, and does seem to be describing a new species. Smith (1824: ii. 203) cites Withering (1787: 362) and Ray (1724: 257) (who has it as a variety, as 'Plantago aquatica longifolia'), as well as Petiver. Bennett (1893) notes that 'Withering's plant seemed to be a dwarf form of the plant with lanceolate leaves'.

Alisma plantago-aquatica

'of plantyn or weybrede ... The greater is larger with brode leves lyke unto a bete ... the rootes ar ... of the thiknes of a finger; it groweth in myri places in hedges and in moyst places'. - Turner (1562: 94).   **1562**

Clarke & Druce both cited 'Plantago aquatica ... about the brinkes of pondes and ditches almost everywhere'. - Gerard (1597: 338), as the first record. Certainly Chapman *et al.* (1995) have no problems with the 1562 reference, where there is also a figure clearly depicting this species.

But see also Turner (1538: A. ij) whose 'Alisma ... [is known] nostratibus Water plantane or Water waybrede' seems to imply familiarity with a wild plant.

Alliaria petiolata

* 'Alliaria est herba passim in sepibus proveniens ... Sauce alone ... Jak of the hedge'. - Turner (1538: A. ij).   **1538**

Allium ampeloprasum

* 'Allium montanum major Anglicum *Newtoni*. In parva insula Holms dicta supra Bristolium in Sabrinae aestuario [Steep Holm, in the Severn Estuary] copiose provenit'. - Ray (1688: 1125, 1688a: 2).   vc6   **1688**

Clarke notes that a specimen collected by Newton in this locality is in Hb. Sloane, 152. f. 153, **BM**.

Allium oleraceum

'Allium montanum bicorne, an Ampeloprassum? ... this I observed on the Scars of the Mountains above Settle in York-shire'. - Ray (1670: 13).   vc64   **1670**

| | vc64 | 1660 |

' … this I observed on the scars of the mountains above Settle in Yorkshire'. - see notes of Ray's 1660 Itinerary in Raven (1950: 115 and note).

Both Clarke & Druce have as their first record 'Allium sylvestre bicorne flore ex herbaceo albicante, cum triplici in singulis petalis stria atro-purpurea … Invenimus inter segetes Notleiae in Essexia'. - Ray (1688: 1119). Yet as pointed out by C.D. Preston (*in litt.*) there has been confusion over the forms of this variable plant, with the Settle plant described as var. *complanatum* or *A. carinatum* Sm.

Allium schoenoprasum

| | vc1 | **1820** |

'Pursuing our journey over the downs towards Mullion [from Kynance] at a distance from any habitation we found *Allium Schoenoprasum*. This plant is extremely rare in a wild state in this kingdom'. - Jones (1820: 26).

The first record given in Clarke was from Lightfoot (1777: 180), who was copying the record from Hope ('In a Park on a mount near Fase Castle [Berwickshire]', May 22nd 1765'. - John Hope's ms 'List of Plants growing in the Neighbourhood of Edinburgh', see Balfour (1907)). Chives had been cultivated for centuries, with no wild locations known, and that record is now thought to be a relic of cultivation, as are later discoveries in Northumberland. Smith (1824: ii. 138) lists other sites in Argyll (also ex Lightfoot) and in Westmorland, but none of these are now thought to be native.

Allium scorodoprasum

| | vc69 | **1690** |

* 'Allium sylvestre amphicarpon foliis porraceis floribus & nucleis purpureis … In montibus Westmorlandicis observavit D. Lawson. In Troutbeck-holm by Great Strickland'. - Ray (1690: 165).

| | vc69 | 1688 |

Lawson's letter was sent to Ray on 9th April 1688 (Lankester, 1848; Whittaker, 1986).

In the copy of Merrett (1666: 6), believed to have been annotated by him, is written (per Lees (1888: 446)) 'Ampeloprasum, sive porrum sylvestre … at Sh[k]ire thorn in Craven, going from Mawater Tarn to Perolic Bridge before you come to ye River yt maketh Wharf River'. Lees equates that with this species, but Foley (2006a) has a slightly different transcription and interprets as *A. ampeloprasum*. Another early record is a specimen sent from James Sutherland to Richardson, from 'the Rocks of Edinburgh Park'. See Dandy (1958a: 218). Sutherland died in 1719 - it is possible that the record dates from c1700.

Allium sphaerocephalon

' ... first noticed in Jersey by Mr J. Woods in the summer of 1836 ... in sandy ground adjoining St Aubin's Bay ... '. - E.B.S. 2813 (1838).

vc113 **1838**
vc113 1836

The only record other than in the Channel Islands, though in a site where it is possibly alien, is 'St Vincent Rocks [Bristol]'. - Stephens (1847).

Allium ursinum

* 'Allium ursinum ... Rammes or Ramseyes groweth in woddes about Bath'. - Turner (1551: B. v.).

vc6 **1551**

There are earlier references in Turner (1538 & 1548) but these cannot be definitely of plants in the wild.

Turner, W. 1538.
Libellus de re Herbaria Novus.

Allium vineale

* 'Allium ... the seconde kynde ... crowe garlike ... groweth in the fieldes'. - Turner (1548: A. vj. verso). 'Allium sylvestre ... In a field called the Mantels on the versoside of Islington by London'. - Gerard (1597: 142).   1548

Alnus glutinosa

* 'Alnus ... an alder tree ... groweth by water sydes and in marrishe middowes'. - Turner (1548: A. vij).   1548

Alopecurus aequalis

* as var. 4 of *A. geniculatus* 'In a marshy place by the stews in Edgbaston Park [Birmingham]'. - Withering (1796: 121).   vc38   1796

Withering notes that 'the anthers are a fine orange, so that the flowering plant may be distinguished at some distance'; an excellent observation.

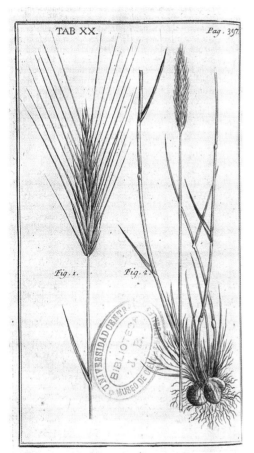

Alopecurus bulbosus

* 'Gramen myosuroides nodosum. Found by Mr. Jam. Sherard'. - Ray (1724: 397), with a figure, but no locality. 'In the first Field next the Road before you go into Northfleet [Kent]. Dr. Wilmer'. - Blackstone (1746: 32).   1724

Ray, J. 1724. *Synopsis Methodica Stirpium Britannicarum*, ed 3.

Alopecurus geniculatus

* 'Gramen fluviatile spicatum'. - Gerard (1597: 13).   **1597**

Alopecurus magellanicus

* 'Mr. G. Donn has favoured us with this new species of *Alopecurus*, discovered by himself on mountains about Loch Nagore [Lochnagar] in Aberdeenshire'. - E.B. 1126.   vc92   **1803**

'Brown discovered the plant in August, 1794, as stated on the ticket accompanying his specimens in Herb. Mus. Brit.'. - Clarke (1900).   vc92   1794

See Smith (1824: i. 80). 'Mr. R. Brown … informs me that he communicated it to Mr. G. Don', and also Mabberley (1985: 26) stating that the correct credit should to be given to Brown.

Alopecurus myosuroides

'Gramen alopecurinum minus. In the moist furrowes of fertill fields'. - Gerard (1597: 9-10). 'Gramen aquaticum spicatum *Lob. ico.*'. - Ray (1660). 'Near Paddington, BuddIe'. - Ray (1724: 397).   **1597**

Gerard (1597: 9-10) illustrates and describes four grasses - Gramen alopecuroides majus and minus (great and small foxe-taile grasses) and G. alopecurinum majus and minus (great bastard and small bastard foxe-taile grasses). He then gives the habitat of the latter but not the former. Clarke and Druce confound the name of the former with the habitat of the latter.

Alopecurus pratensis

* 'Gramen alopecuroides majus … In the moist furrowes of fertill fields'. - Gerard (1597: 9-10).   **1597**

See note above, under *A. myosuroides*, and note that Ray (1670) amended what he had written in his 1660 work, citing this Gerard reference. For a full discussion see Oswald & Preston (2011: 210n).

Althaea officinalis

* 'Althea … marish mallowe or water mallowe … groweth in watery places'. - Turner (1548: A. vij). 'Very plentifully in the marshes both on the Kentish & Essex shore alongst the river of Thames about Woolwich, Erith, Greenehyth, Gravesend, Tilburie, Lee, Colchester, Harwich, and in most salt marshes about London'. - Gerard (1597: 789).   **1548**

Rydén *et al.* (1999) interpret Turner's 'Althea … our countrymen [call it] Holy Oke' (1538: A.ij) as this species. Earle (1880: 27) cites 'merse-mealewe' from an eleventh century ms. Stearn (1965) and earlier editors equate it with *Alcea rosea*.

Ammophila arenaria

* 'Spartum … a kinde of sea bente or sea rishe whereof the frayles are made that figges and rasines are caried hether in out of Spayne. The same bent or sea rishe have I sene in Northumberland besyde Ceton Dalavale [Seaton Deleval], & ther they make hattes of it'. - Turner (1562: 144 verso).   vc67   **1562**

Raven (1947: 50) and Stearn (1965: 242) both agree that this is the 'Spartum herba' of Turner (1548: G. iiij. verso), interpreted by Britten (1882) as *Stipa tenacissima*, and thus ignored by Clarke. However Turner's 1548 record cites it from Friesland, but gives no English locality, and though he coins the English name 'Frailbente', this does not seem to be conclusive proof that he found it here. Presumably a 'frayle' is the basket for importing the produce from Spain.

Anacamptis laxiflora

'St Ouen's Pond, Jersey, June 9, 1838'. - Babington in E.B.S. 2828.   vc113   **1839**

'Grand-mare, Guernsey. Gosselin, J., c.1790'. - see McClintock (1982: 165).   vc113   1790

Plate 2828 Smith J. and Sowerby, J. 1838. *English Botany supplement*, vol 3.

Anacamptis morio

'Cynorchis morio foemina. The female Foole-stones ... do grow naturally to their best liking in pastures and fields'. - Johnson (1633: 208).

1633

Clarke and Druce both cite Johnson (1634: 54) 'Orchis, sive Cynosorchis morio foemina *Lob. Ger.* ... In pratis. The female Foole-stones. But Ray (1660), Curtis (1787) and many others cite the 1633 reference.

Anacamptis pyramidalis

* 'Orchis sive Cynosorchis purpurea spicâ congestâ pyramidali ... In many places, as in a chalkie close at Hinton near where they burn lime'. - Ray (1660: 109).

vc29   **1660**

Anagallis arvensis

* 'Anagallis ... anglice Pympernell'. - Turner (1538: A. ij. verso). 'It groweth commonly amonge the corne. The male hath a crimsin floure, & the female hath a blewe floure'. -Turner (1548: A. viij).

1538

Anagallis tenella

* 'Nummularia flore purpurascente ... I first found it anno 1626 in the Bishopricke of Durham, and in two or three places of Yorkshire ... also on the bogges upon the heath neare Burnt wood [Brentwood] in Essex'. - Johnson (1633: 630).

vc66, 18   **1633**
vc66   1626

There was a typing error in Clarke corrected in Clarke (1909).

Anchusa arvensis

* 'Buglossum siI. parvis floribus'. Chalkedale near Dartford. - Johnson (1629: 9). 'Upon the drie ditch bankes about Pickadilla'. - Johnson (1633: 799).

vc16   **1629**

Pulteney refers to Lyte (1578: 5) ' The wilde kind of Buglosse ... with small bleu fleures ... groweth in moste places of this countrie, in barren soyle, and gravelly grounde'. This seems quite possible, but is 'this country' here?

Andromeda polifolia

* 'Rosmarinum sylvestre ... groweth in Lancashire in divers places, especially in a field called Little Reede, amongst the Hurtleberries, neere unto a small village called Maudsley [Mawdsley, N of Ormskirk]; there founde by a learned Gentleman often remembered in our History (and that woorthily) master Thomas Hesketh'. - Gerard (1597: 1110).

vc59   **1597**

Anemone nemorosa

* 'Ranunculus .. the fourth kinde .. with a white floure … groweth in woddes and shaddish places in April'. - Turner (1562: 114b, with a good figure).

Chapman *et al.* (1995: 542 n.251) make this *Ranunculus planifolius*. This seems extremely unlikely.

1562

Angelica sylvestris

* 'The wilde Angelica that groweth here in the lowe woodes and by the watersydes'. - Turner (1568: 6).

1568

Anisantha sterilis

* 'Bromos sterilis. Barren Otes'. - Gerard (1597: 69).

Pulteney cites 'Bromus or Havergrasse … it groweth in borders of feeldes, upon bankes & …. , and alongst by way sides'. - Lyte (1578: 505). This description and the accompanying drawing are quite possible, if indeed Lyte was referring to England.

1597

Anogramma leptophylla

'The President [Mr Newman] has heard, through the kindness of his friend Mr Henry Hagen, that this pretty little species has been found growing on a bank in Jersey'. - Newman (1853a).

vc113   **1853**

'Miss J. Marett of La Haule first found it in Feb 1852, and sent it to Dr Lindley [See also Gardener's Chronicle, 69, 1853] of University College, London for identification'. - Le Sueur (1984: 4).

vc113   **1852**

Newman (*op.cit.*) goes on to say that Mr W. Christy, who took great pleasure in raising this plant from seed [*sic*], year after year, resided for some months nearby.

Antennaria dioica

* 'Gnaphalium montanum alb. *Lob. Ger.* Mountaine Cudweed or Catsfoote'. - Johnson (1641: 22). 'In Scosby Leas neer Donkester, and in Sherewood Forrest neer Bescot Park path and no where else, Mr. Stonehouse'. - How (1650: 48).

1641

Turner (1548: H.iiij) refers to 'Pilosella … has purple floures mengled with whyte altogether, and thys groweth in heathes where as Ling or heath groweth …'. This could well refer to this species but he might be describing a plant he had seen in Germany.
Parkinson (1629: 375. 6) describes 'Gnaphalium montanum flore albo & flore purpureo' as a plant very difficult to grow in gardens (as opposed to their wild habitat), and figures it in his *Theatrum* (1640: 690) as 'pes cati'.

Anthemis arvensis

* 'White Ox-eye. Lond. Peckham Fields'. - Petiver (1713: xix. 8).　　vc21　　**1713**

Oswald & Preston (2011: 188n) make a persuasive case for identifying this with Ray's (1660: 41) 'Cotula non foetida', but note that Ray omitted two other mayweeds, so there cannot be certainty in his record. Hanbury & Marshall (1899: 195), as does Edgington (2011), cite L'Ecluse (1601: 337) 'Parthenium pleno flore. Londini in Anglia primum mihi conspecta est haec planta, anno salutis humanae 1579', as referring to this species, but C.D. Preston, *in litt.*, points out that the illustration, coupled with the phrase 'pleno flore', suggests rather *Tanecetum parthenium*. Druce (1928: 474) in his note on Du Bois' herbarium, notes this as 'Found by Mr. Buddle, near Greenwich'. Buddle died in 1715.

Anthemis cotula

* 'Cotula foetida … In corne fieldes, neere unto pathwaies and in the borders of fields'. - Gerard (1597: 618).　　**1597**

'Dogge-fenell and mathes is bothe one, and in the commynge up is lyke fenell and beareth many white floures, with a yelowe sede: and is the worst wede that is, except terre [tare, *Vicia hirsuta*]' - Fitzherbert (1523).　　1523

Fitzherbert is cited by both Grigson (1958) and Kay (1971) as to referring to this species and Grigson also identifies it with the Anglo-Saxon 'maegthe'. Note also Pulteney's citing of Lyte (1578: 184.2) 'Cotula foetida … most commonly in this countrie in every corne field'. Lyte contrasts this with another species, only found in France and Germany, though in the gardens of apothecaries here. Turner (1548: F. i. verso) has 'Parthenium to be stynkyng maydweede', which is probably this species.

Anthoxanthum odoratum

* 'Gr. Anthozanthum spicatum *J.B. tom. 2, pag. 466.* locum non memini'. - Merrett (1666: 48). 'Gramen vernum spica brevi laxa'. - Ray (1696: 252).　　**1666**

Anthriscus caucalis

* 'Myrrhis sylvestris nova Aequicolorum *Col.*'. Hampstead Heath. - Johnson (1632: 32)　　vc21　　**1632**

See also Johnson (1633: 1038.5).

Anthriscus sylvestris

* 'Myrrhis … called in Cambrygeshyre casshes … groweth in hedges in every countrey'. - Turner (1548: E. v. verso).　　vc29　　**1548**

Presumably vc29.

Anthyllis vulneraria

* 'Anthyllis leguminosa ... Upon Hampstead Heath neer London, right against the Beacon, on the right hand as you go from London, neere unto a gravell pit; they also growe upon blacke Heath, in the high way leading from Greenwich to Charleton, within halfe a mile of the towne'. - Gerard (1597: 1060).   vc16   **1597**

This record is mentioned in Hanbury & Marshall (1899), but not by Kent (1975). However Kent's predecessors, Trimen & Dyer (1869), suggest that the Hampstead locality is unlikely, and that Gerard might have intended *Ornithopus perpusillus*.

Apera spica-venti

* 'Agrorum venti spica, *Lob.*'. Near Canterbury. - Johnson (1632: 22).   vc15   **1632**

Johnson (1633: 6) relates that it is called 'Windle strawes', and used to 'adorne our chimneys in Sommer time'.
Pulteney cites Gerard (1597:5) as the first record, but though the illustration is that used in Johnson, I feel that Gerard's account was confused, and prefer the gloss in Johnson.

Aphanes arvensis s.l.

* 'Percepier Anglorum ... At Angliae tamen Bristoiae, arbitramur, frequentissimo apud mulierculas usu receptu est'. - L'Obel (1571: 324).   vc34   **1571**

Aphanes arvensis s.s.

Walters (1949b).   **1949**

'Gog-Magog Hills, Henslow, 1825, **CGE**'. - Walters (1949b).

There are several specimens in Hb Sloane, **BM**, of which that from H.S 84:10, from around 1700, appears to be this segregate. Kent (1950) notes Johnson's 1633 record from Tothill Fields, but we have no idea which of the two species this was.   vc29   1825

Aphanes australis

Walters (1949b).   **1949**

'Guernsey. Gosselin, J., c.1790'. - see McClintock (1982: 83).   vc113   1790

Apium graveolens

* 'Elioselinum ... Smallage ... groweth in watery places and also in gardines'. - Turner (1548: C. viij.). 'Apium palustre'. Stoke. - Johnson (1629: 8).   **1548**

Turner (1538: A. ij. verso) also gives this as 'Apium nostrates vocare solent Smallage'.

Apium inundatum

* 'Sium pusillum foliis variis, nondum descriptum, in aquosis. Small water Parsley' - Johnson (1641: 33). 'In Surry near Purbright'. - Merrett (1666: 114).   **1641**

Apium nodiflorum

* 'Sium umbellatum repens, *Matthioli & Italorum*, ut puto'. Between Sandwich and Canterbury. - Johnson (1632: 19).   vc15   **1632**

'Sium repens. 27 Aug.1619….plentifullie by the lakes and rivers at Droxford, Hants'. - Goodyer ms.11, f.82, see Gunther (1922: 114).   vc11   1619

Rose, in Gilmour (1972: 60), claims that 'Sium *Matt. & Italorum*'. Stoke to Cliffe. - Johnson, 1629: 8, is probably this species too.

Apium repens

Riddlesdell (1917), elucidating the native forms of *Helosciadium* [*Apium*].   vc23   **1917**

'First certain record 1858, Port Meadow, W. Holliday, Hb. Druce'. - Druce (1927).   vc23   1858

There are older specimens - 'Cowley Bottom, vc23, 1789, Sibthorp, **BM**' (with the enigmatic note 'I received this specimen of Milne [his gardener]'). Druce (1932: 134) cites 'Sium repens, Peat beds on Bullingdon Green'. - Sibthorp (1794: 97)', but Killick *et al.* (1998) ignore these records.

Aquilegia vulgaris

* 'Aquilina … Columbina anglis … pratensis etiam est Franciae & Angliae'. - L'Obel (1571: 339). 'about Broadsworth, and Hample woods' [nr. Doncaster, S.W. Yorkshire]. - How (1650: 9).   **1571**

Arabidopsis petraea

* 'Nasturtium petreum'. - Johnson (1641: pars alt. p. 8).   vc49   **1641**

Thomas Johnson and party were on Snowdon, Aug 1639. See Raven (1947: 289).   vc49   1639

Arabidopsis thaliana

* 'Pilosella siliquata *Thal.* … ad agrorum margines. Codded Mouse-eare'. - Johnson (1634: 59). 'On the ditch sides in the way to Marybone and in a Close on the left hand of the lane from Islington to Kingsland'. - Merrett (1666: 93).   **1634**

Kingsland was a settlement around Dalston, in E. London.

Arabis alpina

* 'Cuchullin range, Skye'. - Hart (1887).   vc104   **1887**

This discovery was on his honeymoon (Druce & Leach, 1915). His account does not relate whether he was accompanied or not.

Arabis hirsuta

'Cardamine pumila Bellidis folio Alpina. *Ger. emac*. On the rockes nigh the Quarrie by Bath'. - Johnson (1634: 26).   vc6   **1634**

Clarke's entry reads" 'Barbarea muralis *J.B.* ... Upon the walls of the Church of Ashburn [Ashbourne] in the Peak'. - Ray (1670: 38). Ray suggests, very convincingly, that this was the plant mistaken by Johnson (1634: 26) for *Cardamine bellidifolia*'. I have followed that.

Arabis scabra

* 'Cardamine pumila bellidis folio Alpina *Ger. emac*. Nuper in rupe S. Vincentii prope Bristolium in Anglia invenit D. Jac. Newton'. - Ray (1686: 817, 1688a: 3).   vc34   **1686**

Arbutus unedo

* 'Arbutus ... The Strawberry Tree ... hath beene of late dayes found in the West part of Ireland'. - Parkinson (1640: 1490).   **1640**

See earlier references, back to the ancient Irish law code, the Breton Laws, given in Nelson (1979b), who also raises the possibility that Parkinson's record was from Richard Heaton. Scully (1916) cites a MS inquisition of the estates of Rory O'Donohoe made in or about 1584, which includes the following ' ... in the island of Loughleane and elsewhere in the islands, where grow certain trees called Crankany which bear fruit every month throughout the entire year'. Scully notes that the name 'Crankany' is a fair phonetic rendering of the Irish name 'Crouncahinye' for *Arbutus*.

Arctium agg.

* 'Personata ... a Bur ... groweth commonly about townes and villages'. - Turner (1548: F. ij).   **1548**

Arctium lappa

* 'Personatia, Bardana, Lappa Major'. Stoke to Cliffe. - Johnson (1629: 8).   vc16   **1629**

Arctium minus

Babington (1843: 171) was the first to separate this.   **1843**

'Bardana capitulis minoribus non lanuginosis. Found by Mr Buddle near ye town [London]. Bardana minor. From Mr Isaac Rand. Found at Lee in Kent'. - Druce (1928: 475).   vc21   1715

Buddle died in 1715 and Rand in 1743. But see also 'Arctium montanum & Lappa minor Galeni, *Lob*. Button-Burre. Mangersfeild, in Master Langleys yarde'. - Johnson (1634: 20). This is interpreted by many, including Kew & Powell (1932), as this species.

Arctium nemorosum

as *A. intermedium* Lange, under *A. minus* 'near Berwick-on-Tweed'. - Babington (1851: 179).	vc81	**1851**
Mid-west Yorkshire, 1818, **E**. The specimen has not been re-examined, and there are almost certainly earlier records in herbaria.	vc64	1818

Druce also refers to Ray (1724: 217, 2) and Turner (1548: F. ij) as possible earlier records, but the species concept has been muddled all through the ages.

Arctostaphylos alpinus

* as *Arbutus alpina* 'Upon many of the highland mountains … particularly on those to the south of Little Loch Broom in Ross-shire and those in the way between Loch-Broom and Loch-Mari. I found it likewise on a mountain call'd Ben-na-Grian, in Strath, in the Isle of Skye, two miles above McKennan's Castle'. - Lightfoot (1777: 215, with a figure).	vc104, 105	**1777**
'Up the water of Golspie to Benhorn [Ben Horn], near the summit' (later he found it on Ben Valich [Beinn Mhealich]). - Journal of James Robertson for 1767, Henderson & Dickson (1994: 35).	vc107	1767

Lightfoot found this on his 1772 trip. Murray & Birks (2005) suggest the Skye site of Ben-na-Grian [Beinne na Greine] is Beinn Bhuidhe, just to the east, where it was refound in 1990.
There is an entry 'Ben Grihum and several hills in Sutherland' in Hope's 1768 MS, in Balfour (1907) (and cited in Hudson (1778)).

Arctostaphylos uva-ursi

* 'Vaccinia rubra foliis myrtinis crispis. Four miles from Heptenstall near Widdop on a great Stone by the River Gorlpe in Lancashire'. - Merrett (1666: 123); Ray (1670: 309).	vc63	**1666**

The locality in Merrett is actually in Yorkshire.

Arenaria ciliata

* 'Discovered in Sept. 1806 by Mr. J. T. Mackay on the calcareous cliffs of a high mountain adjoining to Ben Bulben, Co. Sligo, Ireland'. - E.B. 1745.	vcH28	**1807**
Dandy (1958a: 156) cites a specimen in Buddle's Herbarium (Hb. Sloane 124, f.6, **BM**) labelled 'A. D. Lhwyd in Hybernia collect. prope Sligo, a D. Richardson habui'. Dandy dates this as probably	vcH28	1700

collected in 1700, as Lhwyd did not reach W. Ireland until that year. See also Sleeman (1870), who adds 'In herbarium as Lychnis alsinoides parva flore albo minimo, vol cxxiv. f. 6'. Druce (1928: 469) notes that 'it is probably one of the plants referred to in Lhuyd (1712 [1711])'.

Arenaria leptoclados

Discussed and separated from *A. serpyllifolia* by Babington (1860b: App. IV), and also in E.B.S. 2972 (1862). Babington (1843) includes it as var. β of the above. — vc29 — **1860**

'Mixed specimens of *Arenaria serpyllifolia* and *A. leptoclados*'. - Druce & Vines (1907: 107). — 1747

Babington (1860b) refers to W. Borrer, as the first to identify this in Britain, with a specimen from Henfield, Sussex, in 1844.
The record from Druce & Vines is dated as 1747, the date of Dillenius' death.

Arenaria norvegica

* 'Unst, in the Shetland islands, first discovered by Mr Thomas Edmonstone, Jun. an enthusiastic naturalist only eleven years of age, and ascertained to be new to Britain, by Dr M'Nab, on his visit to those islands in 1837'. - Hooker (1838: 182). — vc112 **1838** / vc112 1837

Clarke, citing the same reference of Hooker, has 'First gathered on 27th of April, 1837, on a range of serpentine hills ... in Unst ... Shetland by [a] son of Dr. Edmonstone of that place'. But this citation is in fact an abbreviated version of the text in E.B.S 2852 (1841).

Arenaria serpyllifolia s.l.

* 'Alsine minima. Fine Chickweede'. - Gerard (1597: 487) (with a figure). 'Alsine minor *Tab.* minima, *Dod. Lob.*'. - Johnson (1632: 40). — **1597**

L'Obel (1571: 193) whilst giving no English location for 'Alsine minima' does give an English name 'Little Chickweede'.

Arenaria serpyllifolia s.s.

Discussed and separated from *A. leptoclados,* by Babington (1860b: App. IV). — **1860**

'Alsine minor multicaulis', in a collection of plants gathered by Dr. Robert Plot[t], Hb. Sloane 113: 8, **BM**. — 1696

Plot died in 1696.

Armeria alliacea

'The Quenvais, and on the the sand-dunes at St Brelade, Jersey'. - Babington (1839a: 77).	vc113	1839
'In August 1833 Mr W.C. Trevelyan visited Jersey and discovered …'. - Babington (1839a: iii).	vc113	1833

Armeria maritima

* 'Caryophyllus marinus … Arearum margines ornant Belgae et Angli, apud quos in maritimis frequens oritur'. - L'Obel (1571: 189).   **1571**

The subspecies *elongata* was first recorded in 1726, nr. Grantham, vc63, but not satisfactorily identified until 1958 (Gibbons & Lousley, 1958).

Armoracia rusticana

* 'Thlaspi … thys kynde groweth in Morpeth in Northumberland and there it is called Redco'. - Turner (1548: G. vj. verso).   vc67   **1548**

Arnoseris minima

'Hieracium minimum …Londini communicavit 1581 Thomas Pennaeus … frequens in Anglia multis in locis'. - L'Ecluse (1583: 647-50 & 1601: cxlii).   **1583** / 1581

L'Ecluse's trip was in 1581. He inspected Penny's herbarium and took from him several pictures and descriptions (Raven 1947: 169).

Arrhenatherum elatius

* 'Gramen caninum nodosum … In the fields next to Saint James wall as ye go to Chelsey and in the fields as ye go from the tower hill of London to Radcliffe [Rotherhithe]'. - Gerard (1597: 22).   vc17, 21   **1597**

Artemisia absinthium

* 'Absinthium Ponticum … groweth … aboute tounes diches' &c. - Turner (1568: i. 6).   **1568**

Clarke, Druce and Nelson (1959) all cite Turner (1551: A. v), but this detail only appears in the 1568 edition.

Artemisia campestris

* 'Abrotanum campestre … both [this and the next entry, 'Abrotanum inodorum', presumably *A. vulgaris*] found on Newmarket Heath, Mr. Sare'. - How (1650:1). 'At a place called Elden in Norfolk, 12 miles beyond Newmarket in the way towards Lynne, in a hollow bottom, where two great roads cross one another. Th. WiIliseI'. - Ray (1670: 2).   vc28   **1650**

Not only do we know nothing about Mr. Sare, but treatment of his record has been inconsistent. Newmarket Heath is on the borders of Cambs and Suffolk. Hind (1889) does not mention it in his *Flora of Suffolk* and Babington (1860b), in his *Flora of Cambridgeshire*, only brackets it. Does this mean that he disbelieves the record or that he thinks that it was in Suffolk? Ray's 1670 record would be the next.

Artemisia maritima

* 'Absinthium marinum ... is plentuous in Northumberlande by holy Ilande [Holy Island] and in Northfolke beside Lin [King's Lynn]'. - Turner (1548: A. iiij. verso).   vc28, 68   **1548**

Artemisia norvegica

'Discovered by Sir Christopher Cox, in August 1950, on the spur of a mountain (in Wester Ross) between 2350 and 2450 ft'. - Blakelock (1953).   vc105   **1953**
    vc105   1950

See also *The Illustrated London News*, 5th Sept. 1953, p359 'A botanical discovery: first pictures of British *Artemisia norvegica*'.

Artemisia vulgaris

'Artemisia … vulgus hanc herbam ubique vocat Mugwort' - Turner (1538: A. iij).   **1538**

Clarke and Druce cite 'Absinthium ... This comon Mugwurt of ours groweth ... in hedges and amog the corne'. - Turner (1551: E. i. verso) as their first record, but Turner's earlier use of the English name implies familiarity with the plant as a wild species (as indeed is echoed in Earle (1880)).

Arum italicum

* '[Discovered] some three or four years ago … at Steephill [on the Undercliff], Isle of Wight'. - Hambrough (1854).   vc10   **1854**

This is the first record of the native, *A. italicum* subsp. *neglectum*.

Arum maculatum

'Aros [is called] a nostris Cockowpyntell'. - Turner (1538: A. iij).   **1538**

Clarke gives Turner (1548: B. ij. verso) as the first date, but notes the reference in the *Libellus*. Both Stearn (1965) and Rydén *et al.* (1999) accept the 1538 reference.

Asparagus officinalis

* 'Asparagus sativus ... groweth wilde in Essex, in a medowe adjoining to a Myll beyond a village called Thorp, and also Singleton, not farre from Carbie, and in the medowes neere Moulton in Lincolnshire; likewise it groweth in great plentie neere unto Harwich at a place   vc19, 53   **1597**

called Landamer lading, and at north Moulton in Holland, a part of Lincolnshire'. - Gerard (1597: 954).

This is outside of the presumed native range of *A. prostratus* and thus taken to be a garden escape. Gerard initially separates A. sativus, Garden Sperage, and A. palustris, Marish Sperage, but then seems to give only wild sites for the former. As for the locations given only Harwich and Moulton have been traced with any certainty.

Asparagus prostratus

This was finally separated from the archaeophyte *A. officinalis* and raised to species level through the work of Kay (1997).	vc1	**1997**
'Asparagus palustris *Ger.* ... Anno 1667. I found it growing on the cliffs at the Lezard point in Cornwal'. - Ray (1670: 31).	vc1	1667

Pulteney correctly points out that L'Obel (1571: 353) refers to *Asparagus* as occuring spontaneously by the sea, but there is no hint that this is in England. The record in Bowen (2000) citing Turner (1551) is an error.

Asperula cynanchica

'Anglica Saxifraga ... operit Julio et Augusto acclivem cretaceam, & aridum montem, arte militari aggestum [? Silbury Hill] inter Chipnam & Malburu Angliae, Bristoliensi a Londino via'. - L'Obel (1571: 183).	vc7	**1571**

Johnson too (1634: 65) makes L'Obel's plant synonymous with this species, but others have suggested *Scleranthus annuus*. Clarke & Druce had both cited as the first record 'Synanchica *Lugd*'. ... Between Rochester and Gravesend. - Johnson (1632: 30), though Clarke had added the L'Obel record as a possibility.
Another early record is 'Synanchia altera Anglica, sive minor ... Cretaceis gaudet montosis prope Drayton e regione Vectis Insulae'. - L'Obel (1655: 150). Raven (1947: 239) notes that 'About this time [1596] too, he [L'Obel] was certainly at Richard Garth's house at Drayton near Portsmouth'.

Asplenium adiantum-nigrum

Onopteris mas. The male blacke Ferne ... upon trees in shadowie woods ... shadowie sandy banks and under hedges'. - Gerard (1597: 975.1).	1597

Asplenium ceterach

'Asplenum ... Citterach, or Scaleferne ... I heare say it is plentuous in the west countrey here in England'. - Turner (1548: B. iij. verso). '... of Ceterache ... or scaleferne, because it is all full of scales on the inner-syde ... about Bristol'. - Turner (1551: E. iiij. verso).	1548

Asplenium marinum

'Chamaefilix marina anglica, non nisi saxorum rupiumve interveniis aut petrosis asperginibus Cornubiae innatum reperi ad maris alluviones non procul ab aedibus generosi viri D Muyle'. - L'Obel (1576: 474). | vc2 | **1576**

Edgington (2013) plausibly suggests a date of 'by 1571', the year L'Obel left England on his first visit. | vc2 | 1571

Mr Muyle [Moyle]) lived at Bake near St Germans and is also referred to in Parkinson (1640: 1044-5).

Asplenium obovatum

'Filix elegans Adiantheo nigro accedens. On the rocks on North side [of Jersey], W. Sherard'. - Ray (1690: 228). | vc113 | **1690**

Sherard visited in 1689. | vc113 | 1689

There are specimens in Buddle's herbarium at **BM**, from 'around 1700', 'gathered by Mr Thorpe near Tunbridge'.

Asplenium onopteris

as *Asplenium adiantum-nigrum* var. *acutum*, Praeger (1919). | vcH1 | **1919**

'on limestone rocks at Muckross, Killarney, 'a beautiful and delicate variety [of *A. adiantum-nigrum*]". - Mackay (1825). | vcH1 | 1805

Finally raised to species level in Shivas (1969).

Asplenium ruta-muraria

'Ruta muraria, sive Salvia vitae. Wall Rue, or Rue Maiden-haire … upon the wall of the churchyard of Dartford … also upon the walls of the Churchyard of Sittingburne, also upon the Church wals of Railey [Rayleigh, N. of Southend] in Essex'. - Gerard (1597: 983.3). | vc16, 18 | **1597**

This is presumably Turner's (1548: H. iiij.) 'Salvia vita or Ruta muralis … it may be called in english Stone Rue'. But he has no hint of it being an English plant. Lyte too (1578: 409) has 'Reu of the wal is very common in this countrie … upon all olde walles that are moyst'. Again there is no certainty that this refers to England.

Asplenium scolopendrium

'Hemionitis … Hertes tongue'. - Turner (1538: B. ij). 'The common Harts toonge groweth by the waies sides, as yee travell from London to Exceter'. - Gerard (1597: 976). | vc67 | **1538**

Turner notes that this was probably the identity of the plant that he seen in Northumberland (see also Edgington, 2013).

Asplenium septentrionale

'Adianthum ἀκροςιχον seu furcatum *Thal.* Filix saxatilis Tragi *J.B. Park.* On the rocks in Edinburgh Park. T. Willisell'. - Ray (1670: 7).   vc83   **1670**

Raven (1950: 151n.) implies that Willisel's collections were in 1669.   vc83   1669

Parkinson's reference (1640: 1043.8, 1044) of 'Filix saxatilis Tragi … as Lobel saith neere the sea in Cornwall in moyst rockie places' obviously is meant to refer to *A. marinum*, but the illustration is of *A. septentrionale*.

Asplenium trichomanes

'Trichomanes groweth in the same places that Adianthum … hear groweth. … it hath smal leaves of eche syde, growyng in order: in figure lyke unto the leaves of a lentill, one agaynst another upon small twigges shyninge tarte and somthynge blackishe'. - Turner (1562: 157).   **1562**

Turner's illustration is of this species. Gerard (1597: 985) has 'in a shadowie sandie lane in Betsome, in the parish of Southfleet in Kent … walls of her Majesties palace of Richmond'. Johnson (1633: 1146) refers to Goodyer recording it in Jan 1624, 'neere Wollmer Forrest'.

Asplenium viride

'Trichomanes ramosum *J.B.* … In summis rupibus Arvorniæ [Snowdon] copiose oritur … D. Lloyd'. - Ray (1690: 27).   vc49   **1690**

Edgington (2013) suggests that 'it was probably one of the plants that Lhwyd found on Snowdon in the summer of 1688, and sent to Ray the following year'.   vc49   1688

Aster linosyris

* '*Chrysocoma Linosyris* … discovered in September, 1812, by the Rev. Charles Holbech … on the rocky cliff of Berryhead, Devon'. - E.B. 2505.   vc3   **1813**
   vc3   1812

Aster tripolium

* 'Tripolium maius et minus … scatent haec Norbonica, et Anglica littora, & fluminum crepidines'. - L'Obel (1571: 123). 'By the fort against Gravesend: in the Ile of Shepey in sundry places: in a marsh which is under the towne wals of Harwich, in the marsh by Lee in Essex, in a marsh that is between the Ile of Shepey & Sandwich'. - Gerard (1597: 333).   **1571**

Astragalus alpinus

* 'It was discovered on 30 July 1831, on a cliff near the head of the Glen of the Dole, Clova, by Mr. Brand, Dr. Greville, and myself [Dr. R. Graham]'. - E.B.S. 2717.　　vc90　**1831**

Astragalus danicus

* 'Glaux exigua … The true Glaux groweth upon Barton hill fower miles from Lewton [Luton] in Bedfordshire'. - Gerard (1597: 1062).　　vc30　**1597**

Druce cites 'Foenum graecum sylvestre. There is a herbe in England … which aunswereth in many things unto the description of Glaux in Dioscorides, howbeit I think it is not the trew Glaux of Dioscorides wrote of'. - Turner (1562: 12) (given by Druce as 1551 in error). This *may* refer to *A. danicus*, but seems too tenuous to accept here.

Astragalus glycyphyllos

* 'Hedysarum Glycyrrhizatum … In Suffolke … from Sudbury to Corner Church [?Cornard] … in Essex about Dunmow and in the townes called Clare & Hennyngham [Castle Hedingham]'. - Gerard (1597: 1059).　　vc19, 26　**1597**

Pulteney cites L'Obel (1571: 402) 'Glaux vulgaris … Ex Angliae nos litoreis item mediterraneis procul mari, Londini serandam curavimus' which, I agree, refers to this species. Whether this means deliberately sown, or just growing where it seeded, is not clear. Jackson, in Hanbury & Marshall (1899), identifies that record as *Glaux maritima*, an error, I'm sure.

Athyrium distentifolium

as *Psaudathyrium alpestre* 'gathered by Mr Watson, in Canlochan Glen, Forfarshire, in 1846'. - Newman (1851).　　vc90　**1851**

For *A. flexile* 'Herb [J.] Backhouse, 1832, **E**'. - Ingram & Noltie (1981). For *A. distentifolium* 'Collector unknown, 1840, **E**'. - Ingram & Noltie (1981). 'July 1841, Ben Alder'. - Watson (1852).　　vc90　1832

Stace (2010) describes *A. flexile* as a variety of this species, but prior to that was often treated separately. This was described as *Pseudathyrium flexile* from 'Micaceous rocks at the head of Glen Prosen, Clova Mountains, Forfarshire, James Backhouse, James Backhouse, jun., Thomas Westcombe'. - Newman (1853b).

Athyrium filix-femina

'Filix mas non ramosa pinnulis angustis, raris, profunde dentatis. This groweth also in many places in the shade'. - Johnson (1633: 1130).　　vc11　**1633**

Goodyer described this on 4th July 1633 (ms. 11. f. 140) see Gunther (1922: 183) who identifies it as *Dryopteris affinis*), along with three other ferns, all, presumably, from Maple-durham neer Petersfield. Druce and Rose (in Gilmour (1972) identify 'Filix foemina'. - Johnson (1629: 8)' In between Isle of Grain and Rochester, as this species. Edgington (2013) is certain that that last should be *Pteridium aquilinum*.

Atriplex glabriuscula

* as *A. rosea* 'Guernsey'. - Babington (1837). 'I have lately gathered *A. rosea* on the coast of Holy Island (Lindisfarn), Berwick, and the Forth near Newhaven, and Mr Borrer has sent it to me from the Sussex coast'. - Babington (1838).	vc113	**1837**
'A. maritima ad foliorum basin auriculata procumbens & ne vix sinuata, *Pluk. Alm. 61. Totum rubrum variatur 7 spec. folio Isle of Anglesey towards Bolaton* [?] *plentifully.* (The first British record of *A. Babingtonii*, Woods …)'. - Druce & Vines (1907: 56 as species 152.7), and in the Appendix to the above (142, sheet 151) from Portsmouth.	vc52	1726

Moss (1914: 177) has synonymy to Ray (1670: 35) as 'A. maritima nostras', to Morison (1680: ii 607), but these must be to the *A. prostrata* group only. He (Moss) also equates Smith's (1804: 1092) *A. patula* β with this species, but the Babington seems the safer reference. See also Wolley-Dod (1937: xxxvi & 373) for possible earlier records.
The record from Druce & Vines is dated 1726, as it was found on Dillenius' trip to Wales with Brewer and others. Bolaton is very possibly Plas Bodafon, not far from Moelfre (James Robertson *in litt.*).

Atriplex laciniata

* 'Atriplex maritima *J.B.* maritima laciniata *C.B* … Hanc plantam nuperrime in Anglia invenimus in litore maris arenoso prope Holland, vicum 5 m. p. à S. Osithae [St Osyth] oppido remotum in Essexia: D. Dale in Merseia [Mersea] …'. - Ray (1686: 193). 'On the Sea shore near little Holland in Essex: found also by Mr Dale in the isle of Mersey, not far from Colchester plentifully'. - Ray (1688a: 3).	vc18	**1686**
Marginalia in a copy of Du Pinet, *Historia Plantarum*, 1561 in the BM library. Gunther (1922: 235) attibutes these to Dr Walter Bayley and thought most likely to date from 1570-1572.		1572

See the comment under *A. pedunculata* as regards Johnson's 1639 record from near Harlech, which Merrett identified as this species. Indeed, *A. laciniata* is still found at Mochras, just to the south.
Moss (1914: 179) and Pulteney equate Gerard's (1597: 257.4) 'A. marina, Sea Orach' with this species. Gerard does say 'covered with

a certain mealiness' and gives records from Rochester, Reculver, Tilbury and Gravesend, but Hanbury & Marshall (1899) do not cite the record, and the species is not that common in Kent.

Raven (1950: 245) dates Ray's first record to the summer of 1685, on the basis of a ref to *Corr. 140* (ie Lankester, 1848: 140), but there is nothing on that page or any others near it.

Atriplex littoralis

* 'Atriplex maritima altera Osyridis aut Scopariae fol. sive minima ... Salsugine riguis pratis juxta aedes D. Richardi Garth, non procul ab Ostia sive portu vernaculo discursu Portsmout, triplicem atriplicem maritimam collegi'. - L'Obel (1655: 85).   vc11   **1655**

'Halimus or Atriplex maximus, upon hable [R. Hamble] bankes in great quantitie'. - Marginalia in a copy of Du Pinet, *Historia Plantarum*, 1561 in the BM library. Gunther (1922: 235) attributes these to Dr Walter Bayley and thought most likely to date from 1570-1572.   vc11   **1572**

Richard Garth died in 1597.

Atriplex longipes

'… my collection from a stand of *Phragmites australis* in the NE of Wigtown Bay, Kirkcudbright, vc73, 9th September, 1975, **MANCH**'. - Taschereau (1977).   vc73   **1977**

'Brands Bay, vc9, 1969, P. Symmons, **RNG**'. - Bowen (2000).   vc9   1969

The specimen from Dorset was redetermined as this (or possibly with an element of *A. prostrata*) by J.R. Akeroyd.

Atriplex patula

'Atriplex sil. Polygoni aut Helxcines foliis, *Lob.*'. Hampstead to Kentish Town. - Johnson (1629: 12).   vc21   **1629**

But see Turner (1551: E. v. verso) '… wild areche, and it groweth abrode in the corne feldes'. - Chapman & Tweddle (1989: 248) identify this with this species. Clarke and Druce cited 'Atriplex sylvestris vulgaris'. - Johnson (1633: 326, 5), as their first record, but all modern authors have chosen the 1629 record.

Atriplex pedunculata

'Atriplex maritima Halimus dicta, erecta, semine folliculis membranaceis bivalvibus, in latitudinem porrectis, & utrinque recurvis, longo pedunculo insidentibus, clausò. *Phytogr. Tab. 36 fig.*1. Hanc plantam jampridem in palustribus maritimis, prope Scirbeck, juxta Sepulchetum, unum milliare a Boftoniä [Skirbeck, a mile from Boston] oppido remotu, in agro Lincolnienfi copiossimam invenimus'. - Plukenet (1696: 61).   vc53   **1696**

Both Clarke and Druce cite 'A. marina semine lato nondum descripta. Sea Orach with broad seeds'. - Johnson (1641: 16) and 'Neer Harlecham [presumably Harlech, as this is the location on his 1639 trip where Johnson describes finding it on the beach (*op. cit:*11) 'copiose']'. - How (1650: 13). But this species has never been found in Wales and the record is not cited in any later authority. Kew & Powell (1932: 98-99) suggest that Johnson's plant might be *A. laciniata*, and indeed, the figure illustrated in their work is reproduced from plates of Johnson's held in **BM** and annotated in Merrett's hand as that species (Kew & Powell, 1932: 127).

Gibbons (1975) gives the Plukenet date as 1691, which presumably refers to the publication date of the first two volumes of his *Phytographia*. But whilst I can see the drawing there, there is no indication that it was found in Britain.

Atriplex portulacoides

* 'Cepea ... may be called in English see purcellyne ... I found the same herbe of late besyde the Ile of Porbek' [Purbeck, Dorset]. - Turner (1551: I. iij. verso).	vc9	**1551**
Turner's visit to Dorset was in 1550 - see Raven (1947: 102)	vc9	1550
See also Turner (1548: C. i) 'Cepaea Plinij groweth by the sea syde'.		

Atriplex praecox

'Kylestrome, W. Sutherland, vc 108, 4th September 1975, **MANCH**'. - Taschereau (1977).	vc108	**1977**
N. Uist, vc110, 1861, det. Taschereau, **E**.	vc110	1861

Atriplex prostrata

'Atriplex sil: sinuata'. Hampstead to Kentish Town - Johnson (1629: 12).	vc21	**1629**
Johnson's record is identified as *Chenopodium album* by Clarke and Druce, but Rose (in Gilmour, 1972: 127) identifies this as *Atriplex hastata* (*prostrata*), as does Kent (1975).		

Atropa belladonna

* 'Somniferum et Laethalae ... Angl. Dwale, Great Morell ... in Anglia ubique obvio proventu fruticat'. - L'Obel (1571: 102). 'It groweth very plentifully in ... Lincolnshire and in the Ile of Elie at a place called Walsoken neere unto Wisbitch. I founde it growing without the gate of Highgate'. - Gerard (1597: 269).	**1571**

This is also in Turner (1538: C. i. verso) as 'Solanum soporiferum aliqui vocant Dwale', which implies familiarity with the plant, though it might be still as a cultivated plant.

Avena fatua

* 'Aegylops Bromoides Belgarum. In Anglia ... inter hordeum et secale nonnusquam occurrit'. - L'Obel (1576: 21). **1576**

Avenula pratensis

* 'Gramen avenaceum montanum spica simplici, aristis recurvis ... In summis tumulis seu colliculis Bartloviensibus [Barton Hills] manu quoddam aggestis, in ipso limite agri Essexiensis versus Cantabrigiam inventum ad nos attulit D. Dale'. - Ray (1688: 1290). vc19 **1688**

Avenula pubescens

* 'Gramen avenaceum panicula purpuro-argentea splendente *D. Doody* ... In pascuis circa aedes Comitis Cardiganiae ad Twitnam [Twickenham] Middlesexiae vicum'. - Ray (1688: 1909) 'In the pastures about the Earl of Cardigans house at Twittenham in Middlesex'. - Ray (1688a: 9). vc21 **1688**

Baldellia ranunculoides

'Plantago aquatica minor, shown to me by the learned English doctor Thos Penny in watery places in England'. - L'Obel (1581: 370). **1581**

See Raven (1947: 168). Clarke and Druce both overlook the L'Obel record and cite Johnson (1633: 418) 'Plantago aquatica humilis ... I found [this] in the companie of Mr. Will. Broad and Mr. Leonard Buckner in a ditch on this side Margate in the Isle of Tenet'. Did Penny show this to L'Obel before he left Britain in 1571?

Ballota nigra

* 'Ballote . . . marrubium nigrum is named in english stynkyng Horehound or blacke Horehound ... groweth in hedges communely in every countrey'. - Turner (1548: B. iiij. verso). **1548**

Barbarea vulgaris

* 'Barbare herba groweth about Brokes and water sydes ... in englishe wound-rocket'. - Turner (1548: H. j. verso). 'Barbarea'. Between Hampstead and Kentish Town. - Johnson (1629). **1548**

Bartsia alpina

* 'Crataeogonon foliis brevibus obtusis Westmorlandicum ... Prope Orton in Westmorlandia juxta rivulum qui decurrit secus viam qua inde Crosbeiam itur'. - Ray (1670: 86). vc69 **1670**

This was on Ray's 1668 Itinerary. See Lankester (1848: 26) and Raven (1950: 148). vc69 **1668**

The plant still occurs, by Ray's Bridge, below Orton Scar, on the way to Crosby Ravensworth.

Bellis perennis

* 'Bellis ... est ilIa herba quam vocamus a Dasy'. - Turner (1538: A. iij). 'In Northien men call thys herbe a banwurt'. - Turner (1551: F. ij. verso).

Turner (1538) adds that in Northumberland the name 'daisy' only refers to that with a red flower, cultivated in gardens, and that the wild kind is called a Banwort.

1538

Berberis vulgaris

'Oxiacantha … vulgo Berberis dicitur. - Turner (1538: B. iv). 'The Barbery bushe growes of itself in untoiled places … about a gentleman's house called Master Monke, dwelling in a village called Iver [Bucks], two miles from Colebrooke, where most of the hedges are nothing else but Barberrie bushes'. - Gerard (1597: 1144).

Clarke and Druce cite the Gerard record as the first date, but Turner's English name implies familiarity with the plant as a wild species.

1538

Berula erecta

'Water persely or sallat persely'. - Turner (1562: 138).

Clarke cites 'Sium majus angustifolium … This I first found in the company of M. Robert Larkin [Lorkin] going betweene Redriffe and Deptford'. - Johnson (1633: 257), although wondering if the Turner entry refers to it. Druce, Nelson (1959) and Chapman *et al.* (1996) all cite the Turner reference.

Another early record is 'Pastinaca aquatica minor. Sium odoratum tragi. 2 July 1620 … plentifullie in the River by Droxford in Hampshire'. - Goodyer ms.11, f. 82v, see Gunther (1922: 116).

1562

Beta vulgaris

* 'Ad insulam Sheppey … Beta alba'. - Johnson (1629: 5). 'Upon the sea coast of Tenet [Thanet]'. - Johnson (1633: 319).

Turner and Gerard only knew this plant from cultivated forms in the garden.

vc15   **1629**

Betonica officinalis

* 'Betonica … Betony … groweth muche in woddes and wylde forestes'. - Turner (1548: B. v).

See Turner (1538: A. iij) 'Betonica … vulgus nostram vocat Betony', which implies familiarity with the plant as a wild species.

1548

Betula

* 'Byrche ... in Northumberlande'. - Turner (1551: F. v. verso). Turner calls it .. 'this Frenche tre'.   vc67, 68   **1551**

Note also Turner (1548: B. v. verso) 'Birch tree ... it groweth in woodes and forestes'.

Betula nana

* 'In Bredalbane'. - Lightfoot (1777: 575, with a figure).   vc88   **1777**

'In a letter from Mr Pulteney to the Reverend Professor Martyn, Leicester, 17th February 1763, 'I have lately had from Dr Hope ... he informs that they have discovered the *Betula nana* ... to be native in Scotland'. - Gorham (1830: 116). Martyn (1763; final page) also mentions that 'I have recently been informed by my ingenious Friend Mr Pulteney of Leicester that the *Betula nana* ... is found to be a native of Scotland'.   vc106   1762

Sir James Naesmyth in the moors north of Loch Glash [Glass] Ross-shire'. - Hope, 1768, in Balfour (1907). It is slightly odd that Hope does not mention his earlier record here. However a much earlier record of what is almost certainly this species is referred to in Gibson (1722: col 962) 'Betula rotundifolia nana. N.D. On a moss near Birdale [Birkdale, vc65]'. The background is set out in Horsman (1995: 163), and the finder is said to be Christopher Hunter (Horsman, 2011).

In another possible pointer to an early date John Edgington drew my attention to a letter from Richardson to Petiver on 11 September 1702, (Sloane ms. 4063, f.174): 'The Salix pumila folio rotundo JB which you take to be one of Bocconi I find mentioned in an old catalogue of my garden by the name Betula montana rotundifolia repens, Vitis Idaea crenatifoliis. It grows well in my garden and bears Juli [catkins] yearly but the seed is not included in a pappus as in the Salix kind'. He (Edgington) adds 'I'm quite sure that Salix pumila folio rotundo JB is *Salix herbacea*, (which Bocconi called Salix alpina, minima, alni rotundio folio - Bauhin's name describes *S. reticulata*). But Richardson's garden plant sounds very much like *Betula nana* (for which, before I had seen it, I once mistook *S. herbacea*). Of course we don't know the origin of Richardson's plant - maybe another garden, such as Leiden - but it's still interesting to see such an early reference to *B. nana*, if that is what it was'.

Betula pendula

Babington (1843).   **1843**

Babington (1843) refers to E.B. 2198 (1810), where this species is the lower figure and *B. pendula* the upper figure, and his seems the first really satisfactory split and key.   1810

It would be surprising if there were not earlier herbarium records.

Betula pubescens

as var. *glutinosa* (*pubescens*), Balfour *et al.* (1841: 2). **1843**

Oswald & Preston (2011: 169n) point out that Ray's records for Cambridge (1660: 21) is almost certainly this species, as even Babington (1860), two hundred years later, only recorded *B. pendula* as a planted tree. vc29 1660

See above. Druce cites Gerard (1597: 1295) 'Betula' and Ray (1724: 443), but neither convinces that they were aware of the differences. Kent (1975) assumes that Johnson's 'Betula, Matth. Lob.' (1632: 34) is this species, which is probably correct.

Bidens cernua

* 'Foemina Canabina Septentrionalium, stellato & odoro flore … In Anglia ubiq; udorum, praesertim Londini & Auxonę'. - L'Obel (1571: 227). vc21 **1571**

Where is 'Auxonę'? Druce (1927) does not claim it for Oxfordshire. Auxonne in France? Yet the order of words would suggest somewhere in England.

Bidens tripartita

* 'Eupatoreum Cannabinum foem'. Isle of Grain to Cliffe. - Johnson (1629: 8). vc16 **1629**

See Edgington (2007) for a discussion on Goodyer's recognition of this plant, almost certainly before 1629.

Blackstonia perfoliata

* 'Centaureum Luteum … Quamplurimis tum Galliae, tum Angliae locis … collibus urbi Bristoiae eminentibus, inter segetes'. - L'Obel (1571: 173). 'Upon the chalkie cliffes of Greenhithe'. - Gerard (1597: 437). vc34 **1571**

Blechnum spicant

'Lonchitis altera … hath leves lyke unto ceterache which is called Asplenium, but greater, rougher, and much more divided or cut in … in diverse places of Sommerset shyre and Dorset shyre … and may be called combe ferne. It groweth much in darck laynes about bushe rootes and out of the shaddow oft tymes alone'. - Turner (1562: 41b). vc5, 6, 9 **1562**

Turner's visit to Dorset was in 1550 - see Raven (1947: 102). vc9 1550

Blysmus compressus

* 'Gramen cyperoides spica simplici compressa disticha … a D. Newton mihi primum ostensum est a se collectum in aquosis prope Orton Westmorlandiae vicum, necnon circa Chislehurst. vc16, 69 **1688**

Exsiccatum vidi apud D. Plucknet qui primus illud observasse dicitur'. - Ray (1688: 1910 and 1688a: 10).

Blysmus rufus

* As *Schoenus ferrugineus* 'In the Isle of Mull'. - Lightfoot (1777: 86 & 1138).   vc103   **1777**

as *Schoenus ferrugineus* 'in Sky'. - Hope, 1768, see Balfour (1907).   vc104   1768

Clarke adds 'Specimens from Robert Brown are in **BM** labelled 'Arbigtland in Galloway, 1769, Dr. Walker, who thought it was the *Schoen. ferrugineus* Lin. He further added 'It appears under this name in Lightfoot. Dr. Walker was "its original discoverer". - E.B. 1010'.

Bolboschoenus laticarpus

'So far the earliest Somerset specimens thought to be *B. laticarpus* are from Berrow (vc6) and from a ditch near Long Load (vc5), collected in 1881 and 1891 respectively, by R.P. Murray. Records for *Scirpus maritimus* at both of these sites are included in Murray's *Flora of Somerset* (1896)'. - Crouch (2011).   vc5   **2011**

The earliest evidence is almost certainly from Pulteney (1799), who recorded *Scirpus maritimus* from the R. Stour at Durweston [NW of Blandford Forum, about 15 miles from the sea]. All recent inland records on that river have been of *B. laticarpus*.   vc9   1799

The realization that there were two species in Britain arose from the paper by Hroudová, Z. *et al.* (2007). The 1881 record has been ignored here as though determined by Hroudová on the pedunculate form, the fruit is too immature to be certain (F.J. Rumsey, *in litt.*).
Pulteney's work is cited as 1799, as that is the date of the version printed 'for the use of the compiler and his friends'. An augmented version was published in 1813.

Bolboschoenus maritimus

* 'Ad insulam Sheppey … Cyperus rotundus inodorus septentrionalium *Lob.*'. - Johnson (1629: 5).   vc15   **1629**

Botrychium lunaria

'Lunaria racemosa, Ceterach foliis, racemulis Ophioglossi … Lunaria minor … postremo in Angliae collibus, qui Grenvici Regius aedibus [Greenwich, at the Royal Palace], & Tamesi imminent, satis multia reperimus'. - L'Obel (1571: 360).   vc16   **1571**

Turner (1548: H. iij) has 'Lunaria … in English litle Lunary or Maye Grapes … groweth in Middowes and pastur groundes'. See also Turner (1568: 53) 'the lesse Lunarye', with an excellent illustration, but neither reference specifically says that it grows here.

Brachypodium pinnatum s.l.

* 'Gramen spica Brizae majus. In copses and hedges common enough about *Oxford*, D. Bobart'. - Ray (1696: 248).   vc23   **1696**

Brachypodium pinnatum s.s.

Chapman & Stace (2001).   **2001**

'Gramen spica Brizae majus … (In *Woodstock*-Park, *Mr. J. Sherard*. On all the Heaths and Commons for twenty miles on this side *York*; *idem*)'. - Ray (1724: 392).   vc23, 63   1724

This was the first published note in Britain of a split suspected by Dr M.A. Khan in 1981, and included in his 1984 thesis.
The Ray record for York is most likely to be for this segregate.

Brachypodium rupestre

Chapman & Stace (2001).   **2001**

C. A. Stace (*in litt.*) notes 'Jenner's Flora of Tumbridge Wells (1845) says of *B. pinnatum*: "On the sides of the hill between St Clare and Wrotham". That is without any doubt *B. rupestre*'.   vc16   1845

This was the first published note in Britain of a split suspected by Dr M.A. Khan in 1981, and included in his 1984 thesis.
Not enough redetermination of herbarium species has been undertaken to give a definitive first evidence, but without doubt there will be earlier records.

Brachypodium sylvaticum

* 'Gram. spica Brizae majus *Bauh.*'. Highgate wood. - Johnson (1629: 11). 'Gramen avenaceum dumetorum spica simplici … In sepibus & dumetis hac specie nihil frequentius'. - Ray (1670: 140).   vc21   **1629**

This record of Johnson's is accepted by Kent (1975).

Brassica nigra

* 'Sinapi sativum vulgare'. Isle of Thanet. - Johnson (1632: 4).   vc15   **1632**

Clarke gives ''Mustarde,' recorded as growing 'in the corne in Somersetshyre a litle from Glassenberrye' Turner (1562: 137) as possibly referring to this species, as does Druce (though citing the wrong year) and Chapman *et al.*(1995). But see *Sinapis alba*.

Brassica oleracea

* 'Brassica sylvestris groweth in Dover cliffes where as I haue onely seene it in al my lyfe'. - Turner (1548: B. vj).   vc15   **1548**

Brassica rapa

* 'Rapistrum aliud sylvestre non bulbosum ... I found going from Shorditch by Bednal Greene to Hackney'. - Parkinson (1640: 864).   vc21   **1640**

Druce identifies 'Rapum sativum in agris'. Cliffe. - Johnson (1629: 8) as this, but Rose, in Gilmour (1972: 61), thinks the Johnson record was *B. napus*.

Briza media

* 'Phalaris pratensis minor ... non in segetibus, sed herbidis pratensibusque Germaniae, Galliae, Angliae oritur'. - L'Obel (1571: 16).   **1571**

Briza minor

'Gramen tremulum minus, paniculâ amplâ, locustris parvis, triangulis. Mr Sherard first found in Jersey, afterwards in many meadows in France. Small Quaking Grass with triangular Heads'. - Ray (1696: 254).   vc113   **1696**

Sherard's visit was in 1689.   vc113   1689

See also Le Sueur (1984: xxxiv, 195). The first record other than from the Channel Islands is 'Between Penzance and Marketjew, Cornwall, 1774. Lightf. in his herbarium'. - Smith (1824: i. 133). Another is 'Prope Bath, D. Alchorne'. - Hudson (1778: 38), and also in Babington's *Flora Bathoniensis* (1834), apparently on the personal observation of the author. But White (1912) relates that Babington said later that his record was erroneous, though the source of that comment has not been traced.

Bromopsis benekenii

'In the report of the B.E.C. for 1867 (*J. Bot. 6: 71*), it is stated that Herr von Nechtritz, of Breslaii, recognized specimens of '*B. asper*,' from Derbyshire, as this plant, and that plants from North Yorkshire must be referred to the same subspecies. It appears that *B. serotinus* is the plant usually called *B. asper* by British botanists, and that the restricted *B. asper* is a plant of great rarity in this country; indeed, the only English specimen of typical *B. asper* I have seen, is one in Sowerby's herbarium at the British Museum, collected at the Plough, Camberwell, Surrey, where it may perhaps have been an introduction'. - Trimen (1870b).   vc17   **1870**

Sowerby died in 1822   vc17   1822

Trimen is citing the article by Beneken describing, under the name of *B. serotinus*, in the *Botanische Zeitung* for 1845, p. 724, what he considered a new species, found near Naumburg - that is separating what we now call *B. ramosus* from what is now *B. benekenii*.

Bromopsis erecta

* 'Festuca Avenacea sterilis spicis erectis. In the hedges beyond Botley near Oxford'. - Ray (1690: 237). vc23 **1690**

This record was in the appendix, from J. Bobart.

Bromopsis ramosa

* 'Gramen avenaceum lanuginosum glumis rarioribus longis. Hairy Haver-grasse'. - Johnson (1634: 40). 'Gramen avenaceum dumetorum panicula sparsa'. - Ray (1670: 140). **1634**

Bromus commutatus

'Festuca elatior paniculis minus sparsis, locustis oblongis strigosis aristatis purpureis splendentibus D. Doody'. - Ray (1696: 261.7). This is the reference given in E.B. 920, as cited by Babington (1843). **1696**

See *B. racemosus*, which Cope & Grey (2009) combine with this species, surely a more sensible treatment. Druce (1928: 489) cites 'Festuca Avenacea, spicis strigosioribus e glumis glabris compactis. From Buddle', but in his later work (Druce, 1932) assigns this to *B. racemosus*, adding that 'much confusion has been made between the two species'!

Bromus hordeaceus

* 'Gramen Bromoides vernum spicis erectis. Early meadow Drauk or Darnell-grasse'. - Johnson (1641: 22). **1641**

Bromus interruptus

Druce (1895, 1896b). vc22 **1895**

'The plant was first recorded in 1849 by a Miss A.M. Barnard and a collection made from a field in Odsey, Cambridgeshire. These examples were studied by H.C. Watson who named it *Bromus pseudo-velatinus* (Watson 1850)'. - Lyte & Cope (1999). vc20, 29 1849

Druce's report (1895) was just a note - the full paper is Druce (1896b). His record was from Berkshire, but he was then unaware of the earlier gathering.

Odsey is on the Icknield Way, right on the boundary of Cambridgeshire and Hertfordshire, though the village is in the former. Both counties have claimed the record.

Bromus racemosus s.l.

'Festuca Avenacea spicis strigosioribus e glumis glabris compactis. In pratis humidis provenit. D. Dale observavit & attulit'. - Ray (1688: 1909). **1688**

Both Clarke and Druce cite Ray (1690: 191) but the 1688 reference seems acceptable. Druce (1928: 489) cites this polynomial for *B. commutatus* with the reference Ray (1686: 1907).

Bromus secalinus

'Gramen Avenaceum supinum floscuculis secalinus. ... All these grow in the fields of the land either of plowed or fallow fields. Long-winged Oate Grasse flowring like Rye'. - Parkinson (1640: 1150.10), but with no illustration.

1640

Clarke cites Merrett (1666: 49) 'Gr. bromoides latiore pannicula, *P.1150*. ... Woods below Hamsted', and this is followed by Kent (1975), but the habitat seems unlikely. Pulteney also cites Parkinson, but equates it with 'Gramen bromoides segetum latiore panicula' (1640: 1149.2).

Bryonia dioica

* 'Ampelos leuce ... anglis Bryoni aut Wylde nepe'. - Turner (1538: A. ij. verso). 'In many places of Englande in hedges'. - Turner (1548: B. vj. verso).

1538

Bunium bulbocastanum

* 'Found by Rev. W.H. Coleman in 1839 'near Cherry Hinton in Cambridgeshire'. - E.B.S. 2862.	vc29 vc29	**1841** 1839

Dony (1953) and his successors give the first record as 1835 (Coleman, Cherry Hinton, etc), but this is in error.
Turner (1548: B.vij) notes, under Bunium, that 'it is a rare herbe in Englande ... '.

Bupleurum baldense

* as *B. odontites* 'This new addition to the flora of Britain was gathered by the Rev. H. Beeke, D.D. early in July last, on the marble rocks around Torquay, Devonshire. We indeed received a very diminutive specimen ten years ago from Devonshire, by favour of the Rev. Aaron Neck'. - E.B. 2468.	vc3	**1812**
'Sent by the Revrd. Aaron Neck ... Jany. 19th, 1802'. - Note on original drawing for E.B. 2468, see Garry (1903: 84).	vc3	1801

Bupleurum falcatum

* 'Found by Mr. Thomas Corder in 1831 'at Norton Heath between Chelmsford and Ongar, Essex'. - E.B.S. 2763.	vc18 vc18	**1833** 1831

Bupleurum rotundifolium

* 'Of Throwwaxe ... Perfoliata ... In Summersetshire betwene Summerton and Marlock [Martock]'. - Turner (1568: 56).	vc5	**1568**

Bupleurum tenuissimum

* 'Auricula leporis minima ... By the way-side as you ride to St Neotes beyond Elles-ly [Eltisley]. This I have found in sundry other places in England, viz. in the road to Stilton a little beyond Huntingdon. At Maldon in Essex ... near Fullbridge'. - Ray (1663: 3). vc18, 29, 30 **1663**

How (1650: 18) has 'Bupleurum minimum nondum descriptum floribus luteis ... found in Surrey', which is accepted as the first record by Salmon (1931). See also Johnson (1641: 17) 'Bupleurum angustifolium, *Dod. Tab. Ger.* Narrow leafed Hares eares ... are affirmed by Gerard to grow here, as also by Parkins. pag. 578'. Gerard (1597: 485) describes two species, and adds that he had found them 'growing naturally among the bushes upon Beistone castell in Cheshire'. This seems far too tenuous and an odd habitat; see *Alchemilla alpina*, which Gerard recorded from the same site.

Butomus umbellatus

* 'Juncus cyperoides floridus paludosus ... In lacuniis & torpidis aut lentè fluentibus rivulis Belgiae, Angliae & Londini ad arcem Regiam casteriûmque [Tower of London] navium Liburnicarum nascitur'. - L'Obel (1571: 44). vc21 **1571**

Philip Oswald translates for me 'in ditches & sluggish or slowly flowing brooks of Belgium, England & London by the Royal castle/citadel and the brigantines' [galley's] oar-store'. Kent (1975) translates 'Regiam casteriûmque' as 'the Tower of London'. Note that *Scilla autumnalis* was first recorded by Parkinson (1629: 132) 'at the hither end of Chelsey before you come at the King's Barge-house'.

Buxus sempervirens

* 'Buxus ... the Box tree groweth upon sundry waste and barren hils in Englande'. - Gerard (1597: 1225). 'On Box hill in Surrey'. - Merrett (1666: 18). **1597**

Cakile maritima

* 'Cakile Serapionis ... Vulgo Eruca Marina ... Angliae insulá meridionalem Vectim' [Isle of Wight]. - L'Obel (1571: 77). vc10 **1571**

Calamagrostis canescens

* 'The first discovery of this grass is owing to Mr. John Scampton a Curious Botanist who sent it me from Leicestershire'. - Petiver (1716: n. 69). vc55 **1716**

Dandy (1958a: 200) notes that there is a specimen in Buddle's herbarium (H.S.125, f.11) annotated 'a D. Scampton in agro Northamp [sic]: circa Oundle collecta'. Buddle died in 1715. Horwood & Noel (1933: clxxxvi) cite a letter from Petiver to vc55 1695

Scampton, which they date 1695 (supported by the passage in Dandy), asking for specimens of a grass 'which seems different from ye Gramen arundinaceum Raii Syn. 185' as probably relating to this species.

Calamagrostis epigejos

* 'Calamagrostis nostras sylvae St. Joannis ... Reede grasse of St. Johns wood'. - Parkinson (1640: 1180).	vc21	**1640**
'Calamagrostis. 27 Apr. 1622'. - Goodyer ms.11, f.121, see Gunther (1922: 172). However Kent (1975: 587) cites a record from L'Obel (1655: 42) of 'Highgate, c 1600'. L'Obel died in 1616. Johnson (1638) has 'Gramen arundinaceum paniculatum, sive Calamagrostis [Tottenham]', interpreted by Edgington (2014) as this species.	vc21	1616

Clarke, followed by Hanbury and Marshall (1899) cite 'Gramen tomentosum & acerosum, Calamagrostis quorundam Lob.'. Between Gillingham and the Isle of Sheppey. - Johnson (1629: 5). Rose, in Gilmour (1972), finds this unconvincing and suggests *Polypogon monspeliensis*. Oddly, though, he identifies the same polynomial from L'Obel as *Calamagrostis* from the area between Sandwich and Canterbury in Johnson (1632: 22)! I feel that as the grass is really uncommon in E. Kent that the Parkinson is the safer reference.

Calamagrostis purpurea

' … material collected from Rescobie Loch, Angus, compares well with specimens of *C. purpurea*'. - Stewart (1981).	vc90	**1981**
'R. Mackechnie and E. C. Wallace collected it from near Braemar in 1941'. - Wigginton (1999). There is a specimen in **E**, dated 12/7/1941, from Braemar.	vc92	1941

Calamagrostis scotica

* 'Found by Robert Dick at Loch Duran in Caithness'. - Bennett (1885b).	vc109	**1885**
See Smiles (1878: 340), describing a visit in June 1863.	vc109	1863

The specimen is in Thurso, in the Robert Dick Museum.

Calamagrostis stricta

* as *Arundo stricta* 'Discovered by Mr. G. Don in June, 1807, in a marsh called the White Mire, a mile from Forfar'. - E.B. 2160.	vc90 vc90	**1810** 1807

Callitriche agg.

* ' … another herbe of small reckoning that floteth upon the water called Stellaria aquatica or Water Starwoort'. - Gerard (1597: 681). **1597**

This has been interpreted by Clarke and others to be *C. stagnalis* s.l. (see below).

Callitriche brutia s.l.

'Stellaria aquatica longfolia'. - Petiver (1713: vi. 4). '*C. autumnalis* … in fossis etc, frequens'. - Hudson (1762: 2). **1713**

Pulteney's copy of Hudson is in the Linnean Society, and there he has added a gloss equating this with the Petiver reference and with Ray's (1724: 290) 'Stellaria aquatica foliis longis tenuissimis'. Druce, in addition to the Hudson reference, cites a specimen '*C. autumnalis* (not of L.) circa Oxoniam, Sir J. Banks'. - Herb. **BM**. 1768.

Callitriche brutia s.s.

as *C. intermedia* var. *pedunculata* in Pearsall (1935).		**1935**
Pont Seiont, vc49, 1830, Babington, det. Preston, **CGE**.	vc49	1830

Callitriche hermaphroditica

'In Clunie loch, Scotland. Mr Arthur Bruce'. - Smith (1824: i. 10).	vc89	**1824**
Loch of Clunie, vc89, 1792, McRichie, **BM**.	vc89	1792

Callitriche obtusangula

* In marsh ditches, slightly brackish, bordering on Brading Harbour, Isle of Wight, July 1860'. - More (1870).	vc10	**1870**
St Peters Marsh, Jersey, vc113, 1852, J. Piquet, det. Savidge, **OXF**.	vc10	1860

More cites Syme (1868: viii. 122), who has this as a subspecies of *C. verna*, and was anticipating its discovery in Britain.

Callitriche palustris

'In July 1999, whilst on a botanical holiday in western Ireland, John Bruinsma found …'. - Lansdown & Bruinsma (1999).	vcH15	**1999**

The site is Coole, nr. Gort.

Callitriche platycarpa

The first record by this name was 'St Laurence, Jersey, growing on mud'. - Babington (1839a: 36). 'Near Preston Gobald Churchyard; near Oaks Hall near Pontesbury; Sharpestones Hill near Shrewsbury'. - Leighton (1841: 446).	vc113	**1839**
Clarke (1909) has 'more certainly Petiver's 'Stellaria pusilla pal. repens tetraspermos', of which there is a specimen in his English	vc17	1691

Herbarium at **BM** (H. S. 151, f. 36), having on its printed ticket 'The first discovery of this Plant to be a Native of England we owe to the ingenious Physician and Botanist Dr. Hans Sloane who observed it in a Bog on Putney Heath, June 4, 1691'.

Clarke (1909) also adds that 'but the Alsine palustris minor serpyllifolia referred to in Johnson (1633: 615), as found 'betweene Clapham heath and Touting, and betweene Kentish towne and Hampstead' was probably this'.

Callitriche stagnalis s.l.

' … another herbe of small reckoning that floteth upon the water called Stellaria aquatica or Water Starwoort'. - Gerard (1597: 681). **1597**

Dandy (1958a: 178) notes that there is a specimen of Petiver's annotated that it was first found 'in a bog on Putney-heath, June 4.1691' by Sir Hans Sloane (see *C. platycarpa* above).
Recorded, with no further details, in a ms. Flora and Fauna of the Islands, by James Robertson, 1767 - 1771, reproduced in Henderson & Dickson (1994).

Callitriche truncata

* 'Collected by Mr Borrer, on June 6th 1826, growing "completely under water in a deep ditch between Amberley Castle and Wild Brook", Sussex'. - Trimen (1870a).   vc13 **1870** / vc13 1826

This was first illustrated, as *C. autumnalis*, in E.B.S. 2606, 1829, and not properly sorted till Trimen's paper. The 1826 specimen is at **BM**.

Calluna vulgaris

* 'Erice … in English Heth hather, or Ling, it groweth on frith and wyld mores, some use to make brusshes of heath both in England and Germany'. - Turner, (1548: C. viij. verso). 'Irica. Heth hather and lyng … the hyest hethe that ever I saw groweth in northumberland which is so hyghe that a man may hyde hymself in'. - Turner (1551: P. ij). **1548**

Caltha palustris

* 'Chameleuce … called in Northumberlande a Lucken gollande … groweth in watery middowes with a leafe like a water Rose' [Water Lily]. - Turner (1548: C. ij).   vc67, 68 **1548**

Calystegia sepium

* 'Convolvulus … in english withwynde or byndeweede … wyndeth it selfe about herbes and busshes'. - Turner (1548: C. iiij. verso). **1548**

Calystegia soldanella

'Soldanella marina ... grows plentifully by the sea shore in most places of England, especially neere to Lee in Essex, at Mersey in the same countie, in most places of the Isle of Thanet, and Shepie and many places along the northern coast'. - Gerard (1597: 690).   vc15, 18   **1597**

Turner originally described 'Brassica marina ... Sea Folfote ... I have not sene it in England savyng only besyde Porbeck' [Dorsetshire] . - Turner (1551: G. ij. verso) as this species. But it was actually *Cochlearia officinalis* - see Raven (1947: 102), who describes how Turner realised the problem in his later (1568) edition. L'Obel (1571: 263) also describes 'Soldanella. Brassica marina'.

Camelina sativa

'Myagrum. Gold of Pleasure ... likewise wilde in sundry places of England'. - Gerard (1597: 213), How (1650: 78).   **1597**

Campanula glomerata

* 'Trachelium minus ... Mimori Cervicariae ... natales ... Montium Pratorum Germaniae, & Angliae Occiduae sunt'. - L'Obel (1571: 139). 'Upon the chalkie hils about Greenehyth in Kent'. - Gerard (1597: 365).   **1571**

But for Gerard's reference see also *Gentianella amarella*.

Campanula latifolia

* 'Trachelium majus Belgarum sive Giganteum ... In the yeere 1626 I found it in great plenty growing wilde upon the bankes of the river Ouse in Yorkshire as I went from Yorke to visit Selby the place whereas I was borne'. - Johnson (1633: 450).   vc64   **1633**
vc64   1626

Campanula patula

* 'Rapuntium fl. purp. At Effaton [?Adforton] a mile from Wigmore Hereford shire'. - Merrett (1666: 103).   vc36   **1666**

'Campanula media. Ista campanulae rotundifoliae similis, sed .... Provenit copiose inter Herefordiam et Kyneton, ac per totum Wye Flu: tractum [between Hereford and Kyneton (?Kington, 12m. NW. of Hereford) and along the whole length of the River Wye]'. - Ms. notes of William How, see Gunther (1922: 288).   vc36   1656

A full translation of How's notes (which are written in his copy of his *Phytologia*, between the publication of that and his death in 1656) gives little doubt that he is describing this species, especially as this is in the historical heartland of its distribution.
Another early record that might well predate that of How is 'Rapunculus sylvestris flore rubro albescente. '..in the pastures and hedgesides....neer Petersfield' - from species of plants described in Goodyer's MS (he died in 1664), see Gunther (1922: 195) and

Brewis *et al.* (1996: 221). Goodyer's record has been erroneously identified as *C. rapunculus*.

Campanula rapunculus

'Rapunculus esculentus vulgaris, garden Rampions ... Some of them [ie *Campanulas*], as the first [*C. rapunculus*] [are found] in divers places of this land' - Parkinson (1640: 648.1). 'Rapunculus esculentus vulgaris ... In the Lane that leads from Dartford Heath to Bexley, Kent, plentifully'. - Blackstone (1746: 81).

1640

If the Blackstone record is more certain than the Parkinson reference (which is also cited by Ray (1670: 263)) then the first recorded date is Lawson's 'about Hersham in Surrey' and 'in Nonsuch over court'. - Raven (1948: 8). Raven dates Thomas Lawson's notebook, from which these records came, to not later than 1676. Merrett's (1666: 103) 'neer Petersfield, Mr. Goodyer,' has been lastingly controversial. See *C. patula* above.

Preston *et al.* (2002) give 1597, based on Gerard's (1597: 369) 'Rapuntium parvum' which although that is given as a synonym by Ray and others, Gerard gives its habitat as 'woods and shadowie places', whereas he also describes 'R. majus' (the illustration of which is definitely not *C. rapunculus*) as sown in gardens and eaten in salads.

Campanula rotundifolia

* 'Campanula minor, purpurea ... wilde in most places of England, especially upon barren sandie heathes, and such like grounds'. - Gerard (1597: 368).

1597

Campanula trachelium

'Medium, an sit Mariana viola? ... Coventriae Angliae'. - L'Obel (1571: 138). 'Viola mariana ... these pleasant floures grow about Coventrie in England ... [where they] eate their rootes in Salads, as Pena writeth. *Fol.138*' - Lyte (1578: 194).

vc38   1571

See the note in Raven (1947: 201). Unfortunately Louis (1980) gives no hint as to when L'Obel visited Coventry.

Capsella bursa-pastoris

* 'Bursa pastoris ... Shepherdes purse ... groweth by highwayes, almost in every place'. - Turner (1548: H. ij). 'Trachelium majus ... in the lowe woods and hedgerows of Kent about Canterburie ... and also about Watford and Bushey, 8 miles from London'. - Gerard (1597: 365).

1548

Cardamine amara

* 'Cardamine seu Nasturtium aquat. amarum ... in a bog betwixt the Duke of Norfolks garden, and Lambeth Church in the way by Thames side, and in Cornwall'. - Merrett (1666: 20). vc17 **1666**

There are no confirmed records of this species in the West of England.

Cardamine bulbifera

* 'Dentaria bulbifera ... At Mayfield in Sussex in a wood called Highreede'. - Parkinson (1640: 621.1). 'In the old Park Wood near Harefield [Middlesex] abundantly'. - Blackstone (1737: 23). vc14 **1640**

'At Mayfield in a wood of Mr Stephen Penckhust called Highreed and in another wood of his called ffox-holes, 6 August 1634'. - Goodyer ms.11, ff. 53, 62, see Gunther (1922: 186). vc14 1634

Cardamine flexuosa

* 'Cardamine impatiens altera hirsutior … this is very common in ditches and moist places, flowering at the beginning of the Spring'. - Ray (1670: 54). 'Cardamine impatiens altera hirsutior ... very common in Warwickshire in gardens and moist places'. - Ray (1690: 114). **1670**

Wolley-Dod (1937: 320) has 'There is very little doubt that this is the species referred to by Ray in his letter to Courthope, of Danny, dated Ap. 28, 1662, as "Cardamine impatiens, a different sort than that which we sowed in our gardens. I found it all along by the ditches as I rode to London [from Cuckfield]". vc14 1662

Clarke adds 'This is cited in Withering (1796: iii. 578), and in Withering (1787) it was described as *C. parviflora*, a preoccupied name for which *C. flexuosa* was substituted in the 3rd ed. [1796]'. Raven (1950: 121) dates the trip from Danny to 1662.
Ray refers to his belief that this and *C. hirsuta* are two species in a letter to Lhwyd in 1689. See Gunther (1928: 196).

Cardamine hirsuta

* 'Cardamine minor arvensis *D. Lloyd*, quam inter segetes et in hortis passim provenire ait, tum in agri Salopiensi propè Oswaldstry [Oswestry], tum Montis Gomerici propè Llanvylhin [?Llanfyllin, NW of Welshpool]'. - Ray (1690: 114). vc40, 47 **1690**

Clarke adds 'This is according to Stokes (Withering (1787: 688), who first clearly separated *C. flexuosa*'. I believe that the Ray reference is satisfactory - see above.

Cardamine impatiens

'Cardamine impatiens vulgò Sium minus impatiens *Ger*. Among the stones under the scars near Wharf, a village some three miles distant from Settle toward Ingleborough'. - Ray (1677: 51).

vc64   **1677**

Possibly Ray found this on his 1671 trip.
Both Clarke and Druce cite 'Sium minus impatiens ... I found it ... about Bath and other parts of this kingdome'. - Johnson (1633: 261)' as the first record, but nowhere in Johnson (1633) is there any mention of 'Bath' or 'this kingdome' - in fact Johnson has 'growes naturally in some places of Italy'. I suspect the error arises from their using the 1636 reprint, which in this very unusual instance differs from the 1633 edition. Johnson (1634: 26) does cite Bath. However, Preston (*in litt*.) has drawn my attention to Ray's (1670: 53) citation 'Cardamine impatiens ... Impatient Cuckow-flower, or Ladies-smock. Doctor Johnson saith, That he found this growing wild about the Bath: I suspect it was the next following [*C. flexuosa*] that he found there'. This is convincing, and Johnson's ecology 'In rills and ditch sides about Bath' is all wrong.

Cardamine pratensis

'The second kind of sisymbrium is called cardamine also, in English water-cresses ... groweth in the same places that sion or water perselye groweth in ... it hath leaves fyrst round, but after they be grown furth they are indented lyke the leaves of rocket'. - Turner (1562: 139).

1562

This attribution is from Chapman *et al*. (1995). Turner also mentions that it tastes like water-cresses! Clarke's and Druce's first date was 'Cardamine. . . . In moist meadowes . . . called at the Namptwich Cheshire where I had my beginning Ladie smockes'. - Gerard (1597: 203).

Carduus crispus

* 'Carduus Polyacanthus Theophrasti'. Near Rochester. - Johnson (1629: 8).

vc16   **1629**

Carduus nutans

* 'Carduus muscatus ... groweth in the fieldes about Cambridge'. - Gerard (1597: 1012). 'The Muske-Thistle I have seene growing about Deptford'. - Johnson (1633: 1176).

vc29   **1597**

Carduus tenuiflorus

'Cardui polyacanthi secunda species'. - Johnson (1633: 1176).

1633

Clarke and Druce both cite 'Carduus spinosissimus, capitulis minoribus sive Polyacantha *Lob*. Small welted Thistle' - Johnson (1634: 26); How (1650: 22). But see Oswald & Preston (2011:

442) for an illuminating and convincing discussion on Johnson's sorting of Gerard's confusion over his thistles.

Carex acuta

Goodenough (1792). **1792**

Goodenough (*op.cit.*) cites 'C. digyna, L - spicis filiformibus; foemineis inflorescentibus nutantibus: fructiferis erectis, capsulis acutiusculis apice indiviso' from Uvedale's Herb. vol.12. p.61. n. I. This is in Hb. Sloane, **BM**. As *C. gracilis* 'In Battersea Meadows'. - Curtis, *Fl. Lond*. iv. 62 (1783). 1722

Early identifications of these large sedges is particularly problematic; see the discussion in Oswald & Preston (2011: 211n). Note 'gramen palustre maius…the sharpe edge grasse [in Est mallinge, East Malling] growing in Tuffettes …'. - Manuscript notes of William Mount, of c. 1584, in Gunther (1922: 255 - 263), and identified by him as this species, but in view of the note above there can be no certainty here at all. Kent (1975) gives the first record of both this and *C. acutiformis* as in Blackstone (1737), but the same caveats apply.
Uvedale died in 1722.

Carex acutiformis

as *Carex paludosa*, Goodenough (1792). **1792**

'Gramen cyperoides majus angustifolium, *Park.1265*'. - Druce & Vines (1907: 88). 'As *C. acuta*, Curtis, *Fl. Lond. iv. 61* (1783). 1747

Early identifications of these large sedges is particularly problematic; see the discussion in Oswald & Preston (2011: 211n). Note that Ray (1670: 143). has 'Gramen cyperoides. majus angustifolium Park. ... in pratis humidis'. Druce (1928: 487) cites this as the first record, but there is no certainty at all that Ray was properly separating *C. acutiformis* from *C. acuta*. Clarke cited 'Mr. Ja. Sherard first observ'd this in a pond near Eltham in Kent, about the end of May; and Mr. Rand, in the ditches at the 'King's Arms' against Whitehall'. - Petiver (1716: no. 159), but the same lack of any certainty applies to this record too.
The record from Druce & Vines is dated as 1747, the date of Dillenius' death.

Carex appropinquata

* as *C. paradoxa* 'In a boggy wood at Ladiston, near Mullingar [Westmeath]. Mr. D. Moore'. - Babington (1843: 337). vcH23 **1843**

Askham Bog, 1818, Hb. Backhouse, **E**. 'In Heslington fields [York] in April, 1841'. - Spruce (1844: 842). vc64 1818

But later in the same volume of the *Phytologist*, Spruce (1844: 1121) refers to a note on this species in Teesdale (1798) where the latter describes what is almost certainly this species. Spruce also refers to a specimen from Mr Backhouse 'who gathered it in Ascham-bogs some twenty years ago', presumably that cited above.

Carex aquatilis

as *C. stricta* 'My specimens were collected from the side of the river Esk, near Eskmount, three miles from Brichen in Angusshire'. - Don (1806: fasc 8, 192).   vc90   **1810**

Don's fascicle 8, 'although dated 1806, did not appear until 1810, as is shown in a letter from Don to Mr Winch …'. - Druce (1904: 65). Clarke has 'Common on the Clova range of mountains,' Scotland. Found by W. J. Hooker, W. J. Burchell, and R. K. Greville [about 1824]. - E.B.S. 2758.

Carex arenaria

* 'Gramini cyperoidi ex Monte Ballon simile, humilius, in maritimis & arenosis nascens … In arenosis maritimis frequens occurrit'. - Ray (1688: 1297). ' … Observed on the sandy shores in many places by Mr Newton and my self'. - Ray (1688a: 10).   **1688**

Carex atrata

* 'Upon the highland mountains frequent as upon Benteskerney [Beinn Heasgarnich], Mal-ghyrdy [Meall Ghaordie], Mal-nan-tarmonach [Meall nan Tarmachan], mountains in Breadalbane'. - Lightfoot (1777: 555).   vc88   **1777**

From the Breadalbane localities cited, these unacknowledged records in *Flora Scotica* are without doubt attributable to Stuart'. - Mitchell (1986). Though this seems likely, I am not sure how Mitchell decided this. Stuart was accompanying Lightfoot on the latter's 1772 tour.

Carex atrofusca

* as *C. ustulata* 'Gathered in watery places with a micaceous soil on the mountain of Ben Lawers, by Mr. Geo. Don'. - E.B. 2404.   vc88   **1812**

'I possess one of these specimens [from G Don, reputedly from Ben Lawers] labelled 1810'. - Syme (1870: x.137).   vc88   1810

Carex bigelowii

* 'In summo vertice montis Snowdon, Mr. Hudson. In alpicis Scoticis, Mr. Dickson'. - Goodenough (1792: 193).   vc49   **1792**

'Gathered [by Sir J. E. Smith] in 1782 on the top of Ben Lomond'. - E.B. 2047.   vc86   1782

See Mitchell (1986) where he relates that Smith's ascent of Ben Lomond was undertaken on 23rd August 1782 under the personal guidance of the resident Rev. Stuart.

Carex binervis

'Very common on the driest moors about Aberdeen, Prof. Beattie'. - Smith (1800a: 268). — vc92 — **1800**

'Driest moors near Harrogate. R. Teesdale (1792)' - Lees (1888). That that has not been traced but in Teesdale (1792) there is a record for *C. distans*, which seems actually to be this species, 'upon various heaths, woods, &c. in the neighbourhood of Castle Howard'. — vc62 — 1792

Druce cites 'Gramen Cyp. spicis [parvis] longissime distantibus varietas … loco squalido & aquoso … prato quodam juxta lupeletum Danfeldiae in Essexiae'. - Ray (1696: 266). This is possible but the habitat is less likely. Smith, in his *English Flora* does not cite Ray, but this is not surprising if he had only just described it. Clarke notes that Curtis (1782) under No.112 (*C. distans*), gives localities which probably included *C. binervis*.

Carex buxbaumii

* Found in 1835 'on one of the small islands of Lough Neagh, County Derry, D. Moore'. - Hooker (1835a: 307). — vcH40 — **1835**

Carex canescens

* 'Gramen cyperoides elegans spica composita ... In a pool in a grove not far from Middleton towards Coleshill in Warwick-shire, also near Wrexham in Denbigh-shire, and doubtless in many other places'. - Ray (1670: 146). — vc38, 50 — **1670**

Carter (1960) dates the discovery of this in Denbighshire to Ray's visit in 1662. — vc50 — 1662

I cannot see a note of the 1662 discovery in Lankester (1846).

Carex capillaris

* 'On Benteskerny [Beinn Heasgarnich], Craigneulict, and Malghyrdy [Meall Ghaordie] in Breadalbane. Mr. Stuart'. - Lightfoot (1777: 557). — vc88 — **1777**

Stuart was accompanying Lightfoot on the latter's 1772 trip, but there is no evidence to suggest a date for these records.

Carex caryophyllea

* 'Gramen spicatum foliis vetonicae caryophyllatae, *Lob.*'. Hampstead Heath. - Johnson (1632: 33). — vc21 — **1632**

Carex cespitosa

'near West Mill, Buntingford, Herts, J.G. Dony, 1960'. - James *et al.* (2012). — vc20 — **2012**
 — vc20 — 1960

James' paper states that the original discoverer was a J.C. Gardiner, presumably the long-serving BSBI Treasurer of that name, who joined the BSBI in 1949, and went on to be Treasurer of the Linnean Society.

Carex chordorrhiza

'On Aug 4th … nr. the head of the Loch Naver, at Altnaharra, W. Sutherland'. - Marshall & Shoolbred (1897).   vc108   **1897**

Carex davalliana

'On Landsdown hill near Bath, found there in 1800 by Mr. Groult in company with Mr. Lambert'. - Smith (1804: 3 addenda & corrigenda).   vc6   **1804**
vc6   1800

The record from N. Britain given in Smith (1800a: 266) is an error, as is that from Mearnshire [Kincardinshire, vc 91] in E.B. 2123.

Carex demissa

Nelmes (1947).   **1947**

as *C. flava* var. *oedocarpa* in Druce & Vines (1907: 123).   1747

The taxonomy of the *C. flava* group has been endlessly debated. Nelmes seems to have been the first to satisfactorily set out parameters between the four segregates. The record from Druce & Vines is dated as 1747, the date of Dillenius' death.

Carex depauperata

* 'Discovered by Mr. Curt. [Curtis]. Charlton-wood, Kent. Mr. Woodward'. - Withering (1787: 1049).   vc16   **1787**

Curtis (1791: fasc.6, pt 64) gives the following account of the discovery of this sedge 'My much-valued friend the Rev. Dr. Goodenough, of Ealing, has the merit of discovering the *Carex* here figured; we were herborizing together in company with the Rev. Dr. Whitfield in a small wood at the back of Charlton Church, when a single plant of it first caught his eye, and, on further search, we found it in one part of the wood in abundance; Mr. Dickson informs me that he has observed the same species growing wild near Godalming, Surrey; and we are informed that it has also been found by Mr. Sole, of Bath. The late Rev. Mr. Lightfoot [died 20th Feb. 1788], who had seen it growing with me, was pleased to call it *depauperata* from the paucity of its flowers, a name in which we sometime acquiesced; but on maturer consideration we think the name we have now given it ['*Carex ventricosa*'] more expressive of its principal character'.

Carex diandra

* 'Prope Norwich, Dom. Crowe'. - Goodenough (1792: 163).   vc27   **1792**

But Ray may have known it, for, after describing *C. paniculata*, he adds: 'Hujus aliam speciem per omnia similem, sed minorem spicis gracilioribus sparsim nascentem, non in ejus modi densis cespitibus, iisdem in locis observavimus'. - Ray (1690: 197).

Carex digitata

* 'Prope Bath, D. Sole'. - Hudson (1778: 409).   vc6   **1778**

T. Tofield found it on the crags at Roch Abbey [E. of Rotherham, vc63], April 16th,1779 - see Skidmore *et al*. (n.d.).

Carex dioica

* 'Gr. cyperoides minimum Ranunculi capitulo rotundo. Frequently found on the Bogs on the West side of Oxford'. - Bobart in Ray (1690: 235).   vc23   **1690**

Druce (1897: cviii) identifies Merrett's (1666: 52) 'Gr. Cyperoides spica echinata simplici. Two miles southward from Oxford in the boggs', as probably this. Pulteney too identifies Merrett's record as this. But see *C. echinata*.

Carex distans

* 'Gramen cyperoides spicis parvis, longissime distantibus … Hanc speciem primus mihi ostendit D. Martinus Lister, postea ipse observavi loco putrido & palustri prope Molendinum *Machins mill* dictum sesquimilliari a Witham oppido versus Camalodunum [Witham to Colchester]'. - Ray (1688: 1295; 1688a: 10).   vc19   **1688**

Gibson (1862) says that 'Ray's plant may be *C. binervis*, as Buddle and Uvedale's specimens imply; but Petiver's plant is *C. distans*'. I have not been able to trace the source of that comment. Smith (1828: 109) equates L'Obel's (1655: 61) 'Gramen cyperoides gracile alterum, glomeratis torulis, spatio distantibus' with this species. However there is no hint that L'Obel saw it in this country.

Carex disticha

'Gramini cyperoidi ex Monte Ballon simile … in palustribus et aquosis'. - Ray (1660: 67; 1670: 145). 'Near Bungay, Suffolk, frequent'. - Mr. Woodward in Withering (1787: 1029).   vc29   **1660**

Ray (1670) marks the record with a 'C', signifying that it is a Cambridge plant. See Raven (1950: 90n), agreeing this with some hesitation and Oswald & Preston (2011: 212n) with less.

Carex divisa

* 'Gramen cyperoides palustre majus spica divisa *C.B* … Prope Hitham Colcestrensem [The Hythe, Colchester] in Essexia'. - Ray (1688: 1296).   vc19   **1688**

Carex divulsa

* 'Gramen cyperoides spicatum minus spica longa divulsa seu interrupta *Hist nost. p.1297* ... In pratis & pascuis frequens'. - Ray (1688a: 10). 'Harefield' [Middlesex]. - Blackstone (1737: 36).   **1688**

Ray's record was accepted by David & Kelcey (1985), but Nelmes (1947) was the first to clearly elucidate the *C. muricata* group.

Carex echinata

* 'Gramen cyperoides spicatum minimum, spica divulsa aculeata. Locis palustribus solo putrido & spongioso'. - Ray (1690: 199).   **1690**

Clarke erroneously states that 'Mr. Druce suggests that Merrett's 'Gr. Cyperoides spica echinata simplici. Two miles southward from Oxford in the boggs' (Merrett 1666: 52), may be this', whereas Druce was actually suggesting *C. dioica*.

Carex elata

* 'Prope Norwich, D. Pitchford'. - Goodenough (1792: 196).   vc27   **1792**

Early identifications of these large sedges is particularly problematic (see, for instance the discussion in Oswald & Preston (2011: 211n)). Goodenough (*op.cit.*) equated this with a specimen in Buddle's herbarium 'Gr. Cyp. caryophyllus spiciis erectis sessilibus. *R. Syn. 264*'. - Buddle Hort.Sic. p.30.n.2. Having examined the specimen, I feel very doubtful about this identification. Clarke adds 'For earlier (?doubtful) synonyms see Babington (1860b: 260); Ray (1696: 264. 4); and Hudson (1778: 412) (as *C. caespitosa*)'.

Carex elongata

* 'Mr. Jonathan Salt, who discovered this in June, 1807, in a marshy place at Aldwark, near the river Don, below Sheffield'. - E.B. 1920.   vc63   **1808**
   vc63   1807

Lees (1888: 463) has 1803 but gives no reference to support this. However Wilmore *et al.* (2011) give ' " … 1801 - first British record" inscription on the herbarium sheet, sadly now lost'. It seems that both of these should read 1807, as confirmed by Coles (2011).

Carex ericetorum

* 'On the Gog Magog Hills'. - Babington (1862a). 'Found by C.C. Babington and J. Ball in 1838 on the Gogmagog Hills, Cambridge, but remained undistinguished till 1861'. - E.B.S. 2971.   vc29   **1862**

'… a specimen sent in 1829, as *C. pilulifera* L., to Sir John Trevelyan from Mildenhall Heath, W. Suffolk, was found to be *C. ericetorum*…'. - David (1981a).   vc26   1829

David (*op.cit.*) further notes that it is probable that Sir J. Cullum's '*Carex montana*' from Newmarket Heath in 1775 - 76 was also this species, and cites Bennett (1910a) and Hind (1889) in support of this.

Carex extensa

* 'In paludibus prope Harwich - on the marshy part of Braunton Burrows in Devonshire'. - Goodenough (1792: 175). — vc3, 19 — **1792**

'An Gramen palustris aculeatum italicum vol minus. *C.B. Th.110*'. - Buddle Hort. Sic. p.30. n10. 'Kirkcudbrightshire, 1769, Walker, **BM**. 'On Cley Beach, Norfolk, June 18th, 1776, by the Rev. H. Bryant'. - Smith (1804: 992). — 1715

Dandy (1958a: 225) noted that Goodenough described sedges from Uvedale's herbarium at BM, and presumably researched those from Buddle (d.1715) too. I have examined Buddle's specimen there and am happy it is this species.
Clarke adds that "Petiver's 'Cyperoides echinatum majus', which Smith cites as a synonym, was found near Cambridge in 1715; but *C. extensa* is not recorded for Cambs", and this seems to be an error for *C. demissa*. Druce gives the Cley Beach record as 1777.

Carex filiformis

* 'In meadows near Merston Measey [Marston Maisey] Wiltshire, Mr. Teesdale'. - Smith (1800a: 269). — vc7 — **1800**

'He [Mr Teesdale] found it … in June, 1799'. - E.B. 2046. — vc7 — 1799

Carex flacca

* 'Gramen cyperoides foliis caryophylleis vulgatissimum … In pratis humidis verno tempore'. - Ray (1688: 1293). 'Harefield' [Middlesex]. - Blackstone (1737: 35). — **1688**

Druce (1897: cviii) identifies 'Gramen. Cyp. spicatum *Ger.* as *C. flacca* or *panicea*, but this seems too tenuous.

Carex flava s.l.

* 'Gramen palustre echinatum, Hedgehog grasse, in watery ditches, as you may see in going from Paris garden bridge to St. George's fields [Southwark] and such like places'. - Gerard (1597: 16). — vc17 — **1597**

Carex flava s.s.

Distinguished from *C. lepidocarpa* in Sandwith (1935). Keyed in Nelmes (1947). — **1935**

'… collected by J. Dickinson in 1836, somewhere in the Ennerdale area of Cumbria (vc 70)'. - Blackstock & Ashton (2001). — vc70 — 1836

Sandwith's note ends with 'Dr Sledge's plant from Roudsea Wood, N. Lancs (see Pearsall 1931) is surely to be placed under it [*C. flava* s.s.].

Carex hirta

'Gramen Cyperoides Norvegicum parum lanosum ... there have beene two sorts of this kinde of grasse, found nere unto Highgate'. - Parkinson (1640: 1172). 'Gramen Cyperoides Nortvegicum, ima foliorum basi tantillum lanuginosum ... Collegit D. Guil. Boelius, hujus amoenissimi studii peritissimus, palustribus Londinensis agri juxta Altam Portam, vernacule Highgate'. - L'Obel (1655: 51). | vc21 | **1640**
| | vc21 | 1616

L'Obel died in 1616. Edgington (2011) has 'collected by [the Dutch plant collector] Wilhelm Boel from marshy fields near Highgate, probably about 1600'.

Carex hostiana

* as *C. fulva* 'Habitat prope Eaton, juxta Shrewsbury'. - Goodenough (1792: 177). | vc40 | **1792**

There is some doubt over the real identity of Goodenough's plant; see Pryor (1876) for a full discussion. Goodenough cited a specimen in Buddle's herbarium (Buddle Hort.Sic. p.30. n.11), but I have examined this specimen and feel sure it is *C. demissa*. Perring *et al.*(1964) cite Relham (1802).

Carex humilis

* as *C. clandestina* 'In rupe Sancti Vincentii dicta, prope Bristol, D. Sole'. - Goodenough (1792: 167). | vc34 | **1792**

Carex lachenalii

* 'Discovered on rocks in Lochnagar, Scotland, in Aug.,1836, by Mr. Dickie and Mr. Clark'. - E.B.S. 2815. | vc92 | **1838**

Druce cites 'Loch na Gar, 1830, found by Mr Brand'. - *Bot. Mag. iii, 126, 1839*', but this reference has not been found. If he meant *Annals of N.H. 1839, 3: 126*, which seems likely, as that was the successor to *Comp. Bot. Mag.*, then there is a reference there 'Mr. Brand exhibited a specimen of *Carex leporina*, found by him in 1830, during an excursion to Braemar and the mountains of Aberdeenshire'. | vc92 | 1830

Carex laevigata

* 'In a marsh near Glasgow, 1793. Mr. J. Mackay. Marshes near Aberdeen. Professor Beattie'. - Smith (1800a: 272). | vc77, 92 | **1800**

'A variety of the Gramen Cyperoides spicis longe distantibus, with longer spikes, found in the Boggy grounds about Tunbridge Wells'. - Druce (1928: 487). | vc16 | 1706

There is no date in Druce (1928), but Druce (1932) gives 1706; however this record is not in Hanbury & Marshall (1899).

Carex lasiocarpa

* as *C. tomentosa* 'Plentifully at the south end of Air [?Ayr] Links. Dr. Hope'. - Lightfoot (1777: 553).	vc75	**1777**

Carex lepidocarpa

Nelmes (1947).		**1947**
'Gramen palustre echinatum *J.B. Lob. Ger emac*. On the moores about Cambridge'. - Ray (1660: 70). Oswald & Preston (2011) consider that, of the three similar species, this is the most likely that Ray saw.	vc29	1660

The taxonomy of the *C. flava* group has been endlessly debated. Nelmes seems to have been the first to satisfactorily set out parameters between the four segregates.

Carex leporina

* 'Gramen cyperoides spica e pluribus spicis brevibus molliter composita ... In pascuis locis humidioribus copiose'. - Ray (1690: 197). 'In a wood against the Boarded-river [New River, Middlesex]'. - Petiver in Gibson (1695).		**1690**

Carex limosa

* 'In paludibus turfosis in comitatibus Eboracensi, Lancastrensi, Westmorlandico, &c.'. - Hudson (1778: 409).	vc60, 65, 69	**1778**
'Mr Fab[ricius] 1767'. - Hope, 1768, in Balfour (1907).		1767

It could of course have been *C. magellanica*, in which case Babington's note under that species would be the first reference. But note that *C. magellanica* has never been reliably recorded for Lancashire or Yorkshire.

Carex magellanica

* as *C. irrigua*, Edinburgh Catalogue (1841). 'Muckle Moss, Northumberland. ... It was first noticed in the above station by Mr. John Thompson'. - Babington (1842).	vc67	**1841**
Near Dumfries, vc72, 1839, Balfour, **E** & Edmonston, **CGE**.	vc72	1839

Babington was shown this plant by Thompson on July 15th 1842 (Babington (1897: 114).

Carex maritima

* 'This new species of *Carex* was communicated by Dr. Hope ... discovered in deep loose sea-sand at the mouth of the water of Naver [i.e. at Bettyhill] and near Skelherry, in Dunrosness [Skelberry, Dunrossness], in Shetland'. - Lightfoot (1777: 544).	vc108, 112	**1777**
'*Carex, capitata* an nova species - Round-headed Carex, from mouth	vc108	1767

of the Naver grows among the loose sand which it seems very proper for binding, as it has creeping roots & always surmounts the sand blown over it by the wind'. - Journal of James Robertson, 1767. Henderson & Dickson (1994: 47).

There is an entry 'Carex nova species… J.R. [Robertson] at the mouth of the water of Naver' in Hope, 1768, see Balfour (1907). Scott & Palmer (1987) locate the Shetland site at the north end of Loch of Spiggie.

Carex microglochin

'Glen Lyon, found by Lady Davy and Miss Bacon'. - Druce (1923).   vc88   **1923**

Carex montana

* 'in May, 1843, by the road side towards Eridge in Sussex, about a mile south of Tunbridge Wells'. - Mitten (1845).   vc14   **1845**
     vc14   1843

Wolley-Dod (1937: 486) gives 1842, presumably an error.

Carex muricata

Nelmes (1947) was the first to clearly elucidate the *C. muricata* group.   **1947**

'Vale Royal, Cheshire,1736, D. Wood, **K**. (as *muricata* subsp. *lamprocarpa*)'. - David & Kelcey (1985). This is earliest herbarium specimen traced by them. However there is a specimen in Buddle's herbarium (died 1715), cited by Goodenough (1792), which does appear to be this species, 'Gr. cyp. spicatum minus *R. Hist. 1297* desc., Buddle Hort. Sic. p.32. n.3'.   vc58   1736

Clarke cites three early records, but whilst these might be the *C. muricata* group, there is no knowing which species. 'Gramen cyperoides parvum'. - Johnson (1633: 21), and (1634: 41). 'Gramen cyperoides palustre minus *Park.*'. - Ray (1660: 67)). Oswald & Preston (2011: 212) merely assign these to *C. muricata* agg.

Carex nigra

* 'Gramen cyperoides foliis caryophylleis, spicis erectis sessilibus, e seminibus confertis compositis'. - Ray (1696: 264).   **1696**

Other early records in Clarke are 'Gramen caryophylleum, angustissimis foliis, spicis sessilibus brevioribus erectis non compactis, Nobis. . . Hoc primo a charissimo Fratre Tillemano ostensum est, dein variis in locis observatum'. - Morison (1699: iii. 243) and 'Peat Bogs on Bullingdon Green'. - Sibthorp (1794: 31).

Carex norvegica

* as *C. Vahlii* 'Discovered in 1830 by Mr. Balfour and myself [R.K. Greville] growing in moist declivities among some precipitous rocks which surround a small loch above two miles above Loch Callader'. - E.B.S. 2666.   vc92   **1830**

Carex oederi

Nelmes (1947). **1947**

as *C. oederi*, Retz (*C. viridula*, Michx.). 'Shooters Hill [Bexley, Kent]'. - Druce & Vines (1907: 123). vc16 1747

The taxonomy of the *C. flava* group has been endlessly debated. Nelmes seems to have been the first to satisfactorily set out parameters between the four segregates. The record from Druce & Vines is dated as 1747, the date of Dillenius' death. I have examined the specimen in **OXF**, and there is no doubt that it is this taxon. There is also a specimen in **OXF** from 'Sands of Borrie, E. of Dundee, Mr Don, det. Nelmes 1946'. This would date to before 1814.

Carex ornithopoda

* 'Discovered by J. Whitehead, in company with H. Newton & E. Hibbert, May 31, 1874, in Miller's Dale, near Buxton, Derbyshire'. - Babington (1874a). vc57 **1874**

Linton (1903) has 'First found, Salt, May 7th, 1801 (*C. digitata*)'. I am not sure of the significance of the bracketted item, but assume that it means that this was originally described as *C. digitata*. There appear to be two specimens - that at Middlesborough has recently been identified as *C. digitata*; the other, from Salt's herbarium in Sheffield, has been removed relatively recently (G. Coles, *pers. comm.*) has not been re-examined, and is now almost certainly destoyed. The fact that Linton listed it as a first record for *C. ornithopoda* is persuasive.

Lees (1888: 466) has a record from Mackershaw Wood, Ripon; Hb. Borrer [who died in 1862], *teste* W.W. Newbould. This is outside the current known distribution and has been ignored for the moment.

Carex otrubae

* 'Ad insulam Sheppey … Gra. palustre Cyperoides *Lob.*'. - Johnson (1629: 5). vc15 **1629**

Rose, in Gilmour (1972: 56), suggests that this attibutation is possible, but *C. divisa* might be more likely here. But see Oswald & Preston (2011: 212n), citing, *inter alia*, Parkinson's (1640: 1266) illustration of this species.

Carex pallescens

* 'Gramen cyperoides polystachion flavicans spicis brevibus prope summitatem caulis … In pratis circa Middleton Agri Warwicensis'. - Ray (1670: 144). vc38 **1670**

Carex panicea

* 'Gramen cyperoides foliis caryophylleis, spicis e rarioribus & tumidioribus granis compositis'. - Ray (1696: 264). 'On Chiselhurst and other bogs'. - Petiver (1716: 6).   **1696**

Carex paniculata

* 'Gr. Cyperoides maximum spicis pendulis, at Bocknam [?Brookham, ?Great Bookham] in Surrey, in a bog. Hujus radicibus utuntur pro sedilibus in agro Eboracensi'. - Merrett (1666: 51).   vc17   **1666**

Carex pauciflora

* 'We found this new species of *Carex* ... about halfway up the mountain of Goat-field in the Isle of Arran'. - Lightfoot (1777: ii. 543).   vc100   **1777**

This was Lightfoot, in company with Stuart on 21 June 1772, see Pennant (1774) and also Mitchell (1986).   vc100   1772

Carex pendula

* 'Gramen cyperoides spica pendula longiore ... In the great ditch at the end of the little Thicket adjoyning to Teversham moor, and in other great ditches'. - Ray (1663: 5).   vc29   **1663**

Ray's observation was in 1661, see a letter from Trinity College in July 1661 (Gunther 1932).   vc29   1661

See also Raven (1950: 296) re a possible record of this from Ellesmere, Salop, by W. Bowle(s) on his 1632 trip, given in How (1650: 54).

Carex pilulifera

* 'Gramen cyperoides foliis mollibus tenuibus, spicis brevibus coacervatis ... In ericeto Hampstediensi prope Londinum invenit D. S. Doody'. - Ray (1688: 1910, 1688a: 10).   vc21   **1688**

Carex pseudocyperus

* 'Pseudo Cyperus. In ditches and waterie places'. - Johnson (1633: 29). 'In a ditch between the Boarded- river and Islington road'. - Petiver in Gibson (1695).   **1633**

Carex pulicaris

* 'Gr. cyperoides pulicare ... A mile East from Oxford at Hockley of the Hole'. - Merrett (1666: 52).   vc23   **1666**

Ray (1670: 148), in citing the name 'Flea-grass' adds 'This was so denominated by Mr Goodyer, because the seeds (which turn downwards on the stalk) do in shape and colour somewhat resemble fleas'. This observation by Goodyer would predate 1666, but no mention of this species has been found in his mss.

Carex punctata

'in Herb. Smith, Linn. Soc. ... specimens gathered many years ago by Mr. Dawson Turner at Beaumaris ... Mr. Wilson has since found the same plant at Warrington'. - Hooker (1836: 192).   vc52, 58   **1836**

'Guernsey. Gosselin, J., c.1790'. - see McClintock (1982: 170).   vc113   1790

David (1981b) cites an early record from **MANCH**, 1838, Menai.

Carex rariflora

* 'Discovered in 1807, by Mr. G. Don, on a mountain at the head of a glen called the Dell [Doll], among the mountains of Clova, Angusshire, near the limits of perpetual snow'. - E.B. 2516.   vc90   **1813**
vc90   1807

Druce misprints 1807 as 1802.

Carex recta

* as *C. salina* β *Kattegatensis* 'Caithness, August, 1883, J. Grant, who writes that it is plentiful'. - Bennett (1885a).   vc109   **1885**
vc109   1883

A fuller treatment is in Ridley (1895), where the location is given as 'sand-banks along the River Wick'.

Carex remota

* 'Gramen Cyperoides minimum Boelii tenuifolium parvis per caulem distinctis torulis ... Provenit in AngloBritannica'. - L'Obel (1655: 54).   **1655**
1616

L'Obel died in 1616.

Carex riparia

* 'Gramen cyperoides majus latifolium. In our owne land'. - Parkinson (1640: 1265). 'In fossis & vadis amnium pigriorum'. - Ray (1660: 66).   **1640**

Carex rostrata

* 'Gramen cyperoides polystachion spicis teretibus erectis ... In several Pools about Middleton in Warwick-shire'. - Ray (1670: 145).   vc38   **1670**

Carex rupestris

* 'Discovered by Dr. Dickie and Mr. Templeton on shelves of rock extending from the small round lake at the top of Glen Callader eastward to the break-neck fall, growing with *Salix reticulata* and *Carex atrata*'. - Hooker (1836: 191).   vc92   **1836**

Found Aug. 2nd, 1836'. Note on E.B. drawing (Garry, 1904).

Carex salina

'On the saltmarsh at Morvich, at the head of Loch Duich (v.c.105, Wester Ross) on 2nd July 2004'. - Dean *et al.* (2005).   vc105   **2005**
                                                                                                                          vc105   2004

See also Dean *et al.* (2008) for a fuller account. The discoverer was Keith Hutcheon.

Carex saxatilis

* 'In montibus Scoticis D. Dickson'. - Goodenough (1797: 78). 'Found on Ben Lawers in 1793 by Mr. J. Mackay'. - Smith (1804: 989).   **1797**

Included in an unpublished list of plants compiled by John Stuart in September 1776'. - Mitchell (1986). 'Mr Stuart formerly found a single specimen on Ben Teskerny [Beinn Heasgarnich] which he sent to Lightfoot ... ' - Robert Brown ms., cited by Mabberley (1985: 38).   vc88   1776

There is an entry for this species in Hope, 1768, in Balfour (1907), but it is blank. Does this mean there was no specimen, or that the location or collector was unknown? On the other hand, if Hope was following Hudson, then Hudson's *saxatilis* is our *caryophyllea*. Brown and W. MacRitchie found it on Ben Teskerny in Aug. 1792 (Mabberley 1985: 25).

Carex spicata

Nelmes (1947) was the first to clearly elucidate the *C. muricata* group.   **1947**

'Hinton, 1825, J.S. Henslow, **CGE**'. - David & Kelcey (1985). Llandudno, 1825, W. Wilson, det. David, **CGE**.   vc29   1825

David & Kelcey dismiss Druce's attribution to Johnson's (1633) 'Gr cyperoides parvum' as 'purest fantasy' and postulate that if this is the plant that Hudson (who was the first to use that name) described, then he must have recorded it before 1762. But Hudson's habitat was 'in aquosis' which makes it less than convincing.
It is possible, of course, that writers earlier than the first surviving herbarium specimens, were seeing this species.

Carex strigosa

* 'Gramen cyperoides polystachion majusculum latifolium, spicis muItis, longis, strigosis ... Found in a Lane at Black Notley [Essex] by Mr. Dale'. - Ray (1696: 265).   vc19   **1696**

Carex sylvatica

* 'Gramen Cyperoides sylvarum tenuius spicatum'. - Parkinson (1640: 1171). 'In Madingley Wood'. - Ray (1660: 67).   **1640**

Carex trinervis

* '[among] some duplicates from Norfolk, belonging to Mr. H. G. Glasspoole, I found four specimens … gathered in 1869 - 70'. - Bennett (1884). vc27 **1884**
vc27 1869

Preston *et al.* (2002) have '1869, H.G. Glasspoole, Ormesby, Norfolk'.

Carex vaginata

* as *C. Mielichoferi* 'Found by Mr. W. Borrer in August, 1810, on the rocky ledges of Craig Challoch in Breadalbane'. - E.B. 2293. vc88 **1811**

'I discovered this species [as *C. salina*] new to Britain, on rocks on the high mountains of Cairngorm in August 1802. My specimens from Ben Macdhui … collected in September 1802'. - Don (1806: 9. 216). Dated 1806, but not issued until late 1812 or early 1813 (Druce, 1904). vc92 1802

Carex vesicaria

* 'Gramen cyperoides majus praecox spicis turgidis teretibus flavescentibus ... In ambulacris Coll. Aedis Christi [Christ Church college] collectum est'. - Morison (1699: 242). vc23 **1699**

Crompton (2007) cites a record from Granchester Meadow, vc29, by Rev J. Newton, c. 1680. There was a J. Newton, who was a friend of Ray's, but I feel that this is an error for a Newton who recorded plants around Cambridge in the late 18th century, and that the date cited should be 1780.

Carex vulpina

'1881, near Oxford, H.E. Garnsey, (in the herbarium of the the Botanical Institute of the University of Vienna)'. - Nelmes (1939). vc23 **1939**
vc23 1881

See Grose (1957: 590) relating a discovery by E. Nelmes at Kew, from Alexander Prior's herbarium, of an undated specimen from 'Wilts'. Nelmes wrote to Grose that it was the oldest specimen known in Britain, and as Prior lived at Corsham (vc7), around 1840, when he was sending in Wiltshire records to Babington, it has been traditionally dated to then. He died in 1902.

Carlina vulgaris

* 'Carlina sylvestris … In untoiled and desart places and oftentimes upon hils'. - Gerard (1597: 997). 'Upon Blackheath … Kent'. - Johnson (1633: 1159). **1597**

'Carlina sylvestris … groweth in this countrie'. - Lyte (1578: 531).

Carpinus betulus

* 'Betulus sive Carpinus. .. the Hornbeame tree grows plentifully in Northamptonshire, also in Kent by Gravesend, where it is commonly taken for a kinde of Elme'. - Gerard (1597: 1296).   vc16, 32   **1597**

Carum carvi

'Carum sive Careum, *Caraways*, allata sunt mihi femina ex agro Lincolniensi ibidem nata quae nullatenus cedebant Germanicus. *G.1034*'. - Merrett (1666: 22). 'Carum seu Careum *Ger.* ... In palustribus Lincolniensibus ínque pascuis depressis pinguibus propè Hull oppidum ... copiosè, & alibi etiam in pratis'. - Ray (1690: 67).   vc53   **1666**

Gibbons (1975) has '1661, Boston, Ray' and Ray's trip through Lincolnshire was in 1661 (see Lancaster 1846: 135) where he writes 'We observed, in a close by the town [Hull] great store of Carum [*C. carvi* Linn.]. It grows in many places about this town and in some places in the Fens in Lincolnshire'. It is not clear whether this is a reference to a wild or cultivated specimen, and Gibbons' reference to Boston has not been traced. Turner (1538: A. iij. verso) also refers to Caraway, but presumably as a cultivated plant.

Carum verticillatum

* 'Carvi foliis tenuissimus Asphodeli radice. Near Greenock in Scotland, Mr. W. Housto[u]n'. - Martyn (1732: 154).   vc75   **1732**

At Greenock, not Ayr (error by Clarke & Druce per D.E. Allen, *in litt.*, having checked Martyn).
Carter (1986) has a record for Pembrokeshire from Gough's 1789 edition of Camden's Britannia, which from elsewhere in his writings, he implies must have come from Lhwyd's trip there in 1697. In addition Carter adds that 'Newton, in Ray (1688) must have the credit [for the discovery of this plant]'. However I cannot find the reference there, or indeed, in any of Ray's works.

Castanea sativa

* 'Nux castanea ... Chestnuttes growe in diverse places of Englande. The maniest that I have sene was in Kent'. - Turner (1548: E. vj. verso).   vc16   **1548**

Smith (E.B. 886) mentions the chestnut, at Tortworth, Gloucestershire, .. 'known to have been a boundary tree in the time of King John'.

Catabrosa aquatica

* 'Gramen dulce udorum ... Udis Londinensis agri juxta Thamesis amoenissima fluenta'. - L'Obel (1655: 10).   vc16   **1655**
                                                                                                                vc16   1616

L'Obel died in 1616.

Catapodium marinum

* 'Gramen pumilum loliaceo simile ... This was shewn me by Mr. Newton who found it at Bare about a mile from Lancaster, as also nigh the Saltpans about a mile from Whit-haven Cumberland; at Bright-Helmston in Sussex, & alibi in maritimis'. - Ray (1688a: 11).

vc13, 60, **1688**
70

Raven (1950: 245) however equates with this species Ray's 'Gramen exile duriusculum maritimum' (1688: 1287) and the same 'I found it on the Essex shore by Little Holland between Walton and S. Osythe' (1688a: 11). This last is not cited by any writer other than Dale (1730: 374), who gives both as synonyms.

J. Ray. J. 1688.
Fasciculus stirpium Britannicarum, post editum Plantarum Angliae Catalogum observatarum.

Catapodium rigidum

* 'Gramen panicula multiplici. Medow hard grasse with manifold tufts … In Fieldes and Medowes'. - Parkinson (1640: 1157.5).   **1640**

This entry is from Clarke (1909) where he notes 'this is more satisfactory than Gerard's 'Gramen minus duriusculum. In moist fresh marrishes'. - Gerard (1597: 4) and Johnson (1634: 39)'.

Caucalis platycarpos

* 'Caucalis tenuifolia flosculis subrubentibus … In the corn about Kingston wood and elsewhere'. - Ray (1660: 31).   vc29   **1660**

Centaurea calcitrapa

* 'Carduus stellatus … Upon barren places neere unto cities and townes'. - Gerard (1597: 1004).   **1597**

Centaurea cyanus

* 'Cyanus … hanc ego herbam arbitror esse quam northumbria vocat a Blewblaw aut a Blewbottell'. - Turner (1538: A. iiij. verso).   vc67   **1538**

Turner adds that the chilren entwine this in wreaths on the feast-day of John the Baptist - see *Agrostemma*.

Centaurea debeauxii

as *C. nigrescens* 'heads usually radiant … meadows and pastures in the west of England'. - Babington (1847).   **1847**

'Aug. 22. [1836] On our way to Stoke Bishop [from Sea Mills] we gathered *C. nigrescens*'. - Babington (1897: 57).   vc34   1836

Britton (1922) introduces this species (as *C. nemoralis*) but equates it with Babington's *C. nigrescens*. In Smith (1839: 6. 95) he notes that 'radiate forms of *C. nigra* "become *C. nigrescens* of Willdenow"'. Prof. Stace (*in litt*.), in referring me to this, adds that since radiate forms are very much commoner in *C. debeauxii* (*nemoralis*) than in *C. nigra*, it is highly likely that Smith was referring to *C. debeauxii*, but it cannot be certain.

Centaurea nigra s.l.

* 'Jacea nigra' and 'maior', 'in everie fertile pasture'. - Gerard (1597: 590).   **1597**

See also 'Jacea nigra … in meadowes and pastures' - Lyte (1578: 109.4).

Centaurea scabiosa

* 'Jacea nigra' and 'maior', 'in everie fertile pasture'. - Gerard (1597: 590).   **1597**

Clarke notes possibly 'Jacea major folio multum lacinioso ... In Angliae segetibus Coventriae conterminis abundè provenit'. - L'Obel (1571: 234). This record is given by Cadbury *et al.* (1971) as their first record for Warwickshire. See also 'Jacea nigra ... in meadowes and pastures' - Lyte (1578: 109.4).

Centaurium erythraea

* 'Centaurii duo sunt genera ... minus ... angli vocant Centory'. - Turner (1538: A. iij. verso). 'Centaureum parvum ... In great plentie throughout England'. - Gerard (1597: 437). **1538**

Centaurium littorale

* as *Chironia littoralis* 'Sea coast near Hartley; and Links at Bamburgh, and Holy Island. Mr Winch'. - Turner & Dillwyn (1805: 469). vc67, 68 **1805**

Clarke adds 'There are specimens in **BM** collected on Holy Island by W. McRitchie and J. Shepherd in 1794'. vc68 1794

Clarke, citing the same reference of Turner & Dillwyn, has 'Mr. Winch finds it abundantly near Newcastle and Mr. Brodie ... near Brodie House in Elgin-shire'. But this citation is in fact an abbreviated version of the text in E.B. 2305 (1812).

Centaurium pulchellum

* as *Chironia pulchella* 'First found in England by ... James Watt ... on the north coast of Cornwall on the downs at Port Owen near the sea'. - Withering (1796: 255). vc2 **1796**

'Centaurium minimum purpureum, *H. Ox.ii. 566.* Collegi in Insula Mona ex Prestholm Aug. mense (the earliest British record)'. - Druce & Vines (1907: 88). This Anglesey record is also in the Diary, for Aug 14th, 1726, reproduced in Druce & Vines (1907: li). vc52 1726

Port Owen is Port Quin, nr. Port Isaac - see Davey (1909) - and James Watt is the pioneer of the steam engine. Priestholm is the old name for Puffin Is, off Anglesey, formerly Mona. See also 'Guernsey, in light sandy soil. Gosselin, J., c.1790'. - McClintock (1982: 109). But both Carter (1955b) and Wade *et al.* (1994) attribute to Lhwyd a record of this (as Chironia pulchella) from 1697 or 1698 from the 'seashore'.

Centaurium scilloides

'From Mr J.E. Arnett ... N. Pembrokeshire (Precelly) district ... The friend [J.B. Rhys] who brought me the specimen ...'. - Wilmott (1918). vc45 **1918**

Mr J. Arnett, of Tenby ... gathered in N. Pembrokeshire on cliffs about 2 miles from Newport ... the first certain record ... in England [*sic*!]'. - Druce (1919: 291).

Centaurium tenuiflorum

'I gathered specimens lately in the Isle of Wight, on the west bank of the Medina, between West Cowes and Newport ...'. - Townsend (1879). vc10 **1879**

'Collected by Babington at the Braye du Valle [Guernsey] on 14 August 1837 originally labelled *C. pulchellum* ... But this is *C. tenuiflorum* ... and the only record for the island'. - McClintock (1975). 'Babington's specimen labelled *Erythraea pulchella* from Les Quennevais [Jersey] in 1837 has been determined as this species (Hb. Cantab.)'. - Le Sueur (1984). vc113 1837

How strange that Babington should have discovered it on Guernsey and Jersey in the same year and that it has never been seen there again.

Centunculus minimus

* 'Chamaelinum stellatum, Starred dwarf flax. Beyond Redding'. - How (1650: 26). 'Centunculus *Cat. Giss. p. 161* ... in pascuis ante vicum Chisselhurst loco subudo (in a Dale just before the Common)'. - Ray (1724: Add. facing p.1). vc22 **1650**

The discoverer appears to have been John Watlington, an apothecary from Reading; see Druce (1897: cxi). Druce also notes that in Ashmole's copy of How, he annotates the entry 'upon the end of the hills next Chaucer's Copps. E.A [Ashmole], J.W [Watlington]'.

Cephalanthera damasonium

'Helleborine, Wild white Hellebor. *G. 442.* ... on Roe-hill in Kent, not far from Dartford'. - Merrett (1666: 60). vc16 **1666**

'ad collem Rough-hill dictum, a Dartfordia non longe, ut ibi invenirem Helleborinen albam ... T.Willisellus asseruit'. Letter from Ray to Mr Willughby, 1661 or 1662, describing his unsuccessful search for the plant: see Lankester (1848: 4). Raven (1950: 147) prefers to date this letter to 1669. vc16 1662

Clarke and Druce both cite 'Helleborine flore albo vel Damasonium montanum latifolium C.B. In the woods near Stokenchurch, Oxfordshire ...'. - Ray (1670: 339), but the Merrett reference seems fine. Boulger (1900) claims this as one of Gerard's (1597: 358) 'wilde white Helleborines', but this is not followed by anybody else.

Cephalanthera longifolia

'Elleborine minor flore âlbo. The small or wilde white Ellebor with a white flower ... groweth in many places in England and with the same Gentlewoman also before remembered [Mistris Thomasin Tunstall], who sent me one plant of this kinde with the other [*Cypripedium calceolus*, from the HeIkes, which is three miles from Ingleborough]'. - Parkinson (1629: 348). vc64 **1629**

See the account for *Cypripedium* and the recent paper by Foley (2009). Clarke, Druce and others all gave Merrett (1666: 61) as the first record, from the same site, viz 'Helleborine angustifol. fl. albo oblongo, in Helkwood in York-shire, not far from Ingleborough'. - Merrett (1666: 61).

Cephalanthera rubra

* 'Gathered last June on Hampton Common, Gloucestershire, by Mrs. Smith, of Barnham House in that neighbourhood'. - E.B. 437.   vc33   **1797**

Clarke adds 'But in a letter from the Rev. W. Lloyd Baker to Sowerby, dated June 16th of that year, he says that **he** found it .. 'some years ago'. Desmond (1977) too gives him as the discoverer.

Cerastium alpinum

as Cerastium latifolium 'Upon the rocks on the summits of the highland mountains, as upon Ben-Lomond and the mountains about Glencoe, etc'. - Lightfoot (1777: 242).   vc86, 98   **1777**

Druce cites 'Alsine myosotis lanuginosa alpina grandiflora ... in rupe Clogwyn y Garnedd ... juxta Llanberys. - Ray (1690: 147); (1724: 349) and spec in Herb Dill'; and in Druce & Vines (1907: 107). But note that the latter say that the Clogwyn y Garnedd specimens, from Dillenius, are *C. arcticum* [*nigrescens*], not *C. alpinum*! Moss (1920) follows the Ray attribution, and indeed Ray seems to differentiate satisfactorily between *C. alpinum* and *C. arcticum*. However, sufficient doubt remains to warrant choosing the later record of Lightfoot.

Cerastium arvense

* 'Auricula muris pulchro flore albo ... Upon the hill of health, on Newmarket Heath among the bushes, & in the Devills ditch plentifully, as also on the bankes by London-rode side between Trumpington, and Hawkson, and almost on every dry bank about Cambridge'. - Ray (1660: 19).   vc29   **1660**

Druce cites 'Caryophyllus holostius'. - Gerard (1597: 447). Ray (1724: 348) cites Gerard too, but Harvey (1981) does not. It is possible but not convincing enough.

Cerastium brachypetalum

'Wymington'. - Milne-Redhead (1947)   vc30   **1947**

Cerastium cerastoides

* 'On a tour, in 1792 ... '*Stellaria cerastoides* ... Ben Nevis'. - Dickson (1794).   vc97   **1794**
    vc97   1792

Cerastium diffusum

* as *Sagina cerastoides* 'discovered by Mr. James Dickson on the rocky and sandy shores of Inch-Keith and Inch-Combe in the Firth of Forth, as well as on the beach below Prestonpans'. - Smith (1793: 344). vc85 **1793**

'(In the same cover [as *C. semidecandrum*], but which Dillenius evidently distinguished as a different species, are specimens labelled) "fl. 4 an viscosa, found by Dr Manningham on ye coast of Sussex est praecox", (which are perhaps the earliest known British specimens of *C. tetrandrum*'). - Druce & Vines (1907: 107). This is given too in Wolley-Dod (1937). vc13, 14 1747

McClintock (1982) says Curtis separated it in 1797, but he is in fact referring to the fascicle which Stevenson (1961) dates to 1798. The record from Druce & Vines is dated as 1747, the date of Dillenius' death.

Cerastium fontanum

* 'Alsine hirsuta, Myosotis *Ad*. Mouseare-chickweed'. - Johnson (1634: 18). **1634**

But see Rose, in Gilmour (1972: 53), where he suggests *C. fontanum* or *C. glomeratum* for 'Alsine hirsuta minor' of Johnson (1629: 3). Lobel (1571: 193) shows this as 'Alsine myosotis, Small mouse eare', rare other than in Italy and France.

Cerastium glomeratum

'Alsine hirsuta altera viscosa *C.B*. Spuria 4 *Dod*. The broader leaved Mouse-ear-Chickweed'. - Ray (1663: 3). vc29 **1663**

Edgington (2014) has 'Lychnis segetum parva viscosa flore albo. Alsine spuria 4. Dod. Alsine viscaria. Cam. Mouse-eare Campion or Catchfly Campion [Tottenham]. - Johnson (1638: 15) as this species. The same wording is in Johnson (1634: 49) vc21 1638

Ray's record is in a letter from him to Peter Courthope, dated from Trinity College in July 1661; see Gunther (1934). See also Rose, in Gilmour (1972: 53), where he suggests *C. fontanum* or *C. glomeratum* for 'Alsine hirsuta minor' of Johnson (1629: 8), but that polynomial does not appear in any of Johnson's later works. A more tenuous early record, cited by Pulteney, is in Lyte (1578: 53.3) where under Mouse Eare he describes a third species 'covered over with a clammy Downe, or Cotton, in handling as though bedewed or moystened with Honie, and cleaveth to the fingers', which grows 'in this country'. I find this quite plausible.

Cerastium nigrescens

'Alsines myosotis facie Lychnis Alpina flore amplo niveo *D. Lloyd.* Juxta aquas ad latera montis Snowdon copiose'. - Ray (1688a: 2).   vc49   **1688**

Druce & Vines (1907: 349.5) describe specimens from 'Clogwyn y Garnedd'. Clarke originally cited this record for *C. alpinum*, but changed it in his 1909 supplement 'as suggested by Mr Beeby'. The endemic variety *C. n. nigrescens*, formerly treated as a species in its own right, was first reported, as *C. alpinum,* 'Very rare on serpentine [in Unst]; June 1837'. - Edmondston (1839). Lusby & Wright (1996) relate that Edmondston was aged eleven when he discovered it.

Cerastium pumilum

* 'Found by Mr. James Dickson 'on dry banks near Croydon'. - Curtis (1777-98).   vc17   **1794**

Stevenson (1961) dates this fascicle to 1794.

Cerastium semidecandrum

* 'Cerastium hirsutum minus, parvo flore *Cat. Giss.* ... In vicis circa Londinum'. - Ray (1724: 348).   vc21   **1724**

Rose, in Gilmour (1972: 53), prefers *C. fontanum* or *C. glomeratum* for 'Alsine hirsuta minor' of Johnson (1629: 8) which is cited by Druce as the first record, and bracketted as a possibility by Clarke.

Ceratocapnos claviculata

* 'Fumaria alba latifolia'. - Gerard (1597: 929). 'Fumaria alba latifolia … In a corne fielde betweene a small village called Charleton and Greenwich'. - Johnson 1633: 1088).   vc16   **1597**

Gerard seems to give the location in Johnson to another species, but Johnson corrects this. Pulteney cites Lyte (1578: 24.2) 'Capnos Plinii, Hedge Fumeterre … groweth under hedges, in the borders of fieldes, and about olde walls'. The drawing is perfectly satisfactory, but was it seen in England?

Ceratophyllum demersum

* 'Equisetum palustre ramosum aquis immersum. . . In aquis pigrioribus fere ubique'. - Ray (1660: 49). 'On the back-side of the Grange in the ditches beyond Southwark, and betwixt Limehouse-end and Blackwall'. - Merrett (1666: 36).   vc29   **1660**

Ceratophyllum submersum

* 'Hydroceratophyllon folio laevi octo cornibus armato ... In fossis juxta viam quae ab urbe Chichester ad Insulam Selsey ducit observavere D. Manningham et D. Dillenius'. - Ray (1724: 135). 'Near Yarmouth ... Mr. D. Turner'. - E.B. 679 (1800).   vc13   **1724**

Chaenorhinum minus

* 'Antirrhinum minimum repens ... wilde among corne in divers places'. - Johnson (1633: 550). **1633**

'Antirrhinum minus. 20 Junii 1620'. - Goodyer ms.11, f. 83, see Gunther (1922: 115). Assumed to be from Droxford, S. Hants, in Brewis *et al.* (1996)   vc11   1620

Chaerophyllum temulum

* 'Cerefolium sylvestre ... Found in June and July almost in every hedge'. - Johnson (1633: 1037. 2). **1633**

Pulteney equates Turner's (1562: 60) 'mock chervel' with this species.

Chamaemelum nobile

* 'Anthenus [Anthemis] sive Chamaemelo ... Cammomyle ... groweth on Rychmund grene [Richmond Green] and in Hundsley [Hounslow] heth in great plentie'. - Turner (1548: B. j).   vc17, 21   **1548**

Turner (1538: A.iv.) has 'Chamemelon ... vulgo dicitur Camomyle', which implies familiarity with the plant as a wild species.

Chamerion angustifolium

* 'Chamamerion ... In Yorkshire in a place called the Hooke'. - Gerard (1597: 388.4).   vc63   **1597**

Lees (1888: 234) notes 'Hook, near Goole, on the uncultivated moor-borders, near by which places it is still, but decreasingly of course, a conspicuous feature in the vegetation'.

Chelidonium majus

* 'Hirundinaria ... Chelidonion is of ii kyndes ... the greater is called in englishe Selendine. It groweth in hedges in the spring & hath a yealowe iuce'. - Turner (1548: D. v.). **1548**

Druce adds 'It is carved on St Frideswide's shrine in Oxford Cathedral, c. 1289'. Nelson (1959) cites Turner (1538: A. iv) 'Chelidonium [majus] angli vocant Celendyne aut Celidony' as the first record.

Chenopodium album

'Atriplex sylvestris vulgaris and A. s. altera'. - Johnson (1633: 326).   **1633**

The Johnson record is cited by Parkinson, Ray and Smith. Clarke and Druce have 'Atriplex sil. sinuata'. - Johnson (1629: 12.) as this species. But Rose, in Gilmour (1972), identifies this as *Atriplex hastata* (*prostrata*), as does Kent (1975). Pulteney cites L'Obel (1571: 97) 'Atriplex silvestris sinuata. Wilde orrage' as this.

Chenopodium bonus-henricus

* 'Tota bona Spinaciae facie. Bonus Henrius. Angl. English Mercury'. - L'Obel (1571: 97).   1571

L'Obel doesn't actually give a habitat for England - it could have just been cultivated. Nelson (1959) has an interesting note under Turner (1538: B. iij) 'Mercuralis mas', commenting that 'Turner remarks that he had seen this in the gardens of King's College, Cambridge, whence roots were brought into our own gardens. Jackson and others identify this as *Mercuralis annua,*but this, of course, could not be transplanted.

Chenopodium chenopodioides

* 'We are obliged to the accurate Mr. [Lilly] Wigg for pointing out this plant to us, in waste ground where the soil is moist and sandy, near Yarmouth, and we have gathered the same between the cliff and the sea at Lowestoft'. - E.B. 2247.   vc25   1811

'Mr. L. Wigg, near Yarmouth, Sepr. 1808'. - Garry (1903: 156).   vc25   1808

There is an early herbarium specimen purporting to be this 'Kings Lynn, vc28, Smith, J.E., 1779' in **LIV**. This transpires to be *Suaeda maritima*.

Chenopodium ficifolium

* 'Buddle's Fig Blite. Lond. [London]'. - Petiver (1713: viii. 3).   vc21   1713

Edgington (2011) has 'A specimen in Hb. Buddle (117: 38) about 1710 is the first as a British plant.

Chenopodium glaucum

* 'Rand's Oak Blite. London'. - Petiver (1713: viii. 1).   vc21   1713

'Plentifully, just going into Tothill Fields, near the road next Westminster'. - Petiver in Gibson (1695), cited in Kent (1950). 'Chenopodium angustifolium laciniatum minus, *Inst. R.H. 506.* Variis circa Londinum locis (... it is *C. glaucum*, L. which was first found by Mr. Rand in 1705 in London)'. - Druce & Vines (1907: 58).   vc21   1695

Druce notes 'There are specimens from S. Doody in Herb. Du Bois, c. 1704'. There is no justification for Druce's estimated date - Doody died in 1706. Edgington (2011) has 'Tothill Fields, Westminster; shown to Buddle by James Rand, c.1705'.

Chenopodium hybridum

* 'Chenopodio ... Stramonii acutiore folio ... Circa Colcestriam [Colchester] in Anglia inventa est D. Dale'. - Ray (1704: 123). 'Maple Blite. Lond. [London]'. - Petiver (1713: viii. 7).   vc19   1704

Chenopodium murale
* 'Atriplex syl. latifolia acutiore folio, *Bauh*'. - Johnson (1634: 22). 'Atriplex procumbens folio sinuato lucido crasso'. - Ray (1686: 198).   **1634**

Chenopodium polyspermum
* 'Atriplex sylvestris, sive Polyspermon. Neere unto pathwaies by ditchsides and in the borders of fields'. - Gerard (1597: 257), Johnson (1633: 324. 3).   **1597**

Chenopodium rubrum
* 'Atriplex sil: Laciniatis foliis. Pes Anserinus'. Hampstead Heath. - Johnson (1629: 12)   vc21   **1629**

Pulteney cites Lyte (1578: 545) 'Pes Anserinus … in untoyled places, alongst by the way sides' and there certainly is a goodish illustraion of this there.

Chenopodium urbicum
* 'Broad pointed Blite. Lond [London]'. - Petiver (1713: viii. 8).   vc21   **1713**

Chenopodium vulvaria
'The ranke Goate or stinking Motherworte … stinking like rotten corrupt fishe … it groweth in this countrie in sandie places by the way sides'. - Lyte (1578: 548). 'Atriplex olida … upon dung hills, and the most filthy places that may be founde, as also about the common pissing places of great princes and Noblemens houses'. - Gerard (1597: 258) - *see image p.156*. 'Hampstead Heath'. - Johnson (1632: 32).   **1578**

For the allusion to 'Scoggins' in Gerard, see Haslett (1866).

Chrysosplenium alternifolium
* 'Saxifraga aurea major foliis longius incidentibus … Near Hedley [Headley, near Selborne] Hampshire, Mr. Brown'. - Merrett (1666: 109).   vc12   **1666**

'Saxifraga aurea, maior, foliis, pediculis longis insidentibus … Hedley'. - Records of Rev. William Browne entered by How in his copy of '*Phytologia*' between 1650 and 1656, see Gunther (1922: 302).   vc12   1656

Chrysosplenium oppositifolium
* 'Saxifragia aurea … in Angliae, Northmaniae, Belgii humentibus aut Lichenis saxeis … floret'. - L'Obel (1571: 267). 'About Bath and Wels'. - Gerard (1597: 693).   **1571**

Of stinking Orach. Chap. 42.

Atriplex ollida.
Stinking Orach.

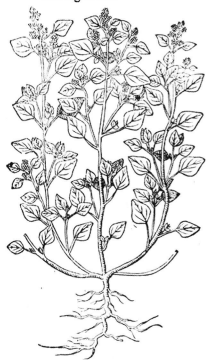

✱ The description.

STinking Orach groweth flat vpon the ground, and is a base and lowe plant with many weake and feeble braunches; whereupon do growe small leaues of a grayish colour, sprinckled ouer with a certaine kinde of durtie mealinesse, in shape like the leaues of Basill : among which leaues heere and there confusedly dispersed bee the seedes as it were nothing but dust or ashes. The whole plant is of a most lothsome sauour or smell, vpon which plant if any should chaunce to rest and sleepe, he might very well report to his friendes that he had reposed himselfe amongst the chiefe of *Scoggins* heires.

✱ The place.

It groweth vpon dung hils and in the most filthy places that may be founde, as also about the common pissing places of great princes, & noblemens houses. Sometime it is founde in places neere brick kils and olde wals, which doth somewhat alter his smell, which is like tostedcheese: but that which groweth in his naturall place smelleth like stinking salt fish, whereof it tooke his name *Garosmus*.

✱ The time.

It is an herbe for a yeere, which springeth vp, and when the seede is ripe it perisheth, and recouereth it selfe againe of his owne seede, so that if it be gotten into a ground, it cannot be destroied.

✱ The names.

Stinking Orach is called of *Cordus Garosmus*, bicause it smelleth like stinking fish, which is called in Greeke γάρον: it is likewise called *Tragium Germanicum*, and *Atriplex fœtida garum olens Penæ & L'Obelij*, for it smelleth more stinking then the rammish male Goate, whereupon some by a figure haue called it *Vulvaria*, and may be called in English stinking Motherwoort.

✱ The nature and vertues.

There hath beene little or nothing set downe of the auncients either of his nature or vertues: notwithstanding it hath beene thought profitable by reason of his stinking smell for such as be troubled with the Mother: for as *Hippocrates* saith, when the Mother doth stifle or strangle, such things are to be applied vnto the nose as haue a rancke and stinking smell.

Chenopodium vulvaria from Gerard, J. 1597.
The Herball, or general history of Plantes.

See species account : Chenopodium vulvaria, p.155

Cicendia filiformis

* 'Centaureum palustre luteum minimum ... versus extrema Cornubiae in solo putrido & lacustri provenit'. - Ray (1670: 63).	vc1	**1670**
'July 1 [1662] ... we rode to the Land's End ... we found another plant on the boggy ground ... it was almost all stalk; it was not above a hand high, it had a yellow flower, but not open in any when we were there, it being a close day'. - Lankester (1846: 189) and see Raven (1950: 128).	vc1	1662

Cicerbita alpina

* 'Discovered on the Aberdeenshire mountain of Loch-nagore by Mr. G. Don, Sept. 1801'. - E.B. 2425.	vc92 vc92	**1812** 1801
Clarke mistakenly gave 1810 for the E.B. plate.		

Cichorium intybus

* 'Intuborum duo sunt genera ... Erraticus intibus dicitur etiam Cichorium ... angli wylde suckery nominant'. - Turner (1538: B. ij).	**1538**

Cicuta virosa

* 'Sium alterum Olusatri facie ... Found by Mr. Goodyer in the ponds about Moore Parke; and by Mr. George Bowles in the ditches about Ellesmere [Salop] And in divers ponds in Flintshire'. - Johnson (1633: 257).	vc20, 40, 51	**1633**
'Sium alterum olusatri facie. 16 Sept. 1625 ... in the ponds about Moore Parke and at Denham in Hertfordshire ...'. - Goodyer ms.11, f. 58, see Gunther (1922: 179).	vc20, 24	1625
Bowles' trip was in 1632, see Raven (1947: 295). Moor Park is in Hertfordshire, but Denham is in Buckinghamshire.		

Circaea alpina

Not clearly separated from *C. intermedia* (= *C. x intermedia*) until Clapham *et al.*(1962). For a more detailed treatment see also Raven (1963).		**1962**
'Lodore Waterfall, Derwentwater, 1806, [D.] Turner. **K**.'. This is the earliest record cited in Raven (1963).	vc70	1806
The first record cited in Clarke and Druce is 'Ad radices montium in Comitatibus Westmorlandico Eboracensi, &c., circa Dallam Tower in agro Westmorlandico'. - Hudson (1762: 10). But Dallam Tower (SD490810) is outside the range of known *C. alpina*, and is almost certainly the hybrid, *C. x intermedia*. See the map in Halliday (1997) which shows a record of the hybrid from the Dallam Tower area, with the species only occurring some way to the north.		

Circaea lutetiana

* 'Circaea lutetiana … Angliae … hortis spontanea'. - L'Obel (1576: 137).   **1576**

Cirsium acaule

'Carduus acaulis septentrionalium. Angliae collibus'. - L'Obel (1576: 480). 'Carduus acaulis *Lob.*'. Between Gravesend and Rochester. - Johnson (1629: 2).   **1576**

Carduus acaulis septentrionalium L'obelii.' 8 July 1620 … on the chalkie downes of Hampsheire plentifullie; and also at Purflett in Essex'. - Goodyer ms.11, f. 105, see Gunther (1922: 118).

Cirsium arvense

* 'Carduus vulgatissimus viarum. By highwaies sides and common paths in great plenty'. - Gerard (1597: 1012).   **1597**

Cirsium dissectum

* 'Cirsium anglicum … provenit in pratis C. viri D. Nicolai Pointz equitis praefecturae Glostriensis in villa vernacule [Iron] Acton nomine'. - L'Obel (1576: 315).   vc34   **1576**

Riddlesdell *et al.* (1948: 290) provides the modern name for L'Obel's site, and it assumed that L'Obel was the finder, not Nicolas Pointz.

Cirsium eriophorum

* 'Tomentosus Carduus Anglicus … Frequens in Angliae collibus strigosis agri Sommerseti juxta aedes … Eduardi Saintloo'. - L'Obel (1571: 370). 'C. eriocephalus … By Pocklington … in Yorkshire; Mr. Goodyer also found it in Hampshire'. - Johnson (1633: 1152).   vc6   **1571**

Cirsium heterophyllum

* 'Cirsium Britannicum … Descriptionem & iconem mihi anno 1581 Londini communicavit C. V. Thomas Pennaeus Londinensis medicus … Provenit in pratis ad radices montis Englebrow [Ingleborough] totius Angliae celsissimi in comitatu Eboracensi'. - L'Ecluse (1583: 655).   vc64   **1583**
    vc64   1581

Cirsium palustre

'Carduus spinosissimus altissimus, forte Carduus palustris, *Bauh.*'. - Johnson (1634: 26).   **1634**

Johnson (1633: 1176) laboured to fit Gerard's plants to those in Dodonaeus' Herbal, and both Clarke and Druce followed his 'third sort … growes on wet heaths …Carduus palustris described in Bauhin'. It seems safer to decide on the less ambiguous description in Johnson (1634)

Cirsium tuberosum

* 'Discovered by A. B. Lambert, Esq. [in 1812] in a wood of his own, called Great Ridge, near Boyton House, Wilts'. - E.B. 2562.	vc8	**1813**
	vc8	1812

Plate 2562, 1813. Smith J. and Sowerby, J. *English Botany*, vol 36.

[2562]

C N I C U S   tuberosus.

Tuberous Plume-thistle.

SYNGENESIA Polygamia-æqualis.

GEN. CHAR. *Cal.* swelling, imbricated with spinous scales. *Recept.* hairy. *Down* feathery, deciduous.

SPEC. CHAR. Leaves with slightly-winged stalks, pinnatifid, lobed, fringed with prickles. Stem unarmed, with about two stalked flowers. Calyx-scales lanceolate, pointed, rather spreading.

SYN. Cnicus tuberosus. *Willd. Sp. Pl. v.* 3. 1680.
Carduus tuberosus. *Linn. Sp. Pl.* 1154.
Jacea tuberosa. *Ger. em.* 728.

WE readily concur with the opinion of Professor Willdenow, first published in his *Prodr. Berolin.* 261, that *Cnicus* is best distinguished by its feathery seed-down from *Carduus*. With this latter genus then will remain our *nutans, t.* 1112, *acanthoides, t.* 973, *tenuiflorus, t.* 412, and *marianus, t.* 976, only, of the British species.

The plant before us, entirely new to Britain, was discovered last year by our excellent friend A. B. Lambert, Esq. in a wood of his own, called Great Ridge, near Boyton house, Wilts, growing plentifully, in one spot only. It flowers the beginning of August, and is perennial.

The root creeps, sending down many oblong perpendicular knobs. The stem is about two feet high, leafy, furrowed, hairy, without spines, simple, except at the top, where it usually bears two flowers, scarcely more, the figure in Gerarde being, as far as we have observed, faulty in this particular. The leaves are pinnatifid, and variously cut, fringed with copious yellowish prickles, slightly hairy, their base running down into narrow winged footstalks. Flowers on long hairy stalks, at first rather drooping, bright purple, with a slightly downy calyx, whose scales have prominent, leafy, minutely spinous, tips. Seed-down feathery.—Linnæus justly says the flowers are like *heterophyllus, t.* 675, but smaller. It is a very distinct species.

Cirsium vulgare

* 'Carduus lanceatus. ... By highwaies sides and common paths in great plenty'. - Gerard (1597: 1012). 'C. lanceolatus, *Tab.*'. Between Sandwich and Canterbury. - Johnson (1632: 20).	**1597**

Cladium mariscus

'Segge or shergrasse ... the edges of this herbe are so sharpe, that they will cut a mannes hand ... it hath a long stalke, and three square, and in the top of that is a sort of little knops instead of seeds, and flowers much like our garden gallyngal ... The people of the fen countries use it for fodder and do heat ovens with it'. - Turner (1551: H. v). 'Cyperus longus inodorus sylvestris. In the watery places of Hinton Moor, and in divers Fen ditches'. - Ray (1660: 43).

1551

Clarke and Druce both gave the Ray record as their first date, and neither Clarke nor Chapman & Tweddle (1989) call Turner's record anything other than a *Carex* species. But it must be *Cladium*, were it not for the three-cornered stems. In their *Dictionary of English Plant-names*, Britten & Holland (1886) say 'it seems likely that *Cladium* is intended' and add that that 'in the Fens *Cladium* is still called Sedge'. Ray (*op.cit.*) also notes that whereas *Cladium* has round stems, Gerard and Parkinson, describing what he assumes to be the same plant, attribute triangular stems. He adds that 'we are not much influenced by the authority of those writers, for perhaps they write thus from conjecture & because they thought it was essential for every Cyperus to have triangular stems'. Alan Leslie (*in litt.*) points out that 'vegetative shoots, which of course are just made up of a large bunch of leaves, are trigonous, in that the way the leaves fold round each other towards the base forms a very definite triangle in section with broad flat faces. This might have led to the mistaken idea that the stem was trigonous in this species'.

Clematis vitalba

* 'Vitis sylvestris solani foliis Heguine (Hedgvine] ... groweth plentuously betwene ware and Barckway in the hedges'. - Turner (1548: G. viij. verso).

vc20   **1548**

Ware and Barkwaye are in E. Hertfordshire.

Clinopodium acinos

* 'Acinos Anglicum ... quam anno M.D.XXCI. mensè Iulio florentem in Angliâ observabam, dum commodum navigandi tempus expectans, ad Ill. Dn. de Cobham, Garterii ordinis Equitem, & quinque portuum Prafectum proficiscerer'. - L'Ecluse (1601: i. 354).

vc16   **1601**
vc16   1581

A translation of L'Ecluse's text might be 'which in the year 1581 in the month of July I was observing flowering in England while, awaiting a favourable time to take ship, I was travelling to [visit] the Most Illustrious Lord of Cobham, Knight of the Order of the Garter & Warden of the Cinque Ports'. He lived at Cobham Hall in Kent. I am grateful to Philip Oswald for this translation. Britten (1887) and Druce (1897: xciv) both interpret Turner's "Corne mint," (1548: B. vii and 1551: 109) *as* this species, but Clarke, Druce and Stearn (1965) prefer *Mentha arvensis*.

Clinopodium ascendens

* 'Calamintha ... the firste kynde is lyke wilde Meriorum and it groweth much aboute Syon in Englande ... bush calamint ... or hore calamynt'. - Turner (1548: B. vij). 'Calamintha vulgaris, *Offic*'. In Thanet. - Johnson (1632: 8).   vc21   **1548**

Clinopodium calamintha

* 'Ad insulam Sheppey ... Calamentha pulegii odore *Lob.*'. - Johnson (1629: 5).   vc15   **1629**

Rose, in Gilmour (1972: 56), says still abundant at Key Street and Bobbing. Pulteney identifies as this species Lyte's (1578: 247.2) 'Calaminthae tertium genus ... groweth in this countrie in rest fields, and on certaine small hilles or knappes'.

Clinopodium menthifolium

* On 29 Aug.1843, 'in a wooded valley near Apes down and Rowledge about three miles and a half from Newport towards Yarmouth'. - Bromfield (1843).   vc10   **1843**

But see Pope *et al.* (2003). 'seen by Sir David Brewster and others in 1841'. No source is given for that, but it may have been from Adams (1873: 96), which gives ... 'Brewster and other eminent naturalists ... '.

Clinopodium vulgare

* 'Clinopodium ... I heare saye that it groweth ... about Oxford. It may be called in englishe horse Tyme'. - Turner (1548: C. iij. verso).   vc23   **1548**

Cochlearia anglica

* 'Cochlearia Anglica, Atriplicis folia ... In Anglia ad amoenissimi Tamesis fluenta qua Londinum praeterlapsus est, etiamque in Bristoiae sinu Occidui Oceani'. - L'Obel (1571: 122).   vc16, 34   **1571**

'Crescit copiose ad Thamesem'. - annotations made by T. Penny on C.Gesner's plant illustrations dated as by 1565 in Foley (2006b).   vc16, 18   1565

see also Foley (2005).

Cochlearia danica

* 'Hederaceum Thlaspi. In Portlandiae peninsula, Cornubiae vicino portu Plimmouvve, aliisq: maritimis Angliae cautibus'. - L'Obel (1576: 338).   vc2, 9   **1576**

Cochlearia micacea

* 'Ben Lawers, Perthshire, from 3500 to 3700ft; Am Binnein, Perthshire, from 3200 to 3500 ft; Ben Dothaidh, Argyle, above 3000 ft. ... August, 1887'. - Marshall (1894: 289). vc88, 98 **1894**

Pentland Hills, 1847, det. Rich, **E**. Pentland Hills, source of S. Medwin Water, vc83, 1878, F.M. Webb, det. Rich, **E**. vc83 1847

The 1847 specimen came from the herbarium of C. Murchison (1830-1879), but it is not known if it was collected by him.

Cochlearia officinalis

'Scurby Wede ... sene it in England at Westchester, at Portlande and at Porbeke [Purbeck]'. - Turner (1568: i. 90). vc9, 58 **1568**

The revised first volume of the Herbal was issued in 1568 (see Raven (1947:102 & 124) by which time Turner had changed his mind over the identification of his find (see *Calystegia soldanella*). Raven dates the finding of this to Turner's visit to Dorset in 1550. vc9 1550

Westchester appears to be Chester (both Lhwyd and Ray use the word, and Ray was a friend of the Bishop of Chester). See Gunther (1928: 58), reproducing a letter from Ray 'about the later end of April 1669, being at West-chester with my lord Bishop of that Diocese'. Nelson (1959) states that Westchester was probably Whitchester, near Heddon on the Wall, vc67, but that is inland and very unlikely.

Cochlearia pyrenaica

'On the mountains of Clova and Loch na Gare, in August 1807 by Mr George Don'. - E.B. 2403, 1812. vc90, 92 **1812**

Gerard's (1597: 324.1) 'C. rotundifolia ... it hath beene found of late growing many miles from the sea side, upon a great hill in Lancashire [*sic*] called Ingleborrough hill' must be this species as it is the only species of this critical group that is found there. vc64 1597

Druce had Don's record as *C. alpina*, and E.B. as *C. groenlandica*, but it appears to be *C. pyrenaica* (T.C.G. Rich, *pers. comm.*). Druce & Vines (1907: 94) give a record of this from Caernarvon, vc49, which can be dated to Dillenius' trip there in 1726.
Another early record is in a list of manuscript additions that William How drafted before his death in 1656 (Gunther (1922: 288) 'Cochlearia rotundifolia sive Batava, *Lob*. Round-leaved Scurvygrasse. It groweth nigh unto a castle in ye Peake in Darbyshire which is 30 miles from the sea'.

Coeloglossum viride

* 'Orchis Batrachites, Frog Satyrion: by Barkway [Herts], Dr. Johnson's ms'. - How (1650: 82). 'Orchis palmata flore viridi vc20 **1650**

… in some pastures and closes on the Northside of Chesterton plentifully'. - Ray (1660: 107).

Johnson died in 1644.	vc20	1644

Ray has a brilliant description of this species in Cambridgeshire. Kew & Powell (1932) describe How's access to some of Johnson's manuscripts.

Coincya monensis

* 'Eruca Monensis laciniata lutea … We found it plentifully, going from the Landing-place at Ramsey to the Town'. - Ray (1670: 103).	vc71	**1670**
See notes of Ray's 1660 Itinerary in Raven (1950: 114).	vc71	1660

Coincya wrightii

'Lundy'. - Wright (1936).	vc4	**1936**
'In June 1844 … a singular variety of *Nedyus contractus* … taken in abundance from a species of *Brassica*, which grows in profusion on the eastern side of the island'. - Wollaston (1845). 'Loose and sprawling but bright in hue'. - Gosse (1874).	vc4	1844

Compton & Key (2000) give as early records Gosse (1874) & Wright (1936). Martin & Fraser (1939) state that it was first named by Wright in 1936, but cite records back to 1884. However Pugsley (1936c), reviewing Wright's paper (and the occurrence of two host-specific beetles there) cited a record of 'wild *Brassica*, on which Wollaston detected two peculiar beetles in 1845'.

Colchicum autumnale

* Of Middow Saffrone 'I have sene it growe in the west cuntre besyde Bathe'. - Turner (1551: L. iij. verso).	vc6	**1551**

Note the record 'In the West country of England in a mead a little from Bruton [c20m S. of Bath]'. - Henry Daniel, dated to c1370, see Harvey (1981) and also Allen (2001).

Comarum palustre

* 'Pentaphyllum rubrum palustre … In a marrish ground adjoining to the land called Bourne ponds, halfe a mile from Colchester'. - Gerard (1597: 836).	vc19	**1597**

Lyte (1578: 82.4) mentions 'Pentaphyllon Rubrum … onely founde in ditches, or aboute ditches of standing water'.

Conium maculatum

* 'Cicuta … oure Hemlocke'. - Turner (1548: C. ij. verso). 'Common Hemlock grows plentifully about towne walls and villages …'. - Gerard (1597: 903)		1548

Conopodium majus

* 'Apios ... called in Englishe ... an ernute or an erth nute ... groweth plentuouslye in Northumberland beside morpeth'. - Turner (1548: B. j. verso). vc67 **1548**

Convallaria majalis

* 'Lilium convallium. .. . On Hampsted Heath fower miles from London, in great abundance: neere to Lee [Leigh-on-Sea] in Essex, and upon Bushie heath, thirteen miles from London, and many other places'. - Gerard (1597: 331). vc18, 21 **1597**

Convolvulus arvensis

* 'Convolvulus ... The common bynde aut The lytell Wynde'. - Turner (1538: A. iiij). 'In most parts of England'. - Gerard (1597: 714). **1538**

Corallorhiza trifida

* *as Ophrys Corallorizha* 'In a moist hanging wood near the head of Little Loch Broom on the western coast of Ross-shire'. - Lightfoot (1777: 523). vc105 **1777**

Lightfoot's tour was in 1772. vc105 1772

Cornus sanguinea

* 'Cornus ... the female is plentuous in Englande and the buchers make prickes of it, some cal it Gadrise or dog tree'. - Turner (1548: C. v). **1548**

Cornus suecica

'Chamaepericlymenum. *Park. Ger.* On the Northwest-end of the highest of Cheviot hills, among the rocks on the West-side plentifully'. - Ray (1670: 339). vc68 **1670**

The first date (L'Ecluse 1601) given in Clarke and Druce is an error, for the record there refers to Danzig [Gdansk] in Poland - see the explanation in Raven (1947: 164). No earlier record than than that in Ray has been found in Johnson (1633), How (1650) or Merrett (1666).

Corrigiola litoralis

* 'Found by Mr. Hudson on Slapham Sands beyond Dartmouth and near the Star Point'. - Withering (1787: 322). vc3 **1787**

Clarke adds 'There is in Herb. Banks in **BM** a specimen sent to Banks by Hudson in 1784'. vc3 1784

Corylus avellana

* 'Corylus ... The hasell is so well knowen that wee nede not any description of it'. - Turner (1551: M. iij).		**1551**

Corynephorus canescens

* 'Gramen maritimum vulgatissimo pratensi Gramini conjener, aut simile. Hoc gramen radice capillata crassiore maritimis litoreis Kantiae oritur'. - L'Obel (1655: 8). 'L'Obel first discovered this on the coast of Kent, and Mr. Buddle since him in Suffolk'. - Petiver (1716: 126). 'On Yarmouth Denes it abounds'. - E.B. 1190. vc15 **1655**
vc15 1616

L'Obel died in 1616. There is no confirmed record of this species from Kent. It is not mentioned in Hanbury & Marshall (1899) though it is quite possible that it was once on the coast from Deal to Sandwich.

Cotoneaster cambrica

* 'On the limestone cliffs of the Great Ormshead, Carnarvonshire, in various places. Mr. W. Wilson, 1825'. Smith (1828: 268). vc49 **1825**

'Mr W. Wilson, about the year 1821'. - E.B.S. 2713. vc49 1821

The E.B.S. reference adds 'but Mr J.W. Griffith, of Garn, is said to have seen it so early as 1783, but "unfortunately he laid it by, instead of describing and communicating it to Sir J.E.Smith"'. The source for this statement is given as Williams (1830).

Crambe maritima

* The 'wonderful great Cole ...groweth at dover harde by the see syd'. - Turner (1551: G. ij. verso). vc15 **1551**

L'Obel later found this at Portland (1570: 92), recognising that it was Turner's plant, and noting that he had received seeds from Turner. See Raven (1947: 104).

Crassula aquatica

'While looking for plants at Adel, near Leeds, on Sept 1st [1921] … it is associated with *Limosella aquatica* and *Polygonum minus*, and it grows in abundance on drying mud'. - Butcher (1922). vc64 **1922**
vc64 1921

Crassula tillaea

* as *Tillaea procumbens* Found on Drayton Heath and several other places near Norwich, in great plenty. First examined and ascertained by the Rev. Mr. [Henry] Bryant in 1766'. - Rose (1775: app. 450). vc27 **1775**
vc27 1766

Hind (1889) gives a record 'Lackford and Culford, Dill. Ray in Hort. Ox.'. This has been interpreted as dating to 1724, the date of Dillenius' revision, but, as Hind makes clear, it refers to an annotated copy of this work which he dates to c. 1780.

Crataegus laevigata

Babington (1843) gives the two varieties and postulates that they are 'distinct species', a distinction formalised in all his later editions. **1843**

Horwood & Noel (1933) cite Pulteney (1749) as the first record 'This I observ'd in Hollinghall Wood [in Charnwood Forest] but sparingly the present year 1749' as 'Mespilus apii folio sylvestris spinosa foliis et fructu majore' from Ray (1724: 454.4). Pulteney separated it from *C. monogyna*. vc55 1749

In fact Ray (*op.cit.*) gives two locations 'found by Mr James Sherard in the Orchard Hedge at Mr Maidwell's at Gadington, Northamptonshire', and 'In Ricot Park and elsewhere in Oxfordshire; *Merr.*'. Druce (1930a) in his *Fl. Northamptonshire*, only tentatively accepts the first record, and does not mention the second in his *Fl. Oxfordshire* (1927).

Druce cites Smith (1824: ii. 359) as his first published record, but all that Smith records is that he is not convinced by Jacquin's arguments (for two species). Clarke does not separate them.

Crataegus monogyna s.l.

* ' … Mattiolus holdeth that our haw tre or whyte thorne tre is Oxycantha … Spina alba … Our comō hawthorn'. - Turner (1562: 73b). 'Oxyacantha … Angli May dicunt'. - L'Obel (1571: 443). **1562**

Crataegus monogyna s.s.

Babington (1843) gives the two varieties and postulates that they are 'distinct species', a distinction formalised in all his later editions. **1843**

There is a specimen of this in Merrett's collection in Hb. Sloane 14: 44, **BM**. 'Mespilus Apii folio sylvestris spinosa, sive Oxycantha, *C.B. Pin.* 454 … About London full grown trees. Hyde Park'. - Druce & Vines (1907: 130). 1695

Merrett died in 1695, though the record is probably earlier. The record from Druce & Vines is dated as 1747, the date of Dillenius' death. There may well be earlier records that are outside the range of *C. laevigata* and thus can be confidently identified as this species.

Crepis biennis

* 'Hieracium maximum asperum Chondrillae folio *C.B* … a D. Newton in Cantia inventum est'. - Ray (1688: 1857, 1688a: 12). vc15, 16 **1688**

This is presumably James Newton, known to Ray as a competent and travelled botanist. See Raven (1950: 218).

Crepis capillaris

* 'Hieracium Aphacoides Succorie Hawkweede … In untoiled places … '. - Gerard (1597: 234). **1597**

Crepis foetida

* 'Hieracium minus Cichorei vel potius Stoebes folio hirsutum'. - Ray (1660: 75). vc29 **1660**

Dunn (1905) has 'Native of SE Europe…...though often styled a native in the SE counties, it is usually recorded expressly from artificial habitats', and thus would make Ray's record alien. See also a long discussion in Oswald & Preston (2011: 220n) who decide that the occurence in Cambridge was not entirely natural. Within the core range the earliest record traced is one from Chelsea, S. Doody (Gibson, 1695). Another is Charlton chalk-pits, dated from c.1700 (Buddle ms., cited in Hanbury & Marshall (1899: 213).

Crepis mollis

* On a tour through the Highlands, in 1789, '*Hieracium molle*. In sylvis Scotiae australis'. - Dickson (1794). **1794**
1789

Crepis paludosa

* 'Hieracium montanum latifolium glabrurn minus *C.B., Park.* ... In montosis Septentrionalibus Angliae'. - Ray (1677: 161). **1677**

Crepis praemorsa

'In July 1988 … in northern Westmorland … on a low bank at the edge of a hayfield'. - Halliday (1990). vc69 **1990**
vc69 1988

Crithmum maritimum

* 'Crithmus ... Sampere ... groweth much in rockes and cliffes beside Dover'. - Turner (1548: C. v. verso). vc15 **1548**

Cruciata laevipes

* 'Cruciata. Crossewoort ... In the Churchyarde of Hampsteed neere London ... also in the Lane or highway beyond Charleton a small village by Greenwich'. - Gerard (1597: 965). vc16, 21 **1597**

Cryptogramma crispa

'Filix petraea nodum descripta. Rocke Ferne'. - Johnson (1641: 21). 'Filix Dauci foliis. On old walls a mile from Haslenden [Haslingden, Lancashire] going over the hill towards Goodschaw Chappel'. - Merrett (1666: 39). vc49 **1641**

as *Allosurus crispus*, Thomas Johnson and party, on Snowdon, Aug 1639. See Raven (1947: 289). vc49 1639

Druce has 'Filicula montana florida perelegans. At the bottom of stone walls made up with earth in Orton Parish…Westmorland, plentifully'. - Ray (1670: 115). But, if it is correct, it seems he missed Johnson's reference.

Cuscuta agg.

'Cassutha … Doder growth aboute Flax, Tares, Nettels, Tyme, Savery, Hoppes, and many other such lyke, it is much more plentuous in Germany then it is in Englande'. - Turner (1548: B. viij. verso). **1548**

Probably Turner (1538: A.iij. verso) 'Cassutha … angli [call it] Dodder', which implies familiarity with the plant as a wild species. Stearn (1965) and earlier editors equate this with *C. epithymum*, but Rydén *et al.* (1999) with *C. europaea*, which seems ambitious, though Nelson (1959) and Stearn (1965) lend some support to this. The reality is that no early writers were aware that there were two wild species.

Cuscuta epithymum

E.B. 378, where the separation of the two species was first clearly set out. **1797**

Cuscuta europaea

* 'Mr. Sowerby last autumn received wild specimens of the real *europaea* from Mr. Alexander Smith of Aberdeen, and others from the Rev. Mr. Hemsted [from Cambridgeshire], which have at the same time verified the plant as a native of Britain'. - E.B. 378. 'Near Newmarket, Rev. J. Hemsted' - Turner & Dillwyn (1805). vc29 **1797**

' … very plentifully in Summerset shire upon Nettles'. - Gerard (1597: 462). 'Cuscuta major. Commonly growing among beans'. - Ray (1660: 42). vc6 1597

Trail (1923) sets out details of the confusion over the Aberdeen record, and, on balance, it seems best to ignore it.
The reference in Gerard to nettles, which Druce takes as referring to this species, seems acceptable, but of course Gerard did not know of two species. Hanbury & Marshall (1899) identify Johnson's 1632 Kent record of 'Cuscuta, Matth. Androsace, Tragi' as this species.

Cynodon dactylon

* 'Gramen dactyloides radice repente *Ger.*, repens cum panicula Graminis Mannae *J.B* ... *Found by* Mr. Newton *on the sandy shores between* Pensans *and* Marketjeu Cornwall, *plentifully*'. - Ray (1688a: 10). vc1 **1688**

The plant is still there, though much reduced at present.

Cynoglossum germanicum

* 'Dwarfe Houndstoonge ... groweth very plentifully by the waies side as you ride Colchester highway from Londonward, betweene Esterford [Kelvedon] and Wittam [Witham Priory] in Essex'. - Gerard (1597: 659). vc19 **1597**

Cynoglossum officinale

* 'Cynaglossus ... Houndes tong ... groweth in sandy groundes and aboute cities & townes'. - Turner (1548: C. vj. verso).   **1548**

Cynosurus cristatus

* 'Gramen Cristatum *Ch. Bauhini* ... quo nihil frequentius variis editis pratis Londinensis agri, iuxta Hackneum (alibique multis in locis transmarinis)'. - L'Obel (1605: 467).   vc21   **1605**

Cyperus fuscus

'On the margins of a Peat pond on Shalford Common, about two miles from this place [Godalming]' [discovered 16th September 1846]. - Salmon (1847).   vc17   **1847**
    vc17   1846

Clarke and Druce both cite 'A. H. Haworth, Esq ... found it in a low marshy meadow [Eelbrook] scarcely half a mile from his late residence in Little Chelsea'. - Graves & Hooker (1821: t. 85) as the first record. This is followed by Kent (1975), who also gives an earlier record of Haworth's of 1819 (**CYN**). But the Eelbrook locality is probably an introduction, see Trimen (1871a) who relates that 'Mr Haworth made no secret that he had sowed it there from seeds which he had obtained from Swiss specimens purchased from Mr Thomas of Bex'.

Cyperus longus

* 'Cyperus longus *Ger.* ... D. Newton in Insula Purbeck dicta Dorcestriae Angliae eum invenit ... à latere Portlandiam insulam spectante'. - Ray (1688: 1299). '*The ordinary Sweet Cyperus or English Galingale. Found by* Mr Newton *in the Isle of* Purbeck Dorsetshire*, near a Chappel, on the side that looks towards* Portland Island'. - Ray (1688a: 5).   vc9   **1688**

The location of the Dorset record has never been satisfactorily identified, though there is a very long-standing site at Ulwell, near Swanage. But that is hardly near to what is presumably the Chapel on St Aldhelms Head that does indeed look along the coast towards Portland.

Cypripedium calceolus

* ' ... Elleborine major, sive Calceolus Mariae. ... in Lancashire neere upon the border of Yorkeshire, in a wood or place called the Helkes, which is three miles from Ingleton, as I am informed by a courteous Gentlewoman, a great lover of these delights, called Mistris Thomasin Tunstall who dwelleth at Bull-banke, near Hornby Castle in those parts'. - Parkinson (1629: 347-348).   vc64   **1629**

Helks Wood is in Yorkshire. Clarke cited Parkinson (1640), but corrected this in his 1909 supplement.

Cystopteris alpina

'Teesdale, Durham, 1872, J. Backhouse'. - Syme (1886: xii. 104).   vc66   **1886**

Edgington (2013) describes plausible records from Snowdonia, with specimens in the Morisonian herbarium at **OXF**, collected by Lhwyd, presumably in 1688. See also Morison (1699: 581) 'Filicula Cambrobritannica pinnulis Cicutariæ divisura donatis. Ad Onopterin majorem, *Tab. p. 474.* quodammodo spectat … E rupibus celsissimi montis Snowdon, loco Clogwyn du yn yr Ardhu ab incolis dictu, D. Lhwyd'.   vc49   1688

The plant was first described, as *Cyathea incisa*, as an alien from Walthamstow. - E.B. 163 (1794), where it had been known since 1778.

Cystopteris diaphana

'In February 2000, Matt Stribley … on a damp woodland bank at Polbrock Bridge, E. Cornwall, (SX06)'. - Murphy & Rumsey (2005).   vc2   **2005**
   vc2   2000

This ignores the undoubtedly alien record from Penjerrick garden (vc1) in 1961. But there is a record in Hanbury & Marshall (1899), under *C. fragilis*, 'Gathered by Mr. Thorpe near Tunbridge (i.e. Tunbridge Wells) in Herb Buddle iii. 31, **BM**, which has been determined (Rumsey 2007b) as this species. Whether this was from a native or cultivated specimen is not known. Buddle died in 1715.

Cystopteris dickieana

' … in a cave near the sea, at Aberdeen, Dr. Dickie'. - Sim (1848).   vc91   **1848**

Marren (1983) refers to Dickie's *Flora Aberdoniensis* (1838) recording *C. fragilis* (*sic*) 'in a coastal cave …rare'.   vc91   1838

Edgington (2013) points out that Dickie denied that he was the first to recognise it 'it was no original discovery of mine, the late Professor Knight having been in the habit of showing it to his pupils [Dickie was one]'. Lusby & Wright (1996) agree with this. See also Moore (1849), for a fuller account and synonymy. Marren adds that the type collection is in **BM**, 1842. The Sim reference is a letter to the magazine, in which he says 'it will prove a valuable addition to our ferneries, if not to the British Flora'.

But note that there is an unresolved taxonomic problem here with inland records from Ben Lawers and the Cairngorms of a form that has been assigned to this species by Stace (2010), and as probable *C. fragilis* by Dyer *et al.*(2000). The earliest such record is 'Ben Lawers, foot of, Linseed Mill, Robert Brown, 8/1794, **BM**'.

Cystopteris fragilis

'Filix saxatilis caule tenui fragili … Upon old Stone-walls and rocks, in the mountains of the Peak in Darby-shire and the West-riding of York-shire, and in Westmorland'. - Ray (1670: 114).   vc57, 64, 69   **1670**

Edgington (2013) suggests that Ray 'surely saw it in1658 during his first visit to Derbyshire, and again in 1660 when he toured the northern counties with Willughby'.

Cystopteris montana

'Discovered by Mr Wilson in Scotland in 1836'. - Newman (1843a).	vc88	**1843**
	vc88	1836

Druce's (1932) citation actually says 'Ben Lawers', rather than 'Scotland', citing Newman, but no site is given there. Druce adds 'Sprengel says it was found in Wales and figured in *Phytographica, t, 89, f. 11*', but any Welsh record is an error.

Cytisus scoparius

* 'Genista ... Broume groweth in al countreis of England where as I haue ben'. - Turner (1548: D. ij. verso).   **1548**

Daboecia cantabrica

Petiver (1703: tab. 27, fig. 4).	vcH16	**1703**
In Connemara in 1700, Lhwyd noted 'a ... heath, so common that ye people have given it ye name of *Prych dabeog* i.e. *Erica (s'ti) Dabeoci ...*'. - Nelson (1979a). See also 'Erica Dabeoci in Mayo'. - Lhwyd to Richardson, 21st October 1700, British Library: Sloane ms. 4063, f.48.	vcH16	1700

Nelson (*op.cit.*) says that this was published by Petiver, and later by Ray (1704: Dendr. 98) as '30. Erica S. Dabeoci Hibernis D. Lhwyd'. Ray too refers to it in a letter to Lhwyd dated 11th June 1701. See Gunther (1928: 280).

Dactylis glomerata

* 'Round tufted Reede grasse Calamagrostis torosa panicula ... by the hedge sides in many Countries of the Kingdome'. - Parkinson (1640: 1182).   **1640**

L'Obel (1655: 43), with the same polynomial, 'frequens Anglo-Britanniae'.   1616

Parkinson (1640: 1161) also has 'Gramen pratense spica multiplici rubra', which later writers equate to the same species.

Dactylorhiza

'Palma Christi mas,' and 'Palma Christi foemina ... The royal Satyrions do grow for the most part in moist or fennie grounds ... I have found them in many places especially in the middest of a wood in Kent called Swainescombe neere to Gravesend, by the village Swainescombe, and likewise in Hampsteed wood fower miles from London'. - Gerard (1597: 170).   vc16, 21   **1597**

The history of the concepts and nomenclature is probably best summed up in Pugsley (1935), tracing the developments from Dodoens (1568: 216) onwards. Pugsley summarises all the pre-Linnean works, concluding that by then only two species were understood - The Male Satyrion Royal or Marsh Orchid, with unspotted foliage (which he equates to *D. incarnata* or *D. praetermissa*), and the Female Satyrion Royal or spotted Orchid, with spotted foliage (which he identifies with *D. maculata*).

Dactylorhiza ebudensis

Lowe (2003), Foley & Clarke (2005).   vc110   **2003**

as *Orchis majalis* subsp. *occidentalis* 'Lingay Strand to Newton'. - Campbell (1937), detailing a find in 1936.   vc110   1936

as *Orchis occidentalis* (Pugsley) Wilmott, *comb. nov.*, in Campbell (1938).

Dactylorhiza fuchsii

'*Orchis fuchsii* Druce' - Druce (1915b), Godfery (1923)   **1915**

'Orchis palmata foemina … frequent in meadows & moist woods'. - Ray (1660: 107) and interpeted as this species in Oswald & Preston (2011: 250). 'In ye wood near Upcerne in Dorsetshire, Rev. W. Stonestreet'. - Druce (1928: 485).   vc29   1660

There were 'wars' between Godfery and Druce, but the papers cited seem the first clear separation. Clarke and Druce both cited for this species 'Palma Christi mas', and 'Palma Christi foemina . . . I have found them in many places especially in ... a wood in Kent called Swainescombe wood neere to Gravesend, and likewise in Hampsteed wood fower miles from London'. - Gerard (1597: 170).

In view of the comments by Pugsley (1935), we need to go to the time when *D. fuchsii* and *D. maculata* were properly distinguished. There might well be earlier evidences, but the two cited are unequivocally of this species. Stonesteet died in 1716.

Dactylorhiza incarnata

'Orchis latifolia … We need go no further than Battersea-meadows to find this plant in tolerable abundance'. - Curtis (1788: Fasc., v, 65).   vc17   **1788**

Stevenson (1961) dates this fascicle to 1787-88. But see Gerard (1597: 171.3) for 'Serapias palustris latifolia' with what seems an excellent depiction of this species. This (and/or *D. praetermissa*) were noted by Ray (1660: 106) and discussed in Oswald & Preston (2011: 249n), but the record cannot be confidently assigned to either species. It was noticed by Babington in Cambs in 1833, 'but not then distinguished from *O. latifolia*'. - Babington (1860b: 225).

Dactylorhiza kerryensis

as *D. majalis* γ *occidentalis* 'I discovered last year [1933] in five locations in counties Clare and Galway'. - Pugsley (1935: 586).	vcH9, H16	**1935**
Praeger (1939) refers to a specimen from Ardrahan (vcH15) collected in 1893, cited by Pugsley.	vcH15	1893

Dactylorhiza maculata

Druce (1915b).		**1915**
Note that the enlarged single figure of the flower in E.B 632 (1799) appears to be this species, see Druce (*opp. cit.*).		1799
See the comments under *D. fuchsii* above. There are very few records of this species in county floras before 1900, but an exhaustive search of herbaria might produce more satisfactory or earlier records than that given here.		

Dactylorhiza praetermissa

'White Water-side, Northants, 1878'. - Druce (1914b).	vc32	**1914**
'An excellent example of *O. praetermissa*, of uncertain origin, is H.S. 312/119/1, [**BM**] similarly labelled among Uvedale's exsiccatae'. - Pugsley (1935).		1722
Uvedale died in 1722. Druce also cites (under *Orchis latifolia*) Webster (1886: 56) as an early reference.		

Dactylorhiza purpurella

'In two or three bogs on the fells ... between Borrowdale and Watendlath [Cumberland], ... identified by Mr. R.A. Rolfe as O[rchis] *cruenta* Muhl.'. - Goss (1899).	vc70	**1899**
Little Harle, vc67, from a painting by Henrietta Lorraine, 5th June 1824, identified as this species by John Richards (*pers. comm.*). Braemar, 1855, A. Croall, det. Summerhayes, **BM** & **K**.	vc67	1824

Dactylorhiza traunsteinerioides

as *Orchis majalis* subsp. *traunsteinerioides*, from Newcastle, Co. Wicklow ' ... was first noticed in the herbarium of the Dublin Museum ... and in the spring of 1935 Mr J.P. Brunker supplied me ... material from Wicklow'. - Pugsley (1936b). As '*Orchis Francis-Drucei*, (23 June 1935) on a hill slope above Loch Maree, West Ross'. - Wilmott (1936).	vc105, H20	**1936**
'in the marsh south of the Station road, Newcastle, 1932'. - Brunker (1950). Brunker adds that the **DBN** specimen was dated 1934.	vcH20	1932
Wilmott's site has been identified as from Slioch (D. Donald *pers. comm.*). Perring *et al.* (1964) refer to specimens collected from Dernford Fen, Cambs, in 1913, **CGE**, but it is not clear whether they (or indeed Crompton (2007)) accept the determination.		

Damasonium alisma

* 'Plantago aquatica minor stellata. ... I found [this] a little beyond Ilford in the way to Romford and Mr. Goodyer found it also growing upon Hounslow Heath'. - Johnson (1633: 418).   vc18, 21   **1633**

'Plantago aquatica stellata. 30 Julii 1618 … in Hounslowe Heath'. - Goodyer, ms.11, f. 56v, see Gunther (1922: 110).   vc21   **1618**

Danthonia decumbens

* 'Gramen avenaceum minus procumbens paniculis non aristatis'. - Ray (1670: 141). 'On Harefield Common' [Middx.], and near Battle's Well'. - Blackstone (1737: 34).   **1670**

Daphne laureola

* 'Daphnoides ... in englishe Lauriel, Lorel or Loury, groweth plentuously in hedges in England'. - Turner (1548: C. vij). 'St. Albans'. - Coles (1657: 311).   **1548**

Turner (1538: A. iv. verso) as 'Daphnoides … [called by] vulgus autem Laury aut Lauriell aut Lowre', which implies familiarity with the plant as a wild species.

Daphne mezereum

* 'Woods near Andover in Hampshire'. - Miller (1752: under Thymelaea. No.3).   vc12   **1752**

Daucus carota

* 'Ye wild carot is found abrode in ye feldes'. - Turner (1562: 80).   **1562**

See also Turner (1538: A. iv. verso) 'Daucus creticus ... mihi videtur, anglis esse Wylde Carrot'. Stearn (1965) equates this with either this species or *Pastinaca sativa*. See also Turner (1548: F. i. verso) 'Pastinaca … Carettes growe in all countreis in plentie'.

Deschampsia cespitosa

* 'Gramen segetale … neere unto hedges, and in fallow fields'. - Gerard (1597: 5). 'Gramen segetum panicula speciosa ... In the borders of corne fields, and grounds that have beene plowed'. - Parkinson (1640: 1157).   **1597**

Clarke has the Gerard record as a probable, and Druce agrees. Ray (1660) too, as 'The fair panicled Corn-grasse', cites the Gerard and Parkinson synonyms, but the habitat seems unlikely.

Deschampsia flexuosa

* 'Gramen paniculatum locustis parvis, purpuro-argenteis majus & perenne D. Doody ... a D. Doodio observatum & nobis ostensum est'. - Ray (1696: 258).   **1696**

Deschampsia setacea

as *Aira setacea* 'in ericeto Strattoniensi [Stratton Strawless] in comitatu Norfolk. D. Stillingfleet'. - Hudson (1762: 30).   vc27   **1762**

'It was sent to him [Hudson] by Benj. Stillingfleet who found it at Stratton Strawless in 1755'. - Petch & Swann (1968).   vc27   1755

Smith (1824: i. 104) gives *Aira setacea* as var. β of *A.*[= *Deschampsia*] *flexuosa*. I have found no confirmaton for the date given in Petch & Swann, though Bob Ellis (*in litt.*) notes that Stillingfleet was certainly spending time with Robert Marsham in Stratton Strawless that year as that was when he wrote his *Calendar of Flora*.

Hudson, W. 1762. *Flora Anglica*.

Descurainia sophia

* 'Sophia Chirurgorum ... in parietinis & areis prope urbes ... Galliae, Nortmaniae, Belgiae, & Angliae'. - L'Obel (1571: 328).   **1571**

Dianthus armeria

* 'Ad insulam Sheppey … Armeria sil. altera caliculo foliolis fastigiatis cincto, &c. *Lob*. Cariophyllus montana'. - Johnson (1629: 5). vc15 **1629**

See Britton (1892) for an interesting account of the English name, but also concerning early records.

Dianthus deltoides

* 'There is a Wilde creeping Pinke which groweth in our pastures neere about London and other places, but especially in the great field next to Detford by the path side as you go from Redriffe [Rotherhithe] to Greenewich'. - Gerard (1597: 476). vc16, 17 **1597**

See Britton (1892) for an interesting account of the English name, but also concerning early records and about the lasting confusion that Gerard caused. Even Raven (1947: 209) attributes this record to *D. armeria*.

Dianthus gratianopolitanus

* 'Armeriae species flore in summo caule singulari … (… On Chidderoks [Cheddar] in Somersetshire; by Mr. Brewer)'. - Ray (1724: 336). vc6 **1724**

The BM copy of Ray (1724) was owned by Trimen (dated 1862), who added the following note regarding the comment from Doody, in <u>his</u> own copy of Ray (1696: 199) "this way [was?] gathered (as I think, somewhere near the Peake) by Mr du Bois and communicated to me which I gave to Dr Plukenet". Plukenet died in 1706. vc6 1706

Ray (1690 & 1696) both give 'It grows in England, as Mr Doody informs', but with no location. In his *Fl. Bristol*, White (1912), gives 'On Chidderoks etc' (as in Ray (1724) but with the reference 'Ray (1696)'. White has erred there and we have no idea if Doody saw a wild plant.

Diapensia lapponica

'In 5th July 1951 Mr. C.F.Tebbutt … at c2500 feet on hills overlooking Glenfinnan in the Arisaig district of Inverness-shire'. - Roger (1952). vc97 **1952**
vc97 1951

See also Blakelock (1952).

Digitalis purpurea

* 'Much in Englande & specially in Norfolke about ye coney holes in sandy ground … Foxe gloue'. - Turner (1568: 16). vc27 **1568**

Diphasiastrum alpinum

'Muscus clavatus foliis cypressi, *Bauh. Ger. em.* On the top of Snowdon as also upon Cheviot hills in Northumberland'. - Johnson (1641: 26).	vc49, 68	**1641**
Thomas Johnson and party were on Snowdon, Aug 1639. See Raven (1947: 289).	vc49	1639

Diphasiastrum complanatum

'Hartlebury Common, Worcester, 1837, Rev. C. Babington'. - Druce (1882). 'between 1881 and 1884 by the Rev. H.P. Reader of Woodchester, near Stroud in Gloucestershire'. - Edgington (2013) as *D. x issleri* citing Jermy (1989), who described the plant as *D. issleri*.	vc37 vc37	**1882** 1837

The taxonomy of this species is still confused, with records also assigned to *D. issleri* or the hybrid *D. x issleri*.
The finders were Churchill Babington (C.C. Babington's cousin) and a Miss Lea. See Druce (1916b) for a full historical account, and for a modern note, taking into account taxonomic advances, see Edgington (2013). Apparently Miss Lea's herbarium is lost, and all that remains is a sterile shoot in **CGE** which has been identified as this species. Reader's discovery might be a more satisfactory first record.

Diphasiastrum tristachyum

' … found [in 1866] by a woman named Sarah Young, while occupied in cutting heath for broom-making, at Lower Waggoner's Wells, Bramshott, N. Hants'. - Rumsey (2012).	vc13 vc13	**2012** 1866

Rumsey (2012) adds that the original note was in the Gardeners' Chronicle for 11th August 1866, and that there is another specimen in **BM**, from Ingleborough (vc64) from 1816. The provenance of the latter is uncertain.

Diplotaxis tenuifolia

* 'Eruca sylvestris … groweth in most gardens of itselfe. You may see most bricke and stone wals about London and elsewhere covered with it'. - Gerard (1597: 192).	vc21	**1597**

Dipsacus fullonum

* 'Dypsacos latine labru veneris aut lavacru veneris dicitur … anglorum vulgus vocant a Wylde tasyll'. - Turner (1538: A. iiij verso).	**1538**

The cultivated species, *D. sativus*, was first recorded in 1762, but must have been grown as a field crop before that. See Thirsk (1997).

Dipsacus pilosus

* 'Virga pastoris ... In Anglia Cantia secus vias & suburbia Rhiae Sandvicium [?Richborough, near Sandwich]'. - L'Obel (1571: 374).   vc15   **1571**

The suggestion of Richborough is from Hanbury & Marshall (1899).

Draba aizoides

* 'Discovered by Dr Wm. Turton, in March 1803, growing abundantly on walls and rocks about Pennard Castle, 8 miles west of Swansea in South Wales'. - E.B. 1271.   vc41   **1804**

'John Lucas Esq., of Stout Hall near Swansea, informs us he found it in 1795 near Wormshead, 16 miles west of that town [Swansea]'. - E.B. 1338.   vc41   1795

Draba incana

* 'Paronychia Gnaphalii facie ... At Clapdale in Yorkshire in the midway betwixt Setle and Ingleborough hill on the rocks'. - Merrett (1666: 90).   vc64   **1666**

Draba muralis

* 'Bursae pastoriae loculo sublongo affinis pulchra planta *J.B* ... On the sides of the mountains in several places of Craven in Yorkshire'. - Ray (1670: 49).   vc64   **1670**

Draba norvegica

* as *D. stellata* 'Found in 1789 on Ben Lawers'. - Dickson (1790: 29 & 1794: 288).   vc88   **1790**
   vc88   1789

Drosera anglica

* 'Ros solis sylvestris longifolius ... This was sent me by Mr. Zanche Silliard an Apothecarie of Dublin in Ireland, which sort wee have growing by Ellestmere in Shropshire by the waysides (the report of Dr. Coote)'. - Parkinson (1640: 1053. 2).   vc40, H18   **1640**

Clarke adds 'First found by Heaton 'in a Bogge by Edenderry'. - see How (1650: 105)' and Colgan & Scully (1898) give the same details.
See also Raven (1947: 303), where Heaton claims that <u>he</u> gave Silliard the plant. Bowles recorded this plant (or *D. intermedia*) from Ellesmere in August 1632. See Oswald, in Trueman *et al.* (1995).

Drosera intermedia

* 'Rorella sive Ros solis foliis oblongis ... '. - Johnson (1634: 65).   1634
'On Hinton moor'. - Ray (1660: 189).

Pulteney cites L'Obel (1571: 454, quoad fig. saltem) 'in England the administrative districts of Kent & Somerset so greatly abound with this [plant] that not far from the celebrated Abbey & mount named Glastonbury there is enough to suffice to load a Horse, with a fibrous root, in six or eight flexible reddish little tails, an inch and a half long; the oblong, concave little leaves, resembling a small spoon or ear-scoop'. Both species are found on the Somerset Levels, and both were not uncommon in Murray's (1890s) day, though *D. rotundifolia* was the more common. But who knows what the levels were like 300 years before that? *D. intermedia* would like more open ground, which might have been the case before water levels were dropped. Clarke has a query against the Johnson attribution. Bowles recorded this plant (or *D. anglica*) from Ellesmere in August 1632. See Oswald, in Trueman *et al.* (1995).

Possibly this species, Pena, P. & L'Obel, M. de. 1571. *Stirpium adversaria nova*.

Drosera rotundifolia

* 'Rosa solis is a litle small herbe that groweth in mossey groundes   1568
and in fennes and watery mores'. - Turner (1568: 79).

'in Anglia' - annotations made by T. Penny on C. Gesner's plant   1565
illustrations dated as by 1565 in Foley (2006b)

Dryas octopetala

* 'Teucrium Alpinum Cisti flore ... In the mountains betwixt Gort   vcH15   1650
and Galloway [Galway], Mr. Heaton'. - How (1650: 120).

In common with other species first found by Richard Heaton but   vcH15   1641
not published till 1650 by How, this is reckoned by Walsh (1978) to have been discovered by 1641.

Dryopteris aemula

'*Aspidium dilatatum* var. *recurvum* … plentifully in the neighbourhood of Penzance'. - Bree (1831).	vc1	**1831**
More details are in Bree (1843), where he says that he saw it at Penzance in 1817 and on the Irish mountains some years previously.	vc1	1817
Druce notes 'But Newman cites 'filix montana ramosa argute denticulata (Ray, 1690: 27), from the top of the Glyder'. Edgington (2013) points out that these are errors for *D. expansa*.		

Dryopteris affinis s.l.

Camus (1991) credits Lowe (1891) with splitting *D. filix-mas* into our current three species, with this as *Nephrodium paleaceum*. Manton (1950) confirmed this.	vc100	**1891**
'Filix mas vulgaris … In horto Chelseano & alibi in sylvis, v.gr. Canewood ([Kenwood, Middlesex] invenitur cum pinnulis non serratis; D. Doody'. - Ray (1724: 120). This record is dated by Edgington (2013) to 1700-1705. As var. *affinis*, Charlton Wood, Kent, Mr Buddle'. - Druce (1928: 91).	vc21	1706
Newman (1844, 1854) treated *D. borreri* as a subspecies of *D. filix-mas*. Doody died in 1706, Buddle in 1715.		

Dryopteris affinis s.s.

as *D. affinis* subsp. *affinis*, Fraser-Jenkins (1980).		**1980**
Tarbet, vc99, 1855, T. Moore, det. Fraser-Jenkins, **K**.	vc99	1855

Dryopteris borreri

as *D. affinis* subsp. *borreri*, Fraser-Jenkins (1980).		**1980**
'Filix mas vulgaris … In horto Chelseano & alibi in sylvis, v.gr. Canewood [Kenwood] invenitur cum pinnulis non serratis; D. Doody'. - Ray (1724: 120). Edgington (2013) traces this to a herbarium specimen in **OXF**, and confirms this is this species, dating it to 1700-1705. 'Bute, 1827, Greville, **E**'. - Pugh (1953).	vc21	1705
This was first described in Newman (1854: 189) as *D. filix-mas* var. *borreri*, separating it from var. *affinis*, but it was not until Fraser-Jenkins that the true taxonomy was understood.		

Dryopteris cambrensis

as *D. affinis* subsp. *stilluppensis*, Fraser-Jenkins (1980).		**1980**
Dalry, vc75, 1877, R. Kidston, det. Fraser-Jenkins, **G**.	vc75	1877
There might well be earlier records in herbaria.		

Dryopteris carthusiana

'Filix ad ramosam accedens palustris, muscosa lanugine aspersa, pinnulis acutioribus … In locis paludosis præsertim solo putrido, non solum circa Oxoniam sed & aliis comitatibus Angliae'. - Morison (1699: 579). — vc23 — **1699**

Edgington (2013) cites early records (c.1705) including 'Filicis maris ramosæ plantæ juniores Petroselini facie … Tunbridge Wells'. - Druce & Vines (1907: 49).

Dryopteris cristata

Aspidium cristatum … gathered by the Rev. R.B. Francis in the low boggy parts of the heath between Holt and Hempstead, Norfolk'. - E.B. 2125 (non t.1949). — vc27 — **1810**

'the Lows, Holt Heath, 1805, [incorrectly given as 1810]'. - Smith (1828: 289). — vc27 — **1805**

Edgington (2013) relates the recent discovery of a sterile frond of this species on a mixed sheet in Buddle's Herbarium (Sect iii. f.15) at **BM**. This has no location or other provenance, but must date from before his death in 1715.

There are reports of a specimen from Ling Common, vc28, 1779, Sir J.E. Smith, **LIV**, but this has not been traced.

Dryopteris dilatata

'Filix mas ramosa pinnulis dentatis, Durford Abbey, Sussex; Mapledurham, Hants, J. Goodyer'. - Johnson (1633: 1129). — vc11, 13 — **1633**

For a fuller note on Goodyer's superb description, dated 4th July 1633, see Gunther (1922: 68).

Dryopteris expansa

as *D. assimilis,* Walker (1961), following on from the work in Manton (1950). — **1961**

'Filix montana ramosa minor argute denticulata. Ad summitatem montis Glyder, qua Lacui Lhyn Ogwan imminent'. - Ray (1690: 27). This is Lhwyd's record, dated by Edgington (2013) to 1688. — vc49 — **1688**

VC92, Beinn a Bhuird, 3000ft., NO/0898, Greville, 1831, **E**'. - Crabbe, Jermy & Walker (1970).

Dryopteris filix-mas

'Filix mas'. On a journey from Stoke to Cliffe. - Johnson (1629: 8). — vc16 — **1629**

see also Turner (1538, B) 'Filix [the male one] [is called] a Ferne or a Brak or a Bracon by the English'. Nelson (1959) supports this attribution.

Dryopteris oreades

Lowe (1891), describing this species as *Nephrodium propinquum*. Manton (1950) was the first to satisfactorily resolve the taxonomic status.	vc64	**1891**

'The oldest specimen I have been able to find is one [in **BM**] that was among those that was purchased from the Linnean Society in 1963 – it is marked ex Herb. Withering and is of a frond cultivated at Kew Gardens from material from the north of England. The label bears the name of Brown [Robert Brown] and says the plant is distinct and has been regarded as a variety within the Banksian herbarium. I would imagine this must date to c.1800'. - F.J. Rumsey *in litt*.   1800

See Camus (1991), who credits Lowe with splitting *Dryopteris filix mas* into our current three species, and this as *Nephrodium propinquum*. Note that Newman (1844: 202) mentioned 'a plant from Ingleborough from Rev. Pinder identical to De Candolle's *Polystichum abbreviatum*', which had been named as *Lastrea filix-mas* var. *abbeviata* by Moore.

Dryopteris remota

as *Lastrea remota* 'The Galway record, from woods at Dalystown, S.E.Galway, seems to date from 1898'. - Praeger (1909b).   vcH15 **1909**
   vcH15 1898

Preston *et al.* (2002) say that the earlier Dunbarton record, 1894, is suspect. Edgington (2013) gives a full account, and similarly doubts the Dunbarton record.

Dryopteris submontana

as *Asplenium spinulosum* var. γ 'North Yorkshire'. - Hooker (1830: 444). '*Aspidium rigidum*. Rev W.T. Bree … on Ingleborough'. - E.B.S. 2724 (1837).   vc64 **1830**

Lees (1888: 510) gives a record 'On Ingleborough, near the hill foot towards the village, in 1815 - Bree in Francis (1837)', adding 'the village was Clapham; the locality meant either Clapdale or Moughton Scars'. This would be the record cited in Newman (1844).   vc64 1815

Echium plantagineum

'*Lycopsis C.B.* In the sandy grounds near St Hilary [Jersey], W. Sherard'. - Ray (1690: 238).   vc113 **1690**

Sherard's visit was in 1689, and Ray, in a letter to Lhwyd in 1689, refers to this. See Gunther (1928: 197).   vc113 1689

The first record other than from the Channel Islands was 'Wandsworth, 1850'. - Irvine (1859) as a casual. The persistent site near St. Just, nr. Cape Cornwall, vc1, was first discovered by J. Ralfs in 1872 (Lomax (1873) in a 'sandy, weedy field on high ground not far from the sea'.

Echium vulgare

* 'Anchusa … the second kind … groweth in gravylly and sandy places … called in some places of Englande cattes tayles, in other places wylde buglose'. - Turner (1551: C. v). — **1551**

Nelson (1959) is alone in ascribing Turner (1538: A. iv.) 'Cirsion … Cattes tayle' to this species (others choosing *Picris echiodes*), though it does seem odd that those are happy to assign the same English name to this species in Turner's later works. Clarke adds 'Echium is called 'Cats-tail' in Cambs'. See Martyn (1794:136).

Elatine hexandra

* as *E. Hydropiper* 'The Rev. Mr. Williams found it flowering in August 1798 about the eastern shore of Bomere Pool near Condover, Shropshire'. - E.B. 955. — vc40 **1801** / vc40 1798

Elatine hydropiper

* 'Detected by the writer of this, in August last, [1830] on the S.E. side of Llyn Coron, near Abberffraw, Anglesea'. - J.E. Bowman in E.B.S. 2670. — vc52 **1830**

Cuttmill Pond, vc17, 1815, H. Groves, det. Bromwich, **E**. — vc17 1815

Eleocharis acicularis

* 'Juncellus omnium minimus capitulis Equiseti … It grows in Binsey Common in the moist ditches next to the River Isis'. - Plot (1677: 145). 'Hunc a se observatum mihi communicavit … D. Dodsworth'. - Ray (1688: 1306). — vc23 **1677**

Eleocharis mamillata

'Material… was first collected in Britain by Mr N.Y. Sandwith in July 1947 "in a marshy ox-bow of the R. Wharfe below Buckden", Upper Wharfedale, vc 64'. - Walters (1963). — vc64 **1963** / vc64 1947

Eleocharis multicaulis

* 'On a bog near the house of Mr M'Kinnon, at Corryhattachan, Isle of Skye, discovered by Mr. John Mackay in 1794'. - Smith (1800: i. 49). — vc104 **1800**

'An Juncello accedens graminifolia Plantula capituli Armeriae proliferae. Gathered near Tunbridge Wells in Kent, Du Bois'. - Druce (1928: 486). — vc16 1740

Du Bois died in 1740. Corryhattachan is probably Coire-chat-achan, NG6222, near Broadford (S.J. Bungard, *pers. comm.*).

Eleocharis palustris

* 'Juncus aquaticus minor capitulis Equiseti *Lob*. Club Rush'. - Johnson (1633: 35. 5 & 1631, with a figure). **1633**

Eleocharis parvula

Edinburgh Catalogue (1841). vc11 **1841**

'Collected in 1837 at Lymington, Hants, in pits dug for the newly excavated swimming baths, with *Scirpus savii* [*Isolepis cernua*]. - G.E. Smith in **BM**'. - Clarke (1900). vc11 1837

Babington (1842) adds 'Found in Hampshire, and probably overlooked in other places owing to its minuteness'. Townsend (1904) gives the date of the **BM** specimen as 1835, but this is not the case.

Eleocharis quinqueflora

'Gramen cyperoides minimum. Caryophylli proliferi capitulo simplici squamato … in paludis locisque putridis'. - Morison (1699: 245 n.40). **1699**

Druce (1909b) shows that the plant referred to in Ray (1690: 210) as 'Graminifolia plantula Alpina capitulis Armeriae proliferae. In pascuis ad radicem excelsae cujusdam rupis *y Clogwyn du ymhen y Glyder* in agro Arvoniensi. D. Lloyd,' was probably this.

Eleocharis uniglumis

* 'Aberdeen, Dr. Dickie'. - Babington (1847: 349). vc92 **1847**

Shelford Common, J.S. Henslow, 6.6.1825, (as *E. palustris*), det. S.M. Walters, 1956, **CGE**. Fen nr. Clayhithe, W.A. Leighton, 13/5/1834 (as *E. palustris*) det. S.M. Walters, 1948, **BM**. vc29 1825

Babington (1897: 285) refers to this in a letter to Balfour, Feb 8th, 1842, in respect of a specimen from Barvas [Isle of Lewis, Outer Hebrides].
See 'Gramen iunceum marinum dense stipatum, Pusshye grasse: in the sandes by the Castles betweene Sandwyche and Dover'. - Manuscript notes of William Mount, of c.1584, in Gunther (1922: 255 - 263), identified by him as this species.

Eleogiton fluitans

'Juncellus capitulis equiseti fluitans … In aquosis Angliae fluitat, unde D. Gillenius attulit'. - Bauhin (1620: 23). **1620**

Britten (1909) includes all that is known of Arnold Gillen and his visit to England, though that article is mainly concerned with records of Rev. M. Dodsworth.
See also the note under *Littorella* for the identity of a Goodyer record cited in Gunther (1922).

Elymus caninus

* 'Gramen caninum aristatum radice non repente sylvaticum. Found plentifully growing in Stoken-Church Woods'. [Oxon.] - Bobart in Ray (1690: 235).   vc23   **1690**

This record was in the appendix, from J. Bobart.

Elytrigia atherica

* 'Gramen maritimum, spica loliacea, foliis pungentibus nostras *Pluk. Alm. 173* ... Prope Sherness in insula Sheppy & ad Delkey [Dell Quay] prope Chichester'. - Ray (1724: 391).   vc13, 15   **1724**

Elytrigia campestris

as *E. repens* subsp. *arenosa,* Trist (1995).   **1995**

Pagham, vc13, 1875, H.E. Fox, **OXF**.   vc13   1875

Trist (*op.cit*) refers to an earlier reference for its occurrence here in Melderis (1980), but there is nothing there other than an inclusion of Britain in the European distribution.

Elytrigia juncea

'Gramen caninum geniculatum marinum, *Lob.*'. Margate. - Johnson (1632: 3).   vc15   **1632**

Rose, in Gilmour (1972: 104), so interprets this (as *Agropyron junceiforme*). Clarke's first date is 'Gram. spica Tritici mutici, Bauh'. Between Margate and Sandwich. - Johnson (1632: 19), adding that Hanbury & Marshall identify this with *E. juncea*. But Rose (*loc. cit.* p114) suggests that reference is more likely to be *E. atherica*. All slightly unsatisfactory.

Elytrigia repens

* 'Gramen caninum ... In gardens and arable grounds, as an infirmitie and plague, or the fields'. - Gerard (1597: 22).   **1597**

See also 'Gramen canarium, … Couche grasse … groweth in Addington in Kent'. - Manuscript notes of William Mount, of c. 1584, in Gunther (1922: 255 - 263), and identified by him as this species. 'Gramen canarium' is slightly odd - C.D. Preston (*in litt.*) wonders if this is a transcription error.

Empetrum nigrum

* 'Erica baccifera procumbens ... I found this growing in great plenty in Yorkshire on the tops of the hills of Gisbrough, betwist it and Rosemary-topin [Guisborough and Roseberry Topping]'. - Johnson (1633: 1383).   vc62   **1633**

Raven (1947: 274) identifies this with Johnson's trip to Yorkshire in 1626.   vc62   1626

Clarke notes that 'Gerard (1597: 1199) refers to a 'Heath which beareth berries ... in the North part of England', but says the berries are 'red'.

Epilobium alsinifolium

* 'Lysimachia siliquosa glabra minor latifolia ... In the rivulets on the sides of Cheviot-hills'. - Ray (1677: 194).   vc68   **1677**

This was on Ray's trip to the north of England in 1671, with T. Willisel. See Raven (1950: 155) as 'probably' this species.   vc68   1671

Epilobium anagallidifolium

* 'On Ben Lomond'. - Lightfoot (1777: 199).   vc86   **1777**

Assuming this is Lightfoot's record, then he climbed this mountain on 13th June 1772, see Bowden (1989: 164 & 220).   vc86   1772

Lightfoot also recorded it from Little Loch Broom, presumably from his visit there in July 1772 (Bowden 1989: 164 & 223).

Epilobium hirsutum

* 'Lysimachia siliquosa ... Neere the waters (but not in the waters) in all places for the most part'. - Gerard (1597: 388). Stoke to Cliffe. - Johnson (1629: 8).   **1597**

There are two other possibilities. 'Lysimachia...[leaves] are thinne, and in fasshō lyke wylow leves ... the flour is darck rede or of ye color of golde. It groweth in watery and in marrishe and fennish groundes. Thys is a very comen herbe in Germany and England'. - Turner (1562: 43b), identified as this species by Chapman *et al.* (1995: 110). Pulteney cites Lyte (1578: 72) 'Lysimachium purpureum primum, the first purple red, willow herbe, also the sonne before the father'. That last seems confusing, as surely it refers to something like *Tussilago*?

Epilobium lanceolatum

Edinburgh Cat. (1841).   **1847**

'Guernsey. Gosselin, J., c.1790'. - see McClintock, (1982: 87).   vc113   1790

See also Babington (1842), where he equates his record of *E. montanum*, var. γ. *lanceolatum* to this species. McClintock (1982: 87) says 'only split off in 1818', but the source of this reference has not been traced.

Epilobium montanum

* 'Lysimachia siliquosa ... [this] varietas ... In Anglia observatur ... locis ...umbrosis, saxosis aut minus udis'. - L'Obel (1571: 145). 'Hard by the Thames ... as you go from a place called the Devils Neckerchiefe to Redreffe [Rotherhithe]'. - Gerard (1597: 388).   **1571**

The Devils Neckerchiefe was at St Saviour's Dock, on the south shore of the Thames, just downstream of Tower Bridge. It was so named because pirates were hung there.

Epilobium obscurum

* 'Wyken, Warwickshire; Ilfracombe, Devon; Llanthony, Monmouthshire; and Sussex; and I am informed by Mr Borrer that it is found in Herefordshire by Mr Purchas'. - Babington (1856: 243). — vc4, 35, 36, 38 — **1856**

'Hampsted Bog [Middlesex. = *E. obscurum*, Schreb.]'. - Druce & Vines (1907: 97). Druce notes that the specimen might have been transposed with the next, from Isle of Anglesey. — vc21 — 1747

Druce adds 'but see Deakin (1845: 548) *E. virgatum*, marshy places about Lincoln, rare'. This record is cited for this species in Gibbons (1975).

The record from Druce & Vines is dated as 1747, the date of Dillenius' death.

Epilobium palustre

* 'Lysimachia siliquosa glabra minor angustifolia ... On Teversham Moor'. - Ray (1660: 93). — vc29 — **1660**

Clarke adds 'But see Johnson (1633: 479)'. It is not clear to which species there he is referring to, but there is 'Lysimachia siliquosa glabra minor angustifolia' (also in Gerard (1597: 386)) which is added at the end of the account as an unillustrated species. It is possible that this is the species. Pulteney concurs, but then again he might have been describing a Flemish species.

Epilobium parviflorum

* 'Lysimachia siliquosa minor hirsuta'. Stoke to Cliffe. - Johnson (1629: 8). — vc16 — **1629**

Epilobium roseum

* 'In uliginosis; primum in Anglia a cel. Curtisio in Lambeth Marsh in comitatu Surr. detecta, et nuper ad Morton prope Ongar in agro Essexiense invenit Dr. E. Forster jun.'. - Symons (1798: 199). — vc17, 18 — **1798**

'Lysimachia siliquosa latifolia glabra altera minor. Found by Mr Rand in Kent'. - Druce (1928: 472). — vc16 — 1743

Rand died in 1743. There is also a specimen cited in Druce & Vines (1907: 149).

Epilobium tetragonum

* 'Lysimachia siliquosa glabra minor, *Bauh*. In humidis saxosis. Little-codded Willow-herbe' - Johnson (1634: 49). — **1634**

Epipactis atrorubens

* 'Helleborine flore albo-rubente, *Park*. ... On the sides of the mountains near Malham 4 miles from Settle in great plenty, as also in many places thereabout'. - Ray (1677: 157).   vc64   **1677**

This was on Ray's trip to the north of England in 1671, with T. Willisel. See Raven (1950: 154).   vc64   1671

See How (1650: 57) 'Helleborine flora atro rubente, *Park*. Wild white Hellebore with dark red flowers: found by Lysnegeragh, Mr. Heaton'. Raven (1947: 302) accepts this as this species, but cites Colgan & Scully (1898: 342), who are more cautious, writing '[How's record] should, perhaps, be referred to this plant. The locality ... is probably near Roscrea in King's Co. [H18]'. In common with other species first found by Richard Heaton but not published till 1650 by How, this is reckoned by Walsh (1978) to have been discovered by 1641.

Note that Ray (1670: 162) also has an earlier record of 'H. flore atro-rubente. Near Sheffield in Yorkshire'. This same species is referred to in a letter from Ray to Lister on 10th Sept. 1668 as 'Helleborine flore atrorubente variis in locis'. But this 1670 record is doubted in Lees (1888), and not included in Wilmore *et al.* (2012). The nearest recent records to Sheffield are in the Peak District, not that distant.

Epipactis dunensis

as *E. viridiflora* f. *dunensis* 'Southport'. - Stephenson & Stephenson (1918).   vc59   **1918**

'… widely scattered on the sand-dunes from Hall Road, in South Lancashire to South Shore in West Lancashire…'. - Wheldon & Travis (1913).   vc59, 60   1913

There is an earlier specimen 'Southport, 5/8/1869, J. Barrow, **MANCH**', which, although quite possible, has not yet been retraced. But note too Michell & Smith (2012), who argue persuasively for earlier records from the S. Lancashire coast. They note records of *E. latifolia* in Whittle (1831) and Hall (1839), from 'a moist grassy spot, among the sand-hills between Waterloo and Crosby' and whilst these might refer to *E. helleborine* and *E. phyllanthes*, Phil Smith (*in litt.*) argues against both of those on habitat grounds.

Epipactis helleborine

* 'The "Satyrion" described by Dioscorides having a whyte floure like a lyly ... I have sene it ... in England in Soffock'. - Turner (1562: 128b). 'In the woods by Digswell pastures halfe a mile from Welwen in Hartfordshire'. - Gerard (1597: 358).   vc27   **1562**

Both Clarke and Druce incorrectly cite p.158 of Turner.

Epipactis leptochila

as *var. nov. leptochila* [of *E. helleborine*] 'on July 29th, 1918, a few miles from Guildford'. - Godfery (1919).	vc17	**1919**
'Gathered in woods at Bomere pool, Salop, 1835'. - Leighton (1841: 435), as *E. viridifolia*. Cf. Babington (1843: 295) under *E. media*.	vc40	1835

Babington and Leighton knew they had something different, but Godfery's descrition seems the first satisfactory description. Leighton's record is not mentioned in Sinker *et al.*(1985) but is included in Lockton & Whild (2015), who also give a record from the same site for 1832.

Epipactis palustris

* 'Tertiae Clusii Helleborines similem facie, riguis pratis pagi Mary-cray [St. Mary Cray] vacati erui elapsa aestate 1601 septem Londino miliaribus Anglicis'. - L'Obel (1655: 94).	vc16 vc16	**1655** 1601

Raven (1947: 292) claims that Robert Abbot had found this species at Hatfield, giving a reference to Gerard (1597: 175.4) 'was found (by a learned preacher called master *Robert Abbot*, of Bishops Hatfield) in a boggie grove where a conduite head doth stand, that sendeth water to the Queenes house in the same towne'. The illustration there is satisfactory, the text just about so. None of the Floras of Hertfordshire mention this record.

Epipactis phyllanthes

'Gerald Smith in the Gardeners' Chronicle 1852, p.660 'This pretty plant was gathered from a single locality, upon the elevated part of Phillis Wood, near Westdean, Sussex … in Sept 1838. Original spec at **K**'. - Young (1952).	vc13	**1952**
Waverley [nr. Farnham], 15/8/1838, Miss Parker, det. D.P. Young and P.D. Sell, **CGE**'.	vc13, 17	1838

Young seems to have been the first to sort out the confusion in a satisfactorily manner. Wolley-Dod (1937: 424) gives an 1839 record.

Epipactis purpurata

* '*E. purpurata* … Parasitical on the stump of a maple or hazel in a wood near the Noris farm at Leigh, Worcestershire, in 1807, Rev. Dr. Abbot'. - Smith (1828: 41).	vc37	**1828**
'An Helleborine latifolia montana *C.B.* In the woods at Tunbridge Wells'. - Druce (1928). This is from Du Bois' herbarium and Druce adds 'If, as I think it is, correctly identifed, it is the first Kentish record, and one of the earliest British examples'. Du Bois died in 1740.	vc16	1740

Maskew (2014) states that it is more likely that Abbot's site was in Norrest Woods, vc37.

Early records of this species are mired in confusion. Note particularly the discussion under *Orobanche purpurea*, where the attribution of Goodyer's record (Johnson (1633: 228)) to this species seems far more satisfactory were it not that that it was noted as growing on the edge of a field. Gunther (1922: 47) sets out both sides of the argument, and at this distance it cannot be resolved.

Other records include Clarke, who gives 'Wood near Tring, Herts, June, 1808. - Dickson, BM'; Raven (1947: 210), who identifies one of the three 'Wilde white helleborines' that Gerard (1597: 358) has 'in a woode five miles from London … Lockridge' as this species, and Raven again (1950: 115), in notes on Ray's 1660 Itinerary 'H. flore atro-rubente … Meadows and pastures around Sheffield'. This last is discussed under *E. atrorubens*.

Epipactis sancta

'*E. peitzii* var. *sancta* var. *nova* … Britannia, Cheviotland, Holy Island, alt.s.m. 5m,12.VII.1994'. - Delforge (2000). Raised to a species in Delforge & Gevaudan (2002).	vc68	**2000**
'… coastal dunes of Holy Island … belonging to *E. dunensis* … A. J. Smith, 1958'. - Swan (1993).	vc68	1958

Epipogium aphyllum

* as *E. gmelini* 'communicated to me on the 9th of this month [September], as found a few weeks since by Mrs. W. Anderton Smith from the Rectory, Delamere, Bromyard [Herefordshire]'. - Hooker (1854).	vc36	**1854**

See also Crotch (1855), where he says he received a specimen on 29th July 1854.

In his letter Rev Anderton Smith added 'I was fortunate in finding a considerable mass of it … as the banks were very much trampled … I decided on digging it up, and planting it in a similar spot in our own grounds'. Both Clarke and then Druce gave 1842 as the first record, citing Hooker. There is no justification for this.

Equisetum arvense

'Equisetum segetale. Corne Horse-taile … which for the most part groweth among corne'. - Gerard (1597: 956).	**1597**

See also Turner (1548: F. iiij. verso) 'Polygonum … short Shaue grasse … groweth in many places by water sides, & some time amōg the corne'. That last can only be *E. arvense*. Druce cites the wrong page of Gerard and the wrong edition!

Equisetum fluviatile

'Equisetum nudum laevius nostras … this is the naked species which grows commonly in England'. - Doody in Ray (1690: 244).

1690

Kent (1975) cites a record from Tottenham, from Johnson (1638), who listed two horsetails 'juxta Tottenham', one of which was 'Equisetum nudum *Ger. junceum Trag.* Naked Horse-taile'. This was the same polynomial as in his 1634 work.

vc21   1634

Although this was included by Gerard (1597: 951) as 'E. nudum, unlocalised', in Johnson (1634, 1638) and in other 17th century works, it was always confused with *E. hymale*. These include 'Equisetum nudum *Ger.*'. - Ray (1660: 48). Doody was the first to satisfactorily distinguish between the two. For an elegant exposition of the problems see Oswald & Preston (1998).

Equisetum hyemale

'Equisetum primum *Math.* quo utuntur arcuarii ad sagittas poliendas, At Skippon and Craven in Lancashire [*sic*], and in Rigby Woods'. - Merrett (1666: 36).

vc64   1666

Edgington (*in litt.* 16/3/10 & 2013: 63) gives the earliest localised record from Merrett (actually Willisel's) from the Craven area (but see Lees, 1888, who interprets this record as *E. telmateia*). Gerard's (1597: 956) 'E. nudum' was cited by Druce as his first record but seems most unlikely - see also Oswald & Preston (1998).
Turner (1538: B. iij) has 'Hippuris is called Shauynge gyrs [Shaving Grass] in English … Horsetayle, Dyshewashyges' and Jackson (1959) notes the rough stalks of *E. hymale* were bound together for use as scrubbing brushes - hence the vernacular name. This might indicate an earlier record, but more likely one of cultivation. However de Winter (2015) cites references of the use of 'shave grass' from English sources back to the 14th century.

Equisetum palustre

'Equisetum palustre, *Lob. Ger.* … in palustribus solo putrido & spongioso'. - Ray (1686: 129.4, 1688a: 5).

1686

A very interesting possible early record is in L'Obel (1571: 355) 'Hippuris minor fontinalis … uti in Anglia passim vivis ipsis scaturiginibus trans undantem aquam eruperis prope Castell Cary'. Edgington (2103: 66) discusses this, suggesting that it could be either *E. palustre* or a branched form of *E. fluviatile*.
Gerard (1597: 956) had 'E. palustre, Water horse-taile … that growes by the brinks of rivers and running steams … often in the middest of the water', but it was not until Ray that the differences were clearly set out - he listed five differences from *E. arvense*.

Equisetum pratense

'*E. Drummondii* … Scotland; banks of the Isla and the Esk, in Forfarshire'. - Hooker (1830: 454).	vc90	**1830**
'ex herb [R.K.] Greville, 1824. **E**'. - Ingram & Noltie (1981).	vc90	1824

Equisetum ramosissimum

S. Lincolnshire (vc53): near Boston, in long grass by river. Seen along a limited stretch only, 24th July,1947, H.K. Airy Shaw'. - Alston (1949).	vc53	**1949**
'Equisetum nudum ramosum, Buddle, I gathered it on Hounslow heath'. - Herb. Sloane 117: 11 in **BM**, det Acock & Rumsey.	vc21	1715

This is from his herbarium in the Sloane collection. Buddle died in 1715. For a full account see Rumsey & Spencer (2012).

Equisetum sylvaticum

'Equisetum sylvaticum. Wood Horse-taile … that growes in woods and shadowie places'. - Gerard (1597: 957).	**1597**

Another early record is 'Equisetum minus omnium tenuifolium. In collibus Altæ Portæ, vulgo Highgate, tribus aut quatuor a Londino miliaribus Anglicis in pratis occurrit, etiamque sylvosis asperginibus'. - L'Obel (1655: 149). L'Obel died in 1616.

Equisetum telmateia

'Equisetum majus. Great Horse-taile … a cubite high'. - Gerard (1597: 955).	**1597**

Equisetum variegatum

'Mr George Don, who gathered it, in July 1807, on the sands of Barry, on the sea coast of Angusshire'. - E.B. 1987.	vc90 vc90	**1809** 1807

Ray (1724: 130), citing a record from Thos Robinson (in an *Essay towards a Nat. Hist. of Westm. and Cumberl. p92*), has 'E. nudum minus variegatum Basileense. This was first shew'd to Mr Lawson at Great Salkeld… on the Banks of the River Eden [Cumberland]'. This is possibly this plant, but Newman (1843b: 728) identifies this as *E. arvense* which does seem much more likely, and is followed by Whittaker (1986).

Erica ciliaris

* 'Sent from a bog near Truro by the late Rev. J. S.Tozer to Dr. Greville, 1828'. - Lindley (1829: 174).	vc1 vc1	**1829** 1828

Long known to Sir Charles Lemon [of Carclew, Cornwall] on a heath at Carclew, nr. Penryn, and also on a heath in the parish of St. Agnes'. - E.B.S. 2618, 1830. [Druce also cites E.B.S. 2618, but dates it 1812]. Miller (1768) includes it in the four British Heaths.

Erica cinerea

'Erica Coris folio sexta ... In Anglia supra Windesoram [Windsor]'. - L'Ecluse (1576: 114). 'E. tenuifolia ... Hampstead Heath'. - Gerard (1597: 1199).	vc22	**1576**
Pulteney (1799) states that L'Ecluse saw this at Windsor on his visit in 1571, and, indeed, this was his only known visit before 1576. Druce (1897: 330 & xcvii) cites L'Ecluse's record, but refers only to the 1601 edition.	vc22	**1571**

Erica erigena

'in Connemara [Urrisbeg Mountain]'. - Mackay (1830).	vcH16	**1830**
There is a specimen of this plant in the Morisonian Herbarium in the University of Oxford. It is labelled 'Erica foliis juniperinis D. Lhwyd ex Hibernia". - Nelson (1979a). Lhwyd's visit to Connemara was in 1700.		1700

Erica mackayana

* 'Discovered by the son of the innkeeper (Macalla) at Roundstone, Cunnemara and communicated to Mr Mackay'. - Hooker (1835b: 158).	vcH16	**1835**
'… gathered by the writer [C.C. Babington] on Sept. 2, 1835, by the side of the road between Crushtour and Ballinaboy … Galway'. This from Garry (1903-4: 120).		

Erica tetralix

* 'Vulgatior Ericae folio Myricae. Pumilla caliculato Unedinis flora ... saxosis montibus Angliae occiduae ad Bristoiam exilior … fruticat' - L'Obel (1571: 447).	vc34	**1571**

Erica vagans

* 'Erica foliis Corios multiflora ... By the way-side going from Helston to the Lezard-point in Cornwal, plentifully'. - Ray (1670: 101).	vc1	**1670**
'on Goonhilly Downs near the Lizard Point is a kind of heath which I have not seen elsewhere in England'. - Lankester (1846: 190, n) in a note on Ray's 1667 Itinerary. See also Raven (1950: 144).	vc1	1667

Erigeron acris

* 'Coniza caerulea acris'. Between Faversham and Gravesend. - Johnson (1632: 30). 'I first observed it ... by Farningham in Kent'. - Johnson (1633: 485).	vc16	**1632**
'Conyza coerulea acris. On the walls of Winchester in Hampshire'. - Goodyer ms.11, f.95, see Gunther (1922: 162).	vc12	1621

This is recorded as 'Conyza maior?' at Llewenny, in a list of annotations of plants found his native county by Sir John Salusbury, of Denbighshire, in his copy of Gerard (1597), dated to between 1606 & 1608. See Gunther (1922: 243).

Erigeron borealis

* On a tour through the Highlands, in 1789, 'on Ben Lawers'. - Dickson (1790: 29 & 1794: 288).	vc88	**1790**
'this was found in the clefts of a high rock on the East Side of Ben-Lawers, by Mr Stuart, who communicated fair specimens 1787'. - Mitchell (1992) citing Lightfoot's annotated copy of his *Flora Scotica*, now at RBGE.	vc88	1787

see also 'Mr Stuart from high rocks East-side of Ben Lawers, 87'. - Hope, 1768, in Balfour (1907). Fletcher (1959) dates this as found by 1768, surely an error, presumably taking the date of the original compilation rather than the additions made subsequently - see introduction.

Eriocaulon aquaticum

* 'It was found, September 1768, in a small lake in the island of Skye, by James Robertson'. - Hope (1769).	vc104	**1769**
'In a small lake by the road side leading from Sconsar to Giesto in Skye, 11 Sept.1764. Sir John Macpherson, who, indeed, first noticed it, leaped from his horse, waded into the lake, and brought it out'. - Dr. Walker in Hooker (1821: i. 270). However Mabberley (1985: 66) gives the date as 1768 in a letter from W. Wright to Robert Brown, 15th Feb. 1801.	vc104	1764

Sconser is just east of Sligachan, and there is a Gesto House, at Bracadale some way to the west. The lochan is given as as Loch-na-Caiplich [L. Mòr na Caiplaich, by the road just west of the Sligachan Hotel] in Hooker (*op.cit.*).
There is an entry for this species from J.R. [J. Robertson] for 1768 in Hope, 1768, see Balfour (1907).

Eriophorum angustifolium

* 'Gramen tomentarium … Upon a bog at the further end of Hampsted Heath … it groweth likewise in Highgate parke neere London'. - Gerard (1597: 27).	vc21	**1597**

Eriophorum gracile

* 'Near Halnaby, Yorkshire'. - Woods (1835a: 290).	vc65	**1835**

Halnaby is NE of Richmond.

Eriophorum latifolium

* as *E. polystachion* 'I found this first in bogs in Northamptonshire'. - Dickson (1794).	vc32	**1794**
'Linagrostis panicula minore *Tourn. 664*. From Mr Stonestreet'. - Druce (1928: 487).		1716

Stonestreet died in 1716. Dickson's record was first noted in 1792 (Druce, 1930a). Desmond (1994) states that A. Bruce was the discoverer of this (as *E. pubescens*) but his herbarium sheet is dated 1793.

Eriophorum vaginatum

* 'Gramen junceum montanum subcaerulea spica. Mossecrops'. - Johnson (1641: 23). 'Idem [i.e. Gramen junceum] cum cauda leporina Bauhini'. - How (1650: 53). 'Gramen plumosum elegans ... In Ellesmer Moores [Ellesmere, Shropshire], Dr. Bowle'. - How (1650: 54).		**1641**
Bowles' only known trip to Shropshire was in 1632.	vc40	1632

Erodium cicutarium

* 'Geranium … Pinke nedle or Cranes byl'. - Turner (1548: D. iij).	**1548**

Clarke refers to this as a possibility, but Druce and Stearn (1965) are more happy. Clarke's first date was 'Geranium Cicutae folio inodorum'. - Gerard (1597: 800, fig. 8).

Erodium lebelii

Baker & Salmon (1920).		**1920**
Seaforth Common [Crosby], July 1866, H.S. Fisher, det G. Hutchinson, **MANCH**. 'Penally Burrows, Tenby, 1873, C. Bailey (Hb. Manchester)', cited in Baker & Salmon's paper.	vc59	1866

Erodium maritimum

* 'Geranium Betonicae folio, Over against Saint Vincents Rocks on the farther side of the River, and at Bass Castle [Boscastle] in Cornwall, and betwixt Milford Haven and Haverford West, on old walls by the way side'. - Merrett (1666: 46) - *see image p.196*.	vc2, 6, 45	**1666**
'In insulis Monensi et Prestholmensi'. - see notes of Ray's 1660 Itinerary in Raven (1950: 114n).	vc52	1660

Priestholm is the old name for Puffin Is, off Anglesey, formerly Mona.
Another early record of interest is in Petiver (1710), who records 'Small Sea Cranesbill … the first discovery of this plant is owing to Dr. Morison … who found it in stoney places about Chadder in Somersetshire'. See Morison (1680: sec. 5, 512). But Cheddar is an inland site and the record is unlikely to be before Ray's of 1660. White (1912) merely cites the Petiver record with no comment.

(46)

Geranium malacoides fol. diffectis minimum, *P*. 707.

Geranium fanguinarium, G. 945. On Saint *Vincents* Rocks, and *Ingleborow* hill.

Geranium Rupertianum, *Herb Robert*, G.939. Geran. Robertianum vulg. *P*.710. in umbrofis & fepibus, variat flore albo.

Geranium fexatile, G. 938 *P*. 707. in muris antiquis & faxorum fiffuris. Betwixt *Branford* and *Sion-houfe* plentifully.

* *Geranium* Betonicæ folio, Over againft Saint *Vincents* Rocks on the farther fide of the River, and at *Bafs Caftle* in *Cornwall*, and betwixt *Milford Haven* and *Haverford* Weft, on old walls by the way fide.

Gladiolus paluftris Cordi, *Water fword flag*, G. 29. Juncus floridus, *Flouring rufh*, P. 1197.

Glandes terræ, peas *Earth-nut*, G. 1236. Lathyrus arvenfis, five terræ glandes, *P*. 1061.

Glaſtum fativum, *Garden woad*, G. 491. *P*. 600. feritur in lætioribus folis.

Glaux Diafcordis, Diafc. his *Milk-tare*, G. 1242. Glaux Hifpanica Clufii, *P*. 1095. in montofis.

Glaux exigua maritima, *Black faltwork*, G. 562. *P*. 1283. In the *Downs*. vid. Icon. Dod.

Glychyrhiza echinata Diafc. *Hedge-bogg licorice*, G. 1302. *P*. 1099. 'I is planted in many places.

Glychyrhiza vulg. G. 1302. Glychir. vulgaris filiquata, *P*. 1099. in quibufdam fylvis, Common in *Yorkſhire*.

Gnaphalium

Erodium maritimum from Merrett, C. 1666. *Pinax Rerum Naturalium Britannicarum*.

See species account : Erodium maritimum, p.195

Erodium moschatum

'Geranium moschatum ... grow[s] wilde in many places of this land'. - Parkinson (1640: 709). 'Geranium moschatum *Ger. Park.* ... In Craven-Common, and near Bristow on a little Green you pass over going thence to S. Vincents Rock'. - Ray (1670: 132).

1640

Lees (1888:182) casts doubt on Ray's Craven record, pointing out that he did not see it himself (it seems to have come from Dr Lister, who said it was very common) and that it is odd that it was abundant at that time and never seen again. Lees (above) adds though that there is a specimen in Hb. Buddle which is genuine. The Bristol record is not mentioned in Ray's 1662 Itinerary 3, but see Raven (1950: 127) dating the discovery to this trip.

Erophila glabrescens

Druce (1930). **1930**

Delver, Fulneck [Pudsey, Leeds], vc63, 1816, det. Rich, **DBN**.   vc63   1816

There were many attempts to identify and fit the British species into the French research, commencing with Baker (1858) but Druce's is the first satisfactory account.

It has not been possible to trace Delver, but the specimen is from the herbarium of Dr S. Litton, who was born in Liverpool and made frequent visits back from Ireland.

Erophila majuscula

Druce (1930). **1930**

Woolwich, vc16, 1724, det. Rich, **OXF**.   vc16   1724

There were many attempts to identify and fit the British species into the pioneering French research, commencing with Baker (1858) but Druce's is the first satisfactory account.

Erophila verna s.l.

* 'Paronychia vulgaris. Common Whitlow Grasse ... Upon bricke and stone wals, upon olde tiled houses, which are growen to have much mosse on them ... upon the bricke wall in Chancerie Lane belonging to the Earle of Southampton in the suburbes of London'. - Gerard (1597: 500). 'Paronychia vulgaris alsine folio. Common Whitlow Grasse'. - Parkinson (1640: 555).   vc21   **1597**

The main Gerard entry seems to be for this species but the location for Chancery Lane could equally apply to this or to *Saxifraga tridactylites*, or both. Kent (1975) gives the location to the latter.

Erophila verna s.s.

Druce (1930). **1930**

'Bursa pastoria minima oblongis siliquis, sive verna loculo oblongo *J.B*', in Hb. Sloane, 123: 2, **BM**.   1715

The Hb. Sloane specimen is in the Herbarium of Buddle, who died in 1715.

Eryngium campestre

* 'Eryngium vulgare, *J.B* ... On a rock which you descend to the Ferrey, from Plymouth over into Cornwall'. - Ray (1670: 105).   vc3   **1670**

'Found by Ray on July 7, 1662'. - Lankester (1846: 195).   vc3   1662

Eryngium maritimum

* 'Eryngium ... sea Holly, it groweth plentuously in Englande by the sea syde'. - Turner (1548: D. j).   **1548**

<1379. 'est half of Britlond on the see sandys'. - Henry Daniel, see Harvey (1981), and also Allen (2001).

Erysimum cheiranthoides

* 'Camelina - Treacle Wormseed ... wilde in sundry places of England'. - Gerard (1597: 213). 'About one mile from Redding'. - How (1650: 19).   **1597**

Erysimum cheiri

* 'Viola alba ... one is called in english Cheiry, Hertes ease or Wal Gelefloure ... groweth upon the walles'. - Turner (1548: G. viij.) 'Leucojum luteum ... Hic in Anglia uti plerasque alias nataliciis christi saepius ... vidimus'. - L'Obel (1571: 141).   **1548**

L'Obel's text suggest it is frequently blooming at Christmas.

Euonymus europaeus

* 'Euonymos ... I have sene it betwene Barkway and Ware in the hedges ... Spyndle tree'. - Turner (1548: D.j).   vc20   **1548**

Pulteney (1799) relates how it was Turner who coined the English name.

Eupatorium cannabinum

* 'Eupatorium vulgare ... water Hemp, because it groweth about watersydes and hath leaues lyke Hemp'. - Turner (1548: H. ij. verso).   **1548**

Euphorbia amygdaloides

* 'The first kind of Spurge' called Characias ... or Amigdeloides 'Thys kynde have I sene in diverse places of England. Fyrst in Suffock in my lorde Wentfurt his parte [parke] besyde Nettelstede [Needham Market, nr. Ipswich] afterward in Sion parke above London ... woode spourge'. - Turner (1562: 153v).   vc21, 25   **1562**

Euphorbia exigua

* 'The fifth [kind of Spurge] is called Cyparissias ... groweth much in the stuble after the corne is caried in ... pyne spourge'. - Turner (1562: 154v (with a figure)). 'Esula exigua Tragi'. Near Rochester. - Johnson (1629: 2).   **1562**

Euphorbia helioscopia

* 'The fourth [kind of Spurge] is called Helioscopius ... groweth most comonly in old wastes and fallen doun walles and about cities   **1562**

... called in diverse partes of England Wartwurt ... it maye also be called son spourge or son folonynge spourge'. - Turner (1562: 154v (with a figure)).

Euphorbia hyberna

* 'Tithymalus Hibernicus'. - How (1650: 121). **1650**

It has always been assumed that How's record must have come from Richard Heaton (see *Dryas octopetala*, *Gentiana verna* and others), and although no record has been found of this, it seems unlikely to have been otherwise. Walsh (1978) argues that all of Heaton's records were made before 1641. Colgan & Scully (1898) note that the earliest definite locality is from Co. Limerick, given in K'Eogh (1735). Note also that Ray (1670: 299) cites a record in Stephens & Brown's *Catalogus horti botanici Oxoniensis* (1658), presumably of a cultivated specimen. 1641

Euphorbia lathyris

* 'Certainly wild, and perhaps indigenous, in several places in, and near the parish of Ufton near Reading, springing up in dry stony thickets, periodically for a year or two after they have been cut, and till choked by briars. Rev. Dr. Beeke'. - Turner & Dillwyn (1805: i. 27). vc22 **1805**

'Mr. Banks found one plant of it upon the island [Steep Holmes] (on Saturday, July 3rd, 1773)'. - Riddlesdell (1905: 300). vc6 1773

Clarke (1909) adds that 'This specimen is in **BM** with the following note in Banks's hand: 'I found this one plant among the *Ligustrum* on the south side of the Steep Holmes island, but being hurried by the tide had not time to search for more'.

Euphorbia paralias

* 'The thyrde kynde [of Spurge] is called Paralius … I have sene the same … in the West countre besyde the sea syde'. - Turner (1562: 154). **1562**

One is tempted to think that this would have been on his 1550 visit to Dorset.

Euphorbia peplis

* 'Tithymali marini species minima ex Cornubia'. - Merrett (1666: 118). 'In arenoso maris litore inter Pensans et Marketjeu copiose in Cornubia'. - Ray (1670: 237). vc1 **1666**

On Ray's Itinerary 'July 1 [1662] … On the beach near Pensance … a kind of sea-pease, species Tithymali, which runs close to the ground, the stalks of it red and round … '. - see Lankester (1846: 191). vc1 1662

Euphorbia peplus

* 'Peplus sive rotunda Esula, Petie spurge ... growe in salt marshes neere the sea, as in the ile of Thanet by the sea side betwixt Reculvers and Margate in great plentie'. - Gerard (1597: 407).   vc15   **1597**

Gerard's record seems an odd location, but the illustration and description are satisfactory. Pulteney cites Lyte (1578: 361.2) 'Pityusa minor ... found in this Countrie in arable fieldes and bankes'. It is possibly this species, but has a poor illustration.

Euphorbia platyphyllos

* 'Tithymalus platyphyllos Fuchsii *J.B* ... Nos in Comitatu Somersetensi non longe ab oppido Kinesham copiosum invenimus, & alibi'. - Ray (1670: 299).   vc6   **1670**

On Ray's Itinerary 'June 20th [1662] ... in the corn fields about Camerton [nr. Bath] ... '. - see Lankester (1846: 180).   vc6   1662

Euphorbia portlandica

* 'Tithymalus maritimus minor Portlandicus ... The Reverend Mr. William Stonestreet ... first discovered this about a year since on a narrow Neck of Land covered with Peebles [*sic*] which joyns Portland with the Coast of Dorset-shire'. - Petiver (1715a: 282).   vc9   **1715**

'Tithymalus maritimus minor, Portlandicus ... Found by the Reverend Mr. Stonestreet in the narrow Neck of Land which joyns Portland to Devonshire [*sic*] 1711'. - Ray (1724: 313).   vc9   1711

Gunther (1922: 52) cites and equates this with this species, L'Obel (1571: 163), as a variety of his 'Arborescens', 'In Australis Angliae Insula Portlandia, pusillam, nó multú diversã ab istis forma, ortu, habuimus: sed quia inter Tithymalos orta'. The accompanying illustration could be this, but looks to me far more like a *Limonium*, which also occurs on Portland.

Euphorbia stricta

* 'Occurs between Tintern and the Wind-cliff, Borrer'. - Hooker (1842: 292).   vc35   **1842**

'Tuesday June 29th. Found near Tintern Abbey a few specimens of *Euphorbia platyphyllos* by the Brook side, going from the Abbey to the Forge where they make wires'. - Riddlesdell (1905), describing a visit to Wales by Lightfoot and Banks in 1773, and identifying the *Euphorbia* as this species.   vc35   1773

Earlier records, of presumed casuals, as *E. verrucosa*, from Essex and York are given in Ray (1696: 183.3).

Euphorbia villosa

as *E. epithymoides* 'In plenty in a lane leading from Prior Park Lodge to Combe down, and also in a wood to the east of the Monument at Prior Park … pointed out to me by the discoverers, Mr. E. Simms and Dr. Heneage Gibbs'. – Babington (1834)	vc6	**1834**
Prior Park, Bath, June 1831, A. Gibbs, **BM**.	vc6	1831

This species has been identified with the 'Esula major Germanica, Turbith nigrum & adulterinum, *Gal*. Angliae frequentissima in sylva D.Ioannis Coltes, prope Bathoniam'. - L'Obel (1576: 194) and with 'Esula major Germanica *Ad. Lob.Ger*. Quacksalvers Turbith. By a wood side, some mile south of Bathe'. – Johnson (1634: 34).

Euphrasia agg.

* 'Euphrasia … These plants growe in drie medowes in greene and grassie waies and pastures standing against the sunne'. - Gerard (1597: 537).		**1597**

Figured and described Turner (1568: 30), but with no indication that it occurs here.

Euphrasia arctica

Yeo (1978).		**1978**
Mid Lothian, 1823, det. Yeo, **E**, **K**.	vc83	1823

The taxonomy of this species does not seem to have been satisfactorily clarified until Yeo's revision.

Euphrasia cambrica

Pugsley (1929), described from a specimen of his own.	vc49	**1929**
'Capel Curig, 1914 (pers.com. Yeo)'. - Ellis (1983: 134)	vc49	1914

The full details were in Pugsley (1930), the paper being read in 1929, and thus the abstract was able to appear in that year.

Euphrasia campbelliae

'in 1939 … from Uig'. - Pugsley (1940a).	vc110	**1940**
	vc110	1939

Euphrasia confusa

'Simonsbath, 1918'. - Pugsley (1919b).	vc5	**1919**
W. Cornwall, 1879, det. Yeo, **CGE**.	vc1	1879

See also Hiern (1909). The herbarium sheet is annotated 'SW42', with no other details

Euphrasia foulaensis

'Foula'. - Druce (1896a)	vc112	**1896**
'Bressay, 1884, W.E. & H. Smith, **BM**'. - Scott & Palmer (1987: 261).	vc112	1884

Euphrasia frigida

as *E. latifolia* 'W of Melvich, July 15th 1897'. - Townsend (1897).	vc108	**1897**

Euphrasia heslop-harrisonii

'Rhum, 1943, Heslop Harrison'. - Pugsley (1945).	vc104	**1945**
The type specimen, from 1943, is in **BM**.	vc104	1943

Euphrasia marshallii

Pugsley (1929), described from a specimen from Marshall.	vc108	**1929**
'Melvich, 1897 [E.S. Marshall]'. - Yeo (1978: 294).	vc108	1897

The full details were in Pugsley (1930), the paper being read in 1929, and thus the abstract was able to appear in that year.

Euphrasia micrantha

as *E. gracilis*, Townsend (1897).	**1897**
'Euphrasia *J.B.* Unlocalised but the earliest British example, collected by Du Bois'. - Druce (1928: 479).	1740

Syme (1866) had included this as *E. gracilis*, but not all his illustrations were of this species. Townsend's paper seem a better first record.
Du Bois died in 1740. There is another specimen cited in Druce & Vines (1907: 90).

Euphrasia nemorosa

Babington (1874b: 260).	**1874**
'E. tenuiore folio vulgaris. From Mr Stonestreet'. - Druce (1928: 479).	1716

Stonestreet died in 1716. See also Rose, in Gilmour (1972: 50), where he essays that the 'Euphrasea vulgaris'. Gravesend to Rochester, of Johnson (1629: 1) would have been almost certainly this.

Euphrasia officinalis s.s.

Pugsley (1930), but see also Townsend (1897: 465).	**1930**
'E. latiore folio, flore majore. Unlocalised'. - Druce (1928: 479).	1740

Druce's record is from Du Bois herbarium. He died in 1740. There is a specimen cited in Druce & Vines (1907: 89) 'E. vulg. cum fl. majore. In collibus cretaceis prope Greenhithe [Kent]'.

Euphrasia ostenfeldii

Yeo (1971).		**1971**
Brough of Birsay, 1888, Trail, det. Silverside, **BM**.	vc111	1888

Salmon had coined *E. curta* var. *ostenfeldii* in 1933, but none of his records were from the British Isles.

Euphrasia pseudokerneri

Pugsley (1929), described from a specimen from C.E. Salmon, Reigate Hill, 1896, **BM**.	vc17	**1929**
Shelford Common, [Cambs], 1826, Henslow, det. Yeo, **CGE**. Bath, [N. Somerset], 1830, det. Yeo, **CGE**.	vc29	1826

The full details were in Pugsley (1930), the paper being read in 1929, and thus the abstract was able to appear in that year.

Euphrasia rivularis

Pugsley (1929), described from a specimen of his own.	vc49	**1929**
'Cadir Idris, 1905, W.A. Vice, det. Pugsley'. (*pers. com.* Yeo). - Ellis (1983: 134).	vc48	1905

The full details were in Pugsley (1930), the paper being read in 1929, and thus the abstract was able to appear in that year.

Euphrasia rotundifolia

Pugsley (1929), described from a specimen from E.S. Marshall, Melvich.	vc108	**1929**
'Melvich, 1897, [E.S. Marshall]'. - Yeo (1978: 295).	vc108	1897

Euphrasia salisburgensis

'on limestone rocks south of Lough Mask, Co. Mayo, on July 15th, 1895 … about two miles south of Clonbur'. - Townsend (1896).	vcH16	**1896**
'Aran Is, Co Clare, 1852, D. Oliver, det. Yeo, **CGE**. This was originally wrongly named as *E. gracilis*'. - Webb & Scannell (1983: 156). Oliver (1852) refers to his collection, and relates that he sent it to Babington, who thought it might be a form of *E. gracilis*, and possibly *E. Salisburgensis*.	vcH9	1852

Clonbur is just south of Lough Mask, and in West Galway. In Moore & More (1866) is a reference to 'a slender form found in Great Arran Island and about Castle Taylor, Galway, is pronounced by M. Boreau to be the *E. cuprea* of Jordon'. Webb & Scannell (*op.cit.*) give this as the first record, but Townsend was the first to clearly identify it.

Euphrasia scottica

as *E. paludosa* 'Around Braemar'. - Townsend (1891).	vc92	**1891**
Orkney, 1888, Johnston, det. Yeo, **K**.	vc111	1888

Euphrasia tetraquetra

as *E. occidentalis* 'St Mary's, Scilly, 1862, det. Wettstein'. - Townsend (1897).	vc1	**1897**
Cardiganshire, 1848, Babington det. Yeo, **CGE**. Pembrokeshire, 1848, Babington, det. Yeo, **CGE**.	vc45, 46	1848

Euphrasia vigursii

'Porthtowan [1905]'. - Davey (1907).	vc1	**1907**
	vc1	1905

Exaculum pusillum

'*Cicendia candollei*, Griseb. I have lately received this plant from Mr Fred. Townsend, who gathered it 'on waste broken ground near Paradis in the island of Guernsey, in company with *Cicendia filiformis* and *Radiola millegrana*". - Babington (1850b).	vc113	**1850**

Fagus sylvatica

* 'Fagus ... Bech trees growe plentuously in many places of England. Two of the greatest that euer I sawe growe at Morpeth on ii hylles right over the Castle'. - Turner (1548: D. j. verso).	vc67	**1548**

Fallopia convolvulus

* 'Elatine ... groweth amonge the corne & in hedges, it maye be named in englishe running Buckwheate or bynde corne'. - Turner (1548: C. viij).	**1548**

Fallopia dumetorum

* 'Found September 20, 1834, in a wood at Wimbledon, by Mr. J. A. Hankey'. - Babington (1836).	vc17	**1836**
	vc17	1834
This species was found by Mr. Luxford on 22 Sep. 1826, near Reigate, Surrey'. Note on drawing for E.B.S. Plate [2811] in Bot. Dept. Brit. Mus'. - Garry (1904: 160). This appears from the text in E.B.S. to be a misreading of the 1836 date. Salmon (1931) gives the Luxford record, but dates it 1837.		

Festuca altissima

* as *F. calamaria* 'In a moist wooded valley at the foot of Ben Lawers, 1793, Mr. Mackay'. - Smith (1800: i. 121).	vc88	**1800**
	vc88	1793

Festuca arenaria

Babington (1874b) has *F. oraria*, which refers to E.B. 2056 [erroneously given as 2058].	**1874**

Garry's note (1904: 254) on this plate of E.B. has 'Yarmouth, Dawson Turner, Esq., 25 June 1809'. vc26 1809

Druce (1908) lists *F. rubra arenaria*, *F. rubra juncea* and *F. dumetorum*.

There is an entry in Sibbald (1710) which reads 'Gramen marinum juncifolium tenuissimum spicam avenaceam. I found it in the sands below Blackness'. Sibbald gave no authority for the polynomial. John Edgington (*in litt.*) pointed me to this, but we have not traced the polynomial. He suggests, with merit, that this might well be this species. Apparently G. Don suggested that it might be *Poa distans* var. *retroflexa* but that too is conjecture.

Festuca armoricana

as *F. ovina* var. γ, from Jersey. - Huon (1970). vc113 **1970**

'… M. Wilkinson … determined as this a specimen collected by Vaniot in 1881 … from the dunes of St Ouen's Bay'. - Le Sueur (1984) (as *F. ophioliticola*). vc113 1881

Festuca filiformis

as *F. capillata*, Howarth (1925). **1925**

'Gramen capillacium Locustellis pinnatis, non aristatis … D. Dale'. - Ray (1688: 1288). There is also a superb specimen in Hb. Sloane, 125.17, **BM** (Buddle's collection). 1688

Although it has been traditional to assign Ray's record to this species, the real concept of the species was not clarified until Howarth (1925), as *F. capillata*. Both Sibthorp (1794: 44), from Bullington Green, vc23, and Babington (1843), as var.γ, described a non-awned form of *F. ovina* as *F. tenuifolia* (= *F. filiformis*).

Festuca huonii

'This was stated (*in litt.*) by Wilkinson to be the common cliff top species of the north coast [of Jersey] in 1983'. - Le Sueur (1984). vc113 **1984**

Dunes, St Ouens, Jersey, 1936, Thompson [F.W.], ?det. Hubbard, **K**. vc113 1936

See Auquier (1973) and Wilkinson & Stace (1991). Another early record is 'Rock crevices, 9th June 1953, Jerburg, Guernsey, Hubbard, det. Wilkinson, **BM**'.

Festuca lemanii

Wilkinson & Stace (1988). **1988**

Seven Sisters, Herefords. Stace, C.A., 1982. Field record. vc36 1982

Clarke, under *F. fallax*, as a synonym of *F. duriuscula* Linn. (= *F. lemanii*), cites 'Gramen pratense panicula sparsa versus unam partem, duriore'. - Ray (1670: 153).

Festuca longifolia

as *F. caesia* 'In almost the last botanical excursion that I enjoyed with my late friend Mr Crowe, when we examined the country around Bury, in June 1804, along with our experienced guide Sir Thomas Cullum, we were much struck with the very glaucous aspect of this grass on the dry barren heaths north of that town'. - E.B. 1917.   vc26   **1808**

   vc26   1804

This is presumably Sir J.E. Smith writing in E.B. See also Trist (1973) and Gibson & Taylor (2005).

Festuca ovina s.l.

'Gramen sparteum montanum spicâ foliaceâ gramineâ majus … The greater grasse-eared Grasse, amongst the rockes at Carnedh Lhewelyn'. - Johnson (1641: 23).   vc49   **1641**

The Johnson attribution is from Druce, and if he is correct then this would have been found on Johnson's trip of 1639.   vc49   1639

Clarke cited 'Gramen capillaceum locustellis pinnatis non aristatis … Observavit et ad me attulit D. Dale'. - Ray (1688: 1288).

Festuca ovina s.s.

Howarth (1925).   **1925**

'from ye House calld Cwm Brwynog where David Griffith liveth at Llandberrys [Llanberis]'. - Druce & Vines (1907: 120). This would be from Dillenius' trip to Wales in 1726.   vc49   1726

Festuca rubra s.l.

'Gramen pratense, panicula duriore, laxa, unam partem spectante … in pascuis'. - Ray (1688: 1285).   **1688**

Ray's English name is 'Meadow grass with harder sparsed panicles'. As Druce has seen Buddle's grass with Ray's epithet (From Mr Buddle'. - Druce (1928), this seems acceptable. Druce cited incorrect page number in Ray.

Festuca rubra s.s.

Babington (1874b).   **1874**

'Gramen pratense, panicula duriore, laxa, unam partem spectante … in pascuis'. - Ray (1688: 1285).   1688

Since Babington was the first to separate *F. rubra* s.s from any others in that aggregate, this has to be the first record, although Ray was certainly describing this taxa.

Festuca vivipara

'Gramen sparteum montanum spicâ foliaceâ gramineâ majus ...   vc49   **1641**
The greater grasse-eared Grasse, amongst the rockes at Carnedh Lhewelyn'. - Johnson (1641: 23). 'Gramen montanum spicâ foliaceâ gramineâ. *P.B. Grass upon grass.* In summis altissimorum Cambriae montium, *Snowdon, Caderidris,* &c, verticibus, inter saxa, ubi nulla ferè alia planta praeter muscum provenit, copiosé'. - Ray (1690: 184).

Thomas Johnson and party were on Snowdon, Aug 1639. See   vc49   1639
Raven (1947: 289).

See *F. ovina s.l.* above and also Wycherley (1953: 44) on early discoveries. Horsman (1995) records a record of Ralph Johnson, from Teesdale, which he dates to 1672.

Johnson, T. 1641. *Mercurius Botanicus pars altera.*

(23)

rum margines. *Great Haver-grasse.*
*Gramen geniculatum, *Tab. Ger.* Paniceum spicâ asperâ, *Bauh.* In pratis humidis.
*Gramen geniculatum aquaticum, *Tab. Ger. Water kneed grasse.* 𝔍𝔫 𝔱𝔥𝔢 𝔐𝔢𝔞𝔡𝔬𝔴 𝔩𝔶𝔦𝔫𝔤 𝔞𝔪𝔬𝔫𝔤 𝔱𝔥𝔢 𝔅𝔯𝔦𝔡𝔤𝔢𝔰 at *Wilton* 𝔗𝔬𝔴𝔫𝔢𝔰 𝔢𝔫𝔡. *Park.* 1177.
Gramen junceum montanum subcærulea spica. *Mosse-crops.*
Gramen junceum maritimum, *Lob. Ger. Sea Rush-grasse.*
Gramen paniceum spica simplici. *Lob. Ger. Single eared Panicke grasse.*
Gramen Parnassi, *Dod. Lob. Ger.* Hepatica alba, *Cordi. Grasse of Parnassus or White Liverwort.*
Gramen sparteum montanum spicâ foliaceâ gramineâ majus. *The greater grasse-eared Grasse.* 𝔄𝔪𝔬𝔫𝔤𝔰𝔱 𝔱𝔥𝔢 𝔯𝔬𝔠𝔨𝔢𝔰 𝔞𝔱 *Carnedh Lhewelyn.*
Gramen sparteum spicâ foliaceâ gramineâ minus; *The lesser grasse-eared Grasse or Grasse upon Grasse.* 𝔒𝔫 𝔱𝔥𝔢 *Fells* 𝔟𝔢𝔱𝔴𝔢𝔢𝔫𝔢 *Pereth* 𝔞𝔫𝔡 *Kendall,* 𝔞𝔰 𝔞𝔩𔰰𝔬 𝔬𝔫 *Snowdon.*

H

*HEdera saxatilis, *Ger.* Asarina, sive Saxatilis hedera, *Lob. Rocke Ale-hoofe.* 𝔍𝔫 𝔰𝔬𝔪𝔢 𝔭𝔩𝔞𝔠𝔢𝔰 𝔬𝔣 *Somersetshire* 𝔞𝔠𝔠𝔬𝔯𝔡𝔦𝔫𔤤 𝔱𝔬 *Lobell. Park. pag.* 677.
*Helleborus niger flo. viridi, *Bauh.* Veratrum nigrum, *Dod. Clus.* Helleborastrum *Ger.*
In

L 4

Ficaria verna

* 'Hirundinaria … Chelidonium minus … Fygwurt, it groweth under the shaddowes of asshe trees. It is one of the fyrst herbes that floures in the spring'. - Turner (1548: D. v). **1548**

Nelson (1959) cites Turner (1538: A. iiij. verso) 'Chelidonium minus', but since Turner does not know of a common name it seems safer to go with the 1548 record.

Filago gallica

* 'Gnaphalium parvum ramosissimum foliis angustissimis polyspermon, *Hist. nost*. Among corn in sandy grounds about Castle-Heveningham [Hedingham] in Essex plentifully. Mr. Dale'. - Ray (1696: 85). vc19 **1696**

Gibson (1862: 172)) says the specimen is in Herb. Dale (at **BM**). Ray (1686: 296) records having a dried specimen of the plant but with no record of where it was collected.

Filago lutescens

* as *F. apiculata* (provisional name) 'Sandy borders of fields, hedge-banks and road-sides, Cantley, Rossington, etc., near Doncaster'. - Smith (1846). vc63 **1846**

Fairmile, 1840, H.C. Watson, **K**. vc17 1840

Dony (1953) gives as the first record 'Abbot, 1798', but this must be an error, since that does not appear in Abbot's flora (1798).

Filago minima

* 'Gnaphalium minimum, *Lob.*'. … Between Sandwich and Canterbury. - Johnson (1632: 23). vc15 **1632**

Rose, in Gilmour (1972: 120), wonders if this might be *Gnaphalium uliginosum*.

Filago pyramidata

* as *F. Jussiaei* 'Gathered four or five years ago, about eight miles from this town [Saffron Walden]'. - Gibson (1848). vc19 **1848**

Salmon (1931: 376) cites 'Gnaphalium prolifer minus. About Epsom'. - Petiver (1713: xviii. 10), and adds that there is a specimen in Herb. Buddle (Hb. Sloane 118, 29), **BM**. vc17 1713

Actually Petiver has this as 'Low Chidding [?Childing] Cudweed', and whilst his record might be the first, there is no indication of taxonnomic differentiation in this critical group. There is another early record cited in Druce & Vines (1907: 143).

Filago vulgaris

* 'Of cottonwede … Gnaphalium … I have sene the herbe … in some places of Englande'. - Turner (1562: 11v).

Clarke adds 'with a figure evidently this'.

1562

Filipendula ulmaria

* 'Medewurte … groweth about water sydes, moist places and fennes … called of some 'Ulmaria'. - Turner (1568: 8).

1568

Filipendula vulgaris

* 'Oenanthe … Filipendula groweth in great plentie besyde Syon & Shene in the middowes'. - Turner (1548: E. v).

vc17, 21   1548

see Turner (1538: C). 'Phellandryon … vulgus Filipendula et Droppewort', which implies familiarity with the plant as a wild species.

Foeniculum vulgare

* 'By the Sea-side in Cornwal towards the lands end plentifully, also upon the chalky cliffs at the West-end of Pevensey marsh in Sussex, and elsewhere'. - Ray (1677: 111).

vc1, 14   1677

'Foeniculum at Rie [Rye] - Leucoii polii Qu. D.'. - W. How, manuscript addition to his copy of Phytologia Britannica, cited in Gunther (1922: 281). How died in 1656.

vc14   1656

The 'D' in How's ms. is assumed by Gunther (1922: 296) to refer to Dr. J. Dale and this is followed by Wolley-Dod (1937).

Fragaria vesca

* 'Fragraria … Euery man knoweth wel inough where strawberries growe'. - Turner (1548: D. ij. verso).

1548

see Turner (1538: B. verso) 'Fragum, non fragrum (ut quidam scioli scribunt), ab anglis vocatur a Strawbery', which implies familiarity with the plant as a wild species.

Frangula alnus

* 'Alnus nigra sive frangula … I found great plentie of it in a wood a mile from Islington in the way from thence toward a small village called Harnsey [Hornsey]… and in the woods at Hampsteed neere London and in most woods in the parts about London,' - Gerard (1597: 1286).

vc21   1597

Frankenia laevis

* 'Polygonum serpillifolium. I found it flouring the third day of September 1621 on the ditch bankes at Burseldon ferrey [Bursledon, on the R. Hamble] by the sea side in Hampshire, Jo. Goodyer'. - Johnson (1633: 567). | vc11 | **1633**

'About this time [1596], too, he [L'Obel] was certainly at Richard Garth's house at Drayton near Portsmouth and was the first to find *F. laevis* in that neighbourhood (see Parkinson 1640: 1485)'. - Raven (1947: 239). | vc11 | 1596

See also Goodyer ms.11, f. 89, in Gunther (1922: 151) giving the 1621 record. Gunther (1922: 148) cites another Goodyer record (ms. 11, f. 103) as this species 'An polygoni marini species. On the sea shoare in the west parte of the Iland of Haylinge in Hampsheire and in other places by sea likewise'. This second record is not dated.

Fraxinus excelsior

* 'Fraxinus ... An Ashe tree ... Asshes growe in euery countrey'. - Turner (1548: D. ij. verso). | | **1548**

Fritillaria meleagris

* 'Fritillaria praecox purpurea variegata *C.B. Pin.* ... In Maudfields near Rislip Common [Ruislip] observed above 40 years by Mr. Ashby of Breakspears'. - Blackstone (1737: 29). | vc21 | **1737**

'In a letter written on 11 December 1736 by John Blackstone to Sir Richard Richardson'. - Kent (2001). | vc21 | 1736

Fumaria

* 'Capnos ... Fumaria, and in englishe Fumitarie ... groweth amonge the corne'. - Turner (1548: B. vij). | | **1548**

Fumaria bastardii

* as *F. agraria* 'Tintagel, Cornwall, Mr. Borrer'. - Mitten (1848). | vc2 | **1848**

'A specimen from "Tintagel, July 2, 1839", W. Borrer, 1848, is in Ed. Forster's Herbarium, in **BM**'. - Davey (1909: 28). | vc2 | 1839

Fumaria capreolata

as *F. pallidiflora* 'from Salcome and Ilfracombe, Devon; Watchet, Somerset; Oystermouth, nr. Swansea, Glamorgan; Caernarvon; Oswestry, Shropshire'. - Babington (1860a). | vc3, 4, 5, 40, 41, 49 | **1860**

Braid Hills, vc83, 1825, G. Lloyd, det. Pugsley, **K**. There are earlier records in the databases but these are not critically determined. | vc83 | 1825

Clarke cites 'Fumaria major scandens flore pallidiore. In hortis & ad muros vel sepes'. - Ray (1670: 122), but since the same wording appears in Ray (1685: 7), and it is not at all clear there that Ray was referring to *F. capreolata* - see Oswald & Preston (2011: 467n) - there must be considerable doubt over what plant Ray was describing in 1670, although, of course, it might be a specimen from a different part of the country than Cambridge.

There is a good illustration in Curtis (1787: f. 6, t47), purporting to be this, but Rose Murphy (*pers. comm.*), identifies this as *F. muralis*. Another early record, which of course cannot be confirmed in the absence of a specimen, is 'At the Hermitage and by waysides', June 2nd 1765'. - John Hope's ms. 'List of Plants growing in the Neighbourhood of Edinburgh', reproduced in Balfour (1907).

Fumaria densiflora

* as *F. calycina*, n. sp. 'Near Edinburgh … and from Dover, gathered by R. M. Lingwood, Esq'. - Babington (1840: 34).	vc15, 83	**1840**
'Swaffham Prior, L. J[enyns], 7.6.1827, **BATHG**'. - Crompton (2007).	vc29	1827

This was figured in E.B.S. 2876 (1844) as *F. micrantha*.

Fumaria muralis

* 'I have seen *F. muralis* from Barnes, Surrey (Mr Pamplin); Shrewsbury, Salop; Wrexham, Denbighshire (J.E. Bowman); Sheffield (Rev. W.W. Newbould)'. - Babington (1860a: 160).	vc17, 40, 50, 63	**1860**
'Guernsey. Gosselin, J., c.1790'. - see McClintock (1982: 40). But see also the Curtis record for 1787 above, under *F. capreolata*.	vc113	1790

Fumaria occidentalis

' … in the late spring of 1902 … looking over specimens in the collection of Mr. A.O. Hume … from Newquay … in June … at Penzance … I met with … identical specimens'. - Pugsley (1904).	vc1	**1904**
Pugsley, in Moss (1920: 172) mentions that 'the earliest known specimen was collected at Newlyn, Cornwall, in 1881, but there is evidence that long previously it was locally known as an unusually beautiful weed of cultivation'.	vc1	1881

Pugsley (1919a: 261) notes that 'This species is perhaps the F. vulgaris corbensis alba (*sic*) of Parkinson (1640: 287) noted as growing in cornfields in Cornwall'.

Fumaria officinalis s.s.

* 'Capnos … Fumaria, and in englishe Fumitarie … groweth amonge the corne'. - Turner (1548: B. vij).		**1548**

This has been traditionally identified as this species, presumably as it is the most widespread.

Fumaria parviflora

'Discovered by the late Mr Jacob Rayer in the corn-fields about Woldham near Rochester, in September 1792'. - E.B. 590. vc16 **1797**
vc16 1792

Clarke, Druce and Hanbury & Marshall (1899) all cite 'Fumaria tenuifolia ... In a corne fielde between a small village called Charleton and Greenwich'. - Gerard (1597: 929.13), but this seems tenuous and is not mentioned by Pugsley in his 1912 mongraph. In Gerard both this and F. latifolia minor (929.6) are recorded from the same location.

Fumaria purpurea

Pugsley (1902), citing specimens from a numbers of locations, and equating with the *F. Boraei* of Babington (1860a) and Syme (1863: i. 106). **1902**

'Fumaria major scandens flore pallidiore ad sepes prope Shrewsbury'. - Druce & Vines (1907: 71), and '26th July 1726' for Dillenius' Journey into Wales (1907: xlvii). vc40 1726

This Druce reference is confirmed in Lockton & Whild (2005) who have examined the specimen in **OXF**. In Druce (1928) he states it was 'in the garden of Du Bois at Mitcham before 1700', but no evidence has been found to support this assertion.

Fumaria reuteri

as *F. paradoxa* 'brought to my notice in October 1904, by Mr F.H. Davey ... from Gilly Tresamble [S. of Truro]'. - Pugsley (1912: 31). vc1 **1912**
vc1 1904

Fumaria vaillantii

* 'I had gathered this plant on Chatham Hill, Kent, about five years ago'. - Henslow (1832: 88). vc15 **1832**

'First found by Prof Henslow in May 1831'. - Babington (1860a). vc29 1831

Gagea bohemica

'…collected by R.F.O. Kemp at Stanner Rocks, nr. Kington, Radnorshire, in April 1965, and recorded as *Lloydia serotina*'. - Rix & Woods (1981). vc43 **1981**
vc43 1965

see also Kemp (1968).

Gagea lutea

* 'Ornithogalon luteum ... Angliae nemorosis Sommerseti collegimus'. - L'Obel (1571: 56). 'In a Cornfield by Winecaunton, Somersetshire'. - How (1650: 85). vc6 **1571**

How's habitat seems strange. See Horsman (1995) for a 1672 record, by Ralph Johnson, from Greta Bridge, vc65 'making a faire show among the Anemones in the skirts of our woods', a much more satisfactory habitat.

Gagea serotina

* 'Bulbosa Alpina juncifolia pericarpio unico erecto in summo cauliculo dodrantali. ... In excelsis rupibus montis Snowdon v.g. Trigvylchau y Clogwyn du ymhen y Gluder, Clogwyn yr Ardhu, Crib y Distilh. &c. D. Lhwyd'. - Ray (1696: 233). vc49 **1696**

Carter (1955a) mentions that 'On Sept 23rd, 1688, J. Bobart wrote to Lhwyd, having received specimens from him including probably *Lloydia* [now *Gagea*]'. This conjecture arises from a letter reproduced in Gunther (1945: 86), where Bobart reports that he cannot find the bulb that Lhwyd had specially packed. It is Gunther who wonders if this bulb was *Lloydia*! Marren (1999) too gives 1688 as the date of discovery.

Galeopsis angustifolia

* 'Ladanum segetum, *Lugd.*'. Gravesend to Rochester. - Johnson (1629: 2). vc16 **1629**

See also Johnson (1633: 699, par. 7), where he discusses this plant further..

Galeopsis bifida

as var. β of *G. tetrahit*, Babington (1847). **1847**

'Cannabis spuria altera flo. purp. Netle Hempe … *Lob. Icon. 527*'. - Goodyer ms. annotation in How's interleaved copy of his *Phytologia*, received by Goodyer in 1659, see Gunther (1922: 194). 1659

Babington (1843) had forecast that this species would be found 'The continental botanists distinguish two plants [from *G. tetrahit* s.l.], both of which should be found in Britain'. The first illustration traced is in Syme (1867: pl1079).

Galeopsis segetum

* 'Sideritis arvensis latifolia glabra, *Ger*. In Occidentali comitatus Eboracensis parte v.g. circa Wakefield, Darfield, Sheffield, &c. Inter segetes'. - Ray (1670: 283). vc63 **1670**

Dated to Ray's 1660 Itinerary, see Raven (1950: 114). vc63 1660

Galeopsis speciosa

* 'Cannabis spuria flore pallido, labro purpureo elegante. In fimetis & ruderibus. Faire-flowere Nattle-hempe'. - Johnson (1641: 18). 'At Marish in Cambridgeshire & in Kighley, Yorkshire, plentifully'. - Merrett (1666: 19). **1641**

Galeopsis tetrahit s.l.

* 'Cannabis spuria ... in the corne fieldes of Kent as about Graves ende, Southfleete, and in all the tract from thence to Canterburie'. - Gerard (1597: 573). vc15, 16 **1597**

Galeopsis tetrahit s.s.

Babington (1847). **1847**

'Juxta Thames, Chelsey, vertic. purpuras'. - Druce & Vines (1907: 78). vc21 1747

Babington (1843) had forecast that this species would be found - 'The continental botanists distinguish two plants [from *G. tetrahit s.l.*], both of which should be found in Britain'. Druce (*op.cit.*) separates the two species, but Kent (1975) seems to distrust early records. The specimen in **OXF** seems to be this species. The record from Druce & Vines is dated as 1747, the date of Dillenius' death.

Galium album

* 'Mollugo vulgatior herbariorum ... Collibus incultis & cretaceis agrorum marginibus ... Angliae plurima'. - L'Obel (1576: 465). **1576**

Galium aparine

* 'Apparine ... vocatur ab anglis Goosgyrs aut Gooshareth'. - Turner (1538: A. ij. verso). **1538**

Galium boreale

* 'Rubia erecta quadrifolia, *J.B* ... Prope Orton, Winander-mere & alibi in Westmorlandia. Apud alios autores nulla ejus mentio quod sciam'. - Ray (1670: 268). vc69 **1670**

In a letter from Ray to Lister, 1668 'Gallium cruciatum *J.B.* in Westmorlandia prope Orton et alibi'. See Lankester (1848: 26-27) and Raven (1950: 148). vc69 1668

Galium constrictum

'Marshes at St Brelade, Jersey, 1924; L'Ancresse, Guernsey, 1906. ... specimens gathered by Miss Todd at Lymington in 1924'. - Druce (1925: 438). vc11, 113 **1925**

'This is what both Babington on 27 June 1838 and Marquand in 1891 collected near Fort Doyle as Heath Bedstraw. Babington however later noted on his specimen '*G. debile* Desv. probably' and so it was'. - McClintock (1975). vc113 1838

Rayner (1929) states that it 'was found by Miss Todd at Lyndhurst, and by the present writer at Hatchett Pond, near Beaulieu, and near Holmsley Bog, in 1924'. We do not know who found it first.

Galium odoratum

* 'Wood rose or wood rowel ... a short herbe of a span long, four square and smal, about ye which growe certaine orders of leaves, certayne spaces goynge betwene, representing some kindes of rowelles of sporres, whereof it hath the name in English. The floures are whyte and well smellinge'. - Turner (1568: 25). **1568**

Galium palustre

* 'Gallium album *Tab*. palustre, *Dod*.'. Between Margate and Sandwich. - Johnson (1632: 16).   vc15   **1632**

Gerard (1597: 967.2) gives 'Galium album, Ladies Bedstraw with white flowers … groweth in hedges among bushes in most places'. Johnson (1633: 1126), using the same name, alters the habitat to 'marish grounds and other moist places'.   1597

Galium parisiense

* 'Aparine minima. Found at Hackney on a wall'. - Sherard in Ray (1690: 237).   vc21   **1690**

Galium pumilum

Goodway (1957), separating the octoploid forms of the south and east of England from what is now described as *G. sterneri* elsewhere in the British Isles.   **1957**

as *G. sylvestre* 'Cheddar. First noticed by Borrer [died 1862]'. - Murray (1896). There is a 1853 record from Steep Holm, vc6, F.M Barton, det. Goodway, **BM**, which seems quite acceptable, and in its native range, but Roe (1981) treats it as a casual from that site, and it is not mapped in Preston *et al.* (2002).   vc6   1862

The separation of the two species in this aggregate had been unclear, although Babington (1874b) probably was describing this species under *G. sylvestre* var β. *nitidulum* and Druce (1900) gave a probably native site in Buckinghamshire. Goodway (1955) previews the results of his 1957 paper, which is based on the work of Sterner in Sweden.

Galium saxatile

* 'Galium album minus *Tab*. In montosis. Little white Ladyes Bedstraw'. - Johnson (1634: 37).   **1634**

Galium sterneri s.l.

* as *G. pusillum* 'In montibus, prope Kendal in comitatu Westmorelandico'. - Hudson (1762: 57).   vc69   **1762**

Galium sterneri s.s.

Goodway (1957), separating the octoploid forms of the south and east of England from what is now described as *G. sterneri* elsewhere in the British Isles.   **1957**

as *G. pusillum* 'In montibus calcareis. prope Kendal in comitatu Westmorelandico'. - Hudson (1762: 57).   vc69   1762

Galium tricornutum

* 'Aparine semine leviore … inter segetes passim'. - Ray (1663: 6).	vc29	**1663**
First mentioned in a letter from Ray to Peter Courthope, dated from Trinity College in July 1661. See Gunther (1934) & Thompson (1974).	vc29	1661

Galium uliginosum

* 'Aparine palustris minor Parisiensis, flore albo ... On the Lower Bog at Chisselhurst; Mr. J. Sherard'. - Ray (1724: 225).	vc16	**1724**
'I found this in ye bogs at Hampstead and it is ye Aparine palustris minor Parisiensis'. - Buddle in Hb. Sloane 121.2 & 10 [a much better specimen], **BM**. 'Aparine palustris minor, Parisiensis flore albo *Tourn*. Found by Mr Buddle in some ditches near Hampstead'. - Druce (1928: 473).	vc21	1715
Both Trimen & Dyer (1869) and Kent (1975) in their floras of Middlesex give 'c. 1700' for this record, but there is no date in Druce (*op.cit.*). Buddle died in 1715.		

Galium verum

* 'Galion ... Named in englishe in the North countrey Maydens heire'. - Turner (1548: D. ij. verso).		**1548**

Gastridium ventricosum

* 'Gramen serotinum arvense spica laxa pyramidali, *Hist. nost. 1288* ... Near Tunbridge Wells in Kent, Mr. Doody'. - Ray (1688a: 11-12).	vc16	**1688**

Gaudinia fragilis

Druce (1908).		**1908**
'New reservoir works towards Molesey, VC17. Wolley-Dod, June 1903, **BM**. First record in a possibly 'native' site 'abundant in a meadow near Ryde, Isle of Wight, J.W. Long,10 July 1917, **BM**'. - McClintock (1972).	vc17	1903
This species was always treated as a casual until the note by McClintock (1972). See also Leach & Pearman (2003) for a full discussion on the status and distribution of this species.		

Genista anglica

* 'Genistella ... I have not sene it in England sauyng once besyde Coome parcke ... Thorn-broume or prickly broume'. - Turner (1548: H. iii).	vc23	**1548**
Clarke, followed by Salmon (1931), claim 'Coome parke' for Surrey. Chapman & Tweddle (1989: 323, n.72) identify this with the home of Sir Thomas Elyot, nr. Woodstock, Oxon. Elyot was a		

correspondant of Turner, and this is more likely (but not given in Druce (1927). See also *Lathyrus linifolius*.

Genista pilosa

* 'Found by Sir John Cullum about Lackford four or five miles from St. Edmund's Bury in July 1774'. - Rose (1775: app. 452).	vc26	**1775**
'First notice 1771. Sir J. Cullum'. - Hind (1889: 105).		
Hind also gives two other entries 'Discovered by Sir Jno. Cullum and Mr Dickson amongst the heath about Culford and Lackford,' *Dill. Ray in Hort. Ox.* and 'Heath West of Icklingham Church, Sir J. Cullum, 1771,' *Dill. Ray in Hort. Ox.* The source of these is given as 'Ms. notes in a copy of *Dill. Ray* [ie Ray (1724)] in the Library of the Oxford Botanical Garden, attributed to Mr Pitchford, 1780'.	vc26	1771

Genista tinctoria

* 'Genistella infectoria. Angl. Dieweed ... Taurinis, Pedemontanis et Angliae occiduae'. - L'Obel (1571: 407).	**1571**

Gentiana nivalis

* 'Ben Lawers'. - Dickson (1794).	vc88	**1794**
Dickson's tour was in 1792	vc88	1792

Dickson, J. 1794. An account of some plants newly discovered in Scotland. *Trans. Linn. Soc.* 2: 286-291

> 290   Mr. DICKSON's *Account of*
>
> 3. The leaves short and flat.—4. The spikes many, upon slender footstalks, and pendulous.—5. The involucrum shorter than the spike.
>
> ERIOPHORUM alpinum. *Linn. Spec. Plant.* 77. found by Mr. *Brown* and Mr. *Don*, in a moss about three miles east of Forfar, in the shire of Angus.
>
> A specimen of this was presented to the Linnean Society some time ago by Mr. *Teasdale*.
>
> GENTIANA nivalis. *Linn. Spec. Plant.* 332. Ben Lawers.
>
> SIUM repens. *Jacq. Fl. Aust.* t. 260. Wet places in the south of Scotland.
>
> SAXIFRAGA cernua. *Linn. Spec. Plant.* 577. Amongst the rocks on the summit of *Ben Lawers*.
>
> STELLARIA cerastoides. *Linn. Spec. Plant.* 604. *Smith's Plant. Ic.* t. 15. Ben Nevis.
>
> ASPLENIUM alternifolium. *Murray, Syst. Vegetab. edit.* 14. *Jacq. Misc.* 2. t. 5. f. 2. Rocks in the south of Scotland.
>
> POLYPODIUM dentatum, nova species*. A figure and description of this will be given in my third Fasciculus. Rocky mountains of Scotland.
>
> I found the following, not described in the *Fl. Scotica*, but in the *Flora Anglica*.
>
> PHALARIS arenaria. Sea-coast, near Prestonpans.
>
> LYSIMACHIA thyrsiflora. Woods in the south of Scotland.
>
> DROSERA anglica. Near Fort Augustus.
>
> BARTSIA alpina. Rocks to the east of Malghyrdy.
>
> * P. dentatum. *Fasc. Pl. crypt.* 3. *p.* 1. *t.* 7. *f.* 1.
>
>   CAREX

Gentiana pneumonanthe

* 'Pneumonanthe ... On a wet moorish ground in Lincolnshire 2 or 3 miles on this side Caster [Caistor] and as I remember the place is called Netleton Moore'. - Johnson (1633: 438).	vc54	**1633**
Raven (1947: 274) identifies this with Johnson's trip to Yorkshire in 1626.	vc54	1626

Johnson (1633) notes that Gerard's records (1597: 355) from Kent were errors. However, the latter's record of *G. verna* (1597: 354) as 'groweth plentifully in Waterdowne forest in Sussex in the way … Eridge house' may be this species - see also Raven (1947: 210) and, with more certainty, Wolley-Dod (1937).

Gentiana verna

* 'Gentianella Alpina verna ... In the Mountaines betwixt Gort and Galloway abundantly. Mr. Heaton'. - How (1650: 46). 'Gathered in April, 1797, in Teesdale Forest, Durham, by Mr. John Binks, and sent us by the Rev. Mr. Harriman, the first botanist who has ascertained it in England'. - E.B. 493.	vcH15	**1650**
In common with other species first found by Richard Heaton but not published till 1650 by How, this is reckoned by Walsh (1978) to have been discovered by 1641.	vcH15	1641

Gentianella amarella

'Gentianella fugax Autumnalis elatior Centaurii minoris foliis … found first by Dr Eales near Welling [Welwyn] in Hartfordshire; then by Mr Dale in some barren layes, at Belchamp S. Paul [S. of Clare], Essex'. - Ray (1696: 156).	vc19, 20	**1696**
A 'kind of Gentian ... groweth in England both in Dorsetshyre [*sic*] upon the playne of Salisberrye, and also in Yorkshyre in bare places'. - Turner (1568: 25).	vc8	1568

The Turner record may well be correct, but as Raven (1947: 278) notes, this species and *G. campestris* were not clearly distinguished till Ray's time. However *G. campestris* has always been very rare in the south of England, and much rarer than this species. See also Gerard's (1597: 354) record of *G. pneumonanthe* 'groweth upon Longfield downes in Kent … upon the chalkie cliffes neere GreeneHythe and Cobham', which Raven (1947: 210) suggests may be an error for this species. However both Johnson (1633: 438) and Clarke identified that as *Campanula glomerata*.

Gentianella anglica

' … appears to have been first observed on Afton Down, Isle of Wight, on May 25, 1864, by Tucker (1870)'. - Pugsley (1936a).	vc10	**1936**
Brighthelmston [Brighton], vc14, Rand, det. Rich, **OXF**.	vc14	1743

Rand died in 1743. Druce cites 'G. fugax minor, fig. 3'. - Johnson (1633: 437) and Ray (1696: 156), which are mere conjecture.

Gentianella campestris

'Gentianella fugax verna seu praecox … found by Mr Fitz-Roberts on the back-side of Halfe-fel-nab [Helsfell] near Kendal'. - Ray (1696: 156).	vc69	**1696**
'G. alpina verna … Waterdowne forest in Sussex … on the plaine of Salisbuie harde by the turning from the saide plaine … Wilton … upon a chalkie bank in the highway between St. Albans and Gorhamberrie [Gorhambury, Herts]'. - Gerard (1597: 354).	vc8, 20	1597

The Waterdowne Forest record may well be *Gentiana pneumonanthe*, but the others seem correct. See Johnson (1633: 437) who attempts to clarify, but the records for this species and *G. amarella* are muddled. See also the note under *G. amarella*. All the Hertfordshire floras accept the Gerard record.

Gentianella germanica

* 'Some years ago Mr. W. Pamplin observed a Gentian with large flowers in the neighbourhood of Tring in Hertfordshire. The Rev. W. H. Coleman subsequently identified this as *G. germanica*'. - Anon (1841).	vc20	**1841**
Dandy (1958a: 219) cites a specimen in Petiver's Hortus siccus Anglicanus (H.S.152, f.61) labelled 'this was gathered near St Albans by Mr Tilden'. He dates this to 1700-1707.	vc20	1707

Pryor (1887) cites under *G. germanica* 'Near the walls of Verulanium [St. Albans], Feilden in herb. [Isaac] Rand'. Rand died in 1743. Dony (1967) 'more certainly Dickson and Anderson 1812'. However McVeigh *et al.*(2005) claim that the St Albans record **is** this taxon.

Gentianella uliginosa

' … collected last September [1923] around Tenby …'. - Pugsley (1924b).	vc41	**1924**
Rich (1996) reports the finding of specimen from Braunton Burrows in **BM** ex herb **E**, Forster (from his herbarium that was acquired in 1849).	vc4	1849

The record in Druce citing Nyman for an early Scottish record is unproven (and his reference 'Trail in *Ann. Scot. Nat. Hist., 173,* 1906' is incorrect). Druce also cites a record of his own, from Golspie, 1913, teste Lindman, but this has not been cited by any later compilation.

Gentianopsis ciliata

'*Gentiana Pneumonanthe* in Bucks. This gentian has been collected during the autumn by a lady (Miss Williams) on a hill not far from Wendover, Bucks'. - Anon (1875). ' ... the specimen in the British Museum Herbarium, sent by Miss Williams from Wendover, represents *G. ciliata*'. - Britten (1879). 'During the summer of 1982, P. Philipson discovered a gentian in chalk grassland near Wendover, vc24'. - Knipe (1988).	vc24	**1875**
'A painting by Mrs E.G. Hood, annotated 'Coombe Downs, Bucks, Oct 8th [18]73". - Pope (2003).	vc24	1873

The earlier references to this plant are filled with confusion. Druce (1926) dismissed the nineteenth century record as a gross error, and it was not seen again until 1982. More recently Edgington (2003) has raised the delightful possibility that James Dickson and William Anderson found it near Tring, Hertfordshire in 1812, five miles from the Wendover locality.

Geranium columbinum

* 'Geranium Columbinum fol. magis dissectis, pediculis longissimis flore magno, In several places of Hampshire, Mr. Goodyer'. - Merrett (1666: 45).	vc12	**1666**
'Geranium columbinum foliis magis dissectis, pediculis longissimis. Aug 1654 … I found it wild … '. - Goodyer ms.11, see Gunther (1922: 191).	vc12	1654

See note under *G. dissectum*. Gerard (1597: 793) has a perfectly good drawng and description of this, as 'Geranium columbinum', ignored by Clarke and others, perhaps because of lack of differentiation from the other similar species.

Geranium dissectum

'Geranium columbinum majus dissectis foliis, *Ger. emac.*'. - Ray (1690: 155.8).		**1690**
'Pes Columbinus, Geranium 2, *Dios*. foliis amplis valde dissectis'. Road-side, Erith. - Johnson (1629: 10). 'Geranium columbinum maximum foliis dissectis. In hedges about Marston, and on that part of Botley-Causey next Oxford, in great plenty'. - Plot (1677: 145).	vc16	1629

Rose, in Gilmour (1972: 64 & 109), points out that Johnson did not differentiate between *G. columbinum* and *G. dissectum* and that this record might well apply to either. Ray (*op.cit.*) was the first to describe both and to clearly separate them.

Geranium lucidum

* 'Geranium Saxatile of Thalius ... Mr. Goodyer found it growing plentifully on the bankes of the highway leading from Gilford [Guildford] towards London neere unto the townes end'. - Johnson (1633: 988).   vc17   **1633**

Geranium molle

* 'Geranium … Pes columbinus of the commune Herbaries … maye be called in englishe Douefote'. - Turner (1548: D. iij). 'Geranium secundum Dios. Pes columbinus'. Between Hampstead and Kentish Town. - Johnson (1629: 12).   **1548**

See notes under *G. pusillum*. Smith (1824: 238), in implying that they were not really satisfactorily sorted until the late 18th century, writes 'For the accurate discrimination of these species, about whch all botanists had been uncertain, I am … indebted to my late friend Mr Davall. They can never more be mistaken'. Davall, by the way, died in 1798.

Geranium pratense

* 'Geranium Batrachiodes. … pratis & glareosis … Angliae lasciuit, floribus caeruleis'. - L'Obel (1571: 297).   **1571**

Geranium purpureum

'From plants observed in Jersey by Mr Sherard 'G. saxatile lucidum foliis Geranii Robertiani. In several places near the shore. I have since found it near Swannigh [Swanage], Dorsetshire'. - Ray (1690: 238).   vc9, 113   **1690**

Sherard's visit to Jersey was in 1689.   vc113   **1689**

Le Sueur (1984) was happy with the attribution of Ray's record. For a very full account of the history of this critical species and its varieties, see Wilmott (1921).

Geranium pusillum

* 'Geranium malachoides sive Columbinum minimum. On the hill of health'. - Ray (1660: 61).   vc29   **1660**

See also *G. molle*. Hanbury & Marshall (1899) quote Johnson's 'Geranium alterum *Dioscor.* sive Columbinum. *Tab.* pes columbinus *Dod.* et ejus altera species foliis roajoribus magis dissectis ' (1632: 18) as including this species. But Rose, in Gilmour (1972: 109), seems unsure whether the species in Johnson is *G. molle* or *G. pusillum*. Oswald & Preston (2011: 207n *et seq.*) think that it is not clear whether Ray correctly separated the two species.

Geranium robertianum

* 'Sideritis … semeth to be the herbe called in englishe herbe Roberte'. - Turner (1548: G. iij). 'Geranium Robertianum, Herba Roberti'. Between Hampstead and Kentish Town. - Johnson (1629: 12). **1548**

Geranium rotundifolium

* 'In muris, tectis et ad sepes circa Bath et Bristol copiose, inter Battersea et Wandsworth'. - Hudson (1762: 265). vc6, 17 **1762**

There is a note against the entry in Druce & Vines (1907: 151), describing a record from near Oxford, 'Although the plant appears to have been known to Buddle … '. Buddle died in 1715. Trimen & Dyer (1869: 68) also mention Buddle's record of 'various places around London' dating it to 'about 1712'. vc21 1715

Hudson gives as a synonym 'G. columbinum majus flore minore caeruleo' - Ray (1724: 358). Ray gives colour varieties, including one from Hackney. Hudson seems to have a better grasp of the species than Ray.

Geranium sanguineum

* 'Geranium haematodes, sive sanguinale. Red or bloody Cranesbill … About S. Vincents rock nigh Bristow'. - Johnson (1634: 38). vc34 **1634**

Geranium sylvaticum

* 'Geranium batrachoides montanum … In pratis montosis & dumetis Agri Westmorlandici et Eborancensis copiose'. - Ray (1670: 130). vc69 **1670**

'frequente in Anglia, in sepibus et sylvis'. - annotations made by T. Penny on C. Gesner's plant illustrations (Vol 8.13 Blatt 431 recto) dated as by 1565 in Foley (2006b). 1565

See also notes of a 1660 Itinerary in Raven (1950: 114).

Geum rivale

* 'Caryophyllata montana purpurea. Found … in Wales … by Mr. Thomas Glynn'. - Johnson (1633: 996). **1633**

Geum urbanum

* 'Geum … Avennes … groweth communely about hedges'. - Turner (1548: D. iij). 'Caryophillata'. Between Gravesend and Rochester. - Johnson (1629: 3). **1548**

Gladiolus illyricus

* '… found by the Rev. W. H. Lucas … at least a mile from any house … in the New Forest, Hampshire, in 1856'. - Babington (1857: 158). vc11 **1857**

'Discovered by a Mrs Phillips in the midst of a wild tract of copse and heath called the "Apse" or "America" woods [Apsecastle Wood, Shanklin, Isle of Wight]. Only one plant was noticed; it was in bud on 7th July 1855'. - More (1862).   vc10   1855

Babington's note should be consulted for very interesting additional details.

Glaucium flavum

* 'Papaver corniculatum ... horned poppy or yealow poppy ... groweth in Dover clyffes, and in many other places by the sea syde'. - Turner (1548: F. i). ' … also in Dorsetshyre'. - Turner (1562: 77).   vc15   **1548**

Glaux maritima

* 'Glaux exigua ... Angliae plerisque mari conterminis'. - L'Obel (1571: 178). 'Between Whitstable and the yle of Thanet in Kent'. - Gerard (1597: 448).   **1571**

Jackson, in Hanbury & Marshall (1899), cites L'Obel (1571: 407) 'Glaux vulgaris … Ex Angliae nos litoreis item mediterraneis procul mari, Londini serandam curavimus'.

Glebionis segetum

* 'Chrysanthemum segetum … segetes frigidiorum tractuum Galliae, Germaniae, & Angliae scatent'. - L'Obel (1571: 237).   **1571**

'in Anglia' - annotations made by T. Penny on C. Gesner's plant illustrations dated as by 1565 in Foley (2006b)   1565

Grigson (1955) cites Fitzherbert in the *Boke of Husbandrie*, 1523 writing '… an yll wede, and groweth commonlye in barleye and pees'.

Glechoma hederacea

'Ground Ivye is very common in all this countrie, and growth in many gardens and shadowy moyst places'. - Lyte (1578: 389). 'Hedera terrestris Ale-hoofe. It is founde as well in tilled, as in untilled places, but most commonly in obscure, base and dark places, upon dunghils and by the sides of houses where eves do drop'. - Gerard (1597: 705).   **1578**

L'Obel (1571: 268) has this too, and whilst he doesn't specify England, gives the English name of 'Ale Hove'. Grigson (1958) too cites the name 'Hofe' in Old English.

Glyceria declinata

'about 1872, by roadside opposite Shedfield House [N. of Fareham]'. - Townsend (1883).   vc11   **1884**

'North of London, 1844. T. Moore, **CGE**'. - Kent (1975). Hampstead, vc21, 1846, T. Moore, det. Hubbard, **K**.   vc21   1844

Townsend refers to his earlier description from St Mary's Marsh, Isles of Scilly as *G. plicata* var. *nana*. (1864: 118). He also wonders if this equates to var. 2 of *G. fluitans* (saltmarsh, leaves blunt ... glaucous; 6-8 fl. (*fluitans* mostly 10fl) in *Withering* ii. 185 [this is from the 1830, 7th edition, where the plant is described from Lymington, vc11].

Glyceria fluitans

* 'Gramen fluviatile ... Flote grasse ... growes everywhere in waters'. - Gerard (1597: 13). 'Gramen aquaticum panicula longissima nondum descriptum ... about London'. - How (1650: 50).	**1597**

Glyceria maxima

* 'Gramen majus aquaticum ... In fennie and waterie places'. - Gerard (1597: 6).	**1597**

Glyceria notata

* [Distinguishing from *G. fluitans*] 'in some meadows north of London'. - Moore (1845).	**1845**
'Gramen aquaticum cum longissima panicula [a specimen from Mr. Brewer of Bangor] no 31 [see Brewer ms.], is the earliest record of *G. plicata*'. - Druce & Vines (1907: 120).	1743

Moore notes that E.B. 1520 ('*Poa fluitans*') represents this species. Druce refers to a specimen in Herb. Du Bois, 1700, but his 1928 account of this does not include this species. The record from Druce & Vines is dated as 1743, the date of Brewer's death.

Gnaphalium luteoalbum

'Gnaphalium ad Stoechadem citrinam accedens *J.B.* On the walls and dry banks, very common [Jersey, Sherard.]' - Ray (1690: 238).	vc113	**1690**
Sherard's visit was in 1689.	vc113	1689

The first record other than from the Channel Islands was from Little Shelford, Cambs, 1802, 'far from any house'. - E.B. 1002.

Gnaphalium norvegicum

* Described, as a variety of *G. sylvaticum* 'sometimes occurring upon the highland mountains'. - Lightfoot (1777: 472).	**1777**

Smith (1800, 1825) and Babington (1847) merely included this species as a form of *G. sylvaticum*, stating that they seemed to be different, but couldn't say why. Yet in the next edition on Babington's *Manual* (1851) he was confident enough to say 'Fl. longer in proportion to the inv. [involucre], and quite distinct from *G. sylvaticum*. — Highland Mountains'.

Gnaphalium supinum

* as *G. alpinum* 'upon the tops of the highland mountains not unfrequent, as on Creg-chaillech [Creag na Caillich], near Finlarig, in Breadalbane, upon Mal-ghyrdy [Meall Ghaordie], and upon the mountains fof Glenlyon, Glenurchy [Glen Orchy], and Glenco, &c. Mr. Stuart'. - Lightfoot (1777: 470, with a figure).   vc88, 98   **1777**

Finlarig is just north of Killin.

Gnaphalium sylvaticum

* 'Centunculus ... Chafweede, it is called in Yorkeshyre cudweede'. - Turner (1548: C. i). 'Gnaphalium Anglicum ... Tertio a Londino miliari opacae sylvae clivus multam alit, cis Tamesim'. - L'Obel (1571: 202).   **1548**

Gnaphalium uliginosum

* 'Gnaphalium vulgare ... Upon drie sandie banks'. - Gerard (1597: 518).   **1597**

This attribution by Clarke and followed by Druce and others, seems less than convincing as regards habitat, but is followed too by Ray (1660).

Goodyera repens

* as *Satyrium repens* 'We found it growing amongst the *Hypna*, in an old shady moist hanging birch wood … facing the house of Mr. Mackenzie, of Dundonald, about two miles from the head of Little Loch Broom ... Ross-shire. It has also been found in a wood opposite to Moy-hall, on the south side of the road to Inverness, as we have been informed by Dr. Hope'. - Lightfoot (1777: 520).   vc96, 105   **1777**

as *Satyrium repens* 'from bridge of Nairn & Dalmagerie [Dalmagarry] … in the wood called Cragenon [Craiganeoin]'. - Journal of James Robertson, May 1767, Henderson & Dickson (1994: 39).   vc96   1767

There is an entry 'in a wood opposite Moy Hall south side of ye road to Inverness 1767' in Hope, 1768, see Balfour (1907). Moy Hall and Bridge of Nairn are on the A9 S. of Inverness.

Groenlandia densa

* 'Tribulus aquaticus minor muscatellae floribus ... In the river by Droxford in Hampshire. John Goodyer'. - Johnson (1633: 824).   vc11   **1633**

'Tribulus aquaticus minor, muscatellae floribus. 2 Junii 1621 … in the river by Droxford in Hampshire'. - Goodyer ms.11, ff.120a, 122, see Gunther 1922: 126) - *see image p.226*.   vc11   1621

> 126   JOHN GOODYER
>
> sight of the plant the 2 of Iune, 1622.—*MS.* ff. 120 a, 122 ; *Ger.* . 823.
>
> [In the fair copy (f. 122) Goodyer altered the name 'floribus uvae' to 'quercus floribus'. The rough copy (MS. f. 120 a) has the notes 'about ye beg. of July, or fortnight before Lamas' and 'both begunn to flower 2 Junij 1621' written in the margin. If the year 1622 given in the text be correct, Goodyer would appear to have been at Durford two years running on the 2nd of June.]
>
> Pondweede. *Potamogeton densus* L.
> Tribulus aquaticus minor, muscatellae floribus.   2 Junii 1621
> This hath not flatt stalkes like the other, but round, kneed, and alwaies bearing two leaves at every ioint, one opposite against another, greener, shorter and lesser then the other, sharpe pointed, not much wrinkled and crumpled by the edges. *Clusius* saith, that they are not at all crompled. I never observed any without crumples and wrinckles. The flowers grow on short small footstalkes, of a whitish green colour, like those of *Muscatella Cordi*, called by *Gerard*, *Radix cava minima viridi flore:* viz. two flowers at the top of every foot-stalke, one opposite against another, every flower containing foure small leaves: which two flowers beeing past there come up eight small husks making six several waies a square of flowers. The roots are like the former. This groweth abundantly in the river by Droxford in Hampshire. It flowereth in June and July when the other doth, and continueth covered over with water, greene, both winter and somer.—*MS.* ff. 120 a, 122 ; *Ger. emac.* 823-4.
>
> [The rough copy (MS. f. 120 a) has the note 'in running brooks' and a reference to Clusius cclij after the name.]
>
> ? *Melilotus indica* L.
> Melilotus Indiae orientalis.   11 Junij 1621
> This is in stalks, branches, leaves flowers and smell altogether like *Melilotus Italica* Camerarii, but smaller more branched & delicate, not growinge so high, & the stalks are greene and have no redd at all. ‖ The seed is also like, but smaller. The root groweth downe right, and is small white, with a very fewe thredds, and perisheth when the seed is ripe the same yere it is sowen. ‖ This hath not beene written of by any that I find. I receaved seeds thereof from Mr. William Coys often remembred.—*MS.* f. 97.
>
> *Epipactis violacea* Dur.
> Nidus avis flore et caule violaceo purpureo colore. an Pseudo-leimodoron Clus. Hist. Rar. Pl. p. 270.   29 June 1621
> This riseth up with a stalke about nine inches high, with a few

Goodyer ms.11, ff.120a, 122, reproduced in Gunther, R.T. 1922. *Early British Botanists and their gardens.*

See species account : Groenlandia densa, p.225

Gymnadenia borealis

Described as a variety of *Habenaria gymnadenia*, from 'Watendlath, Cumberland, 1907'. - Druce (1918: 172). As *G. conopsea* subsp. *borealis*, Rose (1991).   vc70   **1918**

'On a down near Helston'. - Hore (1842).   vc1   1842

G. borealis is not uncommon at the Lizard, and *G. conopsea s.s.* does not occur in Cornwall, so it seems reasonable to attribute Hore's record to this species. But note that Foley & Clarke (2005) consider that it is probably this species referred to in Johnson's (1634: 55) record of the aggregate, see below.

Gymnadenia conopsea s.l.

* 'Orchis palmata minor calcaribus oblongis ... In montosis. Spurre-flowred Orchis, or Red-handed Orchis'. - Johnson (1634: 55).   **1634**

Gymnadenia conopsea s.s.

Druce (1918: 170)		**1918**

'Orchis palmata rubella cum longis calcaribus rubellis, Red-handed Orchis with long spurs', Hb. Sloane 124: 40, **BM**. Cunswick Scar, vc69, 1895, C.H. Waddell, det. Perring, **BFT**.   1715

The Druce reference is not terribly enlightening, other than to show that he was aware that other taxa might be involved. It has not been possible to find earlier determined herbarium records.

The Hb. Sloane specimen is in the Herbarium of Buddle, who died in 1715, and though the specimen has not been expertly determined it seems to be *G. conopsea s.s.* Another from Dodsworth's (d.1697) collection (Hb. Sloane 34: 7) is not named but is possibly this too.

Gymnadenia densiflora

Druce (1918: 170) refers to a record of var. *densiflora* Dietrich from J.W. Heslop Harrison (1917) from Billingham, Durham. This is repeated in Graham (1988).   vc66   **1918**

Borthwick, vc83, 1835, Balfour, det. F. Rose, **E**.   vc83   1835

There is a specimen in **BM** from 'Sowerby's Herbarium' with no date. It is quite possibly that which was used in his illustration in E.B. 10 (1791), though that illustration does not look quite as dense as the herbarium specimen. Lindley (1835) described β *densiflora*, but only from Gotland and Öland, in Sweden.

Gymnocarpium dryopteris

'*Polipodium dryopteris* … It groes in Birks-Wood, North-Dean-Wood, Puttin-Park, and other woods in the neighbourhood of Halifax, plentifully'. - Bolton (1785: 53).   vc64   **1785**

'In divers places in Westmorland, plentifully'. - Ray (1670: 95). Edgington (2013) dates this to his 1661 Itinerary.   vc69   1661

Bolton was the first to distinguish this species. Druce cited 'Dryopteris Tragi' from Gerard (1597: 974) or Johnson (1633: 1135). But Nelson (1959), Stearn (1965) and Rydén *et al.* (1999) all equate Turner (1538, B) 'Driopteris … it grows on trees … the common people call this the brake of the tree' with this species. This seems most unlikely.

Gymnocarpium robertianum

'*Polipodium dryopteris* variety, growing in White Scars near Ingleton and in the Peak of Derbyshire'. - Bolton (1785: 53).   vc57, 64   **1785**

'A specimen of *G. robertianum* in Du Bois' herbarium at **OXF** from 'a wood by Panswick [Painswick] four miles from Gloster', collected by Beller c.1710'. - Edgington (2013: 138). 'Filix ramosa minor,   vc33   1710

Raii Syn. 125, in tolerable plenty about Kilnsay, particularly among loose lime-stones on the right hand side of the Girling Trough near Coniston' - Curtis (1787: no.130) describing a visit in 1782.

Lees (1888: 514) cites the Curtis as the first record, from Conistone and near Kilnsey. Edgington (2013) has 'more securely [as *G. robertianum*] were plants from Faringdon in Hampshire, known before 1654 to John Goodyer (see Gunther, 1922: 189), whose delightful record was added by How to his interleaved *Phytologia*, now at Magdalen College, Oxford'.

Druce (1886) cites a record from Bobart [Morison] (1699), but omits this in his second edition (1927).

Hammarbya paludosa

* 'Bifolium palustre, Marsh Bisoile ... In the low wet grounds betweene Hatfield and S. Albones [St. Albans]; in divers places of Romney Marsh'. - Parkinson (1640: 505.2).   vc15, 20   **1640**

Raven (1947: 261) qualifies this identification 'what seems to be *Hammarbya*'.

Parkinson, J. 1640. *Theatrum Botanicum . . . an Herball.*

Hedera helix s.l.

* 'Hederam greci cisson vocant, Angli Ivy'. - Turner (1538: B. verso).   **1538**

Hedera helix s.s.

as *H. helix* subsp. *helix*, McAllister & Rutherford (1990).   **1990**

'Hederam greci cisson vocant, Angli Ivy'. - Turner (1538: B. verso).   1538

It has been very difficult to find an early record that is unequivocably this taxa, but since Turner was not familiar with parts of Britain where *H. hibernica* is found, it seems reasonable to take his as the first evidence.

Hedera hibernica

as *H. helix* subsp. *hibernica*, McAllister & Rutherford (1990). **1990**

'Tresco Abbey ruins'. - Woodley (1822). vc1 1822

Since *H. helix s.s* barely occurs in Cornwall, the Tresco record is assumed to be this.

Helianthemum apenninum

* 'Chamaecistus montanus Polii folio D. Plukenet … Found by the Doctor upon Brent Downs in Somerset-shire near the Severn Sea'. - Ray (1688a: 4). vc6 **1688**

Helianthemum nummularium

* 'Helianthos … crescit in Anglia'. - L'Obel (1571: 185). 'In Kent'. - Gerard (1597: 1102). **1571**

Helianthemum oelandicum

'Cistus, mas breviore folio *P. 659*. Cistus mas dentatus *G.1276*. Small-leav'd male Cistus, Accepit Dr Bowles insignis Boranicus mihiq; communicavit, ex Cambro-Britannia'. - Merrett (1666: 27). **1666**

Raven (1947: 294-7) gives an account of Bowles' life, and from that it appears his only trip to Wales was in 1632. 1632

Merrett's record is not cited in Smith (1825: 23) but seems satisfactory for the species. Clarke cites 'Chamaecistus sive Helianthemum Alpinum folio Pilosellae minoris Fuchsii *J.B* … Found by Mr. Newton on some Rocks near Kendall in Westmorland'. - Ray (1688a: 4).

Helleborus foetidus

* 'Helleboraster maximus … growes wilde in many woods and shadowie places in England'. - Gerard (1597: 826). 'At Cherry Hinton'. - Ray (1663: 6). **1597**

See also Turner (1538: B) 'Elleborum nigrum … vulgus cantabrigiense [Cambridge], vocat Bearefote'. It is uncertain as to whether Turner was referring to a wild or cultivated plant.

Helleborus viridis

'[of Berefoote] with the broader leaf … in Northumberlande in the west parke besyde Morpeth …'. - Turner (1551: L. vj). vc67 **1551**

Clarke and Druce both cite Turner (1562: 160b), as does Nelson (1959) but the wording in Turner (1551) is identical. See also Chapman & Tweddle (1989: 284).

Helminthotheca echioides

* 'Cirsium ... oure Langdebefe'. - Turner (1548: C. iij). 'Lang de Bese …in great plentye betwene Sion and Branfurd' [Brentford, Middx.]. - Turner (1568: i. 143).   **1548**

Turner (1538: A. iv) 'Cirsion … that well-known prickly herb which we call Langdebefe … and the country people call Cattes tayle'. Stearn (1965) assigns to this species, but Rydén *et al.* (1999) only with a query.

Heracleum sphondylium

* 'Sphondilium ... Cow persnepe, or rough Persnepe. It groweth in watery middowes'. - Turner (1548: G. v).   **1548**

Herminium monorchis

* 'Orchis pusilla odorata … In the chalk-pit close at Cherry-hinton'. - Ray (1663: 7).   vc29   **1663**

In a letter from Ray to Peter Courthope, dated from Trinity College in July 1661, see Gunther (1934).   vc29   1661

Herniaria ciliolata

Babington (1836).   vc1   **1836**

On Ray's 1667 Itinerary as H. glabra, 'in great plenty, on the rocks, Lizard Point, Cornwall'. - See Lankester (1846: 190, n.) and also Raven (1950: 144). 'Herniaria … Glabram invenimus spontaneam in promontorio Cornubiensi The Lizard Point dicto'. - Ray (1686: 214).   vc1   1667

Babington was the first to clarify the taxonomy and referred the Lizard plant to this species.

Herniaria glabra

* 'Herniaria ... in barren and sandie grounds, and is likewise founde in darkish places that lie wide open to the sunne'. - Gerard (1597: 455).   **1597**

Hieracium agg.

* 'Hieracium intybaceum … Endiues Hawke-weed, in untoiled places … '. - Gerard (1597: 234.6).   **1597**

Gerard gives many habitats.

Hierochloe odorata

'About ten minutes' walk from the town of Thurso there is, by the river-side, a farm-house known by the name of the Bleachfield, opposite to which, on the eastern bank of the river, there is a precipitous section of boulder clay; opposite to the clay cliff, and   vc109   **1854** fringing the edge of the stream ... Farther up, between Geize and a section of boulder clay a little below Todholes, the plant may likewise be picked in hundreds. *Hierochloe* has never failed to appear in these localities for twenty years'. - Dick (1854).

At Dr. Hooker's suggestion I send you, for the herbarium, two specimens of a rare British plant (*Hierochloe Borealis*), which, after having been erased from the list, was rediscovered near Thurso by Mr. Robert Dick, who states that it flowers so early in the year as May and the beginning of June, disappearing soon afterwards; so that there was no wonder I and others could not find it in Don's station, Glen Kella, Angusshire, as botanists seldom go there before the end of July'. - Gourlie (1855).

Smiles (1878) reproduces Dick's account and adds more details concerning Dick's sighting of this species over a number of years. The traditional first record is that cited by Clarke and others 'In a narrow mountain valley called Kella, Angus. G. Don'. - Hooker (1821: 28); the illustration in E.B.S. 2641 (1830) notes 'Discovered in 1812'. But this record was doubted in Ingram & Noltie (1981), and is not mapped in Preston *et al.* (2002).

Himantoglossum hircinum

* 'Orchis saurodes, sive Tragorchis Maximus, *Ger*. The Lizard flower, or Great Goatstones. Nigh the highway betweene Crayford and Dartford in Kent'. - Johnson (1641: 27).	vc16	**1641**

This was in fact found by George Bowles, see Raven (1947: 296), citing Ray (1670: 341).

Hippocrepis comosa

* 'Ferrum equinum siliquis in summitate, *Bauh*. Hedysarum minus, *Tab*. Horse-shoe. On the Hils about Bath and betweene Bath and Marleborow'. - Johnson (1634: 35).	vc6, 8	**1634**

Also 'collected by Goodyer on 9th August 1634 at Buttersworth Hill [Sussex], Goodyer ms.11, f.53v'. - Gunther (1922: 187).

Hippophae rhamnoides

* 'Rhamnus primus Dioscoridis L'Obelio sive littoralis. Sea Buckes thorne with willow-like leaves ... In our owne land by the sea coasts in many places'. - Parkinson (1640: 1005). 'Betwixt Sandwich and Deal' [Kent]. - Merrett (1666: 104).		**1640**
'This Rhamnus I founde betweene Dover and Foulkestone by the sea syde under the Clyffes, ao.1582, with reddish beries Orenge colored'. - Manuscript notes of William Mount, of c.1584, in Gunther (1922: 255 - 263), and identfied by him as this species.	vc15	1582

Hippuris vulgaris

* 'Cauda equina foemina. . .. In waterish places'. - Gerard (1597: 957). 'Polygonum foemina semine vidua, *Lob.*'. Between Margate and Sandwich. - Johnson (1632: 19).

1597

Holcus lanatus

* 'Gramen pratense paniculatum molle, *Bauh.* Hoary Meadow-grasse'. - Johnson (1634: 38).

See also Johnson (1633: 30a), where he mentions grasses described by Bauhin that he cannot quite equate with British species.

1634

Holcus mollis

* 'Gramen caninum paniculatum molle … in arvis inter segetes'. - Ray (1688: 1285; 1688a: 11).

1688

Holosteum umbellatum

* 'On the city walls of Norwich … first noticed by Mr. John Pitchford in spring, 1765'. - Rose (1775).

vc27 **1775**
vc27 1765

Homogyne alpina

On the rocks among the Clova mountains are to be found the following interesting plants … *Tussilago alpina* …'. - Don (1813: app.13).

Don's account was just a bare record in a list and was long discounted. But it '… was [re]found in August 1951 in the parish of Cortachy and Clova [Corrie Fee], by A.A.P. Slack'. - Ribbons (1952).

vc90 **1813**

Honckenya peploides

* 'Anthyllis prior lentifolia Peplios effigie, maritima … In Angliae Insulis Australibus, ea praesertim quae Portlandia vocatur'. - L'Obel (1571: 195).

vc9 **1571**

Hordelymus europaeus

* 'Gr. Secalinum maximum, the Greatest Rye grass, *P.1144*. In the woods a mile west from Petersfield'. - Merrett (1666: 57).

Presumably a Goodyer record, if from Petersfield? The vice county could be vc11 or vc12.

vc12 **1666**

Hordeum marinum

* 'Gramen secalinum palustre et maritimum … In palustribus frequentissimum est'. - Ray (1688: 1258, 1688a: 12).

1688

Hordeum murinum

'Phenix is called ... Hordeum murinum by the Romans. By some it is called Avena sterilis. It grows in fields and on tiles recently coated. This herb is called Waybent by some people, because it grows close to paths'. - Turner (1538, B. iiij. verso).

1538

'Hordeum murinum ... The wal Barley whiche groweth on mud walles' - Turner (1548: D. v. verso). This is Clarke and Druce's first date, but Turner (*op.cit.*) seems perfectly acceptable. But Clarke and Druce go on to equate with *Lolium perenne* the rest of Turner's paragraph 'Phenix Dioscoridis semeth to be the herbe which is called in Cabridgeshire Way Bent. It is like unto barlei in the eare'. Stearn (1965) ignores the possibility of two species in Turner.
A really convincing record is 'Hordeum spontaneum spurium Holcus Plinii, Anguillarae'. Near Rochester. - Johnson (1629: 2).

Hordeum secalinum

* 'Gramen secalinum or Rie-grasse ... commonly in our medowes'. - Johnson (1633: 29.4).

1633

Hornungia petraea

* 'Nasturtiolum montanum annuum tenuissime divisum ... brought me by Richard Kayse from S. Vincents Rock near Gorams Chair, in the parish of Henbury, three miles from Bristol'. - Bobart in Ray (1690: 236).

vc34   1690

Hottonia palustris

* 'Viola palustris ... I have not founde such plentie of it in anyone place as in the water ditches adjoining to Saint George [Southwark] his fielde neere London'. - Gerard (1597: 679).

vc17   1597

Gerard describes this species and an aquatic *Ranunculus* on this page. It is not certain which he is referring to in this location.

Humulus lupulus

* 'Lupus salictarius ... Hoppes do growe by hedges and busshes both set and unset'. - Turner (1548: E. ij). 'Lupulus salictarius'. Between Nash and Queakes. - Johnson (1632: 12).

1548

Turner (1538: B. ij. verso) has 'Lupus salictarius ... Anglice Hoppes', but this could as easily be a record of a cultivated plant.

Huperzia selago

'Chamaepeuce *Turneri part 1 pag.129*. Dr Turner's Dwarf pine ... on a very high hill in the North called Boulser'. - Merrett (1666: 25). [Boulsworth Hill near Colne, Lancashire]. 'Muscus erectus abietiformis. Snowdon, Caderidris, and the other high mountains of Wales: also on Ingleborough-hill in York-shire, and the hills of the Peak in Darby-shire'. - Ray (1670: 214).

vc59   1666

'on Aug 20th, 1658 … on the moorish hills and in the pastures hereabout [Buxton], I first observed Vitis idaea thymi foliis, muscus erectus abietiformis'. - Ray's First Itinerary, see Lankester (1846: 125). | vc57 | 1658

Lees (1888: 518) identifies this species with Merrett's (1666: 25) 'Chamaepeuce foemina seu polyspermos at Dovestones in the side of Yorkshire, neer Widdop and a mile this side Lippock in Hampshire', but Townsend (1904) and others equate this with *Lycopodium clavatum*.

Hyacinthoides non-scripta

* 'The comune Hyacinthus is muche in Englande about Syon and Shene and is called in Englishe crowtoes'. - Turner (1548: D. vj). | vc17, 21 | **1548**

Turner (1538: C. verso) 'Vaccinium … I have not found out with certainty what it is called by our people', which implies familiarity with the plant as a wild species.

Hydrilla verticillata

'*Hydrilla verticillata* in England' [from Eshwaite Water, N. Lancs]. - [Tattersall] (1914), followed by Pearsall (1914). | vc69 | **1914**

See also Bennett (1914) and Druce (1915a).

Hydrocharis morsus-ranae

* 'Morsus Ranae … ad arcem Londini regiam & moenia'. - L'Obel (1571: 258). | vc21 | **1571**

Druce seems to be citing the wrong species (*Alba minor*) and the wrong page (257) from L'Obel.

Hydrocotyle vulgaris

* [The] 'shepe kyllinge penny grasse that growth in merishe and waterye groundes'. - Turner (1562: 169). | | **1562**

This is Turner comparing this species with *Umbilicus rupestris*. This is not identified as such in Chapman *et al.* (1995). In fact Fitzherbert (1523) had described it as 'peny-grasse, and growth lowe by the erthe in a marsshe grounde, and hath a leafe as brode as a peny of two pens, and never beareth floure … what thynge rotteth sheepe'.

Hymenophyllum tunbrigense

'Adiantum petraeum perpusillum Anglicum, foliis bifidis vel trifidis … found by G. Daire [Dare], circa Tunbrigiam'. - Ray (1686: 141 & 1696: 47). 'This was first shewn me by Mr Newton, who in company with Mr Lawson, found it on Buzzard rough crag, near Wrenose [Wrynose], among the Moss. Afterward I understood that | vc16 | **1686**

Mr Dare, an Apothecary in London had formerly found it about Tonbridge in Kent'. - Ray (1688a: 1).

Hymenophyllum wilsonii

'Wet rocks. Abundant in the highlands of Scotland and in many parts of Ireland. About Killarney, Mr. W. Wilson'. - Hooker (1830: 450). ' … luxuriantly growing in company [with *H. tunbridgense*] in the rocky woods which border … Upper Lake of Killarney … in the summer of 1829'. - W. Wilson in E.B.S. 2686. vcH1 **1830**

'Adiantum petræum perpusillum Anglicum foliis bifidis vel trifidis … D. Jacobo Newtoni … Hardknot and Wrenose … Iulio mense Anni 1682'. - Ray (1686: 141). vc69 1686

Babington's view (see Lankester (1848: 197)) was that it was this plant, not *H. tunbrigense*, that was reported as 'Adiant. Petr. perpusillum', on Buzzard rough Crag, close to Wrenose, in Westmorland, with Ja. Newton' in a letter from Mr Tho. Lawson to Mr Ray, Apr. 9th 1688. Edgington (2013) agrees and this view seems almost certain to be correct. Druce cites 'Wilson found it on Cwm Idwal in 1828', but this reference has not been traced and is possibly an error, as surely Hooker would have mentioned it.

Hyoscyamus niger

* 'Henbayne … groweth about the sea syde and about guttures and ditches, about townes and cytyes'. - Turner (1551: B. vij. verso). **1551**

This also in Turner (1538: A. ij. verso) as 'Apollinaris … Angli Henbayne', but this could as easily be a record of a cultivated plant, since it was so well known medicinally.

Hypericum androsaemum

* 'Clymenon Italorum … Tutsan … plurima Angliae sylvis, lucis, et nemoribus, praesertim Bristoiae et Glocestriae conterminis'. - L'Obel (1571: 279). vc34 **1571**

'Anglia' - an annotation made by T. Penny on C. Gesner's plant illustrations. Foley (2006b) dated these as by 1565, but did not include this species. For further details see Zoller & Steinmann (1987-91). 1565

Hypericum elodes

* 'Upon divers boggy grounds of this kingdome is to be found … Ascyrum supinum ἑλώδης' - Johnson (1633: 542). 'On a rotten moorish ground not farre from Southampton'. - Johnson (1634: 21). **1633**

The record came from Goodyer, see Edgington (2007).

Hypericum hirsutum

* 'Androsaemum excellentius dictus, seu magnum ... In Anglia aestate superiore prope Bristoiam: In Vincentii praeruptis & sylvosis trans flumen'. - L'Obel (1571: 173).	vc6, 34	**1571**

Clive Lovatt (*in litt.*) wonders if there has been confusion over the early records for this species and *H. montanum*.

Hypericum humifusum

* 'Hypericum minus repens'. Between Sandwich and Canterbury. - Johnson (1632: 21).	vc15	**1632**

Hypericum linariifolium

'In Jersey'. - Babington (1837). This was in a bare list of additions to the flora there.	vc113	**1837**
' … below La Crete guardhouse, St Catherine's Bay, Jersey, in July, 1837'. - C.C. Babington in E.B.S. 2851, 1840.	vc113	1837

The first record away from the Channel Islands is 'Found by the Rev. Thomas Hincks of Cork among granite rocks near the banks of the Teign, Devon in the summer of 1838'. - *Ann. N. H. 1st ser. vi. 76.1840*'. The note in *Annals* came out in October 1840; the plate in E.B.S was published in December 1840.

Hypericum maculatum

* as *H. dubium* 'Discovered in July 1794 by Dr. John Seward of Worcester growing plentifully about Sapey in that County'. - E.B. 296.	vc37	**1796**
'Hypericum quadrangulum flore majore, in sepibus prope Dolgelle'. - Ms diary of Dillenius' Journey into Wales (1726), reproduced in Druce & Vines (1907: xlix).	vc48	1726

Druce claims this as 'the first British record of *H. quadrangulum*, L. (*H. dubium*, Leers)'.
Lower Sapey is in Worcestershire; Higher Sapey in Herefordshire.

Hypericum montanum

* 'Androsaemum Matthioli ... About Bristow and Bath'. - Parkinson (1640: 577). Androsaemum Clampoclarensium, *Col.* On St.Vincent Rocks near Bristol, and on the Hedge sides beyond Lewsham [Lewisham, vc16]'. - Merrett (1666: 8).	vc6, 34	**1640**

Clive Lovatt (*in litt.*) wonders if there has been confusion over the early records for this species and *H. hirsutum*.

Hypericum perforatum

* 'Hypericon ... nonnulli herbam ... Perforatam, vulgus appellat Saynt Johns gyrs'. - Turner (1538: B. verso). 'Ad insulam Sheppey … Hypericon ubiq. in aridis'. - Johnson (1629: 5).	**1538**

Hypericum pulchrum

* 'Hypericum pulchrum, *Tragi*'. Near Canterbury. - Johnson (1632: 25). 'In S. Johns Wood and other places'. - Johnson (1633: 540).   vc15   **1632**

Hypericum tetrapterum

* 'Ascyron is not very common in England, howe be it I sawe it thys last yere in Syon parck ... square saint Johans grasse'. - Turner (1548: B. iij). 'St. Peter's woort, or square St. John's grasse groweth plantifully in the North part of England, especially in Landesdale [Westmorland] and Craven'. - Gerard (1597: 434).   vc21   **1548**

Hypericum undulatum

* 'I first collected the plant on the 7th Aug. 1861 ... at Common Wood, about four or five miles from Plymouth, Devon; and ... Fursdon, Egg Buckland'. - Briggs (1864).   vc3   **1864**

'… collected by myself in July 1857, in Treverry Bog, near Helston, which I then mounted unnamed …'. - Rogers (1891: 99 n.).   vc1   1857

See also Babington (1864a), for a fuller account confirming Briggs' find.

Hypochaeris glabra

* 'Hieracium parvum in arenosis nascens seminum pappis densius radiatis ... On the gravelly heath-grounds near Middleton in Warwick-shire'. - Ray (1670: 167).   vc38   **1670**

'in Anglia' - annotations made by T. Penny on C. Gesner's plant illustrations dated as by 1565 in Foley (2006b)

Hypochaeris maculata

* 'Hieracium montanum caule aphyllo non ramoso flore pallidiore ... On Gogmagog hills and Newmarket heath'. - Ray (1663: 6).   vc29   **1663**

'In agro Cambrigensi in Anglia crescit prope pagum Shelfordii in collibis aridis'. - Notes by Thomas Penny, see Zoller *et al.* (1972-80) and Foley (2005) for the justification for the 1565 date.   vc29   1565

Hypochaeris radicata

* 'Hieracium longius radicatum ... In untoiled places', &c. - Gerard (1597: 234.7). Near Canterbury. - Johnson (1632: 25).   **1597**

Chapman *et al.* (1995: 58) identify Turner's plant (1562: 18) as this species 'Hyoseris is like unto succory… the herbe that I take for this hath a rowghe leafe growinge harde by ye grounde, indented after the maner of succory or dandelion, but the teth are not so sharpe ... It groweth in sandy baron groundes and about casten diches that have muche sand in them'. Pulteney cites Lyte (1578: 566.2) ' Hieracium minus primum ... growing in untoyled places'. This too is possible.

Hypopitys monotropa

* 'Orobanche Verbasculi odore … It grows at the bottom of Trees in the woods near Stoken-Church and we find it mentioned in some ms. notes of the famous Mr. Goodyer'. - Plot (1677: 146).	vc24	**1677**
'Orobanche verbasculi odore. 22 August 1620….in a hedgerowe in a ground belonging to Droxford farme'. - See Gunther (1922: 122), from a ms copy of a lost note of Goodyer written on back of f.249v of Banister's herbarium siccum in Hb. Sloane, **BM**.	vc11	1620

Stokenchurch is in Bucks, see Druce (1932). Vines & Druce (1914: 173) note that Bobart published his record (in Morison (1699)) after Plot, but said that he had known it for 26 years, i.e. since c.1673.

Iberis amara

* 'Iberis Nasturtii foI. *P. 854*. On the Clifs beyond Deal Castle in Kent'. - Merrett (1666: 66). 'In arvis circa Henley et alibi comitatu Oxoniensi'. - Hudson (1778: 285).	vc15	**1666**

Ilex aquifolium

* 'Ruscus … An holy tre … e cujus corticibus ipse admodu puer viscu confeci'. - Turner (1538: C.).	vc67	**1538**

Addyman (2016) translates this as 'I made birdlime from this as a young child', and thus locates it to Morpeth.

Illecebrum verticillatum

* 'Alsine floribus ad instar Polygoni marini … This was sent me from Cornwall. *N.D.*'. - Merrett (1666: 5). 'Polygonum serpyllifolium verticillatum … In humidioribus pascuis & palustribus udis circa Pensans, et alibi versus extremum Cornubiae angulum'. - Ray (1670: 248).	vc1	**1666**
On Ray's third Itinerary 'Sat. June 28th 1662 … between St Columb and St Michael (and in several other places) … Alsine palustris minor serpillifolia'. - Lankester (1846: 186) where Babington feels that *Illecebrum* is intended by this polynomial. See also Raven (1950: 128).	vc2	1662

Is Ray's 'St Michael' referring to St Michael's Mount?
Johnson (1633: 563) has a good picture of this species, as 'Polygala repens … growing commonly …', which must be a misconception; in any event he gives no record of it growing in Britain.

Impatiens noli-tangere

* 'Persicaria siliquosa ... First found to grow in this kingdome by ... Mr. George Bowles ... first in Shropshire on the banks of the river Kemlet at Marington in the parish of Cherberry [Chirlbury]'. - Johnson (1633: 446).　　vc40　　**1633**

This was in August 1632. See P. Oswald, in Trueman *et al.* (1995).　　vc40　　**1632**

Imperatoria ostruthium

'Of great Pilletorye of Spayne … after the Dutch, Maisterwurt ... I never hard that it grew wilde in Englande, savinge aboute Morpeth, in the North parke there'. - Turner (1568: 36).　　vc67　　**1568**

Turner describes it both in part ii of the Herbal (1562: 83) but again, with no reference to the earlier record, in part iii (1568: 36). The later reference is the only one that suggests it is found in England. Druce cites the wrong edition.

Inula conyzae

* 'Baccharis Monspeliensium. Plowmans Spiknard. ... In divers places in the West parts of England'. - Gerard (1597: 647).　　**1597**

Inula crithmoides

* 'Crithmum Chrysanthemum ... In the mirie Marsh in the yle of Shepey as you go from the Kings ferrie to Sherland house'. - Gerard (1597: 428).　　vc15　　**1597**

This is where the main road, A249, crosses onto the island.

Inula helenium

* 'Helenion ... in pratis villarum et praediorum Angliae, Germaniae, Galliae'. - L'Obel (1571: 246). 'In the fieldes as you go from Dunstable to Puddlehill: also in an orchard as yee go from Colbrook to Ditton ferrie, which is the way to Windsore, and in sundrie other places, as at Lidde [Lydd], and Folkestone, neere to Dover by the sea side'. - Gerard (1597: 649).　　**1571**

Inula salicina

* 'In June, 1843, I went to Portumna, Co Galway … on the margins of Lough Derg'. - Moore (1865).　　vcH10　　**1865**
　　vcH10　　**1843**

the actual site is in H10.

Iris foetidissima

* 'I have sene a litle flour delice growyng wylde in Dorset shyre'. - Turner (1562: 23). 'Xyris ... called in the yle of Purbek Spourgewurt'. - Turner (1562: 171). 'I have seene it wild in many places, as in woods and shadowie places neere the sea'. - Gerard (1597: 53).	vc9	**1562**
Turner's visit to Dorset was in 1550 - see Raven (1947: 102 - 103). Both Clarke and Druce incorrectly cite p.32 of Turner.	vc9	**1550**

Iris pseudacorus

'Acorum … In Northumberland a seg, those who live in Ely and in the swampy places near Ely call it a lug. The most widely used names are Gladon, a Flag and a Yelowe floure delyce'. - Turner (1538, A.ij.) .	vc29, 67	**1538**

The Turner 1538 reference seems perfectly acceptable. Clarke and Druce cited 'Acorus [meaning *Acorus calamus*] groweth not in England wherefore they are farre deceyued that use the yelowe flour de luce whiche some call gladen for Acorus'. - Turner (1548: A. v. verso) (misprinted by Clarke as 1546).

Isatis tinctoria

* 'Glastum sylvestre. The wilde kinde groweth where the tame kinde hath been sowen'. - Gerard (1597: 394).	**1597**

Nelson (1959) cites Turner (1538: B. ii. verso) 'Isatis … vulgus herbam allellat Wad'. But this is no indication that it was cultivated or in the wild.

Isoetes echinospora

'In two places near Llanberis in N. Wales. Loch of Park near Aberdeen. Ben Voirlich, Dunbartonshire'. - Babington (1862). 'On August 6, 1845, in company with Dr. Balfour and a small party of students, I visited Loch Sloy and Ben Voirlich, near Loch Lomond, in Scotland, and gathered what I then called *I. lacustris* in a little pool near to the top of the mountain'. - Babington (1863a), where he adds 'the first record of its discovery in Britain is, I believe, contained in a letter addressed by me to the Linnean Society of London, and read at the meeting of March 20, 1862, and published in the Proceedings of the Society for that year, at p. lxiii'.	vc49, 92, 99	**1862**
'Samuel Brewer …. from Llyn Ogwen [Llyn-y-cwn] … in June 1727'. - Druce & Vines (1907: 95). Note however that Edgington (2013) argues persuasively for Richardson's discovery (see note below) to date from his meeting with Lhwyd in 1696.	vc49	1727

In Ray (1724: 307) Dillenius adds 'Subularia fragilis, folio longiore & tenuiore ... Dr Richardson', thus adding another species to that described by Ray as *I. lacustris*. But it was Babington who first satisfactorily clarifed matters.

Isoetes histrix

'This little plant I found in damp spots on L 'Ancrisse Common [Guernsey] … in June 1860'. - Wolsey (1861).	vc113 vc113	**1861** 1860

See also 'My first information of its discovery was contained in a letter from a very intelligent and obliging gardener in Guernsey, Mr. G. Wolsey, dated 15th October, 1860. It contained a bit of the *Isoetes*, asking its name, and mentioning that it was found on L'Ancresse Common, in Guernsey, in June of that year'. - Babington (1863a). For the first record other than from the Channel Islands, from the Lizard, in Cornwall, see Robinson (1919).

Isoetes lacustris

'Subularia lacustris … Aizoides Fusiforme Alpinorum lacuum, D. Lloyd'. - Ray (1690: 210).	vc49	**1690**
In a letter to Edward Lhwyd, 3rd November 1689 thanking him for the account. See Gunther (1928: 201) and Raven (1950: 247).	vc49	1689

See Camus (1991) where she mentions Newman,1840, discussing two forms of quillwort that may correspond to this species and *I. echinospora*.

Isolepis cernua

'Gramen junceum maritimum exile Plimostii … found as well at Plimmouth as Dover, in their wet grounds'. - Parkinson (1640: 1271.9). As *Scirpus savii* 'Renoyle in the West of Ireland … Anglesea, W. Wilson'. - Hooker, in E.B.S. 2782 (1832).	vc3, 15	**1640**
See also Briggs (1880: xxvii), where he cites L'Obel (1655: 67) 'Gramen Juncoides exile omniumque tenuissimum Pleymuense & montosis udis juxta Plymouth in Cornubiae [*sic*]'. As Briggs points out this might refer to either species, but when taken in conjunction with the Parkinson record, *I. cernua* seems more likely. L' Obel died in 1616.	vc3	1616

Parkinson described both species, and a late 18C gloss in the Wellcome Library's copy (Shelfmark: 4833/D) has this so annotated.
Merrett (1666: 55) has 'plentiful where a small dril issueth out betwixt the rocks, near the South-east point, in the Isle of Wight', cited in Bromfield (1856: 544). Clarke has 241Anglesea, July 8, 1828. - Wilson (1831) (as *S. setaceus*). *S. Savii* [*I. cernua*] was, however, intended'.
Although there is no doubt that these early records describe this species, probably Hooker was the first to really appreciate there were two species and to describe their differences.

Isolepis setacea

* 'Juncellus omnium minimus Chamaeschoenos. Dwarfe - rush'. - Johnson (1634: 45).   1634

Pulteney cites L'Obel (1571: 44) as his first record. Here, under 'Juncellus', L'Obel has ' … seen everywhere on dry gravels and sterile northern heaths'.

Jasione montana

'Scabiosa montana minima, *Lob.*'. Hampstead Heath. - Johnson (1632: 37).   vc21   1632

'Scabiosa minima hirsuta. At Sheete [N. edge of Petersfield]'. - Goodyer ms.11, f. 57v, see Gunther (1922: 110).   vc12   1618

Rose, in Gilmour (1972: 62), suggests that 'Scabiosa montana minima. Near Dartford. - Johnson (1629: 9)', attributed by Hanbury & Marshall (1899) to *Jasione* was probably a slip for *Scabiosa columbaria*, as Johnson expressly says the locality is chalky. Johnson may have found *Jasione* later on Dartford Heath, and confused the two records'. This seems quite possible, but it seems safer to go for Johnson's definite later record. Pulteney points out that Lyte (1578: 109.3) includes 'Sheepes scabious … groweth in the fieldes and by the waysides'.

Juncus acutiflorus

* 'Gramen junceum sylvaticum, *Tab. Ger.*'. Hampstead. - Johnson (1634: 34).   vc21   1632

Gunther (1922: 255 - 263) in listing manuscript notes of William Mount, of c.1584, interprets 'Ponde grasse…in a Ponde in Snodelande…'. as this species. This is not convincing enough.

Juncus acutus

'Juncus major durior'. Between Margate and Sandwich. - Johnson (1632: 16).   vc15   1632

Rose (in Gilmour, 1972: 114-115) thought this referred to *J. acutus*, not *J. maritimus*, as was taken by Clarke. His seems a convincing argument. Clarke's first date was 'Juncus maritimus acutus sive capitulis Sorghi, Bauh. ... Pricking Sea Rush:' - Johnson (1641: 24), but Parkinson (1640: 1193.4) predates this, referring to 'Juncus maritimus capitulis Sorghi', distinguishing it from his 1194.7, 'Juncus acutus maritimus Anglicus' (*J. maritimus*)'. This is followed by Bicheno (1818) and others. Note that Johnson's 1641 record was made on his 1639 trip, and the record can be located to Harlech, in Merioneth (Johnson 1641: 11).

Juncus alpinoarticulatus

'About ten days after this [August 4th 1887] Mr. James Brebner sent me some rushes from near Killin, and amongst these were three panicles of what appeared, without doubt, to be *Juncus alpinus*. … Finally on August 20th, Mr. Barclay and I botanised the neighbourhood of Blair-Athole, and found a small bed of the the long-looked-for rush'. - White (1887).   vc88, 89   **1887**

'Glen Doll, 1870, J.C. Hutcheson'. - Ingram & Noltie (1981).   vc90   1870

Druce says 'possibly found by Mr Hutcheson in 1870 in Glen Dole, Angus'. There is a specimen in **BM** recently determined by Jaroslaw Procków, a researcher from Poland, from Duddingston, vc83, 1864, an unlikely location unless from the adjacent Arthur's Seat.

Juncus articulatus

* 'Gramen junceum aquaticum Bauhini, *Park.*'. - Ray (1660: 68).   vc29   **1660**

Juncus balticus

* as *J. arcticus* 'Sands of Barry near Dundee, Mr. [T.] Drummond'. - Hooker (1821: 104),   vc90   **1821**

'There are specimens in Don's herbarium from the Sands of Barrie and near Montrose, labelled *J. filiformis*'. - Druce (1904). Don died in Jan. 1814.   vc90   1814

Juncus biglumis

* 'On the top of Mal-ghyrdy, a mountain lying betwixt Glenlochy and Glenlyon, and upon Ben Teskerney, both in Breadalbane. Mr. Stuart'. - Lightfoot (1777: 1100).   vc88   **1777**

'Mr. John Stewart'. - Hope, 1768, in Balfour (1907).   vc88   1768

Ben Teskerney [Beinn Heasgarnich], Mal-ghyrdy [Meall Ghaordie].

Juncus bufonius s.l.

* 'Gramen junceum. Rush grasse groweth in salt marshes neere unto the sea, where the marishes have been overflowen with salt water'. - Gerard (1597: 4). 'Lane going by Totenam Court toward Hampsted'. - Johnson (1633: 4).   **1597**

Juncus bufonius s.s.

Cope & Stace (1978) were the first to properly separate these species, reporting it from every vice-county in the British Isles.   **1978**

'Gramen junceum parvum sive Holosteum Mathioli … Toad-grass' in Hb. Sloane 125: 38, **BM**'. 'Blackheath'. - Druce & Vines (1907: 126).   1715

The Hb. Sloane specimen is in the Herbarium of Buddle, who died in 1715.

Juncus bulbosus

* 'Gramen junceum capsulis triangulis minimum ... In boggy places by rills of water on Hampsted-heath, and else-where'. - Ray (1670: 150).   vc21   **1670**

Clarke adds 'But Johnson's 'Gramen junceum ... the lesser varietie ... upon the bogs upon Hampstead Heath' (Johnson (1633: 4) was probably this'. Edgington (2011) agrees.

Juncus capitatus

as *Schoenus minimus* [from] T.F. Forster jun. 'Insula Sarnia sed rarissime in loco infra pagum Bovet dictum inter Fort-George et Fernain Bay. ... Nova Species'. - Symons (1798: 197).   vc113   **1798**

McClintock (1975: 222) points out 'But it was in fact found (in Guernsey) by Capt. J. Finlay, c1780, who described it in detail in Latin in a letter to Banks dated 22nd Feb 1787, and supported by four specimens (now in **BM**)'.   vc113   1787

The first record other than from the Channel Islands is 'Land's End, Cornwall, on an extensive peat moor near the sea. Here it was in profusion, being by far the most abundant plant on the moor ... '. - Beeby (1872). He also recorded it from Kynance the same month.

Symons, J. 1798.
Synopsis plantarum insulis Britannicis indigenarum.

Juncus castaneus

* as *Juncus Jacquini* 'Primus in Scotia invenit D. Dickson'. - Symons (1798: 87). **1798**

White (1898) has G. Don, Ben Lawers, 1794 as first record. This agrees with Druce (1904: 138, under 'discoveries'), and he also refers to an entry under *J. supinus* for this species. vc88 1794

First found on Ben Challum, Perthshire by Dr John Stuart'. - Bicheno (1818: 322). See also Mitchell (1986), recognising Stuart as the true discoverer. But neither of these sources give the date of his discovery.

Juncus compressus

* 'Gramen junceum maritimum, *Lob. Ger.* Sea Rush-grasse'. - Johnson (1641: 23). 'In meadows and by footpaths'. - Ray (1660: 68). **1641**

Juncus conglomeratus

* 'Juncus laevis glomerato flore, *Lob*. Smooth round-headed Rush'. - Johnson (1634: 45). **1634**

Juncus effusus

* 'Juncus laevis vulgatus … Pratis udis deterioribus … in Anglia potissimum'. - L'Obel (1571: 44). **1571**

Juncus filiformis

* 'Juncus parvus calamo supra paniculam compactam longius producto … In Westmorlandia non procul ab oppido Ambleside dicto invenit D. Newton'. - Ray (1688: 1305; 1688a: 13). vc69 **1688**

Juncus foliosus

Simpson & Walters (1959). **1959**

Barmouth, 1833, J. E. Bowman, det. Cope & Stace, **BM**. vc48 1833

Juncus gerardii

* as *J. coenosus* 'In salsis copiose'. - Bicheno (1818: 309). **1818**

'Gramen junceum maritimum vel palustre cum pericarpis rotundis. From Adam Buddle'. - Druce 1928: 485). 1715

Mr Buddle died in 1715. There is another specimen cited in Druce & Vines (1907: 125).

Juncus inflexus

* Juncus acutus … Upon drie and barren groundes'. - Gerard (1597: 31). **1597**

Pulteney points out that Lyte (1578: 510.2) refers to the 'our common harde rushes' and accurately describes the pith.

Juncus maritimus

'Juncus acutus maritimus Anglicus English Sea hard Rushes ... Differs little growing from the other Sea Rushes, but that the pannickle is longer and slenderer ... groweth on many of our English coasts'. - Parkinson (1640: 1194.7). **1640**

Clarke, following Hanbury and Marshall (1899) used 'Juncus major durior'. Between Margate and Sandwich. - Johnson (1632: 16) as his first date, but Rose (in Gilmour, 1972:114-115) thought this referred to *J. acutus*, not *J. maritimus*. Rose's seems a convincing argument.

Ray (1688: 1303) cites Parkinson, adding a record from 'circa Camalodunum in Essexia', and this is followed too by Bicheno (1818).

Juncus pygmaeus

* 'Mr Beeby sent another little rush ... in a damp hollow on the top of the downs near Kynance Cove'. - Trimen (1872a).   vc1   **1872**

Remarkably, Beeby's discoveries of this species and *J. capitatus* appeared on the same page of the *Journal of Botany*. See also Trimen (1873) for a full discussion of Beeby's find.

Juncus ranarius

Cope & Stace (1978) were the first to properly separate this from *J. bufonius s.l.*   **1978**

'Hampstead Heath, vc21, 1819, det. D. McKean, **E**'. 'Little Chelsea, 1842, det. Cope, **MANCH**' - Kent (2000). 'Goodworth Chatford, Andover, Hants, 2 Sept. 1861, det. Cope & Stace, **K**'. 'Deal Sandhills, Aug. 1862, J.T.Syme, det. Cope & Stace, **BM**'.   vc21   1819

See also Edgington (2007) who, citing Johnson (1633: 5) '..a variety of this (*J. bufonius*) ... groes on the bogges upon Hampstead heath', wonders if this might be the first record for this species. However, Oswald & Preston (2011: 467-8n) find this implausible, not least because *J. ranarius* is a plant of brackish habitats. However, although Hampstead Heath is not on the coast, Mark Spencer (*in litt.*) notes that there were a few anomalous records of coastal species from that general area in the 19th & early 20th century.

Juncus squarrosus

* 'Oxyschoenos sive Juncus acutus CambroBritannicus, Welsh hard or sharpe Rushes ... Found by Dr. L'Obel in his life time upon a high hill in Wales called Bewrin [Berwyn], in sundry wet and moorish groundes in many places thereabouts'. - Parkinson (1640: 1194.2).   vc50   **1640**
   vc50   1616

L'Obel died in 1616. Berwyn is SW of Llangollen. See also Raven (1947: 239).

Juncus subnodulosus

* 'Juncus nemorosus folio articuloso ... In Peckham-field inveni cum glumis albis; *Dood. Not.*'. - Ray (1724: 433).   vc21   **1724**

'Gramen junceum aquaticum magis sparsa panicula. From Mr Adam Buddle'. - Druce (1928: 485).   1715

Buddle died in 1715. Clarke adds 'but see Ray (1696: 276), where under 'Gramen junceum aquaticum' he adds a 'Secunda specie elatius & majus est' ; and Davies (1810) using '*subnodulosus*' for the first time. Oswald & Preston (2011: 213n) conjecture that Ray may have been describing this species in his Cambridgeshire Flora (1660: 68).

Juncus trifidus

* 'I found it upon the summits of the highland mountains to the south of Little-Loch-Broom in Ross-shire, and on Ben-na-scree [Beinn Sgritheall], above Arnesdale, on the side of Loch-Urn [L.Hourn], in Invernessshire &c.'. - Lightfoot (1777: 183).   vc97, 105   **1777**

Lightfoot's tour was in 1772.   vc105   1772

Juncus triglumis

* 'Juncello accedens graminifolia plantula capitulis Armeriae proliferae D. Lhwyd ... On the mountains of Wales'. - Ray (1696: 275).   vc49   **1696**

Carter (1955a) includes this plant from Caernarvonshire in his list of first British records for Lhwyd.

Juniperus communis

* 'Juniperus ... groweth ... in many places in Englande'. - Turner (1548: D. vij). 'Most plentuouslye in Kent, it groweth also in ye bysshopryche of Durram and in Northuberlande'. - Turner (1562: 25).   **1548**

'Tetbyri [Towbury] Castle. It is 2 miles from Theokesbyri [Tewkesbury] …. in the parish of Twyning. Leland VI, 71' - Riddlesdell (1948). Leland would have visited here by 1545, almost certainly before 1543.   vc33   1545

<1379. '... in many places in Kent, …and … near a town they call Chatham'. - Henry Daniel, see Harvey (1981), and also Allen (2001). Heaton (see Walsh,1978) was the first to record the subsp. *nana*, by 1641, near Gort, H15.

Kickxia elatine

'Elatine altera … Both these plants I have founde … about Southfleete in Kent'. - Gerard (1597: 501) [this and *K. spuria*]   vc16   **1597**

Clarke and Druce both cite L'Obel's record for *K. spuria* for this species too, but I can see nothing in his text to justify this.

Kickxia spuria

* 'Elatine Dioscori. Veronica foemina *Fusch.* ... Rarius ... Anglia comparet [with other European countries]'. - L'Obel (1571: 197). **1571**

See also 'Both these plants I have founde ... about Southfleete in Kent'. - Gerard (1597: 501) [this and *K. elatine*]. However Nelson (1959) refers to Turner (1548: H. v.) 'Veronica ... Fluellyng' as this species, citing Lyte (1578: 26) 'as stating that 'The floures be yellow with small croked tayles' with an appropriate figure of a *Kickxia*. Clarke and others identify that plant as *Veronica officinalis*.

Knautia arvensis

* 'Scabiosa ... Scabius ... groweth amongst ye corne is rakest [rankest] of al other'. - Turner (1568: 68). **1568**

Kobresia simpliciuscula

* as *Schoenus monoicus* 'The honour of making this singular plant known is due to Mr. Dickson, who gathered it in the County of Durham in 1799. The Rev. Mr. Harriman had indeed found it in 1797; but not being aware of its novelty he liberally disclaims the merit of its discovery ... finds it wild on the mountain of Cronkley, Durham; also near Widdy bank in Teesdale forest'. - E.B. 1410.   vc65   **1805**

Horsman (1995: 199) cites from a letter from Harriman to Smith in 1803 'I could wish you would say in the letter press that Mr. Dickson found it upon Cronkley, **Yorkshire** [emphasised] in 1799.   vc65   1797

Koeleria macrantha

* 'Gramen pumilum hirsutum spica purpuro-argentea molli ... Observatur nobisque communicata D. Dale ... in montosis et campestribus sed rarius'. - Ray (1688: 1265).   **1688**

'On Banstead Downs, M. Doody'. - Ray (1688a: 11).   vc17   1688

Koeleria vallesiana

'I went down on Oct.16th to the Uphill habitat'. - Druce (1905), after discovering Dillenius' specimens in **OXF**.   vc6   **1905**

'on the Uphill Church (Hill), gramen incognitum, spica sparti, foliis reflexis, glaucus striatis, radice fungosa, foliis figura quodammodo Statices, July 16th, 1726. Dillenius' Diary of Journey into Wales and also The Synopsis (Appendix) 412.17'. - Druce & Vines (1907: xlv &120).   vc6   1726

There are specimens of both records in **OXF**.

Koenigia islandica

'collected by Mr. H.M. Montford and Dr. Montford (Dr. Nellie Carter) near the summit of the Storr in the Isle of Skye, on August 31st 1934 … its most southerly station in Europe'. - Burtt (1950). vc104 **1950**
vc104 1934

Lactuca saligna

* 'Lactuca sylvestris laciniato minima, nondum descriptum … This was found on a bank and in a ditch by the side of a kind of drove or lane leading from London road [in Cambridge] to the river, just at the water near a quarter of a mile beyond the spittle-house end'. - Ray (1660: 83). vc29 **1660**

Lactuca serriola

'Lactuca sylvestris foliis dissectis'. - Johnson (1633: 309). **1633**

'Lactuca agrestis, 13th Sept. 1621'. - Goodyer ms.11, f.100, see Gunther (1922: 159). 1621

Rose, in Gilmour (1972), suggests Johnson's record should be *L. virosa*, but Edgington (2010) agrees with Clarke. But for a masterly account see Oswald (2000), who convincingly argues both that this was the plant that Johnson found, and that he found it, not on Hampstead Heath, as Clarke and Druce cite, but 'betweene London and Pancridge [Old St. Pancras] Church'.

As for Goodyer's record, although we think of *L. agrestis* as a modern synonym for *L. virosa*, he describes that latter species too (Goodyer ms.11, f. 99) and to my mind, and to that of Philip Oswald (*in litt.*), satisfactorily separates them.

Lactuca virosa

* 'Lactuca agrestis opii odore vehementi, soporifero & viroso … in Anglia frequens'. - L'Obel (1571: 89). 'Ad insulam Sheppey'. - Johnson (1629: 5). **1571**

Lamiastrum galeobdolon

* 'Lamium luteum … Hampsteed, neere London, Lee [Leigh-on-Sea] in Essex, neere Watford and Bushie in Middlesex'. - Gerard (1597: 567). vc18, 21 **1597**

The diploid *L.* subsp. *galeobdolon*, confined in Britain to N. Lincs. was not collected until 1969 and was elucidated in Wegmüller (1971).

Lamium album

* 'Lamium … dead nettle or whyte nettle … groweth commonly in hedges'. - Turner (1548: D. vij. verso). **1548**

But see Turner (1538: A.ij. verso) 'Anonium … et Anglice Archangell aut Dede nettell', which implies familiarity with the plant as a wild species.

Lamium amplexicaule

* 'Alsine Hederula altera ... In gardens among pot herbes, in darke shadowie places, and in the fieldes after the corne is reaped'. - Gerard (1597: 493).

1597

Lamium confertum

* 'May 12th 1836 ... Dr. N. Tyacke exhibited specimens of *L. intermedium* lately found by him in several localities in the neighbourhod of Edinburgh and pointed out the differences between it and *L. amplexicaule*, *incisum* and *purpureum*'. - Bot. Soc. Edinb. (1837: 27).

vc83 **1837**

Druce cites Don (1813) as the first record for this species. The relevant text reads '*Lamium* novaspecies which I propose to call *intermedium*: it is perfectly distinct from all the known species: in cultivated fields'.

1813

See also E.B.S. 2914.

Lamium hybridum

* 'Lamium rubrum foliis profunde incisis, In the Kings new Garden near Goring-house'. - Merrett (1666: 69).

vc21 **1666**

Buckingham Palace is on the site of Goring House

Lamium purpureum

* 'Galeopsis ... red Archaungel ... It is lyke Archaungel, but it hath a purple floure, and lesse leaves and shorter. It groweth in hedges'. - Turner (1548: D. ij. verso).

1548

Clarke adds 'Mr. Britten makes this *Stachys sylvatica*, but the 'Red Archangel' of Gerard and Ray is *Lamium purpureum*'. Turner (1538: B. verso) 'Galeopsis ... vulgo Rede archangell' seems acceptable, though Stearn (1965) interprets as either this species or *Stachys sylvatica*. Stearn also has the 1548 reference as probably *Stachys*, but Turner's description seems to favour *Lamium*.

Lapsana communis

'Lampsana ... groweth most commonly in all places, by high way sides, and especially in the borders of gardens amongst wortes and potherbes'. - Lyte (1578: 560). 'Lampsana. Dock Cresses ... Upon walles made of mudde or earth, and in stonie places'. - Gerard (1597: 199).

1578

Lyte has a good illustration, but Gerard's figure is of another plant. Johnson (1633: 255) substitutes a correct one.

Lathraea squamaria

* 'Dentaria maior *Matthioli* ... I found it growing in a lane called East-lane, upon the right hande as ye go from Maidstone in Kent unto Cockes Heath, ... it doth grow also in the fieldes about Croydon ... and also in a woode in Kent neere Crayfoord ... it groweth likewise neere Harwood in Lancashire, a mile from Whanley [Whaley, NW of Blackburn]'. - Gerard (1597: 1388).

vc16, 17, 59   **1597**

Lathyrus aphaca

* 'Aphaca *Lob. Dod*. Elatine 3. *Tab*'. Near Gravesend. - Johnson (1632: 29). 'I have found it in the corne fields about Dartford in Kent and some other places'. - Johnson (1633: 1250), with a good illustration.

vc16   **1632**

See Edgington (2007) for details of a record from John Goodyer sent to Johnson to incorporate into his 1633 work.

Johnson, T. 1632. *Descriptio Itineris Plantarum Investigationis.*

> stis, Dod. Sylvestris, Trag.
> **Polypodium.**
> Glaux vulgaris, Lob. Glycyrhiza sylvestris, Gesn. Fænugræcum syl. Trag.
>
> Sylva relicta, gratiisq; amico actis, ingressi viam quæ Cantuaria Londinum ducit, qua potuimus celeritate per Sittingburne ad Rochester, sicq: tandem ad Gravesend venimus, & præter prius recensitas plantas, sese nostro conspectui hæ offerebant.
>
> *Muscus aquaticus denticulatus.*
> *Alopecuros maxima Anglica paludosa, Lob.*
> Ruta muraria, Dod. Capillus Veneris, Trag. Salvia vitæ, Lob. Ad. Lugd. Ad. Lugd. Adianthum album. Tab. paronychia, Matth.
> *Calamentha pulegii odore. Lob.*
> Quinquefolium peiræum majus, Tab. Tormentilla facie, Ger. Pentaphyllum album, Matth. exiguum alterum, Tragi.
> *Pilosella repens vulgaris.*
> Phyllitis, Matth. Dod. Lingua cervina, Offic. Scolopendria vulgaris, Trag.
> *Aphaca, Lob. Dod. Elatine 3. Tab.*
> *Antirrhinum medium.*
> Tanacetum vulgare, Matth. Dod. Artemisia Diosc. Tab.
>
> 8   *Castanea*

Lathyrus japonicus

* 'Pisa in littore nostro Britannico ... carta quodam in loco Suffolciae inter Alburnum et Ortfordum [AIdborough and Orford] oppida ... autumnnali tempore anni 1555 sponte sua nata sunt adeo magna copia sufficerint vel millibus hominum'. - Joannis Caius (1570: ii. 29).   vc25   **1571**
vc25   1555

It transpires that Caius had previously told Conrad Gesner of this occurrence, and Gesner had apparently included it in Vol. 4 of his *Historia animalium* (1558: 256), thus pre-dating the Caius record. The full details of this 'miraculous' appearance in Suffolk are set out in Halliday (1999).

Lathyrus linifolius

* 'Astragalus ... peaserthnut ... I haue sene thys herhe of late in Coome parke'. - Turner (1548: B. iiij). 'in Come Park and on rychemunde heth'. - Turner (1551: E. v. verso).   vc23   **1548**

Clarke, followed by Salmon (1931), claim 'Coome parke' for Surrey, probably influenced by the additional site of 'rychemunde heth [Richmond Heath]'. Chapman & Tweddle (1989: 323, n.72) identify this with the home of Sir Thomas Elyot, nr. Woodstock, Oxon. Elyot was a correspondant of Turner, and this is more likely (but not given in Druce (1927). Raven (1947: 72) also seems to favour this. See also *Genista anglica*.

Lathyrus nissolia

* 'Ervum sylvestre *Dod*.'. Hampstead Heath. - Johnson (1632: 36).   vc21   **1632**

Pulteney cites Lyte (1578: 507.1) 'Catanance ... groweth in copses and in pastures', with a perfectly good drawing, but is it from England?

Lathyrus palustris

* 'Lathyrus flore ex caeruleo et rubro mixto. In a wet marsh ground on the left hand of Peckham Field from London'. - Merrett (1666: 70). 'Vicia Lathyroides purpuro-caeruleis floribus. Found and brought to us by Th. Willisel in Peckham-field, on the back of Southwark, in a squalid watery place'. - Ray (1670: 316).   vc17   **1666**

Lathyrus pratensis

* 'Vicia siI: *Tabernaemontani*, A. phacoides'. Near Cliffe. - Johnson (1629: 8).   vc16   **1629**

Lathyrus sylvestris

* 'Lathyrus Angustifolia - there groweth great store thereof in Swanescombe wood a mile and a halfe from Greene-Hithe in Kent as you go to a village thereby called Betsome [Betsham]'. - Gerard (1597: 1054).	vc16	**1597**

Clarke's date of 1579 is a transposition error

Leersia oryzoides

* 'I have found it in three places in the Henfield level ... I first observed it on the 24th of September last [1844]'. - Borrer (1844).	vc13	**1844**
'Kew, September'. Specimen at **K** in Rev S.C. Goodenough's herbarium.	vc17	1805

Goodenough died in 1827, but left Kew for Rochester in 1802. The latest date in his herb is 1805 ... so probably collected before then (Tom Cope, at Kew, *in litt*.).

Legousia hybrida

* 'Speculum Veneris ... I found it in a field among the corne by Greenhithe'. - Gerard (1597: 356). 'Among the corn in Chelsey field'. - Johnson (1633: 440).	vc16	**1597**
'in Anglia' - annotations made by T. Penny on C. Gesner's plant illustrations dated as by 1565 in Foley (2006b)		1565

Pulteney cites Lyte (1578: 171) 'Auicularia ... groweth in goode grounde, in fields amongst wheate, or where as wheate hath growne'. There is, however, no specific indication that this was in England.

Lemna gibba

* as *L. minor* β *gibba* 'in stagnis'. - Hudson (1778: 399).		**1778**
'Lower Bishop's pool, Northwick, near Worcester, and in a pool near the east side of Malvern Chase. Sep. 8, 1776. ST. [Stokes]'. - Withering (1787: 1020).	vc37	1776

Lemna minor

'Lens palustris is believed to be the herb which people call Duckes meat. It grows on the surface of pools or other stagnant and motionless waters'. - Turner (1538: B. iij).	**1538**

Clarke and Druce both used 'Lens palustris ... duckes meate ... well knowen in England and specially of them that have pondes'. - Turner (1562: 33 verso), but the earlier reference seems perfectly acceptable.

Lemna trisulca

* 'Hederula aquatica ... I found it once in a ditch by Bermondsey house neer to London and never else where'. - Gerard (1597: 681).   vc17   **1597**

Leontodon hispidus

* 'Hieracium Dentis leonis folio hirsutum. *Ger. em.* Dandelion Hawke-weed. Meadowes'. - Johnson (1634: 43).   **1634**

Turner's 'Hiosyris ... roughe Dandelion' (1562: 18) was probably this, though Clarke and Druce are doubtful. Johnson (1633: 303) also pictures this, but he does not seem certain that he has seen it in Britain.

Leontodon saxatilis

* 'Hieracium pumilum saxatile asperum praemorsa radice *C.B.* ... Found on the banks of New Parks and divers other places about Oxford'. - Bobart in Ray (1690: 237).   vc23   **1690**

Clarke wonders whether 'Hieracium montanum saxatile, Columnae'. Thanet. - Johnson (1632: 10) is this species. Rose, in Gilmour (1972: 106), follows that.

Lepidium campestre

* 'Thlaspi vulgatissimum ... provenit cultis et incultis arvis Angliae'. - L'Obel (1571: 73).   **1571**

Lepidium coronopus

* 'Coronopus Ruellii ... In some places of England they call it Swynescressis'. - Lyte (1578: 95). 'In Tonthill fielde neere unto Westminster'. - Gerard (1597: 347).   **1578**

Lepidium heterophyllum

* 'Thlaspi Vaccariae incano folio perenne ... in montosis Cambro-britanniae & alibi observavi'. - Ray (1670: 296).   **1670**

'Thlaspi Vaccai folio incano perpetuum. in ye Ile of man'. This is a herbarium specimen which Preston (2016) suggests was collected by Willughby when he visited there with Ray in 1660.   vc71   1660

Preston (*op.cit.*) dates Ray's Welsh record to 1662.

Lepidium latifolium

* 'Lepidium ... called wyth a false name Dittany ... groweth in Morpeth in Northumberlande by a water called Wanspeke' [River Wansbeck]. - Turner (1548: E. j).   vc67   **1548**

The Northumberland site is probably an escape - the species was much used as a hot condiment before Horseradish came into favour - and the first record from a possibly native site seems to be 'Dittander is planted in gardens, and is to be found wild also in England in sundrie places, as at Clare by Ouenden in Essex'. - Gerard (1597: 187).

Lepidium ruderale

* 'Thlaspi minus, Bowyers Mustard ... In England in sundrie places wilde'. - Gerard (1597: 204.4). 'Near the sea in many places v.g. Maldon in Essex, Lynne in Norfolk, Truro in Cornwall, &c'. - Ray (1670: 296).   **1597**

Leucanthemum vulgare

* 'Bellis major ... Ang. Greate Daysie … margines agrorum & pratorum familiaris'. - L'Obel (1571: 200).   **1571**

Leucojum aestivum

* 'Found undoubtedly wild betwixt Greenwich and Woolwich ... also ... in the Isle of Dogs'. - Curtis (1777-1798: v. 23).   vc16   **1788**

This part was issued in Nov. 1788, see Stevenson (1961).

Leymus arenarius

* 'In littoribus maritimis frequens'. - Hudson (1762: 45). 'In Insula Bute; et prope Exmouth in Devonia'. - Hudson (1778: 56).   **1762**

Ligusticum scoticum

'Imperatoriae affinis, Umbellifera maritima Scotica. Scotish Sea-masterwort'. - Sutherland (1683: 366), Sibbald (1684: ii. 32). 'On a certain sandy & stony hill six miles from Edinburgh towards Queensferry in Scotland'. - Ray (1688a: 13).   **1683**

According to Desmond (1977), Sir A. Balfour was the discoverer of this plant.

Ligustrum vulgare

* 'Cypros ... Ligustrum ... Pryuet groweth very plentuously in Cambrich shyre in the hedges'. - Turner (1562: 38).   vc29   **1562**

Perring *et al.*(1964) give Ray (1660) as the first record in Cambs, but they ignore all of Turner's records.

Limonium auriculae-ursifolium

'Arsene, L. (1930) was the first to recognise a difference between this species and that now known as *L. normannicum*'. - Ingrouille (1985).   vc113   **1930**

as *L. lychnidifolium* 'Cliffs at Crabbe [Rouge Nez], Jersey, Sep. 1916, Attenborough'. - Druce (1917).   vc113   1916

Limonium bellidifolium

* 'Limonium minus flagellis tortuosis ... Found on the coast of Norfolk by Mr. Henry Scott'. - Blackstone (1746: 47).	vc28	**1746**

Limonium binervosum agg.

* 'Limonium parvum Rocke Lavander ... Upon the chalkie cliffe going from the towne of Margate downe to the sea side upon the left hand'. - Gerard (1597: 332).	vc15	**1597**

Presumably this would be *L. binervosum* subsp. *binervosum*, the common subspecies in that area.

Limonium britannicum

'Trevose Head, W. Cornwall, 1979'. - Ingrouille & Stace (1986).	vc1	**1986**
	vc1	1979

Limonium dodartiforme

'Durdle Dor [*sic*], Dorset, July 1979'. - Ingrouille & Stace (1986).	vc9	**1986**
	vc9	1979

Limonium humile

* 'Limonium Anglicum minus, caulibus ramosioribus, floribus in spicis rarius sitis ... WaItoniae vico in Essexiae non procul ab Harvico [Harwich] portu prope Molendinum copiosum invenit D. Dale, nobisque communicavit'. - Ray (1704: 247).	vc19	**1704**

Clarke adds that a specimen collected by Dale (c.1700) is in **BM**.

Limonium loganicum

'Logan's Rock, east of Lands End, W. Cornwall, July 1979'. - Ingrouille & Stace (1986).	vc1	**1986**
	vc1	1979

Limonium normannicum

as *L. lychnidifolium* var. *corymbosum*, first described by Salmon (1901).	vc113	**1901**
'Rocks, Alderney. 1900'. - Ingrouille (1985).	vc113	1900

Limonium paradoxum

'St David's Head, Pembroke, A.E. Ellis, ... but found previously in 1905 by E.F. Linton'. - Pugsley (1931).	vc45	**1931**
	vc45	1905

Druce cites incorrect year.

Limonium parvum

'Saddle Point, Pembroke, July 1979'. - Ingrouille & Stace (1986).	vc45	**1986**
	vc45	1979

Limonium procerum

Ingrouille & Stace (1986). | | **1986**
Ingrouille & Stace cite Salmon's records of *L. occidentale* var. *procerum* from Llandudno as the first records - see Salmon (1903). Of the three records given there the earliest is Llandudno, 1832, Jn Roberts, Herb Borrer. | vc49 | 1832

Limonium recurvum

'Portland, Dorset, 1832, Henslow'. - Salmon (1903: 72). | vc9 | **1903**
Babington (1860c) had previously described this as *Statice Dodartii*. | vc9 | 1832

Limonium transwallianum

'Near Tenby, on maritime cliffs [1923]'. - Pugsley (1924a). | vc45 | **1924**
| vc45 | 1923

Limonium vulgare

* 'Limonium … Sea Lavander … upon the walles of the fort against Gravesend [Kent], fast by the King's Ferrey going to the Ile of Shepey, in the salt marshes by Lee in Essex, in the Marsh by Harwich and many other places'. - Gerard (1597: 333). | vc15, 16, 18, 19 | **1597**

Limosella aquatica

* 'Plantaginella palustris *C.B.* … secus vias ubi per hyemem aquae stagnarunt'. - Ray (1663: 8). 'Plantago Aquat. minima *Clus. P.1244*. … On Hunslow-heath, a mile off that Town in the way to Coln-Brook'. - Merrett (1666: 95). | vc29 | **1663**

This is described in a letter from Ray to Peter Courthope, dated from Trinity College in July 1661. See Gunther (1934). | vc29 | 1661

Limosella australis

as *Limosella aquatica* var. *tenuifolia* 'From Kenfig Pool'. - Hiern (1901). | vc41 | **1901**
Kenfig Pool, 1897, Trow, A.H., **NMW**. | vc41 | 1897

Linaria repens

* 'Linaria odorata Monspessulana *J.B.* In Cornubia non longe ab oppido Perin [Penryn] versus Occidentem in sepibus'. - Ray (1670: 195). | vc1 | **1670**
'on July 2nd, on Ray's 1662 Itinerary'. - Lankester (1846: 191). | vc1 | 1662

Linaria vulgaris

* 'Osyris … groweth plentuously in Englande … Lynary or todes flax'. - Turner (1548: E. viij). | | **1548**

Linnaea borealis

* 'Specimens from Prof. James Beattie, of Aberdeen, were presented by the President, and among them *Linnaea borealis*, discovered by that gentleman, for the first time in Britain, in an old fir wood at Mearns near Aberdeen'. - Linnean Soc. Minutes (1795).   vc91   **1795**

There is also a reference in Dickie (1860) 'Woods of Inglesmaldie, parish of Fettercairn … in 1795, by Prof. J. Beattie, of Aberdeen'. Fettercairn is in the area known as Howe of the Mearns.

Linum bienne

* 'I have sene flax or lynt growyng wilde in Sommersetshyre wythin a mile of Welles [Wells]'. - Turner (1562: 89b). 'Linum sylvestre tenuifolium'. 'Yle of Shepey'. - Gerard (1597: 447).   vc6   **1562**

Turner goes on 'but it has fewer bowles in the top then the sowen flax hath…'. Nelson (1959), however, thinks that the Turner record is more likely *L. usitatissimum*.

Linum catharticum

* 'Chamaelinum perpusillum … In many places of the Yle of Shepey … also between Quinborow [Queenborough] and Sherland house'. - Gerard (1597: 447). 'Linum sil. pusillum candidis floribus'. Between Chatham and Gillingham. - Johnson (1629: 4).   vc15   **1597**

'in Anglia' - annotations made by T. Penny on C. Gesner's plant illustrations dated as by 1565 in Foley (2006b)   1565

Linum perenne

* 'Linum sylvestre caeruleum … Found on Newmarket Heath [Cambs] Mr. Sare'. - How (1650: 69).   vc29   **1650**

Liparis loeselii

* 'Orchis lilifolius minor sabuletorum Zelandiae et Bataviae *J.B.* … In the watery places of Hinton & Teversham Moors'. - Ray (1660: 106).   vc29   **1660**

Druce cites this as 1640.

Lithospermum arvense

* Of a kind of Lithospermon 'We have in England growyng among the corne'. - Turner (1562: 41). 'Anchusa degener facie milii solis … In the yle of Thanet neere Reculver'. - Gerard (1597: 487).   **1562**

Lithospermum officinale

* 'Lithospermon … Grummel … Gray myle … groweth plentuously aboute woddes and busshes'. - Turner (1548: E. j. verso). 'In the street at Southfleete in Kent'. - Gerard (1597: 487).   **1548**

Lithospermum purpureocaeruleum

* 'Lithospermum majus Dodonaei flore purpureo, semine Anchusae, *J.B.* On the top of a bushy hill near Denbigh-town, on the North-west-side, also in Somerset-shire, not far from Taunton, &c'. - Ray (1670: 197).	vc5, 50	**1670**
'May 18th, 1662', on Ray's third Itinerary. - See Lankester (1846: 167).	vc50	1662

Littorella uniflora

* 'Holosteum minimum palustre capitulis longissimis filamentis donatis ... On Hinton-moor, near Cambridge, on the moist fenny grounds about Glastenbury, Pensans, &c'. - Ray (1670: 169).	vc1, 6, 29	**1670**
Ray was in Cornwall in 1662 and 1667, passing through Somerset and Penzance on each occasion.	vc1, 6	1667

Crompton (2007) cites as the first Cambridgeshire record Merrett (1666: 63) 'Holosteum repens juncifolium... in standing waters in and about Stretham Ferry'.

Merrett's account also gives 'At the bottom of the Moor on the East side of Peters-field', which sounds like a Goodyer record, and indeed there is a record from How (from a manuscript addition to his copy of *Phytologia Britannica*, cited in Gunther (1922: 293)) of 'Holosteum perpusillum growes in ye same lake [as *Ludwigia palustris*] in ye East part of ye said heath [neere Petersfeild in Hamshire] green all ye winter under water, and flowers when ye water is vanished in August, and sometimes much sooner. I first observed this plant in a pond neare Holburie in ye new Forest in Hamshire. J. Goodyer'. These is also a gloss to this entry, seemingly from Goodyer, that he found it on 2nd June 1656, and How died the same year. I have not traced a later reference to the polynomials of How, but Dillenius, in his *Indiculus plantarum dubiarum* in Ray (1724), includes Merrett's polynomial (though with a scrambling of How's ms polynomials and sites). Thus there is no confirmation of the identity of Goodyer's plant. It is conceivable that the record refers to *Eleogiton fluitans*, which is described as 'Gramen junceum clavatum minimum, seu Holosteum palustre repens' in both Morison (1699) and Plukenet (1692), and indeed there is a specimen of this species in the Dillenean Herbarium from this site, but unless there is something hitherto undiscovered in the Goodyer papers, there can be no certainty.

Lobelia dortmanna

* 'Gladiolus lacustris *Clusii* ... In a Pool or lake called Hullswater [Ullswater] that divides Westmorland from Cumberland, 3 miles from Pereth, plentifully'. - Ray (1677: 132).	vc69, 70	**1677**
This was on Ray's trip to the north of England in 1671, with T. Willisel. See Raven (1950: 154).	vc69, 70	1671

Lobelia urens

* 'Supra Shute Common inter Axminster et Honiton D. Newbery'. - Hudson (1778: 378).	vc3	**1778**
'Habitat solum humidum et spongiosum Ericeti cujusdam Shute Common dicti Devoniae, inter padodem turfosam et viam publicam qua ab Axminster ad Honiton ducit; plures plantas primum invenit *Dom. Newbery* & Coll. Exon: Oxon. Circiter finem Octobris, Anno 1768' - Bowden (1989: 132). 'Lightfoot'. - **BM**, Herb. Banks.	vc3	1768

There is a specimen in Herb. Banks, **BM**, and the plant was still to be found at the Devon site recently.
Note that Clarke has 'First found by Lightfoot near Axminster in 1774. - see Letter from Pulteney to Martyn, 3 Oct. 1775, in Gorham (1830)'. But note too that Edwards (1862) relates that Mr Newbery collected it between 1762 and 1778.

Loiseleuria procumbens

* as *Azalea procumbens* 'The Rev. Mr. Lightfoot found it near Arnisdale in the Highlands of Scotland, Pennant's Tour, 1772'. - Withering (1776: 122).	vc97	**1776**
'on Ben Horn [behind Golspie], Ben Valich [Mhealich], Scaraben and Ben Wyvis'. - Journal of James Robertson, 1767, see Henderson & Dickson (1994: 20).	vc107	1767

There are also records of specimens labelled 'plentifully on a hill in Glen Criven on Scaraber, Caithness on Benevalich, Sutherland &c.' in Hope, 1768, see Balfour (1907).
Following Stace (2010) this should now be cited as *Kalmia procumbens*.

Lolium perenne

* 'Hordeum murinum … Phenix Dioscoridis … called in Cabridgeshire Way bent'. - Turner (1548: D. v. verso).	vc29	**1548**

See entry under *Hordeum murinum* for confusion between Clarke, Druce and Stearn (1965).

Lolium temulentum

* 'Lolium … Darnel groweth amonge the corne and the corne goeth out of kynde into Darnel'. - Turner (1548: E. ij). 'Lolium sive Triticum temulentum *Lob.*'. Between Sandwich and Canterbury. - Johnson (1632: 21).		**1548**

The long description in Turner (1538: B. ij. verso, under 'Lolium … angli Darnell') seems a probable reference to an English record.

Lonicera periclymenum

* 'Periclymenum ... wod bynde and Honysuccles ... Wodbyne, is commune in every wodde'. - Turner (1548: F. ij).   **1548**

This is also referred to in Turner (1538: B. iv) as 'Periclimenon … angli vocant Wodbynde' and seems a probable first reference.

Lonicera xylosteum

'from Mr W. Borrer, junior "growing plentifully, and certainly wild, in a coppice called the Hacketts, to the east of Houghton bridge, 4 miles from Arundell, Sussex"'. - E.B. 916.   vc13   **1801**

Druce also cites an indisputedly alien record from 1770 'Under the Roman wall near Shewing sheels [Northumberland]'.

Lotus angustissimus

* as *L. diffusus* 'Among the rocks near Hastings. Mr. Dickson. At Kingsteington and Bishopsteington, Devonshire. Rev. Mr. Beeke'. - Smith (1800: ii. 795).   vc3, 14   **1800**

'Guernsey. Gosselin, J., c.1790'. - see McClintock (1982: 76).   vc113   1790

Dickson's specimen is in **BM**, 29th May, 1798.

Lotus corniculatus

* 'Trifolium siliquosum minus ... In most fertill fields of England'. - Gerard (1597: 1022).   **1597**

Lotus pedunculatus

* 'Trifolium corniculatum majus hirsutum'. - Johnson (1632: 21).   vc15   **1632**

This was between Ash and Canterbury, see Rose, in Gilmour (1972: 119).

Lotus subbiflorus

* as *L.angustissimus* β. *major* 'Cornwall, near the Lizard and near Penzance. H. C. Watson'. - Hooker (1831: 330). 'Found by Mr. J. Woods near the Castle at Dartmouth in June 1828; and in July 1836 in Jersey'. - E.B.S. 2823 (1838).   vc1   **1831**

'Guernsey, roadside between the Ponchez and les Pelleys. Gosselin, J., c.1790'. - see McClintock (1982: 76).   vc113   1790

Lotus tenuis

'Trifolium corniculatum minus angustioribus foliis fruticosius. Among standing corn & in moister places'. - Ray (1685: 18).   vc29   **1685**

Accepted by Babington (1860b) and Oswald & Preston (2011), but ignored by Clarke and Druce, who cited '*L. decumbens* ... Near Mr. Sloper's Farm not far from Tonbridge. A new species; first found at Hastings near Bulverhithe'. - Forster (1816: 86).

Ludwigia palustris

* 'Anagallis aquat. flore parvo viridi caule rubro, in a great Ditch neer the Moor at Peters field Hamshire, Mr. Goodyer'. - Merrett (1666: 7). 'Mr. Borrer found it ... at Buxtead, Sussex ... July, 1827'. - E.B.S. 2593 (1829). vc11 **1666**

'Herba aquatica rubescens, facie Anagallidis flore luteo. 19 Aug.1645 ... in the rivulett on the east side of Petersfield'. - Goodyer ms.11, f. 141, see Gunther (1922: 188). vc11 1645

Also as a ms addition in How's own copy of P*hytologia Britannica* (1650), dated to between then and 1656 (his death). See Gunther (1922: 293).

Luronium natans

* 'Alisma repens foliis gramineis et subrotundis. ... In the great lake below the old castle at Llanberys; Mr. Brewer'. - Martyn (1732: i. 17). vc49 **1732**

'July 21st [1727]. Took a walk down to the meadows below the old castle [Llanberis], ans in the great and lowermost lake ... a plant with a white three-leaved flower and in fruit'. - Samuel Brewer's diary, reproduced in Hyde (1931). vc49 1727

Clarke adds 'see Sam. Brewer's ms. account of his journey in Wales in 1726 in Bot. Dep. Brit. Mus', but the diary is dated 1726 - 1727, and the entry is clearly for the latter year when Brewer was living in N. Wales. For further details see extracts from letters from Dillenius to Brewer, in Sept 1727 and March 1728, cited in Druce & Vines (1907: lxiii, lxviii).

Luzula arcuata

* 'On the most stony and barren summits of the Cairngorum and others of the Grampian mountains. *Professor Hooker*'. - Smith (1824: ii. 183). **1824**

'Discovered by George Don in 1812 on 'Ben Mac Dowie [Macdui]' ... specimens sent to the Countess of Aylesford are now in my herbarium'. - Druce (1932). vc92 1812

'Don refers to this in a letter to Mr Booth in 1812'. - Druce (1904). See *Salix lanata*.

Luzula campestris

* 'Gramen exile hirsutum ... In watery ditches ... going from Paris garden bridge to Saint Georges fields' [Southwark]. - Gerard (1597: 16). vc17 **1597**

Though this is a very odd habitat, the attribution was accepted by all my predecessors, including Ray (1660), who was the first to describe *L. multiflora* (see below). See also 'Gramen hirsutum nemorosum, Hearye or hoarye Grasse in my orcharde'. - Manuscript notes of William Mount, of c.1584, in Gunther (1922: 255 - 263), and identified by him as this species.

Luzula forsteri

* 'Mr. Edward Forster first observed this *Juncus* in 1795, growing intermixed with *J. pilosus*, under trees between Hoghill and Collier-row in Hainhault forest, Essex'. - E.B. 1293, and Smith (1804: 1395). vc18 **1804**
vc18 1795

Druce erroneously notes the date of the drawing as 1790.

Luzula multiflora

*'Gramen hirsutum majus panicula juncea compacta … In the pastures about Gamlingay, near the boggy ground by Sir Roger Burgoynes park, where they dig turves'. - Ray (1660: 68 & 1688: 1291). vc29 **1660**

Fomerly described as a var. of *L. campestris*, but as a separate species (*L. congesta*) in Forster (1816).

Luzula pallescens

'Woodwalton Fen, 13th June 1907, coll. E. W. Hunnybun'. - Druce (1908a: 312). vc31 **1908**
vc31 1907

Luzula pilosa

* 'Gramen hirsutum nemorosum … In woods … or shadowie places'. - Gerard (1597: 17). **1597**

Luzula spicata

'Upon Ben-na-Scree, on the north side of Loch Urn, on the western coast of Inverness-shire, &c'. - Lightfoot (1777: 187). vc97 **1777**

'The ascent of Beinn Sgritheall was made on on 6th August 1772 … 'Messrs Lightfoot and Stuart sallied out …'. - Mitchell (1986). vc97 1772

In fact Pennant (1774) notes that they sallied out 'in high spirits' after being plied with several glasses of rum cordialized with bilberries.

Luzula sylvatica

'Gramen hirsutum Nemorum latioribus majoribusq; foliis praecox Vernum. Repertu facile silvis & silvosis marginibus altae Portae [Highgate], quatuor Milliaribus Londino dissitis: folia altero tanto latiora …'. - L'Obel (1655: 39). vc21 **1655**
vc21 1616

L'Obel died in 1616. Both Clarke & Druce cite 'Gramen nemorum hirsutum majus ... about Highgate and other places. - Parkinson (1640: 1184-6)' as the first record, but I think that refers to *L. pilosa*, though Ray and other writers do point out that the illustration is correct. A ?late 17th century gloss in the Wellcome Library copy of Parkinson (Shelfmark: 4833/D) also identifies his plant as *L. pilosa*.

Lycopodiella inunduata

'Muscus alius terrestris, lycopodium *Hist. lugd. pag.1325*. On Putney-heath on a bog near the wind-mill'. - Merrett (1666: 79) and 'Chamaepeuce foemina seu polyspermos at Dovestones in the side of Yorkshire, neer Widdop and a mile this side Lippock [Liphook] in Hampshire'. - Merrett (1666: 25).

vc12, 17, 64   **1666**

Townsend (1904) and others equate the second Merrett reference with *Lycopodium clavatum*, but Edgington (2013) argues otherwise convincingly: in fact Edgington goes further, making the very plausible suggestion that the Liphook record must be one of Goodyer's, and, if so, would most likely date from the 1640s or earlier.

Merrett, C. 1666. *Pinax Rerum Naturalium Britannicarum*.

Lycopodium annotinum

'Lycopodium elatius Juniperinum clavis singularibus sine pediculis nondum descriptum. Above Lacum Lhin-y Cwn [Llyn - Cwn, just N of the Llanberis Pass]'. - Ray (1690: 16). 'Muscus terrestris foliis retro reflexis *J.B.* Club-moss with reflexed leaves, and single heads without footstalks ... in monte Rhiwr Glyder, supra lacum Llyn y cwn. D. Lloyd'. - Ray (1690: 18).

vc49   **1690**

The record came from Edward Lhwyd, and Edgington (2013) gives the full story and suggests a discovery date of 1688.	vc49	1688

Ray refers to this in a letter to Lhwyd on 26th September, 1689. See Gunther (1928: 199)

Lycopodium clavatum

'Chamaepeuce … there is a much lesse kinde … which groweth in the mountaines of Germany and Wales, and it creepeth hard by the ground, al rough and full of small leves … out of the end of the stalke cometh two rough fruytes, much like unto the longe blomes that come forth of the Haselnut tree in winter, but they are a great deale smaller and yellower'. - Turner (1568: 1. 130).	1568

See also 'In Angliae … Frequens'. - 'L'Obel (1576: 645) and 'Muscus clavatus … which I have not else where found then upon Hampstead Heath, neere unto a little cottage'. - Gerard (1597: 1373.11).

Lycopodium lagopus

Rumsey (2007).		**2007**
Garbh Bheinn, Fersit Forest, v.c.97, 1896, E.S. Marshall, det F. Rumsey, BM.	vc97	1896

A disputed taxon, which might be a subspecies of *L. clavatum*.

Lycopus europaeus

* of Sideritis 'Marrubium palustre Tragi, that is water horehound. That herbe groweth alwayes … about water sydes, and hath a stinking smell of garleke'. - Turner (1562: 135b).	1562

Chapman *et al*. (1995: 570) make this a species of *Stachys* and certainly the 'smell of garlic ' is hardly a characteristic of *Lycopus*. Lyte (1578) gives 'Marrubium palustre, water Horehounde' and Gerard (1597: 565) 'Marrubium aquaticum … on the brinks of water ditches' (although the illustration was not this species, but corrected in Johnson (1633: 700)), and this is followed by Ray (1660). Grigson (1955) notes the fast black dye given by this plant, and this accords with Lyte's (and Gerard's) description of 'the Egyptians herbe, because the Rogues and runnegates, whiche call themselves Egyptians, do colour themselves blacke with this herbe'. Perhaps Lyte (1578) would be a better first record.

Lysimachia nemorum

* 'Anagallis lutea, … in Angliae nemoribus, locisque opacis … in quadam densa et amoena Sylva Coventriae proxima'. - L'Obel (1571: 194).	vc38	1571

Lysimachia nummularia

* 'Centimorbia ... Nummularia ... Herbe ii pence or two penigrasse ... groweth in moyste groundes'. - Turner (1548: H. ij. verso).   **1548**

Lysimachia thyrsiflora

* 'Lysimachia lutea flore globoso *Ger. Park*. ... Nuperrimè peritissimus Botanicus D. Dodsworth, in Anglia, Comitatûs Eboracensis orientali parte hanc invenit'. - Ray (1688: 1023). ' ... in the East riding of Yorkshire'. - Ray (1688a: 15).   vc62   **1688**

In Raven's paper on Thomas Lawson's note-book (1948: 6) he has 'nigh York by a friend of J. Newton'. Raven dates this note-book to around 1676, but writes that this particular note is later since Newton only began botanising in 1680. Whittaker (1986) dates these Yorkshire records to 1680-1682.   vc62   1682

Robinson (1902: 141) cites 'J. Ray, 1685, Gough's "Britannia"' [*sic*], adding 'In the E. Riding, but not seen by R. Teesdale, nor, to our knowledge, by anyone since'. Crackles (1990: 114) states that Ray's record is from Leckby Carr, vc62, (though gives no source for that) and certainly there is no known record for vc61. The record given by How (1650: 71), under the polynomial cited by Ray, from 'Kings-Langly in Hartfordshire', is treated by Pryor (1887) as a double-flowered form of *L. vulgaris*.

Lysimachia vulgaris

* 'Lysimachia ... groweth by the Temes syde beside Shene ... yealow Lousstryfe'. - Turner (1548: E. ij. verso).   vc17   **1548**

Lythrum hyssopifolia

* 'Gratiola angustifolia ... Found by my friend Mr. Bowles at Dorchester' [Oxon]. - Johnson (1633: 582).   vc23   **1633**

See Raven (1947: 295) dating Bowles' tour to 1632.   vc23   1632

Lythrum portula

* 'Alsine aquatica foliis rotundioribus, sive Portulaca aquatica'. Between Canterbury and Faversham. - Johnson (1632: 25).   vc15   **1632**

Lythrum salicaria

* 'Lysimachia purpurea ... groweth by water sydes ... purple losestryfe'. - Turner (1548: E. ij. verso). 'Under the Bishops house wall at Lambeth neere the water of Thames'. - Gerard (1597: 388).   **1548**

Maianthemum bifolium

'Ken Wood, 1813'. - Hunter in Park (1814: 28).   vc21   **1814**

Clarke and Druce, and many others, cite as the earliest record 'Monophyllon groweth in Lancashire in Dingley wood, sixe miles from Preston in Aundernesse; and in Harwood neere to Blackburne'. - Gerard (1597: 330). The species is a very doubtful native especially in Gerard's sites, which have never been seen again, though his illustration is correct. Could he have muddled his species, or perhaps his correspondant, presumably Hesketh, did so? Of the records accepted in Preston *et al.* (2002), the earliest is Ken Wood but even there considerable doubts exist over its nativeness. See Jackson (1913) for a full history of the species not only in London but elsewhere in Britain, but note that he cites the wrong edition of Park.   vc21   1813

Malus pumila

'Malus hortensis, in English an Appel tree'. - Turner (1562: 47b).   **1562**

Turner (1548: E. iij) also refers to 'Malus … in englishe an Apple tree', but without specific mention of cultivated or wild occurrences.

Malus sylvestris

* 'Malus sylvestris [called] in ye South countre a Crab tre, in ye North countre a scarbtre'. - Turner (1562: 47b).   **1562**

Malva arborea

* 'Malva arborea marina nostras … In an island called Dinnie three miles from Kings Roade and five miles from Bristow as also about the cottages neere Hurst Castle over against the Ile of Wight'. - Parkinson (1640: 306).   vc11, 35   **1640**

Denny Island is in the Severn Estuary, and in vc35 (Monmouthshire).

Malva moschata

* 'Malva verbenaca … By the ditch sides on the left hand of the place of execution called Tyborne: … a village fourteene miles from London called Bushey … Hackney … Bassingburne in Hartfordshire [actually Cambridgeshire], three miles from Roiston'. - Gerard (1597: 786). 'Very familiar in Kent'. - Parkinson (1640: 306).   vc20, 21, 29   **1597**

Pulteney cites Lyte (1578: 584) 'Alcea' with 'leaves more cloven and divided into sundry partes … is not very common in this Countrie'.

Malva neglecta

* 'Malva sylvestris pumila … Wilde dwarfe Mallow - among potherbes by highwaies and the borders of fieldes'. - Gerard (1597: 786).   **1597**

Malva pseudolavatera

as *Lavatera sylvestris* 'We are indebted to Mr Curnow … first observed it in July 1873 [in the Isles of Scilly]'. - Trimen (1877a).	vc1 vc1	**1877** 1873

Lousley (1971) notes that J. Ralfs and R.V. Tellam were with Curnow on that trip.

Irvine (1859) gives a record from Wandsworth, but with a query against it, and Leslie (1987) shows this only as a bracketted entry. Even if correct it would be only as a casual.

Malva setigera

* 'In arvo prope Cobham in Cantia anno 1792 invenit J. Rayer habitationum stirpium clarissimus indagator'. - Symons (1798: 200).	vc16 vc16	**1798** 1792

Malva sylvestris

* 'Malva … mallow … groweth wilde about Townes and hygh wayes'. - Turner (1562: 45).	**1562**

Marrubium vulgare

* 'Marrubium … Horehound … groweth aboute townes and villages'. - Turner (1548: E. iij). 'Groweth plentifully in all places of England'. - Gerard (1597: 562).	**1548**

This species is widespread as an alien arising from past medicinal use, and it is difficult to separate casual and native sites. The earliest probable native site traced so far is 'Abundant on all the downs west of Calbourne [Isle of Wight]'. - Snooke (1823). Hind (1889) gives a Suffolk record from 1773, and although it is deemed to be native in the Breckland, he does not give the location.

Matricaria chamomilla

* 'Chamaemelum, sive Anthemis vulgatior, *Lob.*'. Hampstead Heath. - Johnson (1632: 36).	vc21	**1632**

Matthiola sinuata

* 'Leucojum marinum purp: L'Obelii … gathered by Mr. George Bowles upon the rocks at Aberdovye in Merionethshire'. - Johnson (1633: 461).	vc48	**1633**
'Purple sea stock Gilloflower … by seaside by Sir John Salusbury his weare theare at Lland[d]ulas'. - Annotation in a copy of Gerard (1597), dated to 1606 - 1608. See Gunther (1922: 243). There are no modern records from N. Wales, but see Wynne (1993) for an old record from Prestatyn, 14km to the east.	vc51	1608

For the dating of Bowles' record to 1632, see Raven (1947: 295).

Meconopsis cambrica

* 'Argemone Cambro-Britanica lutea - Yellow wild Bastard Poppy of Wales. In many places of Wales … Abber … Denbigh to Guider … Balam … Banghor … Anglesey'. - Parkinson (1640: 370).   vc49, 50, 52   **1640**

See Raven (1947: 264), where he notes that Parkinson explicitly states this was 'found … by L'Obel in his life time'. L'Obel died in 1616.   1616

Medicago arabica

* 'Medica minor fructo cochleato aspero: Tribulus terrestris minor repens, *Lugd.*, ut etiam ejus varietas foliis maculatis, Medica Arabica, *Cam.*'. Hampstead Heath. - Johnson (1632: 33).   vc21   **1632**

Medicago lupulina

* 'Medica semina racemoso'. Near Rochester. - Johnson (1629: 1).   vc16   **1629**

Rose, in Gilmour (1w972: 50 & 105), identifies this and a later record as *Trifolium dubium*. Both Clarke and Druce ignore Johnson (1633: 1186). '*Trifolium luteum lupulinum*', which Smith (1825: 318) equates with this species.

Medicago minima

* 'Trifolium echinatum arvense … In an old gravell-pit in the corn field near Wilborham [Wilbraham] church; also at Newmarket [in Suffolk] where the Sesamoides Salamanticum [*Silene otites*] grows'. - Ray (1660: 166).   vc26, 29   **1660**

Medicago polymorpha

* 'Trifolium cochleatum modiolis spinosis. Hedge-hog-trefoil, with a small fruit, somewhat like the Nave of a Cart-wheel. At Orford in Suffolk, on the Sea-banks close by the Key, plentifully'. - Ray (1670: 302).   vc25   **1670**

'Orford'. In a letter from Ray to Peter Courthope. See Raven (1950: 130), describing Ray's third Itinerary of 1662.   vc25   1662

Clarke adds 'But Hanbury & Marshall (1899: 87) refer Merrett's (1666: 76) 'Medica marina Trifolium cochleatum marinum. Sea-medick. At Rumney, betwixt the Town and Cony-Warren' to this species.

Medicago sativa

* 'Trifolium syl. luteum siliqua cornuta, vel Medica frutescens, *Bauh.* … In montosis. Yellow horne Trefoil'. - Johnson (1634: 74). 'Between Linton and Bartlow'. - Ray (1660: 167).   **1634**

Melampyrum arvense

'Melampyrum purpurascente coma *C.B. Pin. 234*. In the Corn on the right Hand just before you come to Lycham [Litcham] in Norfolk; Mr. J. Sherard'. - Ray (1724: 286*).	vc28	**1724**
Writing to Dr. John Thorpe 'a physician in Rochester', on June 28th, 1716, Petiver says:- 'Betwixt this city [Lynn] and Norwich in a corn field we discovered a beautifull and new plant not found in England before, viz. Melampyrum comâ purpureâ, in flowre (Sloane ms. 3340, f. 255)'. - Britten (1915).	vc28	1716
Another early specimen is in Uvedale's herbarium in the Sloane collection in BM; see Dandy (1958a: 225). Uvedale died in 1722.		

Melampyrum cristatum

* 'Melampyrum cristatum flore purpureo … In Madingley and Kingston woods, and almost in all woods in this county plentifully, likewise it overspreads all the pasture or common grounds you ride through going out of Madingley to dry Draiton'. - Ray (1660: 95).	vc29	**1660**
'Melampyrum cristatum wild in our woods, not described in Gerard, Parkinson. v. Joan. Bauhinum'. This is in a ms. list of Cambridge plants from Samuel Corbyn, dated 20 May,1657. - see Preston (in press).	vc29	1657

Melampyrum pratense

* 'Crataeogonon … In sylvosis collibus, & devexis, umbrosis Norbonae, Angliae, Subaudiae'. - L'Obel (1571: 11). 'Hampsted heath neere London, among the Juniper bushes and bilberry bushes'. - Gerard (1597: 84).		**1571**

Melampyrum sylvaticum

* 'In woods [in Scotland] but not common'. - Lightfoot (1777: 325). 'In the way from Taymouth to Lord Breadalbane's cascade, observed by Mr. Yalden, who communicated specimens. We were informed also by Mr Stuart that it grows about Finlarig, at the head of Loch Tay'. - Lightfoot (1777: 1126 app.). 'In woods around Barnard Castle'. - Robson (c.1780).		**1777**
Hope, 1768, in Balfour (1907) lists both *M. pratense* and *M. sylvaticum*, the latter 'from the wood of Strath Spey'.	vc95	1768
Taymouth Castle and Finlarig are at opposite ends of Loch Tay. Lightfoot's herbarium (Bowden 1989: 177) contains a specimen 'On the road going from Taymouth to the Hermitage, July, 1775'. There is apparently no specimen for Robson's record from the Tees in County Durham, but it was subsequently reported from various woods upstream up to the well-known site at Winch Bridge, about 7 miles away.		

'In Achendenny wood [Auchendinny, Penicuik], June 16th, 1764'. - John Hope's ms. 'List of Plants growing in the Neighbourhood of Edinburgh', reproduced in Balfour (1907), but I think this is the common species.

Melica nutans

* 'Gram. aven. locustis rubris montanum *C.B.* ... This was sent Mr. Pettiver out of the North and by him communicated to us'. - Ray (1696: 262). Ray adds a note (1696: 325) 'Hoc a D. Fitz-Roberts circa Kendalium Westmorlandiae oppidum primarium collectum & ad me transmissum est'. — vc69 — **1696**

In the Ashmole archives at the Bodleian is a letter of Richardson's of 1 March 1691 to Edward Lhwyd (Mss. Ashmole 1817a, f.235) where he offers Lhwyd specimens of plants he found in Yorkshire the previous season. Most are in Ray (1690), which he had with him in the field, but, as he says, "you may find some not mentioned by Mr Ray". Amongst these is 'Gramen montanum avenaceum locustis rubris *C.B.Prodr*'. — 1690

Foley (2008) confirms that it was FitzRoberts who sent Petiver the specimen.

Melica uniflora

* 'Gramen avenaceum rariore grano nemorense, *Lob.*'. Woods near Faversham. - Johnson (1632: 28). — vc15 — **1632**

Pulteney cites L'Obel (1605: 465) as a first record, but this reference contains only a record from Denmark by an Englishman.

Melilotus altissimus

* 'Melilotus Germanica ... No part of the world doth enjoy so great plenty thereof as England doth and especially Essex'. - Gerard (1597: 1034). — vc18 — **1597**

Gerard gives many other sites in an arc from Sudbury in W. Suffolk throughout Essex. See also 'Melilotus Germanica … groweth in this countrie in the edges and borders of fieldes, and medowes, alongst by diches, and trenches'. - Lyte (1578: 498.2).

Melittis melissophyllum

* 'Melissa Moldavica, *Matth.* … found in Mr. Champernons wood by his house on the hill side neere Totnes in Devonshire. Mr. Heaton'. - How (1650: 74). — vc3 — **1650**

Walsh (1978) suggests that Heaton was in Devon in 1645. Ray visited the site during his 1662 itinerary (Lankester, 1846: 195)

Mentha aquatica

* 'Sisymbrium … the red Mynt that groweth by water sydes and is called of some Horse Mynt'. - Turner (1548: G. iij. verso).   **1548**

Mentha arvensis

'Calamintha … The seconde kynde groweth much in England among the corne … in English cornemint'. - Turner (1548: B. vij. & 1551: G. v).   **1548**

There has been confusion over Turner's plant, with Clarke and Druce opting for 'Calamintha aquatica Belgarum … in Anglia'. - L'Obel (1571: 217) as their first date. Britten (1887) refers the Turner species to *Calamintha arvensis* (*Clinopodium acinos*), as does Druce (1897: xciv), but Stearn (1965) favours *M. arvensis*.

Mentha pulegium

'Pulegium ... Penyryal or puddyng grasse ... groweth in such diches and watery places as are ful of water in wynter and are dyred up in the begynnyng of Summer'. - Turner (1548: F.v. verso). 'Penny ryall groweth much with out any settyng besyde hundsley [Hounslow] upon the heth, beside a watery place'. - Turner (1562: 107).   **1548**

Also in Turner (1538: B.iv) as 'Origanum … quam vulgus appellat call Peny ryall, aut puddynge gyrse', but with nothing to say it was wild. Note that Gerard (1597: 545.1 & 2) differentiates wild and garden pennyroyals - the latter upright in his garden, and Turner's 1538 reference might have been to a cultivated plant.

Mentha spicata

* 'Mentastrum … Horse Mint … groweth in divers wet & moist grounds'. - Johnson (1633: 685).   **1633**

Mentha suaveolens

* 'Mentastrum folio rugoso rotundiore spontaneum flore spicato, odore gravi *J.B.* … Rarius occurrit sponte. I observed it growing by the Rivers side at Lydbrook, near Rosse in Herefordshire, plentifully; and in some places of the West-countrey, which I do not now remember'. - Ray (1670: 207).   vc36   **1670**

Ray's visit to Herefordshire was in the summer of 1667, see Raven (1950: 144).   vc36   1667

Guernsey, Grand-mielles. Gosselin, J., c.1790'. - See McClintock (1982: 124). Purchas & Ley (1889) suggest Ray's locality is in vc34.

Menyanthes trifoliata

* 'Trifolium paludosum … frequens paludosis riguisque pratis Angliae'. - L'Obel (1571: 382).   **1571**

Mercurialis annua

* 'Mercurialis ... ab anglis Mercury ... duo sunt genera, mas & foemina, mas haud temere apud nos reperitur, vidi tamen cantabrigie in horto aulae regiae [Cambridge, in the garden of King's Hall], unde radicem in nostrum transtulimus hortum'. - Turner (1538: B. iij). 'French Mercury ... I found under the dropping of the bishops house at Rochester, from whence I brought a plant or two into my garden, since which time I cannot rid my garden from it'. - Gerard (1597: 262).   vc29   **1538**

Perring *et al.* (1964) ignore all of Turner's records and Nelson (1959) notes that one wouldn't replant an annual, and that the plant was more likely to be *Chenopodium bonus-henricus*. However Oswald & Preston (2011: 174, n.157) keep with the *Mercuralis* identification.

Mercurialis perennis

'Cynocrambe ... que passim in Europa tota secus umbrosas vias & in Sylvis udisue'. - L'Obel (1571: 100). 'Cynocrambe ... Dogs Mercury ... about Greene-Hith, Swanes-combe village, Graves-ende and South-fleete in Kent; in Hampesteede woode'. - Gerard (1597: 263).   **1571**

For an eloquent plea for identifying the reference to this species in Turner (1548: E. iiij.) see Nelson (1959), citing 'The herbe which is communely called in englishe mercury hath nothing to do wyth mercurialis [ie *Chenopodium bonus-henricus*] whereof I spake nowe ... and let the commune mercury [ie *M. perennis*] alone'. Note also, as Cynocrambe, listed from 'Oxon', in an anonymous ms. dateable to 1570-72, see Gunther (1922: 236).

Mertensia maritima

* 'Buglossum dulce ex Insulis Lancastriae ... groweth in one of the Iles about Lankashire, there found by Mr. Thomas Hesket [Hesketh]'. - Parkinson (1640: 767.5)   vc69   **1640**

Thomas Hesket is assumed to be Thomas Hesketh of Clitheroe, who died in 1613.   vc69   1613

See also Johnson (1641: 20) 'By the salt pans bewteene Barwicke and the Holy Island'.

Mespilus germanica

* 'Mespilus ... Often-times in hedges among briars and brambles'. - Gerard (1597: 1266). 'Mespilus sylv. spinosa ... In the Hedges betwixt Hampsted-heath and Highgate and in a Holt of Trees three miles westward from Crediton, Devonshire'. - Merrett (1666: 77).   **1597**

Turner (1562: 56) mentions Medlar 'Medler tree or an open arss tre... It hath leves lyke an apple tre…a round apple, good to be eaten, with a larger navel. … it is comen in Italy, Germany and England, and is comenly called a medler'. There is no real indication that this is in the wild. It is similarly referred to in his earlier work, Turner (1538: B. iij).

Meum athamanticum

* 'Meum … I never sawe thys herbe in Englande sauynge once at saynte Oswarldes [St. Oswald, in Lee, near Hexham] where as the inhabiters called it Speknel'. - Turner (1548: E. v.); and see also ' … in the bisshoprik of Durram in wild mores called felles'. - Turner (1562: 57).

vc67   **1548**

Mibora minima

* as *Agrostis minima* 'in Wallia, D. Stillingfleet invenit'. - Hudson (1762: 28).

vc52   **1762**

This record is assumed to be from Anglesey, as the only Welsh site known before the 20th century.

In the Linnean Society's copy of Hudson (1778) there is a ms. addition '*Hist. Ox. p200, t5 f14*' and a further addition 'vide Ray (1724: 411)', which presumably refers to 411.12 'Gramen arvense panicula crispa longiore nostras *Pluk. Alm. 176*', but these seem too vague to be accepted.

L'Obel's (1655: 20) 'Gramen minimum Anglo-Britanicum. Arenoso solo versus, ceanum aliquot, miliaribus a Lio [Leigh], prope Thamesis ostia oritur', was identified by Smith (1824: 84) (as *Knappia agrostidea*) as this species. But Gibson (1862: 360), refutes this, suggesting *Agrostis pumila* (= *A. capillaris*).

Microthlaspi perfoliatum

* 'Thlaspi arvense perfoliatum minus *C.B.* Among the Stone-pits between Witney and Burford in Oxfordshire'. - Bobart in Ray (1690: 236).

vc23   **1690**

Druce and Marren (1999) both cite Ray (1688a) as well, but I cannot find a reference there. There is a specimen in the Morisonian Herbarium 'T. arvense perfoliatum minus, *C. B. P.* Neer 3 miles beyond Witney by the stone pits next the road towards Burford at a round Hillock about 6 or 8 foot high', cited in Vines & Druce (1914: 28), but with no date.

Milium effusum

* 'Gramen miliaceum … in fennie and waterie places'. - Gerard (1597: 6). Between Sandwich and Canterbury. - Johnson (1632: 21).

1597

Johnson (1633: 6) corrects the habitat to 'medowes, and about hedges', which is more satisfactory.

Milium vernale

'Petit Bot'. - Andrews (1900).	vc113	**1900**
Andrews collected the species in April 1899, but probably not at Petit Bot. See Tutin (1950) for the full story.	vc113	1899

Minuartia hybrida

* 'Alsine tenuifolia, *J.B.* ... In the corn fields on the borders of Triplow heath, and in divers other places'. - Ray (1660: 9). 'I have not yet met with it in any other county of England'. - Ray (1670: 18).	vc29	**1660**

Minuartia recurva

'Caha Mountains, Kerry/Cork border … 1964'. - Moore (1966).	vcH1, H3	**1966**
	vcH1, H3	1964

Minuartia rubella

* as *Alsine rubella* Wahlenb. 'Near the summit of Ben Lawers … sent by the late Mr. J. Mackay, in 1796'. - Smith (1828: 267).	vc88	**1828**
'I found the plant upon Ben [Lawers] in Bredalbane, and I never observed it anywhere else. I believe it to be new to Britain. I first found it 1793 in company with Mr. John Macay' [Mackay] . - G. Don ms. note on specimen in **BM**.	vc88	1793
For the 1793 record see also Druce (1904: 128). Smith adds that he had originally referred these specimens to diminutive *M. verna*.		

Minuartia sedoides

'Upon a mountain in Rum called Baikeval,' &c. - Lightfoot (1777: 232).	vc104	**1774**
There is an entry 'J.R. [presumably Robertson] 1766' in Hope, 1768, see Balfour (1907). 'From Ben Klibreck and Ben Greem [Griam]'. - Journal of James Robertson, 1767, Henderson & Dickson (1994: 20).	vc107	1766
Pennant (1774: t33) has a plate of this but seemingly no description. Hudson (1778: 193) cites 'in montibus in boreali parte Scotiae, D. Hope'.		

Minuartia stricta

* 'Found in 1844 by 'a little band of botanists upon the banks of a little stream on Widdybank Fell, not far from Langdon foot-bridge'. - E.B.S. 2890 - *see image p.276*.	vc66	**1844**
Graham (1988) gives the finders as G.S. Gibson, J. Backhouse snr. & jnr., A. Tatham and S. Thompson.		

> 2890.
>
> **ARENARIA** uliginosa.
>
> *Bog Sandwort.*
>
> ———
>
> DECANDRIA *Trigynia.*
>
> GEN. CHAR. *Calyx* of five leaves. *Petals* 5, undivided. *Stamens* 10. *Styles* 3. *Capsule* 1-celled, many-seeded.
>
> SPEC. CHAR. Perennial. Stem ascending, branched, leafy, terminating in elongated, slender, erect peduncles, with 1–3 flowers; when 3, the lateral pedicels with 2 small bracteas. Leaves subulate, semiterete, obtuse, glabrous. Calyx-leaves ovate, acuminate, 3-nerved, about as long as the petals and "rather longer than the ripe capsule."
>
> SYN. Arenaria uliginosa. *Schleich. Cent.-Exsicc.* 1. no. 47. *DeCand. Fl. Fr. v.* 4. 786. *Ic. Pl. Gall.* 14. *t.* 46.
>
> Spergula stricta. *Swartz Vet. Ac. Handl.* 1799. 235–239. *t.* 3.
>
> Alsine stricta. *Wahl. Lapp.* 127. *Fl. Suec.* 279.
>
> Sagina *ramis erectis bifloris. Linn. Lapp. no.* 158.
>
> Alsinanthe stricta. *Reich. Icon. cent.* 14. 29. *f.* 4935.
>
> ———
>
> FOR this highly interesting addition to the British Flora we are indebted to a little band of botanists* who made an excursion to the Yorkshire mountains in the summer of 1844, and detected it in a very circumscribed locality, and in very small quantity, upon the banks of a little stream on Widdybank Fell, not far from Langdon foot-bridge. We have had the opportunity of comparing the first specimens that were gathered
>
> * Mr. John Tatham, jun., of Settle, Messrs. James Backhouse, sen. and jun., of Darlington, Mr. Silvanus Thompson of York, and Mr. G. S. Gibson of Saffron Walden.

Plate 2890 Smith J. and Sowerby, J. 1844. *English Botany supplementvol, vol 4.*

See species account : Minuartia stricta, p.275

Minuartia verna

* 'Auricula muris pulchro flore, folio tenuissimo, *J.B.* ... On the mountains about Settle in Yorkshire plentifully'. - Ray (1677: 35).   vc64   **1677**

'nr. Settle'. - see Raven (1950: 154) describing Ray's Itinerary of 1671.   vc64   1671

Misopates orontium

* 'Much in England in ye corne feldes & in fallowed landes ... Yelow calfes snowte'. - Turner (1551: D. ij. verso, under 'Antirrhinum').   **1551**

It is not clear why Turner called this 'Yelow', since elsewhere in his description he calls the flowers purple.

Moehringia trinervia

* 'Alsine media Middle Chickweede ... bushes and briers, old wals and gutters of houses'. - Gerard (1597: 489). 'Arenaria fontana credita flosculorum foliolis non divisis'. Balsham. - Ray (1663: 3).   **1597**

Moenchia erecta

* 'Holosteum minimum tetrapetalon, sive Alsine tetrapetalos caryophylloides ... Vere floret, in glareosis sterilioribus frequens'. - Ray (1670: 168).   **1670**

Ray (1677: 163) marks this as a Cambridge plant, and adds the locality Gamlingay Park in his 1685 appendix.

Molinia caerulea

* 'Gr. pratense spica Lavendulae, Below the Park on the farther side of Micham Common by the foot way to the River leading to Calalton [Carshalton, Surrey?]'. - Merrett (1666: 57).   vc17   **1666**

Salmon (1931) spells Merrett's Calalton as Casalton.

Moneses uniflora

* as *Pyrola uniflora* 'Never supposed to grow in our island till James Brodie, Esq., of Brodie House in Scotland [Brodie Castle, W. of Forres], found it in that neighbourhood last summer when also Mr. James Hoy, F.L.S., sent it to the Linnean Society from near Gordon Castle'. - E.B. 146.   vc95   **1793**

There is an entry for this species in Hope, 1768, see Balfour (1907) but it is blank. Does this mean there was no specimen, or that the location or collector was unknown? D.E. Allen (*in litt.*) surmises it was a specimen sent by Robertson from the north of Scotland in 1767.

Lusby & Wright (1996) note that there are specimens in the herbarium of Sir J. E. Smith labelled 'from the Western Isles of Harris and Berneray gathered in 1783 by James Hoggan', and assume that this is a mislabelling.

Montia fontana

* 'Alsine aquatica surrectior ... On the boggy grounds about Gamlingay'. - Ray (1663: 3).   vc29   **1663**

Raven (1950: 121) explains that this was a record from 1662.   vc29   1662

Muscari neglectum

* as *Hyacinthus racemosus* 'Fields at Hengrave; and plantations at Cavenham. Sir J. G. Cullum'. - Turner & Dillwyn (1805: 548).   vc26   **1805**

'Sir J.C. Cullum, 1776'. - Trist (1979).   vc26   1776

Turner & Dillwyn add a note that 'We have given a place to this plant because Sir T. Cullum finds it plentifully in the habitats here mentioned, and considers it "at least equally entitled to a place in the British flora as *Tulipa sylvestris* and many other naturalised species'.

Mycelis muralis

* 'Sonchus laevis muralis … Upon walls and in wooddy mountainous places'. - Johnson (1633: 295). 'Sonchus alter folio sinuato hederaceo *Lob*. ... In umbrosis & muris'. - Johnson (1634: 70).		**1633**

Myosotis alpestris

* as *M. rupicola* 'Found 'long ago' from the Highlands of Scotland, specimens of this plant, gathered there by Mr. G. Don, the late Mr. J. Mackay, and other friends'. - E.B. 2559.		**1813**
Clarke adds 'but previously published in Don's Herb. Brit. 205 (1804) as *M. alpina*'. This is in Fascicle 9, no 205, as *M. alpinus*, from Ben Lawers. It is dated 1806, but not issued before 1812, probably either at the end of 1812 or early in 1813. See Druce (1904). Clarke and Druce cite the wrong date for Don, which is odd in so far as Druce had already (*op.cit*.) published the correct version.	vc88	1812

Myosotis arvensis

* 'Myosotis scorpioides … upon most drie gravely and barren ditch bankes'. - Gerard (1597: 267). 'In the backe close of Sir John Tunstall his house a little beyond Croydon'. - Parkinson (1640: 692).		**1597**

Myosotis discolor

* 'Myosotis scorpioides minor flosculis luteis *Park*. ... In siccioribus & arenosis'. - Ray (1670: 217).	vc29	**1670**
'Echium scorpiodes minus flosculis luteis *Bauh. Pin.* 254 growes within 3 miles of Redding plentifully'. - Records of Rev William Browne entered by How in his copy of '*Phytologia*' between 1650 and 1656, see Gunther (1922: 302).	vc22	1656
Ray (*op.cit*.) gives the mark 'C', i.e. that he has seen this in Cambridgeshire, but it is not in his works of 1660 or 1663. Merrett (1666: 82) has 'Myosotis scorpioides hirta minor. In muris & locis siccioribus', which might well refer to this species.		

Myosotis laxa

* as a var. of *M. palustris* 'On Waterdown Forest and near the great Rocks, Tunbridge, Kent'. - Forster (1816: 25).	vc16	**1816**
'In palustribus ericeti Hounsley [Hounslow] ubi subinde aqua deficit'. - Druce & Vines (1907: 75).	vc21	1747

Smith (1824: i. 250) is the first to include this as a distinct species. See also 'Guernsey, Grande Mare. Gosselin, J., c.1790'. - McClintock (1982: 111). The record from Druce & Vines is dated as 1747, the date of Dillenius' death.

Myosotis ramosissima

as *M. arvensis,* E.B.S. 2629. - Borrer (1830), correcting the identification in E.B. 2558.		**1830**
as *M. collina* in Druce & Vines (1907: 75).		1747

Borrer's correction was 'in a note after Sir J.E. Smith's death'. - Garry (1904).
The record from Druce & Vines is dated as 1747, the date of Dillenius' death.

Myosotis scorpioides

* 'Another sort of Scorpion grass ... In almost every shallowe gravely running streame'. - Gerard (1597: 266).		**1597**

Myosotis secunda

* as *M. palustris* 'Moist hills about Glasgow, D. Don; and Ochil hills, G. & D. Don'. - Hooker (1821: 67).		**1821**
'Guernsey. Gosselin, J., c.1790'. - see McClintock (1982: 111).	vc113	1790

McClintock (*op.cit.*) notes that this speices was split from *M. scorpioides* in 1836, but his source has not been found.

Myosotis sicula

' … at the east end of of St Brelades Bay [Jersey] on the fixed dune between the strand and the pond [in 1922]'. - Wilmott (1923).	vc113 vc113	**1923** 1922

Myosotis stolonifera

'Cross Fell, 1919'. - Salmon (1926).	vc70	**1926**
'It was first collected in Britain by A. Wilson nr. Cautley Crag, Sedburgh, [6th June] 1892. **YRK**'. - Halliday (1997). Also Rosgill Moor, Shap, vc69, 1892, C. Bailey, **NMW**.	vc65, 69	1892

Myosotis sylvatica

* 'Myosotis scorpioides latifolia hirsuta. In Charlton and many other woods in Kent'. - Merrett (1666: 82).	vc16	**1666**

Myosoton aquaticum

* 'Alsine major … the great Chickweede … bushes and briers, old wals and gutters of houses'. - Gerard (1597: 487). 'Alsine major glabra'. Gravesend to Rochester. - Johnson (1629: 3).	**1597**

Myosurus minimus

* 'Mouse-taile groweth upon a barren ditch banke as you go from London to a village called Hampsteed; in a field as you go from Edmonton, a village neere London, unto a house thereby called Pims, Woodford Rowe in Waltham forest ... and in other places'. - Gerard (1597: 345).   vc18, 21   **1597**

Pulteney cites Lobel (1571: 187) 'Cauda muris', but with no indication that it is British, save for giving an English name 'Blod strange'!

Myrica gale

* 'Myrtus ... Some abuse a litle shrub called Gal in englishe whiche groweth in fennes and waterish mores for myrto, but they are far deceyued'. - Turner (1548: E. vj). 'called in Cambridge shyre Gall, in Summerset shyre Goul or Golle'. - Turner (1568: 47). 'In the Ile of Ely'. - Gerard (1597: 1228).   **1548**

Grigson (1955) notes that 'Leland, the King's antiquary, when he came to Axholme in Lincolnshire in the fifteen-thirties, wrote that the fenny part of the district "berithe much Galle, a low frutex, swete in burning"'.

Myriophyllum alterniflorum

'in ditches to the north of Ivy Castle, Guernsey'. - Babington (1839a: 36).   vc113   **1839**

'M. aquaticum pennatum ... In fossa prope Lodden-Bridge haut longe a Redinga Oppido J. Bobart observavit'. - Morison (1699: iii, 622), identified by Druce (1897) as 'may well belong to this species'. Druce adds that there is a poor specimen in the Morisonian herbarium, **OXF**.   vc22   1699

Babington was the first to clearly separate the species, and his specimen was gathered on July 26th 1838 (Babington (1897: 74). Clarke also points out that it was not clearly known as a separate species until much later, and cites 'in a pond by the side of the Canal near Whixall Moss [Salop]' - Baxter (1840: 376). Druce, erroneously, cites 1680 for the Morison.

Myriophyllum spicatum

* 'Millefolium aquaticum pennatum spicatum [with a figure] ... In our owne land'. - Parkinson (1640: 1257). 'In the river [Cam] about Stretham ferry'. - Ray (1660: 99).   **1640**

Myriophyllum verticillatum

* 'Millefolium aquaticum minus. In the rivulet Stoure by the little Islet which it makes above the Paper milIs, and in divers other places'. - Ray (1660: 99).   vc29   **1660**

Najas flexilis

'A few scraps of which I found last month [August], in a pond near Roundstone, Connemara'. - Oliver (1850).   vcH16   **1850**

This publication predates that cited in Clarke and Druce by a few weeks. 'On Oct. 11, 1850, at the Botanical Society of London, Mr. Daniel Oliver, jun., exhibited specimens discovered by him in a pond near Roundstone. Connemara, Ireland, in August last'.- G.E.D. in *Phytol. iii. 1088*.

Najas marina

* 'on the 21st of July last, while examining the aquatic vegetation of Hickling Broad …'. - Bennett (1883a).   vc27   **1883**

Bennett (1883b) described the find more fully a few months later.

Narcissus pseudonarcissus

* 'Narcissus herbaceus … is after my judgement our yealowe daffodyl'.- Turner (1548: E. vj).   **1548**

Turner, W. 1548. *The Names of Herbes*

55

shrub called Gal·¹ in englishe, whiche groweth in fennes and waterish mores for myrto, but they are far deceyued.

Napus.

Napus is named in greeke Bounias, in duche Stekruben, in french Rauonet or naueau, I haue hearde sume cal it in englishe a turnepe, and other some a naued or nauet, it maye be called also longe Rape or nauet gentle, as a rape hath a round roote, so hath a nauet a longe roote and somthynge yealowishe. Thys herbe groweth plentuously at Andernake in Germany.

Narcissus.

Narcissus is of diuerse sortes. There is one wyth a purple floure, whiche I neuer sawe, & an other wyth a white floure, which groweth plētuously in my Lordes gardine in Syon, and it is called of diuerse, whyte Laus tibi, it maye be called also whyte daffadyl. Plenie ² maketh mention of a kynde called Narcissus herbaceus, whiche is after my iudgement our yealowe daffodyl.

Nardus.

Nardus is named in greeke Nardos, in englishe Spyknarde, the Potecaries name it Spicam Nardi, it groweth not in Europa that I haue heard tel of. It is hote in the fyrst degree and dry in the seconde.

³ *Nardus celtica.*

Nardus celtica, otherwyse called Saliunca, is in great plentie growyng in the alpes. The Germanes cal it mariend magdalene kraut, it may be called in englishe frēch spiknarde, when the indish spiknard is olde and dusty and rotten, it is better to vse thys in medicines then it.

Nasturtium.

Nasturtium is called in greeke Cardamon, in englishe Cresse or Kerse, in duche Cresuch, in frenche Cresson, Aleuois, and

¹ E. vj.   ² Plinie.   ³ E vj, back.

Nardus stricta

'I and Mr William Broad were at Chissel-hurst … observed this small Spartum whose figure I here give'. - Johnson (1633: 1630).   vc16   **1633**

This attributation of Johnson's record is argued convincingly in Edgington (2010), where he refutes the suggestion of Rose, in Gilmour (1972: 66), who cites 'Gram: Sparteum capillaceo folio minimum'. Hampstead Heath. - Johnson (1629: 11) as probably attributable to this species.

Narthecium ossifragum

* 'Asphodelus, acorifolius, palustris, luteus ... Angliae Flandriaeque; humentibus paludisque'.- L'Obel (1571: 46). 'In moist and marish places neere unto the towne of Lancaster'. - Gerard (1597: 88).   **1571**

Nasturtium microphyllum

The first convincing separation and key is in Howard & Manton (1946).   **1946**

'Nasturtium aquaticum foliis minoribus praecocius … observed by Mr Dale'. - Ray (1696: 172). 'On another sheet 'from Hokley in ye hole [near Oxford] loco rivuloso Maii fine 1744…"'. - Druce & Vines (1907: 93).   1696

Druce & Vines (*op.cit.*) add that Dale's specimen probably came from Essex. See also, from the list of herbarium specimens seen by them, 'Sunning Hill, Berks, 1773, herb Banks, **BM**'. - Howard & Lyon (1950).

Nasturtium officinale s.l.

* 'Cardamine ... Nasturtium aquaticum anglis Water cresses'. - Turner (1538: A. iij. verso). 'The true water Cresse ... groweth muche in brokes and water sydes'. - Turner (1548: G. iij. verso).   **1538**

Nasturtium officinale s.s.

The first convincing separation and key is in Howard & Manton (1946).   **1946**

'Sisymbrium Cardamine seu Nasturtium aquaticum. *JB 2. 884*', in Hb. Sloane 123: 13, **BM**. ' ... Lambeth, a little further of from the water'. - Druce & Vines (1907: 93).   1715

The Hb. Sloane specimen, which is excellent, with two clearly visible rows of seeds, is in the Herbarium of Buddle, who died in 1715. The record from Druce & Vines is dated as 1747, the date of Dillenius' death.

Neotinea maculata

* 'Miss More ... found it early in May, growing in calcareous pastures, at Castle Taylor, county of Galway, Ireland'. - Moore (1864).   vcH15   **1864**

Moore, D. 1864. *Neotinea intacta* Reichb., a recent addition to the British Flora. *J. Bot.* 2: 228-229.

> **NEOTINEA INTACTA, Reichb., A RECENT ADDITION TO THE BRITISH FLORA.**
>
> By D. MOORE, Ph.D., F.L.S.
>
> On the 27th of June I had the pleasure of submitting to the Royal Irish Academy *Neotinea intacta*, Reichb., an Orchid new to the British flora. This pretty little plant, hitherto only known from Greece, Malta, Madeira, Algiers, south of Germany, and Portugal, was discovered by Miss More, the sister of that eminent British botanist Mr. A. G. More. She found it early in May, growing in calcareous pastures, at Castle Taylor, county of Galway, Ireland. After making myself pretty certain of the species, I sent a specimen to Professor Reichenbach, of Hamburg, and in a letter dated June 20, he informs me that the new Irish Orchid " is indeed *Neotinea intacta*." An excellent figure of the wild plant will be found in Reichenbach's Icones Flor. Germanicæ, vol. 13–14, t. 500; and a cultivated specimen is figured by Lindley in the ' Botanical Register,' t. 1525, under the name of *Aceras secundiflora*. In the dry state the plant looks very much like *Habenaria albida*, and it would be well worth while to inquire whether it has ever been passed over elsewhere as that species.
>
> Miss More may be justly proud of her discovery, one of the most important made for many years. It is a highly interesting addition. It is well known that in the south-western counties of Ireland several

Neotinea ustulata

* 'Orchis sive Cynosorchis flor. Purpurascente, Pannonica 4. *Clus.* in montosis pratis. Litle purple-flowred Dogges-stones'. - Johnson (1634: 54). ' Cynorchis minor Pannonica, Clusii. On Scosby-Iease [Scawsby Leys, near Doncaster], Mr. Stonehouse'. - How (1650: 33).   **1634**

Neottia cordata

* 'Bifolium minimum *J.B. I. 31. pag 534.* neer the Beacon on Pendle Hill in Lancashire'. - Merrett (1666: 15).   vc59   **1666**

On Ray's second Itinerary 'August 12th [1661] ... on this moor [Whitby to Gisburgh], not far from Freeburgh Hill [5 miles S of Saltburn], we found Bifolium minimum *J.B.*'. - see Lankester (1846: 149).   vc62   1661

Raven (1950: 116) suggests that Ray, with Willughby, might have visited Pendle Hill in 1660.

Neottia nidus-avis

* 'Satyrium abortivum sive Nidus avis ... It is reported that it groweth ... neere Knaesborough ... I found it growing in the middle of a wood in Kent two miles from Gravesend'. - Gerard (1597: 176).   vc16, 64   **1597**

Neottia ovata

* 'Martagon ... in many places of Englande in watery middowes and in woddes'. - Turner (1548: H. iij. verso). 'Southfleet in Kent'. - Gerard (1597: 326).

1548

Nepeta cataria

* 'Catmynt ... groweth farre from cytyes, and townes, in hedges and in stony groundes'. - Turner (1551: G. vj. verso).

1551

Turner (1538: B. iij. verso) has 'Nepeta … Nepe or catmynte' but with no indication that it was a wild plant.

Noccaea caerulescens

* 'Thlaspi minus *Clus, G.268*. Thlaspi perfoliatum minus, In the Pastures above the Ebbing and flowing Well two miles from Giflewick [Giggleswick] in stony ground amongst the grass, Yorkshire'. - Merrett (1666: 118).

vc64   **1666**

Nuphar lutea

* 'Nymphea is of ii sortes the one hath a whyte flowre and the other hath a yelow flour, they grow both in meres, loughes, lakes and in still or standyng waters'. - Turner (1562: 65b).

1562

see also Turner (1548: E. vij) 'Nymphea … Boeth kyndes of water Roses growe in standyng waters'.

Nuphar pumila

* 'Discovered by Mr W. Borrer in the lake on the Highland mountain Ben Cruachan'. - E.B. 2292.

vc98   **1811**

'Discovered in 1809 by Mr. Borrer in a pool near the farm of Corrie-chastel, at the foot (not the summit) of Ben Chonachan; and in Loch Baladren'. - Graves & Hooker (1828: t. 165).

vc98   1809

Loch Baladren is presumably L. Belladeran (?L. Puladdern, see below), in Moray, where it was collected by Brodie (see Webster, 1978: 84). In fact she cites two records, both from specimens in **E**: 'at Brodie House, to which it was brought from L. Puladdern at Aviemore', but not dated, and 'collected again [by Mr Borrer] in 1811'.

Nymphaea alba

* 'Nymphea is of ii sortes the one hath a whyte flowre and the other hath a yelow flour, they grow both in meres, loughes, lakes and in still or standyng waters'. - Turner (1562: 65b).

1562

This seems a better first record than Turner (1538: B.iij. verso) 'Nymphea … [vocatur] nostris Water rose'. See also note under *Nuphar lutea* above.

Nymphoides peltata

* 'Nymphea lutea minor Septentrionalem ... juxta amoenissima Tamesis fluenta, in udis scrobibus & lacustris pratensibus'. - L'Obel (1571: 258). vc24 **1571**

Druce (1926) claims this record for Buckinghamshire.

Odontites vernus

* 'Crataeogonon Euphrosine ... In fertill pastures and bushie copses'. - Gerard (1597: 85). **1597**

Oenanthe aquatica s.l.

* 'Cicutaria palustris ... In most places of England: it groweth very plentifully in the ditches by a causey as you go from Redreffe [Rotherhithe] to Detforde neere London'. - Gerard (1597: 905). vc16, 17 **1597**

This could, of course, refer to *O. fluviatilis*, but the habitat suggests this species, and this is followed by Salmon (1931) in his *Flora of Surrey*.

Oenanthe aquatica s.s.

Babington (1843: 131) **1843**

'Potteric Carr [S of Doncaster]. In the drains about Rosington. In a pool of water near Mixborough Ferry. T. Tofield, <1747'. - Skidmore *et al.* (c1980). vc63 1747

First separated from *O. fluviatilis* at this date; Babington collected it on June 22nd, 1840 at Menham Bridge, on the R. Waveney [E. Suffolk].
O. fluviatilis is a very rare species in vc63, and Tofield's records are likely to be *O. aquatica*.

Oenanthe crocata

* 'Saxifragia ... Grene Marke ... groweth muche by the Temmes syde about Shene'. - Turner (1548: H. v. verso). vc17 **1548**

Oenanthe fistulosa

* 'Filipendula aquatica ... neere the river of Thames or Tems about the Bishop of Londons house at Fulham and such like places'. - Gerard (1597: 902). vc21 **1597**

Pulteney cites 'Oenanthe vulgaris ... aestate in convallibus herbidis aut verrucis pratensibus Angliae' - L'Obel (1571: 325), which might be this species.

Oenanthe fluviatilis

as *O. phellandrium* β *fluviatilis* 'probably a distinct species … swift streams, Hertfordshire, Rev. W. H. Coleman '. - Babington (1843: 131).	vc20	**1843**
Clarke, with hesitation, and Druce, with none, cited 'Cicutaria palustris tenuifolia, *C.B.P.*'. - Morison (1699: 291) as the first record. There is indeed a specimen in the Morisonian Herbarium at **OXF**, which seems to be this species, although the fruit is poor, with the inscription ' Phellandrium vel cicutaria aquatica quorundam *JB183*. In rivulis & aquis caenosis ... frequens nascitur haec cicuta'.		1699

Dandy (1958a: 182) notes that in Plot's herbarium is a specimen of this species. Plot died in 1696. See also Coleman (1844). Note that Oswald & Preston (2011: 182n) comment that it seems very likely that Ray (1660: 34) encountered both species (this and *O. aquatica*) around Cambridge. Druce also cited a slightly later record 'In rivulo inter Woodstock et celebrem illum pontem Ducis Marlborugii juxta Blenheim, with two sheets of the leaves'. - Druce & Vines (1907: 72). Nevertheless the confusion between the two species was not settled until Babington's note.

Oenanthe lachenalii

* 'Oenanthe Staphylini folio aliquatenus accedens, *J.B.* ... In parochia Quaplod [Whaplode], agri Lincolniensis, non procul ab oppido Spalding'. - Plukenet in Ray (1690: 241).	vc53	**1690**
Gibbons (1975) gives Plukenet's find as 1688 (although she gives no explanation for this).	vc53	1688

Whaplode is just east of Spalding.
Johnson (1632: 19) has 'Oenanthe Angustifolia, *Lob*'. Between Margate and Sandwich. Rose, in Gilmour (1972: 113), agrees with Clarke that Johnson was probably describing this species, but there is no evidence that he was aware of other species. Similarly Gunther (1922: 115) also cites Johnson's polynomial for Goodyer's record 'this 19 of May 1620 I found this wild in East Hoo in ye parish of Subberton (vc11) about 7 miles from Petersfield in Hampshire in a hedgerowe ...'. This record is repeated in Merrett (1666: 84), but does not sound a very likely habitat for this species.

Oenanthe pimpinelloides

* 'Distinguished from *Oe. Lachenalii* and proved to be British ... gathered in a dry meadow on red marl near Forthampton, Gloucestershire, by Mr Edwin Lees'. - Ball (1844).	vc33	**1844**
'By the side of rills, ascending the north side of Breedon Hill'. - Nash (1781).	vc37	1781

The same paper by Ball was also printed in *Annals of NH* (4: 4 -7). Although early records might have been confused with *O. lachenallii*, the habitat of the Bredon Hill record seems reasonable.

Oenanthe silaifolia

* as *O. peucedanifolia* 'Banks of the Isis beyond Ifley - Peat bogs under Headington - Wick Copse'. - Sibthorpe (1794: 98).	vc23	**1794**
'Oenanthe pratensis Asphodeli radice ... prope Blaiden [Bladon] in pratis elatis inter Oxford et Blenheim'. - Druce & Vines (1907: 71).	vc23	1747

The record from Druce & Vines is dated as 1747, the date of Dillenius' death. Townsend (1904: 179), cites a record of Goodyer written in his copy of Merrett (1666), see below, 'first found this plant on the 19th May 1620, it flowered on the 18th day of June'. This is too uncertain and there are no other records for Hampshire. Merrett (1666: 84) has 'Oenanthe aquat. Selini folio, N.D.' and this is possibly the source of Townsend's comment, although that location in Merrett is actually against Oenanthe angustifolia *Lob. P894.*, which polynomial Gunther and Druce have identified with *O. lachenallii*.

Onobrychis viciifolia

* 'Onobrychis sive Caput Gallinaceum ... Upon Barton hill fower miles from Lewton [Luton] in Bedfordshire'. - Gerard (1597: 1064).	vc30	**1597**
'Provenit in agro Cantabrigensi in Anglia circa agrorum margines'. - Notes by Thomas Penny, see Zoller *et al.* (1972-80). The actual reference is: Vol. 6, t. 16 (pp. 42-43). See also Foley (2006b) for the justification of the 1565 date.	vc29	1565

This was grown as a crop, but a probable wild reference is 'Caput gallinaceum Belgarum. *Ad. Lob.* Onobrychis. *Dod. Clus. Ger.* Medicke Fitchling, or Cockes-head. On the farther end of Sarisbury Plaine'. - Johnson (1634: 25).

Ononis reclinata

* 'Dr Graham, of Edinburgh, whilst on a botanising excursion in Galloway, discovered the *Ononis reclinata* in considerable abundance'. - Hooker (1835a: 119).	vc74	**1835**

Fletcher (1959) relates that 'there is a fragment in the Edinburgh Herbarium dated 1833 and carrying the name of Dr Balfour'.

Ononis repens

* 'Anonis called also Ononis ... resta bovis ... groweth in many places about cambryge'. - Turner (1548: B. i).	vc29	**1548**

Ononis spinosa

* 'Ononis aut Anonis Asinaria, sive nocua Offic. Aresta bovis. … Varietas. Quaedam occurrit in pratis maritimis & udis Angliae, praesertim ad Bristoiam & prope Londinum … spinis armatur'. - L'Obel (1571: 378).   vc21, 34   **1571**

Onopordum acanthium

* 'Acantha leuke … spina alba … Besyde Sion in England'. - Turner (1562: 146).   vc21   **1562**

Ophioglossum azoricum

as var. *microstichum* 'Much smaller than the normal plant … reaches maturity in September … found at Swanbister, Orkney, by Mr. J.T. Syme'. - Moore (1859: 341).   vc111   **1859**

as *O. vulgatum* subsp. *polyphyllum* 'St Mary's, Isles of Scilly'. - Millett (1853).   vc1   1852

This does not appear to have been recognised at specific level by British authors until in Jermy *et al.* (1978).

Ophioglossum lusitanicum

'Discovered by Mr George Wolsey … Not far from Petit Bot Bay'. - Newman (1854).   vc113   **1854**

'First found in 1853 by Miss M.E. Guille'. - McClintock (1975: 48).   vc113   1853

McClintock (*op.cit.*) relates how Wolsey sent in the record without mentioning the prior claim of his neighbour, Miss Guille. The first record other than from the Channel Islands was in 1950, St. Agnes, see Raven (1950).

Ophioglossum vulgatum

'Lingua serpetina groweth in many places of England … in englishe Adders tonge. It groweth plētuously in middowes where as Lunary groweth'. - Turner (1548: H. iij).   **1548**

see also 'Adders tonge … it groweth in moyst and medowes in the end of April and in the beginninge of May, and shortely faydeth awaye'. - Turner (1568: 51) with an excellent picture.

Ophrys apifera

* 'Melittias Orchis … The Bee the Fly and the Butterfly Satyrions grow upon barren chalky hils … adjoining to a village in Kent named Greenhithe, upon Longfield downs by Southfleet,' &c … 'likewise in a field … half a mile from S. Albons.' - Gerard (1597: 163).   vc16, 20   **1597**

Ophrys fuciflora

* 'Plentiful on the southern acclivities of the chalky downs near Folkestone, Kent. The conical hill which forms the north-west boundary of the Cherry-garden, near that town, abounding in its upper half with this species, and in the lower half with *O. apifera*. *Mr. Gerard E. Smith*'. - Smith (1828: 273). vc15 **1828**

Mr. Andrew Matthews collected *O. arachnites* several years since at Ospringe ...'. - Smith (1829: 58). This would predate 1828. In the same work there's also a reference to a hybrid between it and *O. fuciflora*, collected by Mr. Lee, but unfortunately not dated.

Ophrys insectifera

* 'Orchis Myodes ... The Bee the Fly and the Butterfly Satyrions grow upon barren chalky hils ... adjoining to a village in Kent named Greenhithe, upon Longfield downs by Southfleet,' &c ... 'likewise in a field ... half a mile from S. Albons'. - Gerard (1597: 163). vc16, 20 **1597**

Ophrys sphegodes

* 'Orchis Arachnitis, Spider Orchis ... upon an old Stone pit ground, which is now green, hard by Walcot a mile from Barneck [Barnack], as fine a place for variety of rare plants as ever I beheld. Dr. Bowle'. - How (1650: 82). 'Near Shelford'. - Ray (1663: 7). vc32 **1650**

Orchis anthropophora

* 'Orchis anthropophora oreades *Col.* ... flore nudi hominis effigiem representans ... Found by Mr. Dale in an old Gravel-pit at Dalington [Ballingdon] near Sudbury [Essex] ...'. - Ray (1690: 171). vc19 **1690**

Clarke notes that Merrett's (1666: 85) 'O. antropophora autumnalis *Col*.', found between Wallingford and Reading, is now thought to have been most probably *Habenaria viridis* [*Coeloglossum*]. See Druce (1897: 477).

Orchis mascula

* 'There are divers kindes of orchis ... one kinde ... hath many spottes in the leafe and is called adder grasse in Northumberland'. - Turner (1562: 152). vc67 **1562**

But see Turner (1551: N. vi. verso) 'Adders gras ... grow[s] plentuously in the myddowes in every quarter of Englande'. Chapman & Tweddle (1989: 300), equate this with *O. mascula* though it is odd that the same species is in two volumes of the Herbal, and it could well be *O. morio*.

Orchis militaris

* 'Orchis Anthropophora autumnalis *Col.*, Orchis Anthropophora Oreades altera *Col. pag. 318*' and 'Orchis Oreades trunco pallido, brachiis & cruribus saturate rubescentibus. These three Satyrions were found on several Chalkey hills, near the highway from Wallingford to Redding on Bark-shire side the River by Mr. Brown'. - Merrett (1666: 85). vc22 **1666**

'Orchis Antropophora autumnalis, Oreades altera, *Col. 318*. The Man Orchis, found on Chalkey hills near the highway from Wallingford to Redding on Barkshire side of the river by Mr Brown[e]'. - Records of Rev William Browne entered by How in his copy of *Phytologia* between 1650 and 1656, see Gunther (1922: 302). vc22 1656

Orchis purpurea

* 'Orchis magna latis foliis, galea fusca vel nigricante *J.B. II.759*. At Northfleet near Gravesend. Mr. J. Sherard'. - Ray (1724: 378). vc16 **1724**

Both Clarke and Druce cite, with some hesitation, the record 'Orchis militaris polyanthos, On Gads-hill in Kent. - Merrett (1666: 87)'. Hanbury & Marshall (1899) also cite that, but add the Ray record as an 'alternative'.

Orchis simia

'Orchis anthropophora foliis angustioribus, spica longiore et tenuiore, nobis … circa Henley, Readingum …'. - Morison (1699: 493). vc22, 23 **1699**

Foley & Clarke (2005) 'An early reference to what may be *O. simia*, along with the Military Orchid [might be] Merrett (1666) [see *O. militaris* above], whilst a more definite one is provided by Morison (1699) for Henley. Druce (1927) dates a specimen in herb. Bobart, **OXF**, to around 1690.

Oreopteris limbosperma

'Varietas illa Filicis [Filix minor palustris] maris vulgaris a Petivero observata in ericeto Dunsmore, prope Rugby'. - Ray (1724: 122). vc38 **1724**

'D. Jac. Petiver had found a variety of Marsh Fern (Filix maris vulgaris varietas) on Dunsmore Heath near Rugby'. - Doody [paraphrased] in Ray (1696: 341). vc38 1692

Babington (1860b), Perring *et al.* (1964), Crompton (2007) and Oswald & Preston (2011) all equate the 'Filix pumila saxatilis altera' of Ray (1670, 1677 & 1685) with this species, but Edgington (2013) prefers *Thelypteris*. I find his reasoning convincing. Edgington also notes that Dillenius, in Ray (1724) was the first to note the distinctive arrangement of the sori, and also clarifies that Petiver's record was in 1692, when he was touring the Midland counties.

Origanum vulgare

* 'Origanum … Our commune organ ... called in some places of England wylde mergerum'. - Turner (1548: E. vij. verso).   **1548**

Ornithogalum pyrenaicum

*'Ornithogalum angustifolium majus floribus ex albo virescentibus, *Bauh*.... Onion Asphodill, Greene starre flowre. It growes in the way betweene Bathe and Bradford not farre from litle Ashley'. - Johnson (1634: 55).   vc8   **1634**

Ornithogalum umbellatum

'Ornithigalum …… I neuer sawe it in Englande, sauyng onely besyde Shene herde by the Temmes syde'. - Turner (1548: E. vij. back)   vc17   **1548**

O. umbellatum has been grown for centuries as a garden plant, and as suggested by Pearman (2013), if it is a native in Britain, then it is as subsp. *campestre*. The core range of that is in Breckland, where the first record is 'Barton, 1772, Sir J. Culham'. - Hind (1889).

Ornithopus perpusillus

* 'Ornitopodium sive Pes avis. varietas [with a figure]. ... Oritur in marginosis herbidis Grinwicii [Greenwich] Regie, ad Tamesim flumen non procul Londino'. - L'Obel (1571: 403).   vc16   **1571**

Ornithopus pinnatus

as *Arthrolobium ebracteatum* 'Guernsey' - Babington (1837). '*O. ebracteata* … last May discovered in Scilly by Miss White'. - Penneck (1838). 'Discovered [in April, 1838] on Tresco, one of the Scilly islands, by Miss White of that place'. – Babington (1838).   vc113   **1837**

'Aug.14. [1837] … Beyond Vale Church ... At about half-way to Mont Guet we found a leguminous plant, which appears to be *Arthrolobium ebracteatum*'. - Babington (1897: 67). 'first found in the British Isles at the Vale [Guernsey] on 14th August 1837 by Babington' - McClintock (1982: 110).   vc113   1837

Also in E.B.S. 2844,1840.

Orobanche alba

* 'Discovered by John Templeton, Esq ... at Cave-hill near Belfast in August, 1805'. - E.B. 1786.   vcH39   **1807**

'Cave Hill pre-1793'. - Hackney (1992: 269).   vcH39   1793

Templeton's ms. *Flora Hibernica*, from which this record came, is traditionally dated to c.1793. Lees (1888: 348) has an early record 'under the rocks on the east side of Malham Tarn, in 1801'.

Orobanche caryophyllacea

* 'In hedges in our area [Folkestone]'. - Smith, G. (1828). See also Smith, G. (1829: 34 - 6). vc15 **1828**

Orobanche elatior

* 'Orobanche sive Rapum Genistae. *Ger. Park.* ... It grew in barley on the right hand of the way between Cambridge and Grantchester: In a corn field nigh the church at Cherry-Hinton'. - Ray (1660: 110). vc29 **1660**

Ray's entry obviously covers both this species and *O. rapum-genistae*. Oswald & Preston (2011: 252, n.500) discuss this fully, including the possibility that it was in fact *O. minor*. Both Sutton (1798) and Babington (1860b) accept this record.

Orobanche hederae

* as *O. barbata,* Babington in E.B.S. 2859. **1841**

'Orobranca or Kepi… upon roots of Ive at Chateau des Marais, Guernsey.1726'. - Thos. Knowlton's ms. diary, cited in McClintock (1975). vc113 1726

Babington lists 10 sites, including Guernsey, where he collected it in 1837. Clarke notes that this was most probably the plant referred to under *O. rapum-genistae* by Curtis (1781: iv. 44), as 'a small Orobanche observed by Mr. Thomas White growing on walls, &c., in Pembrokeshire, the decayed floor of an old castle being almost covered with it'. But see also Raven (1950: 127) where he suggests that an *Orobanche* found by Ray, on his 1662 Itinerary, on Torr-hill, Glastonbury, and which he described in detail but failed to identify (see Lankester 1846: 181), could well be this species.

Orobanche minor

* 'Orobanche flore minore *J.B.* … found by Mr. Rand in a field of Oats two miles beyond Rochester, on the left hand going to Horns-place'. - Ray (1724: 288*). vc16 **1724**

But see Turner (1551: P. v) 'Chokeweed at Morpeth, about the roots of Broom'. Chapman & Tweddle (1989: 312) identify this as this species, but it seems more likely to be *O. rapum-genistae*. This latter is covered in Turner (1562: 71b) - see below. On the other hand, 1724 seems very late for the first record of *O. minor*, though that species may not be convincingly native in many areas (C.D. Preston, *pers.comm*.)

Orobanche picridis

'Observed by myself growing very abundantly … on July 9th, 1844, upon a ledge of the Freshwater cliffs, called by the cliffs-men Rose Hall Green, but supposed to be only *O. minor* at the time'. - Bromfield (1849). vc10 vc10 **1849** 1844

Orobanche purpurea

* 'A single specimen was found in 1779 by a Mr. Scarles near Northreps (Norfolk) and several more last year near Sheringham by Mr. W. Skrimshire'. - E.B. 423. (But in Sutton (1798) the first finding is credited to Mr. Pitchford.)   vc27   **1797**
                                                          vc27   1779

See also Johnson (1633: 228) where is described a plant found by Goodyer in Hants which may have been this. D.E. Allen writes 'Goodyer's plant, found in 1621, has been lastingly controversial'. Druce followed J.E. Smith in referring it to this species, an interpretation Brewis *et al.* (1996) consider 'possible'. But Townsend (1904) thought it more likely to be *Epipactis purpurata*, a view in which he was backed by O.Stapf'.

Orobanche rapum-genistae

* 'Orobanche ... I never sawe it in al Englande, but in Northumberlande ... newe chappel floure'. - Turner (1548: E. viij). 'In many places of England bothe in the Northe countre besyde Morpethe ... and also in the South countre a lytle from shene [Shene, Middx.] in the broum closes'. - Turner (1562: 71b).   vc67   **1548**

Orobanche reticulata

'Found by N.[H.]E. Craven in 1907, about 12 miles from Leeds'. - Druce (1909a: 334).   vc64   **1909**

'There is a fine early specimen in **BM** received from Botanical Record Club, and labelled *O. minor*. Lotherton, M.W. Yorks. G. Webster. 7/1881'. - Pugsley (1940b).   vc64   1881

Lees (1941) gives Mr Craven's initials as H.E. and states that he saw an earlier specimen from J.F. Pickard in 1902.

Orthilia secunda

* 'In shady birch woods among the moss ... near Little-Loch-Broom in Ross-shire, and about Loch-Mari, in the same county, and in the birch-woods of Troschraig, Craig-loisgt, and Coille-mhor, about Loch Rannoch, in Perthshire, where all three species [of *Pyrola*] are found'. - Lightfoot (1777: 219).   vc105   **1777**

as *Pyrola secunda* 'May 1767, from bridge of Nairn & Dalmagerie ... in the wood called Cragenon (Craiganeoin)'. - Journal of James Robertson, Henderson & Dickson (1994: 39).   vc96   1767

There is an entry 'in a wood opposite to Moy Hall, Inverness and in several places in Ross-shire' in Hope, 1768, see Balfour (1907). Lees (1888) and Clarke state that Ray's 'P. folio mucronato serrato' (1690: 176) was not this species - probably *Pyrola minor*.

Osmunda regalis

'Osmunda, Filix florida ... Belgia & Anglia plurima'. - L'Obel (1571: 363). 'Osmunda regalis. Water Ferne, or Osmund Roiall or Osmond the water-man … It groweth in the midst of a bog at the further end of Hampsted heath from London and neere unto Burntwood in Essex'. - Gerard (1597: 969). **1571**

Oxalis acetosella

* 'Oxys ... vulgus etiam vocat Alleluya, wodsore, & cuckowes meat'. - Turner (1538: B. iiij). **1538**

Oxyria digyna

* 'Acetosa Cambro-britannica montana ... A gentleman of Anglesea called Mr. Morris Lloid of Prislierworth [Tre-Iorwerth] found it on a mountaine in Wales'. - Parkinson (1640: 745). vc49 **1640**

Parkinson's entry continues ' … and showed it to Dr Bonham in his life'. Dr Bonham died in 1630. See Raven (1947: 264). vc49 1630

It is assumed that the reference is to Caernarvonshire.

Oxytropis campestris

* 'Discovered by Mr. G. Don, in the summer of 1812, on a high rock, on one of the mountains at the head of Clova, Angusshire, near the White Water'. - E.B. 2522. vc90 **1813**
vc90 1812

This is also in Don (1806: Fasc. 9, no 213). However this was not issued before 1812, probably either at the end of 1812 or early in 1813. See Druce (1904), and also a letter to Mr D. Booth (Druce *op.cit*: 67).

Oxytropis halleri

'On the sea-coast, at the bay of Farr, I had the good fortune of discovering a new species of British pea-grass …'. - Robertson (1768). vc108 **1768**

'From Findhorn to Nairn'. - Journal of James Robertson for 1767, Henderson & Dickson (1994: 31). vc95 1767

There is an entry 'At Cromarty at Farr at Glen Criven on a hill' in Hope, 1768, see Balfour (1907). See also 'Astragalus uralensis ... Upon Carn-dearg, one of the lower heads of Ben-Sguilert ... Mr. Stuart. It has also been discovered at the bay of Farr on the eastern coast, and in a rocky soil at Cromarty by Mr. Robertson. See 'Scotch Magazine' for July, 1768, with a figure of it'. - Lightfoot (1777: 401).
Clarke adds 'I am informed by Mr. Symers Macvicar that it was found by Dr. John Walker near Loch Leven in 1761', and this is repeated in Fletcher (1959) and Lusby & Wright (1996) but no source has yet been found for this.

Robertson, J. 1768. An account and a print, of a new species of *Astragalus*, a plant discovered by James Robertson in 1767.

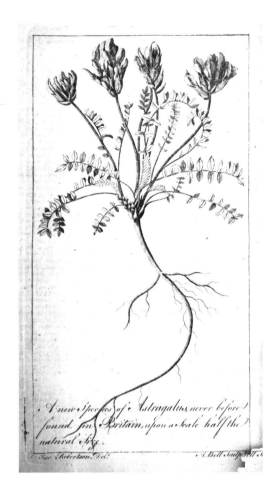

Papaver argemone

* 'Argemone capitulo longiore & A. capitulo torulis canulato ... In Angliae Sommerseti ... segetibus'. - L'Obel (1576: 144). 'These plants do grow in the corne fieldes in Somersetshire and by the hedges and high waies ... as yea travall from London to Bathe'. - Gerard (1597: 301).   vc6   **1576**

Gerard (*op.cit.*) adds that 'L'Obelius found it growing in the next fielde unto a village in Kent called Southfleete, myselfe being in his companie, of purpose to discover some strange plants, not hitherto written of'.

Papaver bivalve

'Papaver corniculatum violaceum ... In the corn fields beyond Swafham as you go to Burwell'. - Ray (1660: 111).   vc29   **1660**

The British species, which is now extinct, belonged to subsp. *hybridum*.

Papaver dubium s.l.

* 'Argemone capitulo longiore glabro ... In agro Cantabrigiensi observavit et ad nos ejus capitula transmisit *D. Pet. Dent*'. - Ray (1686: 856).	vc29	**1686**

Edgington (2014) identifies as this species Johnson's (1638: 19) 'Papaver erraticum alternum, *Fuch. Dod.* Erraticum minus. *Tab.* Little red Poppy, or Corne Rose [Tottenham]'. Johnson (1633: 372.3) describes 'a small kind of red Poppy growing commonly wilde together with the first [*P. rhoeas*] which is lesser in all parts, and the floures are of a fainter or overworne red, inclining somewhat to orange'.

Papaver dubium s.s.

The aggregate was first separated by Babington (1860b: 300).	vc29	**1860**
'Argemone Pavio. 9 Sept. 1621 … the heads or seed vessells are smooth three quarters of an ynch longe … this herbe is like *Papaver Rhoeas* in leaves, stalks, flowers and [white] milkie iuyce, but the stalk are longer, the flowers much paler & the seed vessels longer'. - Goodyer ms.11. f. 113, see Gunther (1922: 155).		1621

Goodyer's description suggests *P. dubium*, not *P. argemone*. This is presumably a Hampshire record.

Papaver hybridum

* 'Argemone capitulo longiore & A. capitulo torulis canulato … In Angliae Sommerseti … segetibus'. - L'Obel (1576: 144). 'These plants do grow in the corne fieldes in Somersetshire and by the hedges and high waies … as yea travall from London to Bathe'. - Gerard (1597: 301).	vc6	**1576**

See notes on *P. argemone*, above.

Papaver lecoqii

The aggregate was first separated by Babington (1860b: 300).	vc29	**1860**
Cherry Hinton, 1836, Babington, det. Kadereit, **CGE**.	vc29	1836

See also McNaughton & Harper (1964) for a fuller account.

Papaver rhoeas

'Under Papaver sativum 'Papaver erraticum redecorne rose, aut Wylde pappy vocatur'. - Turner (1538: B. iiij. verso).	**1538**

Clarke and Druce both cited 'Papaver erraticum … in englishe Redcorn-rose or wylde popy'. - Turner (1548: F. j) and 'With us it groweth much amongst the rye and barley'. - Turner (1562: 77) as the first records, but Turner's earlier record seems quite acceptable, and is followed by Nelson (1959).

Papaver somniferum

'Papaver spontaneum syl. *Lob.*'. Stoke through ... Cowling, and ... to Cliffe. - Johnson (1629: 8).   vc16   **1629**

This is cited in Hanbury & Marshall (1899) and followed in Gilmour (1972). Jackson (1959) prefers Turner (1538: iiij. verso) 'Papaver sativum … Poppi'.

Parapholis incurva

as *Rotbollea incurvata* 'on the sea-coast in several places to the east of North Berwick Law'. - Don (1806: 8.177). This is dated 1806, but as Druce (1904) explains, was not issued until 1810.   vc82   **1810**

'On the beach at Dunwich, 1796. Hb. Davy'. - Simpson (1982). Davy's specimens are incorporated into a 'Hortus Siccus' made by William Kirby which is in Ipswich Museum herbarium – the 1796 specimen is labelled '*Rottbollia incurvata*'.   vc25   1796

The specimen in Don, is definitely *P. incurva* (H. Noltie, *pers. comm.* from the volume at **E**, based on anther lengths). Druce identifies this as *Lepturus filiformis* var. *incurvatus*. Whether Don really had this from N. Berwick, or muddled it from another source, or indeed if this is a garden specimen grown on from wild stock (which it probably was), is, of course, unknown. North Berwick is north of the assumed native range of the species, but there are many other records of it as a ballast plant.

E.B. 760 purports to be this species, but is surely *P. strigosa*. There was much confusion between *P. strigosa* var. *incurvata* and this species - see Hudson (1778: 441), Hooker (1830: 455 var.2) and others, and this was not finally resolved until Hubbard (1936).

Don, G. 1806. *Herbarium Britannicum*. Fasciculi 5 to 9.

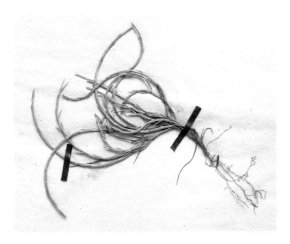

Parapholis strigosa

* 'Gramen parvum marinum spica loliacea'. Near Margate. - Johnson (1632: 31). vc15 **1632**

Rose, in Gilmour (1972: 104), suggested that this was *Elytrigia atherica* (as *Agropyron pungens*). I agree with Hanbury & Marshall's (1899) identification as this species, though Edgington (2010) suggests it is describing *P. incurva*.

Parentucellia viscosa

* 'Euphrasia lutea latifolia palustris ... Primo observavi circa oppidum S. Columbi [St. Columb], & deinceps in Occidentalioribus Cornubiae plurimam'. - Ray (1670: 107). vc2 **1670**

'St Columb, Cornwall, by the wayside in several places' on Ray's 1662 Itinerary, see Lankester (1846: 186) and Raven (1950: 128) - *illustrated opposite, p.299*. vc2 1662

Clarke also cites Merrett's (1666: 31) record of 'Crataeogonum cubitalis altitudinis flora luteo. In the Isle of Wight, Mr. Cole, and in the Kings meadows at Godstone in Surrey', and thinks that this probably refers to this species. This is followed by Pope *et al.* (2003). Mr Cole (or Coles) is probably William Cole, author of *Adam in Eden*, who died in 1662, and, if so, this record would predate that from Cornwall.

Parietaria judaica

* 'Helxine or pardition is called in englishe Pelletorie of the wal … It groweth on walles'. - Turner (1548: D. iiij. verso). From the church wall, Gillingham. - Johnson (1629: 4). **1548**

See Turner (1538: B. verso) 'Helxine dicitur, vulgo Paritory', which implies familiarity with the plant as a wild species.

Paris quadrifolia

* 'Aconitum … Libardbayne or one bery. It is much in Northumberland in a wodd besyde Morpeth called Cottingwod'. - Turner (1548: A. v. verso). vc67 **1548**

Parnassia palustris

* Gramen Parnassi hederaceum recentiorum ... in pratis & udis pascuis Angliae ad Oxoniam'. - L'Obel (1571: 263). vc23 **1571**

Unfortunately Louis (1980) gives no hint as to when L'Obel visited Oxford.

Lankester, E. ed. 1846. *Memorials of John Ray*. This page also covers the discovery of *Illecebrum verticillatum*.

See species account: Parentucellia viscosa, p.298

> 186   MEMORIALS OF RAY:
>
> Saturday, June the 28th, we travelled on to Truro, sixteen miles. By the way we passed St. Columb,* an old town. The churches in Cornwall, for the most part, have good tower-steeples of free-stone; the churches are made up of three rows or ridges of building, of an equal height, and sometimes length too, and covered with slate. Near St. Columb, by the way side, are found in several places, *Euphrasia pratensis lutea, C. B.*† Between St. Columb and St. Michael (and in several other places), a plant, which we guess to be *Alsine palustris minor serpilli folia.*‡ It hath long, weak trailing branches; the stalk is round and red, the leaves of a pale green, growing by pairs, the flowers grow *verticillatim* about the stalk, at every joint; each particular flower is compounded of five, as it were tubuli, in figure like the seed-vessel of larkspur; it grows in watery places near springs. Nothing more common than *Osmunda regalis* about springs and rivulets in this country. Camomile [*Anthemis nobilis*, Linn.] grows in such plenty along the way sides, that one may scent it as one rides. Truro is a pretty town, the second in Cornwall, and is governed by a mayor and four aldermen, with their four assistants; the lord Roberts hath an house there, but it is a small one; the church is handsome and large, and hath two monuments in it, one of them of the three children of the Michells,
>
> * Mr. Ray, in the year 1667, took another tour to the Land's-end, and from thence went to St. Columb. "By the way (saith he) at a place called Baldieu, we saw a tin mine. The load or vein, for the most part, runs east and west, and deepens north. The load is not a vein, but a floor or bed. The load, both above and beneath, is covered with a crust or stony substance, which hath no tin in it, which they call Country, the uppermost they call the North Country, the nethermost South Country. In the mine they sometimes find spar, which is nothing but a flour; some white and hexangular in diamonds. St. Columb is one of the best parsonages in Cornwall, the yearly value between 300 and 400 pounds."
> † This is the *E. lutea*, Linn.; a plant not found in England, and therefore not included in Ray's 'Synopsis.' It is probable that the *Bartsia viscosa, Trixago viscosa, R.* of Bab. Man. (*E. major lutea latifolia palustris*, Ray,) is the plant referred to in this place.—C. C. B.
> ‡ I believe that the *Illecebrum verticillatum*, Linn., is intended by this name, which is not to be found in Ray's 'Synopsis,' where the *Illecebrum* is called *Polygonum serpillifolium verticillatum.*—C. C. B.

Pastinaca sativa

* 'This wild persnepe groweth pletuously besyde Cambrydge in a lane not far from Newnam Milles'. - Turner (1562: 80b). 'Of Persnepes and Skirwurtes, Sisarum sativum magnum'. - Turner (1562: 138b).   vc29   1562

Clarke and Druce both interpret that Turner is describing two species in his first account - see also *Daucus carota*. But both Perring *et al.*(1964) and Crompton (2007) identify this reference with *Daucus carota* only. The illlustration in Turner's later reference is more convincing, for what it is worth.

Pedicularis palustris

* 'Pedicularis *Tragi 97*. cubitalis est altitudinis & foliis Quercinis. Two miles East of Croydon below the wind mill nigh the fish pond'. - Merrett (1666: 91).   vc17   1666

'Lowzye weede…..in black brookes in Estmallinge [East Malling] in Kent'. - Manuscript notes of William Mount, of c.1584, in Gunther (1922: 255 - 263), and identified by him as this species.   vc16   1584

Clarke (with hesitation) and Druce (with none) feel that Gerard's description (1597: 913) 'in moist and moorish medowes' might encompass both this species and *P. sylvatica*.

Pedicularis sylvatica

* 'Pedicularis. Lousewort or Red Rattle ... within verie moorish medows they grow a cubite high or more but in moist and wet heathes and such like barren grounde not above a handfull high'. - Gerard (1597: 913).   1597

Pulteney identifies as this species Lyte's (1578: 516.1) 'Fistularia, Reede Rattel ... groweth in moyst medowes, and is very noysome to the same'.

Persicaria amphibia

* 'Potamogeton angustifolium … In standing waters pooles, ponds, and ditches, almost everywhere'. - Gerard (1597: 675).   1597

Gerard's habitat description covers both this species and *Potamogeton natans*.

Persicaria bistorta

* 'Bistorta ... called in some places of England Astrologia and in some places Pationes'. - Turner (1568: 12). 'The great Bistort groweth in moist and watery places'. - Gerard (1597: 323).   1568

Persicaria hydropiper

* 'Crataeogonum ... Englishe men cal Arssmerte ... groweth in watery & moyst places'. - Turner (1548: C. v).   1548

Turner (1538: B. iv. verso) 'Persicaria, Angli vocant Arsmert or Culerage', which implies familiarity with the plant as a wild species. Nelson (1959) is sure it is.

Persicaria lapathifolia

* 'Persicaria mitis major foliis pallidioribus ... In the lands and furrows of Hedington Field above S. Bartholomews Hospital half a mile from Oxford. ... a D. Jacobo Bobarto observata & descripta est'. - Ray (1696: 58).   vc23   1696

Persicaria maculosa

* 'Persicaria ... Arsmert whiche hath the blacke spotte in it ... groweth most commonlye in moyst places'. - Turner (1568: 62).   1568

Persicaria minor

'Persicaria pusilla repens *Lob*. In barren gravelly and wet places'. - Johnson (1633: 446).

Ray (1660: 116), as interpreted by Oswald & Preston (2011: 259), Smith (1824: ii. 235) and others support this attribution. Clarke (1909) cites Morison (1680: 589) 'Persicaria minor seu pusilla procumbens nobis ... in Angliae pratis' as his first record.

1633

Persicaria mitis

* 'About London; Lagasca [La Gasca] and Mr. Borrer. Near Cambridge, Mr. Babington'. - Hooker (1838: 165). 'First detected [in 1826] by Prof. La Gasca, of Madrid, by the road side at Chelsea'. - Babington in E.B.S. 2867 (1843).

vc21   1838

'Persicaria mitis maculosa et non maculosa *C.B.* Unlocalised, but one of the earliest examples known of this species'. - Druce (1928: 483). Druce's record is from Herb. Du Bois, who died in 1740.

1740

Babington's text in E.B.S. gives an excellent history and synonomy. There is also a record 'Insipida, Chelsea' in Druce & Vines (1907: 54) which they claim as the first British record, but this cannot be dated as before Dillenius' death in 1747. But Druce (1932) also cites a plant in Petiver (1695: i, 13, no. 90) 'Persicaria angustifolia ex singulis geniculis florens, an Pers. angustifolia *C.B. 101? & pr* [*Bauhina prodromus*] *43?* Narrow-leaved Lake-weed. This I found the last Autumn with the common Arsemart on the ditch-banks in the Meadowes beyond Lord Peterborough's house at Westminster'. However, Kent (1975) identifies this with *P. minor*.

Persicaria vivipara

* 'Bistorta minor ... In Westmerland, at Crosby, Ravenswaith'. - Gerard (1597: 323).

vc69   1597

Petasites hybridus

* 'Petasites ... a butter bur northumbrienses vocant an Elden'. - Turner (1538: B. iiij. verso). '... called around Morpeth Eldeus'. - Turner (1548: F. ij).

vc67   1538

Petrorhagia nanteuilii

'Caryophyllus sylvestris, prolifer *C.B.* ... Found by the Rev. Mr. Manningham in Selsey Island, Sussex. Merret relates it to grow in the meadows betwixt Hampton-Court and Tuddington'. - Ray (1724: 337).

vc13   1724

Clarke and Druce both cite 'Armeria prolifera, *Ger*. Chiding sweet Williams in the grounds 'twixt Hampton Court and Tuddington'. - How (1650: 10), which is Merrett's record too. Kent (1975) wonders if this refers to *Dianthus deltoides*, but either way, it is too much of a conjecture to use as a first record.

Petrorhagia prolifera

'Cockford Heath, 30 September 1835, K.Trimmer, **CGE**'. - Akeroyd & Beckett (1995)

vc28 **1995**
vc28 1835

Akeroyd & Beckett trace the history of this plant, mentioning earlier authorities. The record 'in the grounds 'twixt Hampton Court and Tuddington'. - How (1650: 10) could refer to this species or, more likely, to *P. nanteulii*. Kent (1975) wonders if it refers to *Dianthus deltoides*. Either way, it is too much of a conjecture to use as a first record.

Petroselinum crispum

'The Garden Parsley (*Apium petroselinum*) … may have naturalised itself in a few places along the southern coast'. - Jones & Kingston (1829). 'Is in many places, especially in the South of England. Our specimens are from St. Vincent's Rock, upon the ledges of which the plant has altogether a wild apprearance'. - E.B.S. 2793 (1835).

vc3 **1829**

Druce cites Culpeper (1652: 439) 'Frequent on St Vincent's Rocks' as the first record, presumably from White (1912), who gives the same details. But, thanks to research from Clive Lovatt (*in litt.*) it transpires that White was certainly citing the wrong edition of Culpeper and probably the wrong species. There are no records from St. Vincent's Rocks before that in E.B.S.

This is a very late first date for an archaeophyte; in S. Devon and Cornwall, as well as in the Avon Gorge it gives the appearance of having been established for a very long time.

Petroselinum segetum

* 'Sium terrestre'. Stoke through … Cowling, and … to Cliffe. - Johnson (1629: 8).

vc16 **1629**

'Sium siifoliis, Hone-wort. 18 Aug.1620. … 'I took the description of this herb the yere 1620 but observed it long afore'. - Goodyer ms.11, see Gunther (1922: 121) and also Johnson (1633: 1017)..

1620

Peucedanum officinale

* 'Peucedanum or Hogs Fennell groweth on the South side of a wood belonging to Waltham at the Naze in Essex … at Whitstable in Kent … at Feversham in Kent'. - Gerard (1597: 897).

vc15, 19 **1597**

It is interesting that Gerard gives almost all the sites where it has ever been recorded.

Phalaris arundinacea

* 'Gramen arundinaceum acerosa gluma nostras. Great Reed Grasse with chaffy heads. In the low moist grounds by Ratcliffe neere London'. - Parkinson (1640: 1273.2).

vc21 **1640**

How's gloss in L'Obel (1655: 44) equates L'Obel's 'Calamagrostis aquatica Cambro-Britannica aceroso pluma … provenit non procul Londino, via qua itur Ratleam [Ratcliffe]' with Parkinson's species. L'Obel died in 1616.   vc21   1616

Ratcliffe was on the Thames between Shadwell and Limehouse, downstream from Tower Bridge.

Phegopteris connectilis

'Filix minor Britannica pelliculo pallidiore … Ex borealibus Comitatibus Angliae … Tho. Lawson & Ed. Lhwyd'. - Morison (1699: 575, n.17).   **1699**

Whittaker (1986) points out that Ray (see Gunther (1928: 187)) mentions receiving this fern some time before June of 1689. But Edgington (2013) notes that Ray (1670: 114) was describing this same plant as 'Filix pumila saxatilis altera … in montibus saxosis, Derbiensibus, Eboracensibus & Westmorlandicis copiose', and that he [Ray] had seen it, probably as early as his first visit to Derbyshire in 1658.   vc57, 69   1658

Parkinson (1640: 1043.2), citing the same polynomial as Ray, has 'these all doe grow in rockey and stony places' but with no hint that this is a British plant.

Phleum alpinum

* 'Said to be found on Craigneulict above Killin'. - Lightfoot (1777: 1133, under 'doubtful natives or such as have not yet come under the author's inspection'). 'In montibus prope Garway Moor'. Scotland. - Dickson (1794).   vc88   **1777**

Lightfoot's location has not been traced, although Bowden (1989: 74) notes that Lightfoot and Pennant 'At Killin, climbed Sron a' Chlachain (521m)'. Garway Moor is probably Garvamore (NN5294), on the upper R. Spey - there are recent records of this species from Carn Dubh, three miles to the south-west.

Phleum arenarium

* 'Gr. Typhoides maximum alterum. Betwixt Deal and Sandwich'. - Merrett (1666: 59). 'Gramen typhinum maritimum minus. Sea-Cat's-tail-grass'. - Ray (1670: 157).   vc15   **1666**

Hudson (1778: 23) refers this to Parkinson's (1640: 1170.4) 'Gramen typhinum Danicum minus, Sea Canary-grass', but Parkinson did not know it from Britain.

Phleum bertolonii

'Gramen Typhinum minus'. Around Margate. - Johnson (1632: 5).   vc15   **1632**

See the note under *P. pratense* and also Rose, in Gilmour (1972: 106), and Hanbury & Marshall (1899).

Phleum phleoides

* as *Phalaris phleoides* 'On Newmarket Heath'. - Relhan (1786: 23).   vc29   **1786**

'c1760, when Banks came to Cambridge to meet [Mr I.] Lyons and collected the *Senecio integrifolia*'. - Crompton (2007). 'First discovered in Great Britain by Mr. Woodward & Mr. Crowe near Swaffham, Norfolk, in 1780'. - Withering (1787: 66).   vc29   1760

Crompton (*supra*) gives much more and convincing detail, citing Banks' herbarium sheet (in **BM**) labelled 'Cambridgeshire on the banks of the Devils ditch between the turnpike and stichworth Lyons'. Lyons died in 1775.

Phleum pratense

* ' ... described by *Bauhine*, who first gave the figure and description thereof in his *Prodomus* [sic], *pag*.10 called Gramen Typhoides maximum spica longissima ... by the bridge entring into Chelsey field, as one goeth from St James to Little Chelsey'. - Johnson (1633: 11).   vc21   **1633**

'Gramen cuspidatum Alopecuroides latiore maximaque panicula Avenacea gluma ... inter Islington, et altam Portam, vernacule Highgate'. - L'Obel (1655: 38).   vc21   1616

There seems to have been a satisfactory separation of *P. pratense* and *P. bertolonii* at a very early stage, though it was probably done on size! Raven (1947: 247) identifies L'Obel's grass as a large specimen of this species. L'Obel died in 1616, though this record (*pace* Raven) probably dates to c1605.

See also 'Gramen typhinum, an other Sedge Grasse in watery moyste places in my Alderes & ?muche'. - Manuscript notes of William Mount, of c.1584, which Gunther (1922: 255 - 263) identifies as this species. Druce gives the wrong date (1597) for the Gerard/Johnson record.

Phragmites australis

* 'A kind of Reede ... called phragmitis ... groweth ... muche in England'. - Turner (1551: E. ij).   **1551**

Phyllodoce caerulea

* as *Menziesia caerulea* '... Discovered at Aviemore, in Strathspey, and in the western isles of Shiant ... we have received specimens from Mr G. Don and Mr P. Neill ...'. - E.B. 2469. 'Near Aviemore   vc96   **1812**

in Strathspey, where it was first noticed by Mr Brown of Perth. *Mr Patrick Neill*. In the western isles of Shiant. *Mr G Don, and Dr. De Ramm*'. - Smith (1824: ii. 222).

'G.D's [George Don] Letter 26 Aug't [18]10 discovered by Messrs Jas. & Robert Brown Nursery Men Perth'. - a handwritten annotation on a specimen of this species in herbarium of the University of Glasgow, cited by Nelson (1977). vc96   1810

The entry in Smith (*op.cit*.) is ambiguous. Mr Brown of Perth is either Jas. Brown or Robert Brown, both nurserymen of Perth; see Nelson (1977) and Hardy (2011). Desmond (1994) notes that Neill was a friend of Don, so is it likely that Neill would have collected it in Strathspey and Don in Shiant? Does the Aviemore location actually refer to the site on the Sow of Atholl, 30 miles to the south, that was the only known site for 150 years, or was there a separate site known there in the 19th century (see Webster (1978) who lists herbarium specimens from 'near Aviemore' up to 1835)? No other sites were found before 1966, and none of those are near Aviemore. The first record traced that actually refers to the Sow of Atholl is in **DBN** by William McNab '2nd Aug. 1824, near Dalnacardoch (a village near the mountain' - Nelson (*op.cit*.) ' Hooker (1838) too gives 'Heathy moor in the 'Sow of Athol', Dalnaspidal, Perthshrie. Mr Brown of Perth'. But that too is unsatisfactory, since in earlier editions he merely cites 'Heathy moor near Aviemore, Mr Brown'. How did he suddenly decide it was the Sow of Atholl? However, both Nelson and Hardy (as above) tend to think that, on balance there might well have been two different sites.

In Henderson & Dickson (1994: 20) they note 'that in his 1771 Journal Robertson recorded (*op.cit*. p. 171) "A [Andromeda] …c..., leaving Pitmain [nr Kingussie] for Strath Dearn, at the head of the water of Findhorn, not mentioned by Linnaeus or any other botanist". There is no specimen listed for Hope's herbarium, so there will never be certainty, but the area is the only place the plant occurs in Britain now'. Elsewhere they write of the competition to find new plants, and the poaching of others' records, explaining the use of codes for secrecy. Marren (1990) accepts this theory as a probability, and the headwaters of the R. Findhorn would be near to the untraced site near Aviemore referred to above.

Physospermum cornubiense

* 'Cornwal Saxifrage'. - Petiver (1713: xxvi. 9). vc2   **1713**

In Ray (1724: 209) there is a more accurate drawing. A specimen is in Buddle's herbarium, 'a D. Stephens e Cornubia missum'. - Hb. Sloane, cxx. 37, in **BM**. Buddle died in 1715, and possibly Stevens' specimen predates Petiver's record, as Davey (1909) relates that Stephens was the vicar of Menheniot, outside Liskeard, the centre of the British distribution of this species, from 1685.

Phyteuma orbiculare

* 'Rapunculus Corniculatus montanus ... Mr. Goodyer ... found it growing plentifully wilde in the inclosed chalky hilly ground by Maple-Durham neere Petersfield in Hampshire'. - Johnson (1633: 455).   vc11   **1633**

'Rapunculum silvestre Trago. 5th Julii 1620 …'. - Goodyer ms.11, f. 84, see Gunther (1922: 116) but seen by him first at Droxford in 1618. ms.11, f. 57, see Gunther (1922: 111).   vc11   1618

Phyteuma spicatum

* 'Rev. Ralph Price met with it, in 1825, near Hadlow Down, in Mayfield, Sussex'. - Borrer in E.B.S. 2598.   vc14   **1829**
   vc14   1825

Wolley-Dod (1937: 276) has 1824 for same record. Parkinson (1640: 648) speaks of this species, as 'Rapunculus spicatus alopecuroides', together with some of the other 17 Bellflowers described on that page, growing 'wilde in divers places of this land,' but no subsequent botanist appears to have noticed it in Britain until Price. It is assumed that either Parkinson was wrong, or was describing a cultivated species.

Picris hieracioides

* 'Hieracium asperum, *Tab. Ger.* ... In montosis pratis. Rough Hawkweede or yellow Succory'. - Johnson (1641: 24).   **1641**

Chapman *et al.* (1995) suggest that Turner's 'Greate hawkewede ... in the medowe a lytle from Shene'. - (1562 :14) may have been this species.

Pilosella flagellaris

'Pastures and outcrops, west side of White Ness, 1962' - Scott (1967).   vc112   **1967**
   vc112   1962

Pilosella officinarum

* 'Pilosella … yealowe Mouseare'. - Turner (1548: H. iiij & 1668: 58).   **1548**

This might be that referrred to in Turner (1538: A. ij) under 'Asine' [*Stellaria media*] as 'not being our Mouseare which grows in rough and untilled places'.

Pilosella peleteriana

'Not uncommon on sandy banks [in all the islands]'. - Babington (1839a: 58).   vc113   **1839**

'Guernsey, Terres Gosselin. Gosselin, J., c.1790'. - see McClintock (1982: 154).   vc113   1790

McClintock (1975: 213) cites a record from St. Martins, 'Gay, 1832'. The first record other than from the Channel Islands is 'Craig Breidden, vc47, 1834, Babington, det. Sell, **CGE**', which

Babington (1897: 275) alludes to in a letter to W.J. Hooker, dated Dec. 1835.

Pilularia globulifera

'Gr. Piperinum, Pepper-grass, Intermiscentur enim foliosis plurima grana magnitudine & figura grani piperis, near Petersfield'. - Merrett (1666: 57).   vc11   **1666**

This must almost certainly be a Goodyer find (he died in 1664) but no mention of this species has been found in Goodyer's papers.

Pimpinella major

* 'Pimpinella Saxifraga maior'. Between Gravesend and Rochester'. - Johnson (1629: 3).   vc16   **1629**

Pulteney cites Lyte (1578: 285.1) 'Saxifragia maior … groweth in high medowes, and good groundes'. This is given with an adequate drawing, but, of course, we do not know if this is a British record.

Pimpinella saxifraga

* 'Pimpinella Germanica … groweth commonlye in Englande in bankes of eche syde of holowe hygh wayes and in manye meaowes and in verye great plentye'. - Turner (1568: 11).   **1568**

See Nelson (1959: 121) pointing out that there is a reference to 'P. germanica' in Turner (1551: 201) that might be this species.

Pinguicula alpina

* 'This plant was picked by the Rev. George Gordon in June, 1831, in the bogs of Auchterflow and Shannon, on the Rose Haugh property, Ross-shire … There are two specimens in the herbarium of Sir J. E. Smith, upon the same paper with *P. lusitanica*, sent to him by Mr. James Mackay in September 1794, from the island of Skye'. - E.B.S. 2747.   vc106   **1832**

The Skye record has never been verified, and is generally ignored, but where did the specimen come from?
Rosehaugh is in the Black Isle.

Pinguicula grandiflora

* ' … found plentifully in marshy ground in the west part of that county [Co. Cork] by Mr Drummond, curator of the botanic garden at Cork'. - E.B. 2184.   vcH3   **1810**

'Islands at the head of Kenmare river … abound with Butterwort'. - Smith (1756: 85).   vcH1   1756

P. grandiflora and *lusitanica* are frequent in Co. Kerry, but *P. vulgaris* is rare, so Smith's record seems an acceptable first evidence. Scully (1916) also cites this as first date.

Pinguicula lusitanica

'He [Richard Garth] told me that this plant both the blue-flowered [*Pinguicula vulgaris*] and the white-flowered [*P. lusitanica*] are found in many parts of England'. - L' Ecluse (1601: 311). 'Pinguicula minima fl. albo. . . In the mid way betwixt Oakhampton and Lanceston, Cornwall, bertwixt a great wood and the River in boggy Meadows'. - Merrett (1666: 94).

1601

L'Ecluse's trip was in 1581. See Raven (1947: 170).

1581

Raven (*op.cit.* 314) points out that Oakhampton is a misprint for Kilkampton. The location is probably Meddon Moor (I.J.Bennallick *pers. comm.*). Ray (1670: 244) described it as new to science as 'Pinguicula flore minima carneo', and he found this on on his 1662 trip to Cornwall 'circa Kilkhamton'.

L'Ecluse [Clusius], C. 1601. *Rariorum Plantarum Historia.*

Pinguicula vulgaris

* 'Pinguicula siva Sanicula Eboracensis … Crosbie, Ravenswaith in Westmerland, upon Ingleborough fels twelve miles from Lancaster … in Harwood … neere to Blackburne … Bishops Hatfield [Hatfield, Herts] … in the fens in the way to Wittles meare from London, in Huntingdonshire'. - Gerard (1597: 645).   vc20, 30, 59, 64, 69   **1597**

'He [Richard Garth] told me that this plant both the blue-flowered [*Pinguicula vulgaris*] and the white-flowered [*P. lusitanica*] are found in many parts of England'. - L' Ecluse (1601: 311).   1581

L'Ecluse's trip was in 1581, and it was T. Penny who passed the information from Richard Garth to him. See Raven (1947: 170) who adds (from L'Ecluse, 1583) that Garth had sent L'Ecluse plants in full flower from the county of Derby and Wybsey not far from the city of Halifure [Wibsey, just outside Bradford, and Halifax]. These must have been *P. vulgaris*, as *P. lusitanica* is not found in those areas.

See also a reference to annotations made by T. Penny on C. Gesner's plant illustrations dated as by 1565 in Foley (2006b: 96) 'Women in England use it to heal the splits and cracks in the udders of cows'. Presumably this would be this species, the larger of the two.

Pinus sylvestris

* 'Abies, the Fir tree … In Scotland … as I have been assured, but not in Ireland or England, that I can heare of, saving where they are planted'. - Parkinson (1640: 1540).   **1640**

Plantago coronopus

* 'Coronopus … called in Cambryge herbe Iue … groweth muche aboute Shene aboue London it loueth wel to growe by the sea banks also'. - Turner (1548: C. v).   vc17, 29   **1548**

Plantago lanceolata

* 'Plantago … ejus duae sunt species major & minor. Majorem vulgus appellat Waybred aut Plantane. Minorem vero Rybwort, rybgyrse aut Lancell'. - Turner (1538: B. iiij. verso).   **1538**

Plantago major

* 'Plantago … ejus duae sunt species major & minor. Majorem vulgus appellat Waybred aut Plantane. Minorem vero Rybwort, rybgyrse aut Lancell'. - Turner (1538: B. iiij. verso).   **1538**

Plantago maritima

* 'Plantago marina … This growes neere unto the sea side in all the places about Englande where I have travelled, especially by the fortes on both sides of the water at Gravesend; at Erith neere   vc14, 16, 18, 34, 58   **1597**

London; at Lee in Essex; at Rie in Kent; at Westchester, and at Bristowe'. - Gerard (1597: 343).

Turner (1562: 212) does mention a third species 'that groweth by the seaside only', and although Pulteney cites this, it seems too tenuous.

Plantago media

* 'Plantago incana Hoarie Plantaine ... almost everywhere'. - Gerard (1597: 339).   **1597**

Platanthera bifolia

* 'Orchis alba bifolia minor calcari oblongo *C.B.* The lesser Butterfly Orchis. In pascuis'. - Ray (1696: 238).   **1696**

Ray adds that this was differentiated from *P. chlorantha* by D. Dale.

Platanthera chlorantha

* 'Orchis ornithophora candida ... which resembleth the White Butter-flie ... North ende of Hampsteed heath ... High-gate ... in the wood belonging to a worshipfull gentleman of Kent named Master Sedley of Southfleete'. - Gerard (1597: 166.11).   vc16, 21   **1597**

Poa alpina

* 'On the sides of Craig-challeach [Creag na Caillich] above Finlarig in Breadalbane, Mr. Stuart'. - Lightfoot (1777: 96).   vc88   **1777**

'Ben Crooken [Ben Cruachan?] by Mr Oaks in 1767, on whose authority I insert it'. - Hope, 1768, see Balfour (1907).   vc98   1767

Finlarig is just north of Killin.

Poa angustifolia

Treated as a subspecies of the variable *P. pratensis* agg. at least since Babington (1843) as var. γ, but raised to a species in Hubbard (1954).   **1843**

Cambridge, J.S. Henslow, 14.6.1825, det. P.J.O. Trist, 1983, **CGE**.   vc29   1825

Ray (1696: 325) has 'Gramen pratense paniculatum maius angustiore folio', from Bobart at Oxford. This is the polynomial followed by Smith (1824) and is probably a good first record.
In Babington (1843) and in Druce (1908) it is treated only as a subspecies. See Barling (1959) for a much fuller treatment. Crompton (2007) gives as her first record '*P. (angustifolia) panicula diffusa*, narrow leaved Meadow-grass. *Habitat* ad sepes'. - Lyons,1763: 8 (15). This follows Hudson (1762: 34), but I still feel more comfortable with treating Babington as the first record.

Poa annua

* 'Gramen minimum album ... Very plentifully among the hop-gardens in Essex'. - Gerard (1597: 3.2).   vc18, 19   **1597**

Poa bulbosa

* 'Prope Clapham in Com. Surriensi'. - Hudson (1762: 34). 'Dom. Stone nuper prope Yarmouth spontaneam invenit'. - Smith (1800: pref. p. 7).   vc17   **1762**

Poa compressa

* 'Gramen pratense medium culmo compresso *Buddle Hort. Sicc* ... Mr. Buddle, that accurate Graminist, observed this on the walls of Malden'. - Petiver (1716: n. 130). 'On the walls about Eltham [Middlesex] and in several other places. Mr. J. Sherard'. - Ray (1724: 409. 5).   vc18   **1716**

Poa flexuosa

* 'On Ben Nevis, Mr. Mackay'. - Smith (1800: i. 101).   vc97   **1800**

Poa glauca

* as *P. pratensis* β *alpina* 'In monticis Westmorlandicis, Cumberlandicis, &c'. - Hudson (1778: 39). 'Crib y Ddeseil [Snowdon], Mr. Griffith'. - Withering (1796: ii. 148).   vc69, 70   **1778**

Carter (1955a) states that Griffith found this on Crib-y-Ddescil in 1778, but gives no source for this.

Poa humilis

'Knochan, W. Ross, 1894'. - Druce (1913).   vc105   **1913**

'Guernsey. Gosselin, J., c.1790'. - see McClintock (1982: 178).   vc113   1790

Druce's article was the first to clearly separate this from *P. pratensis s.s*, although Lindman (1927), a more satisfactory paper, confirms that Smith (in E.B. 1004 (1802)) was correctly describing this species. Smith cites records from Anglesey in 1799, and equates this to Hudson's (1778: 39) *P. alpina* from Westmorland and Cumberland. But Smith and his successors subsequently confused matters, so Druce seems a better first record.

Poa infirma

'Jersey'. - Salmon (1914).   vc113   **1914**

Jersey: sandy places near the sea, March, 1877, J. Piquet, **BM**.   vc113   1877

The first record other than from the Channel Islands was long thought to be 'Scilly, widely distributed'. - Raven (1950), though it had been found as a casual 'Banks of the Gala below Galashiels, Selkirks, October 1916'. - Hayward & Druce (1919). However Margetts & David (1981) cite 'First found in Cornwall in 1876 by

William Curnow'. There is a likely specimen at **BIRM**, collected by Curnow, from Jackdaw Cliff, Rame, E. Cornwall, 1876, but it is too poor to be absolutely certain that it is this species.

Poa nemoralis

* 'In sylvis et umbrosis'. - Hudson (1762: 34).   **1762**

Druce (1897: 577) identifies it with the 'Gramen pratense   vc23   1696
paniculatum majus angustiore folio *C.B. prod.p.5* Oxonio ab Amico benevolo D. Tilleman Bobart ad me [Petiver] transmissum'. - Ray (1696: 325). Druce adds that the specimen in the Morisonian Herbarium is this species. See also 'from Tilleman Bobart (from Oxford) … Bobart appears to be the first to discover this plant in Britain'. - Druce & Vines (1907: 120).

Bobart died in 1719. Druce's attribution seems probable, although it is not referred to in Smith (1824: i.129). Pulteney (1799) notes 'Seems to have been first noticed in England by Mr. Hudson or by myself, who sent it to him in the year 1759'.

Poa pratensis s.l.

* 'Gramen pratense minus. In dry and barren grounds'. - Gerard   **1597**
(1597: 2.2).

See 'Gramen pratense, great leavyd Medowe grasse very vulger'. - Manuscript notes of William Mount, of c.1584, in Gunther (1922: 255 - 263), and identified by him as this plant. Note that Gunther has the polynomial the other way around to Gerard.

Poa trivialis

* 'Gramen pratense groweth everywhere'. - Gerard (1597: 3.2).   **1597**

Gramen minus' - Manuscript notes of William Mount, of c1584, in Gunther (1922: 255 - 263), and identified by him as this plant, with a query. Note that Gunther has the polynomial the other way around to Gerard. Clarke notes that this and *P. pratensis* were first well distinguished by Hudson and Curtis; see especially Curtis (1777: ii. 5, 6).

Polemonium caeruleum

* 'Valeriana Graeca *G.1076. P.123*. On the Rocks betwixt Maw-   vc64   1666
water [Malham] Tarn and Mawanco, where the highest Rock standeth round like a Castle'. - Merrett (1666: 123).

Nelson (1959) follows Britten (1882) and in turn is followed by Stearn (1965) in ascribing to this species Turner's (1548: F. ij. verso) 'Phu … the other kind is called Valeriana greca, and this is our commune Valerian that we use agaynste cutts with a blewe floure'. There is no indication here that this is a description of a plant in the wild. Lees (1888) has Maw-water and Mawam-co.

Polycarpon tetraphyllum

* 'in pratis aridis et locis arenosis, circa Lymston [Lympstone] prope Exeter in Devonia; et in Insula Portlandica'. - Hudson (1778: 60).   vc3, 9   **1778**

'Clark, Jersey'. - Druce & Vines (1907: 109). Le Sueur (1984) dates this record as to between 1730 and 1740.   vc113   1740

It was first found by Lightfoot in Portland in 1774, see a letter from Pulteney to Martyn, 3 Oct. 1775, in Gorham (1830).
Moss (1920: 15) cites Johnson's (1633: 622) 'Anthyllis marina incana alsinifolia … in Shepey, as also in West-gate bay by Margate in the Isle of Thanet' as this species, but there is no mention of this in Hanbury & Marshall (1899) and it seems unlikely.

Polygala amarella

* as *P. austriaca* var. *uliginosa* 'At the back of Cronkley Fell, Upper Teesdale, Yorkshire, at an elevation of about 1500ft above the sea, Yorkshire, on 24th May, 1852 by Messrs. James Backhouse sen. and jun.'. - Babington (1853a).   vc65   **1853**
   vc65   1852

The variety *austriaca*, from Kent and Surrey, was first recorded in 1871, Wye Downs, J.F. Duthie, **BM**.
Babington (1853b:168) mentions a species recorded by Ray (1724: *287) 'Polygala myrtifolia palustris humilis et ramosior', from Croydon (vc17) which has never been satisfactorily identified.

Polygala calcarea

* as *P. amara* 'Found growing abundantly at Cuckstone, Cobham and other parts of Kent in the summer of 1831. It has also been observed in Surrey and Wiltshire'. - D. Don in E.B.S. 2764.   vc16, 17   **1834**

Ightham, W. Kent, 1831, W. Peete, det. D.R. Glendinning, **CGE**.   vc16, 17   1831

Druce notes that there are specimens noted from Halstead in Essex (see Druce & Vines (1907: 90)), which he claims were the earliest British specimens known. But *P. calcarea* does not occur in Essex.

Polygala serpyllifolia

as *P. depressa* 'Bleak Down, Newtown to Niton [Isle of Wight], May, 1846'. - Bromfield (1847).   vc10   **1847**

'Polygala flo. caeruleo, et flo. albo, Amarella *Gesn*. flos Ambervalis *Dod*'. Canterbury to Faversham. - Johnson (1632: 25).   vc15   1632

Rose, in Gilmour (1972: 122), says 'almost certainly this species, not *P. vulgaris*, that Johnson saw in the acid Blean Woods [between Canterbury and Faversham]'. There are specimens from Hampstead and Southgate, Middlesex & Bishop's Castle, Salop, see Druce & Vines (1907: 90), which Druce claimed as the first British records.

Polygala vulgaris

* 'Polygala ... I have sene thys herbe oft in England'. - Turner (1562: 96 verso).   **1562**

Polygonatum multiflorum

* 'Polygonaton ... 'Thys herbe is well knowen ... in England'. - Turner (1562: 98). 'In Somersetshire upon the north side of a place called Mendip, in the parish of Shepton Mallet' [with other localities in Kent, Hants, Wilts and Surrey]. - Gerard (1597: 758).   **1562**

Polygonatum odoratum

* 'Polygonatum latifolium 2 *Clusii* ... groweth in certaine woods in Yorkshire called Clapdale Woods, three miles from a village named Settle'. - Gerard (1597: 758).   vc64   **1597**

Polygonatum verticillatum

* as *Convallaria verticillata* 'Arthur Bruce, Esq., Secretary to the Nat. Hist. Society of Edinburgh, first found it July 1st, 1792, in the Den Rechip, a deep woody valley four miles north-east of Dunkeld in Perthshire'. - E.B. 128.   vc88   **1793**

'found there by G. Don in 1791 who told me [Brown] and I told others ...'. - Robert Brown ms., cited by Mabberley (1985: 26).   vc88   1791

Polygonum arenastrum

as *P. aequale*, Moss (1914: 126).   **1914**

'Polygonum folio rotundo. Thickset Knotgrass'. - Petiver (1713: x f.2). 'Cambridge, 1835, Babington, **CGE**'. - Perring *et al.* (1964).   1713

Moss (*op.cit.*) states that he is following Lindeman (1912) 'the only treatment of real value', though he does cite Syme (1868: viii. 65) '*P. aviculare* f. *arenastrum*'. Both Ray (1724: 147. 4) and Trimen & Dyer (1869) follow Petiver. Preston *et al.* (2004) cited the Cambridge record as their first record.

Polygonum aviculare s.l.

* 'Poligonon ... vulgus appellat Swyne gyrs & knotgyrs' - Turner (1538: B.iiij. verso).   **1538**

Polygonum aviculare s.s.

Moss (1914: 125).   **1914**

'Polygonum mas vulgare', Hb. Sloane 117: 22 & 117: 23, **BM**.   1715

The Hb. Sloane specimen is in the Herbarium of Buddle, who died in 1715.

Polygonum boreale

Styles (1962), noting that whilst Druce used the name *P. aviculare* var. *boreale* for plants gathered at Loch Leven, Fife (vc85) there was great confusion at that time between this species and the plant then known as *P. littorale*. **1962**

N. Uist, vc110, 1841, J.H. Balfour, det. Akeroyd, **E**. 'Burra Firth, Unst (61), 1865, Tate (**OXF**)'. - Scott & Palmer (1987). vc110 1841

Polygonum maritimum

* 'At Christchurch Head on the sandy shore towards Muddiford, Mr. Borrer … I also have it from Herm Bay, Jersey, collected by Mr. W.C. Trevelyan'. - Babington (1836). vc11, 113 **1836**

'noted by [S.H.] Haslam in Guernsey in 1829 and collected by Babington in 1837'. - McClintock (1975: 144). vc113 1829

Druce cites a specimen, marked as 'the first as British', but with no further details in Druce & Vines (1907: 55 under 147.5). This would have been acquired by Dillenius sometime before his death in 1747. However, Druce did not refer to this in his *Comital Flora* (1932), though this might have been just an oversight in that often carelessly assembled work.
See the note under *P. oxyspermum*.

Polygonum oxyspermum

* 'Polygonum marinum *J.B.* ... In arenosis maris litoribus v.g. prope oppidum Pensans in Cornubia, & alibi'. - Ray (1670: 248). vc1 1670

This would have been on Ray's 1662 trip, see Raven (1950: 129), though it is not mentioned in Lankester (1846). vc1 1662

Clarke and others have identified this as this species, but there seems no reason why it should not refer to *P. maritimum*, as both have been recorded (and recorded recently) from the sands between Penzance and Marazion.

Polygonum rurivagum

Moss (1914: 126), where he cites Norman (1862), as *P. microspermum* and Syme (1868: viii. 67) as *P. aviculare* f. *rurivagum*. 1914

'Polygonum vulgare folio oblongo angusto acuto. Doody', in Buddle's herbarium, det. J.R. Akeroyd, Hb. Sloane 117: 22, **BM**. 'Polygonum oblongo angustoque folio *C.B. Pinax 281*. ... About Camberwell, and amongst the corn in Houndfield by Poundersend plentifully'. - Ray (1724: 146). 1705

The Ray reference is given in Salmon (1931). There is also a reference to the same in 'varietas *P.* angustifolii magni prope Camberwell (Surrey) and in agris ante Eltham (Kent), in Druce & Vines (1907: 54). In fact Ray (1690: 33) also mentions this as a variety.

Polypodium cambricum

Shivas (1962) was the first to satisfactorily explain the taxonomy of this group, though Manton (1947) had described the cytology.   **1962**

'Polypodium Cambro-britannicum pinulis ad margines laciniatis. Rupi cuidam adnascitur in sylva prope Arcem Denis Powis dictam tres mille passus à Cardiffa … D. Newton'. - Ray (1686: 137).   vc41   1686

Polypodium Cambro-britannicum … Lacinated Polypody of Wales. On a rock in a wood near Dennys Powys Castle [Dinas Powis, Glamorgan]'. - Ray (1688: 18). But Wade *et al.* (1994: 1) point out that this was in fact discovered by Richard Kayse of Bristol in 1668, and that it is his record that Ray is citing, though C. Lovatt (*in litt.*) feels the 1668 date is probably a slip for a later date (?1688).

Polypodium interjectum

Shivas (1962).   **1962**

In Du Bois' herbarium at **OXF** 'Polypodium pinnulis latioribis, acuminatis. Found at the Entrance to Over [Iver] Heath in the way thither, from Hillingdon, in some Hedges by Stonestreet in about 1690'. - Edgington (2013).   vc24   1716

Stonestreet died in 1716.

Polypodium vulgare s.l.

'Polypodium … nascitur in muscosis petris … vulgus apppellat Brake of the wall' - Turner (1538, B.iiij. verso). 'Polypodium, old rotten trees, old walls and tops of houses'. - Johnson (1633: 1133).   **1538**

Polypodium vulgare s.s.

Shivas (1962).   **1962**

Four specimens in Buddle's herbarium (Sect. iii, f. 6) at **BM**, without provenance but certainly British plants of c.1705, have 12-14 narrow indurated annular cells and are *P. vulgare s.s.*'. - Edgington (2013).   1715

Buddle died in 1715.

Polypogon monspeliensis

* 'Alopecuros altera maxima Anglica paludosa … Riguis herbidis Comitatus Zout-hamptoniae [Southampton] proximé salinas et antiquas aedes Drayton vocatas D. Richardi Garth … cis mare duobus miliaribus Anglicis à Portsmouth ex adverso Vectis Insulae plurima. … Hanc quoque udis fossis lacustribusq. Essexiensis comitatus legi juxta Thamesis amoenissima fluenta'. - L'Obel (1605: 469).   vc11, 18   **1605**

'L'Obel was at Drayton around 1596'. - Raven (1947: 239). Raven adds (1947: 243) that Garth died in 1597 but his widow, Jane, was visited by L'Obel. | vc11 | 1596

Chatters (2009: 182) has 'known from the Solent marshes since 1595 when Richard Garth … recorded it at Drayton', but the source for this date has not been traced. However in Brewis *et al.* (1996) the date is given as 'c1595'.

Polystichum agg.

'Filix mas non ramosa pinnulis latis auriculatis spinosis' - Johnson (1633: 1130.4, with no illustration). 'Filix mas aculeata'. - Parkinson (1640: 1036). 'Prickly auriculate male Fern … per totem Angliam frequens'. - Ray (1690: 24). | vc11 | **1633**

Newman (1844: 169) was the first to satisfactorily separate *P. aculeatum* and *P. setiferum*. Before him all was conjecture and muddle.

Polystichum aculeatum

Newman (1844). | | **1844**

See Gunther (1922: 183) where he identifies Goodyer, ms.11, f. 140, 'Filix mas non ramosa pinnulis latis auriculatis spinosis' as this species. Goodyer writes that 'the leaves are deeper greene than either of the last two species [*Dryopteris dilatata* and *D. filix-mas*] … at Mapleduham neere Petersfield'. This record is given in Johnson (1633: 1130.4), but with no illustration. | vc11 | 1633

Polystichum lonchitis

'Lonchitis aspera *C.B.* Rough Spleenwort with indented leaves … Clogwyn y Garnedh y Grîb Gôch Trygvylchau; D. Lhwyd'. - Ray (1690: 27). | vc49 | **1690**

Edgington (2013) dates Lhwyd's record to 1688. | vc49 | 1688

Polystichum setiferum

as *P. angulare*, Newman (1844). | | **1844**

'A specimen in Hb. du Bois at **OXF** collected by Bobart at Newbury in 1690 is the earliest sure record of this species'. - Edgington (2013). | vc22 | 1690

The separation of *P. setiferum* and *P. aculeatum* was not fully understood until Newman (*op.cit.*), who set out fully the differences, but even after that the confusion persisted, right up to the 1950s.

Populus nigra s.l.

'Populus nigra is not so comon in England as it is in Italy and hygh Germany'. - Turner (1662: 99) — **1562**

Chapman *et al.* (1995: 519) identify this as *P. nigra*. Druce cites Gerard (1597: 1301) as does Smith (1828), also as *P. nigra* 'in low moist places'. Clarke omits the species.

Populus tremula

* 'Populus alba ... dicitur ab Anglis an aspe aut an esptre'. - Turner (1538: B. iiij. verso). — **1538**

Potamogeton acutifolius

* 'Found by Mr. Borrer, who gathered the specimens here figured in marsh ditches at Amberley, Henfield, and Lewes, Sussex, in June, 1826'. - E.B.S. 2609. — vc13 **1829**

'C. Merrett in Herb. Sloane 19: 114', cited in Dandy (n.d.). Deptford, vc16, 1700, det. Dandy & Taylor, **BM**. — 1695

Merrett died in 1695. The **BM** specimen is in Buddle's herbarium, and it is assumed that the Deptford date is estimated.

Potamogeton alpinus

* as *P. fluitans* ' ... found in the river at Scole, Norfolk, by Mr. Woodward and Mr. Turner ... near Beverley by Mr. Teesdale; and in Lilleshall Mill-pool, Shropshire, by the Rev. Mr. Williams'. - E.B. 1286; and Smith (1804: 1391). — vc27, 40, 61 **1804**

R. Colne, Halstead, vc19, 1748, det. Dandy & Taylor, **BM**. — vc19 1748

Potamogeton berchtoldii

Dandy & Taylor (1938a,1940). — **1938**

In Plot[t]'s herbarium, Hb. Sloane 113: 141, det. Dandy & Taylor, **BM**. In Buddle's herbarium, Hb. Sloane, 117: 30, Hounslow Heath, 1705, det. Dandy & Taylor, **BM**. — 1696

Plot died in 1696. See under *P. pusillus* for an earlier record from Cambridgeshire previously assigned to this species. Dandy (n.d.) identifies as this species Petiver's record (in Gibson 1695) of 'P. pusillum, gramineo folio, caule tereti, Ray (1686: 190) ... this, with the last [*P. perfoliatus*], grows plentifully in the New-river-head'. The New River Head was near to Sadler's Wells, in Clerkenwell.

Potamogeton coloratus

* as *P. plantagineus* 'In damp pits from which peat has been taken, in Guernsey. Mr W. Wilson Saunders informs me that he has gathered it in ditches at Ham ponds, near Sandwich'. - Babington (1838). — vc15, 113 **1838**

Sharpside by Frogden, vc80, 1790, W. McRitchie, det. Dandy & Taylor, **BM**. 'Guernsey, Grand-mare. Gosselin, J., c.1790'. - see McClintock (1982: 157). — vc80, 113 1790

'First found [in Cambs] by Rev. Leonard Jenyns in 1827'. - Babington (1860b: 248). This is another early record away from the Channel Islands.

Potamogeton compressus

'P. ramosum caule compresso folio graminis canini nondum descriptum ... In the river Cam in many places'. - Ray (1660: 125). — vc29 **1660**

See Preston (2010) for a clarification of this record, which had previously been assigned to *P. acutifolius* and to *P. friesii*.

Potamogeton crispus

* 'Pusillum fontilapathum, *Lob.* Tribulus aquaticus minor, *Clus.*'. Between Sandwich and Canterbury. - Johnson (1632: 22; 1633: 823). — vc15 **1632**

'Tribulus aquaticus minor, quercus floribus [uvae]. 2 junii 1621.... from fish-ponds adioyning to a disolved Abbey called Durford, which ponds devide Hampshire and Sussex, and in other standing waters elsewhere'. - Goodyer ms.11, ff. 120a, 122, see Gunther (1922: 123). — vc11 1621

Potamogeton epihydrus

'August 1943, in NE angle of Loch Ceann a' Bagh [S. Uist]'. - Heslop Harrison (1948). — vc110 **1948**
vc110 1943

First found, presumably as an alien, in 1907, nr. Halifax, vc63 (Bennett 1908).

Potamogeton filiformis

* 'Lakes in Forfarshire'. - Babington (1843: 326). — vc90 **1843**

Brodie Castle, vc95, 1780, J. Brodie, det. Dandy & Taylor, **E**. — vc95 1780

In Dandy (n.d.) the Brodie specimen in not dated, and he cites as his first record 'Llynian, Llanfihangel-yn-Nhowyn, Llanfair-yn-Neubwll, vc52, 1798, H. Davies, det. Dandy & Taylor, **BM**'.

Potamogeton friesii

'Potamogeton perpulchrum nostras lucens, angustissimis longis & obtusis foliis, pallide virentibus'. - Plukenet (1696: 304). — **1696**

as 'P. gramineum ramosum caulo compresso' in Plot[t]'s herbarium, Hb. Sloane 113: 141, det. Dandy & Taylor, **BM**. — 1696

Plot died in 1696. See Preston (2010) for a clarification of Ray's (1660: 125) record from the River Cam in Cambridge, which had previously been assigned to *P. friesii*.

Potamogeton gramineus

* as *P. palustre* 'ditches near Beverley'. - Teesdale (1798: 43).	vc61	**1798**
Cornard Mere, vc26, 1739, J. Andrews, det. Dandy & Taylor, **BM**.	vc26	1739

Potamogeton lucens

* 'Potamogeiton longis acutis foliis'. - Johnson (1633: 822. 4). 'In many places in the Thames between Fulham and Hampton-Court'. - Petiver in Gibson's Camden (1695).		**1633**

Potamogeton natans

* 'Potamogeiton latifolium ... In standing waters, pooles, ponds, and ditches, almost everywhere'. - Gerard (1597: 675). **1597**

Gerard's habitat description covers both this species and *Persicaria amphibia*. See also 'Potamogeton ... Pondplantayne or swymmynge plantayne, because it swymmeth above ponds and standyng waters'. - Turner (1548: F. v), which Stearn (1965) identifies as this species, but might, of course, also be *Persicaria amphibia*.

Potamogeton nodosus

as x *Potamogeton Drucei* mihi ' ... first found by Mr. G.C. Druce in the River Lodden, Berkshire, August, 1893'. - Fryer (1898: 31). Raised to species level by Fryer (1899) as *P. nodosus*, Dandy & Taylor (1939).	vc22	**1898**
Alderminster, R. Stour, vc38, 1856, W. Cheshire, det. Dandy & Taylor, **WAR**. As *P. fluitans* var.,'Loddon River, 1893'. - Druce (1897: 516).	vc38	1856

Potamogeton obtusifolius

* 'Potamogiton gramineum latiusculum, foliis et ramificationibus densissime stipatis. *Buddl. H.S. Vol. IV.f. 27*. In fossis prope Deptford'. - Ray (1724: 149).	vc16	**1724**
'Goldingham Hall, Bulmer [W of Sudbury, vc19], 1711, J. Andrews, **BM**'. - Jermyn (1974). 'Potamogeton at ye simpling feast, 1705. Herb. Du Bois'. - Druce (1928: 486).	vc19	1711

Druce gives no reason for citing 1705. Buddle died in 1715.

Potamogeton pectinatus

* 'Millefolium tenuifolium ... In the river Cam in many places'. - Ray (1660: 100). vc29 **1660**

Johnson too (1633: 828) has the same polynomial 'Millefolium tenuifolium, Fennell leaved water Milfoile', and gives the locations as 'lakes, standing waters' etc, but his section includes *Hottonia*, a *Ranunculus* and a *Utricularia* and Ray (1660) notes that although the picture is good the description is not.

Potamogeton perfoliatus

* 'Potamogeton 3. Dodonaei'. - Johnson (1633: 822. 3); and Johnson (1634: 61). 'P. perfoliatum … In the river Cam plentifully every where'. - Ray (1660: 124). 'New-river-head' [Middlesex]. - Petiver in Gibson's Camden (1695).   **1633**

The New River Head was near to Sadler's Wells, in Clerkenwell.

Potamogeton polygonifolius

as *P. natans* var. 2 'Boggy ground, on Birmingham Heath'. - Withering (1787: 172).   vc38   **1787**

Dandy (n.d.) gives as his first record 'Herb. of J. Banister (d.1692) Hb. Sloane 168: 280)'. As 'Potamogeton Lapathi minoris foliis pellucidis D. Lhwyd', in a letter published in the Richardson Correspondance, No 100 (Druce & Vines 1907: liii), where Dillenius gives an account of his ascent of Snowdon in 1726. Lhwyd died in 1709, and his last trip to Snowdon was in 1699. Also from Llyn Lladr, south end of Gribggoch, see Druce & Vines (1907: 142).   1692

Carter too (1955a) includes this plant in his list of first British records for Lhwyd. The source of his information has not been traced, although it might be the specimen of 'P. Lapathi acuto folio, found on Phayon Vrach', described by Gunther (1945: 558). Clarke cited '*P. oblongus*', with syn. '*polygonifolius*'. - Lindley (1829: 250) as his first record.

Potamogeton praelongus

* 'Lakes and pools, Berwickshire, Dr. Robt. Thomson. Moss of Litie, Nairnshire, Mr. J.B. Brechan [Brichan], Mr Staples [Stables]. Lochleven, Mr. Arnott, Mr. J. Hooker'. - Hooker (1835c: 77).   vc85, 96   **1835**

Loch Leven, vc85, 1831, det. Dandy & Taylor, **BM**. Moss of Litie, in the summer of 1832'. - Brichan (1842).   vc85   1831

Brichan refers to an entry in Murray (1836: 108), who notes 'It is said, however, that there are specimens of it in the herbarium of Mr Brodie of Brodie, 20 or 30 years old … '. Dandy (n.d.) too mentions the record of Brodie, who died in 1824.
The records for Berwickshire have never been confirmed.

Potamogeton pusillus

Dandy & Taylor (1938a).   **1938**

In Buddle's herbarium, 117: 30, det. Dandy & Taylor, Hb. Sloane, **BM**. Isle of Dogs, 1817, det. Dandy & Taylor, **BM**.   1715

Buddle died in 1715.
Dandy & Taylor were the first to clarify the taxonomy of this species and *P. berchtoldii* - see Oswald & Preston (2011: 267 n.583). Clarke

and Druce both cited as their first records 'Potamogeton pumilum nondum descriptum ... Between Carleton and Wulwich'. - How (1650: 97) and 'P. pusillum gramineo folio, caule rotundo nondum descriptum ... In the rivulet at Hinton Moor'. - Ray (1660: 125). Both of these are possible, of course, though Oswald & Preston note that until recently *P. berchtoldii* was the commoner in the Cambridge area.

Potamogeton rutilus

'The credit [for being the first to collect] belongs to Beeby, who gathered specimens in Bardister Loch [Shetland] in 1890 and recorded them in *Scot. Naturalist, xi. 30* (1891) as *P. pusillus* var. *rigidus...*'. - Dandy & Taylor (1938b).

vc112 **1938**
vc112 1890

Potamogeton trichoides

* 'Our specimens were gathered near Norwich at "Bixley" by Rev. Kirby Trimmer and at Framlingham by Mr J.B. Wilson, but I learn from the latter gentleman that the same spot is intended. Mr Trimmer originally gathered it in 1848'. - Babington (1850a).

vc27 **1850**

In Buddle's herbarium, 117: 28, det. Dandy & Taylor, Hb. Sloane, **BM**.

1715

Buddle died in 1715.

Potentilla anglica

* 'Pentaphyllum reptans alatum foliis profundius serratis ... In the edges of the *corn-fields* between *Hockley* and the woods under *Shotover*-hill'. - Plot (1677: 145).

vc23 **1677**

Potentilla anserina

* 'Wild Tansey ... groweth in colde and watery places'. - Turner (1568: 4).

1568

But the figures in Turner are not that of this species. One is *Tanecetum vulgare*, the other not identifiable. However the description is plausible, and Pulteney and others give Turner the credit of the first record.

Potentilla argentea

'Pentaphyllum erectum foliis profunde sectis, subtus argenteis, flore luteo *J.B.* In arenosis aut glareosis pascuis, sed rarius; nonnunquam etiam in muris'. - Ray (1670: 236).

1670

See Raven (1950: 125) where he notes that Ray saw *P. argentea* in N. Wales in May 1662 but not as a plant which was new to him.

1662

This species is very rare in N. Wales, and we have no indication of where Ray saw it.

Clarke and Druce (and Edgington (2010)) both cite 'Quinquefolium peiraeum [petraeum] majus Tab.'. Near Gravesend. - Johnson (1632: 29) as this species, though Rose, in Gilmour (1972), identifies this with *P. reptans*. Smith (1824: ii. 418) equates *P. argentea* to Johnson's (1633: 988 f. 7) 'Quinquefolium Tormentillae facie. Wall Cinkfoile', as does Ray (1724: 255) (although referring to the original edition of Gerard (1597: 838), where Gerard gave no indication that it was a British plant).

Potentilla crantzii

* as *P. verna* 'in pratis et pascuis sterilioribus, prope Giggleswick in Comitatu Eboracensi'. - Hudson (1762: 197).	vc64	1762

Rather confusingly Hudson (*op.cit.*) referred to Spring Cinquefoil as *P. opaca*.

Potentilla erecta

* 'Heptaphillon ... nostratibus Tormentyll & Tormeryke dicitur'. - Turner (1538: B. verso).	1538

Potentilla fruticosa

* 'Pentaphylloides fruticosum ... ad ripam Meridionalem fluvii Tees [Yorkshire] ... found by Tho. Willisell'. - Ray (1670: 340).	vc65	1670
Morison (1680: 194) has 'Pentaphylloides rectum fruticosum Eboracense, nobis ... ramulos aliquot exsiccatos ad nos attulit Thomas Willesell sedecim abhinc annis [16 years ago]'. Horsman (1995) takes this to be a record from 1664.	vc65	1664

However Ray (1677: 228) says that Ralph Johnson of York was the first to observe it. Ray himself saw it on his trip to the north of England in 1671, with T. Willisel. See Raven (1950: 155)

Potentilla reptans

* 'Quinque folium ... a nostris Synkfoly'. - Turner (1538: C). 'Commune in al places'. - Turner (1548: F. v. verso).	1538

Potentilla rupestris

* 'Pentaphylloides erectum, *J.B.* ... Ad latera montis cujusdam Craig-Wreidhin [Craig Breidden] dicti in Comitatu montis Gomerici Walliae'. - Ray (1688a: 18).	vc47	1688
Oswald, in Trueman *et al.* (1995), notes that '[Ray's record] was probably after a further visit by E. Lhwyd to Craig Breidden in 1682'.	vc47	1682

In Ray (1696: 141) he adds that it was found by Lhywd.

Potentilla sterilis

* 'Fragaria sylvestris minime vesca, sive sterilis ... Rupibus et cautibus Cornubiae ad aedes generosi viri D. Muli frequentissima'. - L'Obel (1576: 396. I). 'Upon Blackheath'. - Johnson (1633: 998).   vc2   1576

Mr Muyle and his descendants lived at Bake, nr. St Germans, East Cornwall.

Potentilla tabernaemontani

* 'Pentaphyllum incanum repens Alpinum *Park*.... Attulit ad nos Th. Willisellus Pentaphylli genus parvum in pascuis circa Kippax agri Eboracensis vicum 3.m. Pontefracto versus Septentrionem distantem'. - Ray (1670: 235).   vc64   1670

Poterium sanguisorba

'The herbe that is named in Englishe Burnet is called of some comon writers Pimpinella … this herbe is so well knowen in all places of England …'. - Turner (1568: 9). 'Pimpinella hortensis Garden Burnet ... wilde upon many barren heathes and pastures'. - Gerard (1597: 889). 'Pimpinella sanguisorba minor'. Near Dartford. - Johnson (1629: 9).   1568

Turner (*op.cit.*) adds comments on it being good in wine, beer and ale, and his Dutch name describes the flower well too. There is a good illustration, though the text mentions a feature of the stipules that is more akin to *S. officinalis.* Chapman *et al.* (1995: 731) identify Turner's record as *P. sanguisorba*. L'Obel (1571: 320), as 'Pimpinella. Anl. Burnet' mentions too how it is added to drinks in England. It is possible that these entries cover only garden plants.

Primula elatior

Primula veris elatior pallido flore. In Kingston and Madinley woods abundantly and elsewhere'. - Ray (1660: 126).   vc29   1660

Turner (1568: 80) has 'Coweslippe ... there are two kindes of them, and the one is redder yelow then the other, and the other paler. They differ also in smel, for the one smelleth better than the other. The one is called in the West Contre of some a Cowislip and the other an Oxislip, and they are both called in Cambridgeshyre Pagles. … of the same kind is our primrose...'. Chapman *et al.* (1995: 716) interpret this as covering *P. elatior* and *P. veris*. It is slightly odd that this is not so interpreted in any of the Cambridge Floras, though, strangely, many of Turner's records have been ignored in these. C.D. Preston (*in litt.*) comments that the eminent Cambridge botanist, David Coombe, also suspected that Turner's text referred to this species.

Primula farinosa

* 'Primula veris flore rubro ... In Harwood neere to Blackburne in Lancashire, and ten miles from Preston in Aundernesse, also at Crosbie, Ravenswaith and Cragge close in Westmorland'. - Gerard (1597: 639).   vc59, 69   **1597**

'in Anglia' - annotations made by T. Penny on C. Gesner's plant illustrations dated as by 1565 in Foley (2006b). Philip Oswald (*in litt.*) notes that the full text of Penny's annotation reads 'It grows in damper meadows along the course of the River Loyne [Lune] between Lancaster and Kirkby [Lonsdale]. I have never seen [it] except in England'.   1565

Gerard adds 'They likewise grow in the medowes belonging to a village in Lancashire neere Maudsley called Harwood, and at Hesketh not farre from thence, and many other places of Lancashire, but not on this side of Trent [?River Trent] that I could ever have any certain knowledge of: *L'Obelius* reporteth that *Doctor Pennie* ... did finde them in these Southerne parts'. 'Aundernesse [Amounderness] was originally the district roughly between the Ribble and the Cocker, the E. boundary being formed by the fells on the Yorkshire border'. - Ekwall (1922).

See also L'Ecluse (1601: 301) 'I learnt from him [Penny] that this plant [*P. farinosa*] sometimes abounds in the the moist pastures of northern England…..Richard Garth.....had brought to me from the county of Derby and Wybsey not far from the the city of Halifax [Wibsey is just outside Bradford] ... before my departure from London'. Clusius's trip was in 1581. See Raven (1947: 170).

Primula scotica

* 'Found by Mr. Gibb, of Inverness, on Holborn Head, near Thurso, in Caithness'. - Graves & Hooker (1819: t.133).   vc109   **1819**

'from Caithness, mistakenly recorded as *P. farinosa*'. - Journal of James Robertson, 1767, Henderson & Dickson (1994: 50).   vc109   1767

There is also a note of a specimen from 'along the coast of Caithness and Strathnaver in moist pastures' - Hope, 1768, in Balfour (1907).

Primula veris

* 'Coweslippe ... there are two kindes of them ... one is called in the West contre of some a Cowislip & the other an Oxislip and they are both call in Cambridge shyre Pagles'. - Turner (1568: 80).   **1568**

Primula vulgaris

* 'Arthritica... ab anglis dicitur a prymerose'. - Turner (1538: A. ij. verso). 'Our prinrose which I never saw grow in any place saving in England & East Freseland, two cold contrees'. - Turner (1568: 80).   **1538**

This is a slightly odd comment of Turner's, as *P. vulgaris* is frequent throughout much of western Europe. But he had travelled in Central Europe and in Italy too, where it is less common.

Prunella laciniata

'Twenty years ago I noticed a cream-flowered *Prunella* in more than one spot in the Mendip Hills … in 1899 Mrs Gregory [showed it to me] … at 500 - 600 ft overlooking the moors between Draycott and Cheddar, North Somerset'. - White (1906)	vc6	**1906**
Clement (1985) cites Mott (1878) describing a white variety of *P. vulgaris* from 'Birstall Hill, Leicester … an old pasture field on the slope of a low ridge of boulder clay'.	vc55	1878

Mott describes the plant so well that Clement, surely correctly, decides this must be this species. This record is not given in any of the 20th century Leicestershire Floras.

Prunella vulgaris

* 'Symphytum petreum … groweth about Syon seuen myles aboue London … unsauery Margerū'. - Turner (1548: G. v. verso).	vc21	**1548**

This record is not cited in Kent (1975).

Prunus avium s.l.

* 'Cerasus sylvestris: The wilde Cherry tree. In a wood by Bathe'. - Johnson (1634: 28).	vc6	**1634**

Prunus avium s.s.

The differences between this and *P. cerasus* was only properly set out by Bromfield (1841).		**1841**
'Cerasus sylvestris fructo rubro J.B. Common wild Cherry'. - Ray (1670: 64), marked as a Cambridgeshire plant.	vc29	1670

Prunus cerasus

* First clearly described as distinct from *P. avium*, from Whippingham. - Bromfield in E.B.S. 2863 (1841).	vc10	**1841**
There are many early records of the plant itself (the dwarf cherry tree), e.g. 'Chamæcerasus *Ger.* Dwarfe Cherry-tree. In some closes of Teversham going from the Church towards Gains'. - Ray (1685: 5). See also Babington (1860b: 67).	vc29	1685

Horsman (1995: 159) cites a letter from Ralph Johnson to Ray, dated March 29th 1672, enclosing a twig of a cherry tree, which he [Horsman] identifies as *P. cerasus*, referencing it to Ray (1676: 62). However Graham (1988) has interpreted it as *P. padus*.

Prunus domestica

* Under 'Prunus', 'Bulles tre … Plenty of thys sorte of bulles trees then in Somerset shyre'. - Turner (1562: 104).	vc5, 6	**1562**

Prunus padus

* 'Cerasus Avium nigra et racemosa ... groweth very plentifully in the north of England, and especially at a place called Heggdale neere unto Rosgill in Westmerland and in divers other places about Crosbie Ravensworth ... in Marrome Park, foure miles from Blackburne, and in Harward neere thereunto; in Lancashire in almost everie hedge'. - Gerard (1597: 1322). vc59, 60, 69 **1597**

Prunus spinosa

* Under 'Prunus', 'Our slo bush or black thorn'. - Turner (1562: 104). **1562**

Pseudorchis albida

* 'Orchis pusilla alba odorata, radice palmata ... This we found on the back of Snowdon-hill, by the way leading from Llanberis to Carnarvan'. - Ray (1670: 227). vc49 **1670**

on Ray's third Itinerary '26 May, 1662 ... Llanberis ... near the stone tower there a species of Orchis palmata with an odorate flower like to Monorchis'. - Lankester (1846: 170) and see Raven (1950: 125). vc49 1662

Welch (1972) gives a photograph of the Ray specimen.

Lankester, E. ed. 1846. *Memorials of John Ray*.

Pteridium aquilinum

Under Filix 'Foemina filix ... a ferne aut a brak aut a bracon'. - Turner (1538: B).

1538

Turner has a good description of the growth. See also Johnson (1629: 8), between Stoke and Cliffe.

Puccinellia distans

* as *Aira aquatica*, β *distans* 'in arenosis prope Exmouth; circa Northfleet; in agro Cantiano; in comitatibus Eboracensi et Lancasterensi passim'. - Hudson (1778: 34). As *Poa retroflexa* 'Hampstead, July, 1786'. - Curtis (1791).

vc3, 16   **1778**

Puccinellia fasciculata

* as *Glyceria borreri* 'Gathered by Mr. Borrer at Gosport, Selsea, Southampton, Stokes Bay, Shoreham and Freshwater in the Isle of Wight. I have myself noticed it at Harwich, and in Canvey Island near the mouth of the Thames'. - Babington in E.B.S. 2797.

vc10, 11, 13, 18, 19   **1837**

'Gr. paniculatum maritimum vulgatissimum. From Mr Stonestreet'. 'The name refers to *Glyceria maritima* Wahl., but the specimen is the earliest British example known of *G. borreri* [= *P. fasciculata*]'. - Druce (1928: 489).

1716

Stonestreet died in 1716.

Puccinellia maritima

* 'Gramen maritimum vulgatissimo pratensi Gramini congener aut similis ... maritimis litoreis Kantiae oritur'. - L'Obel (1655: 8).

vc15   **1655**
vc15   1616

L'Obel died in 1616.

Puccinellia rupestris

* as *Poa procumbens* 'In the Autumn of 1793, having occasion to be at Bristol, I spent great part of a day in examining the plants of St. Vincent's Rock ... at the foot of the rock'. - Curtis (1798: vi. 11). Also E.B. 532 (1798) from Bristol and from Scarborough.

vc34   **1798**
vc34   1793

Curtis adds that it was in seed, and he took it home to London, where he realised the next year that it was a new species. Stevenson (1961) gives the date of publication of this fascicle as 1798.
'About Trow(bridge) by path, Mr Brewer [d.1743]'. - Druce & Vines (1907: 153). Druce does comment that 'this is an unlikely native station for this maritime species', and it seems better to omit this record.

Pulicaria dysenterica

'Coniza media … eodem loci [as the previous species] secus regiam via atq'. - L'Obel (1571: 145). 'Conyza minor ... In everie waterie ditch'. - Gerard (1597: 391). 'In S. James his Parke, Tuthill fields, &c'. - Johnson (1633: 482).	vc21	**1571**
as *Conysa media* listed from 'in ditches evrywhere about Winchester'. Marginalia in a copy of Du Pinet, *Historia Plantarum*, 1561, in the **BM** library. Gunther (1922: 235) attibutes these to Dr Walter Bayley and thought most likely to date from 1570-1572.	vc11	1571

The L'Obel entry translates as 'and in fact [atque] by the same localities along [secus] the Royal Road', and follows the entry for Coniza minima [*P. vulgaris*]. J. Edgington (*pers. comm.*) points me to 'All high Ways leading to cityes and boroughs are Viae Regiae', from *The antiquity, use and privilege of cities* by Francis Tate (1598).

Pulicaria vulgaris

* 'Coniza minima sive Pulicaria ... In Benardgreyn ara et fossis altero a Londino lapide fruticat'. - L'Obel (1571: 145). 'At Islington, by London'. - Gerard (1597: 391).	vc21	**1571**

Kent (1975) identifies this as Benard Green, nr. Paddington. Marren (1999) and Maskew (2014) identify this with Barnard's Green in Worcestershire, a site recorded in Amplett & Rea (1909), which persisted until the 1930s (Marren 1999). However Kent was following the note by Ardagh (1929), which discusses both claims and offers compelling evidence for the site to be in Middlesex. See also the entry for *P. dysenterica* above.

Pulmonaria longifolia

* 'In England ... found out by John Goodier, a great searcher and lover of plants dwelling at Maple-durham in Hampshire'. - Parkinson (1629: 248).	vc11	**1629**
'Mr. Goodyer found this [Pulmonaria foliis Echii] May 25 Anno 1620 flouring in a wood by Holbury house in the New Forest in Hampshire'. - Johnson (1633: 809), and see also Gunther (1922: 115).	vc11	1620

Pulmonaria obscura

'In 1842 a lungwort with unspotted, cordate leaves was discovered in Burgate Wood, E.Suffolk … found by C.J. Ashfield'. - Birkinshaw & Sanford (1996).	vc25 vc25	**1996** 1842

This was collected and published as *P. officinalis*. - see Ashfield (1862).

Pulsatilla vulgaris

* 'Anemone ... groweth ... about Oxforde in Englande as my frende Falconer toulde me'. - Turner (1551: C. v. verso) [with a figure resembling this species].	vc23	**1551**

Another very early record is 'at Shelford, nr. Cambridge' - annotations made by T. Penny on C.Gesner's plant illustrations, which are dated as by 1565 in Foley (2006b).
Druce (1897: xcv) notes that Turner's description points to *Anemone nemorosa* though the plate is of *Pulsatilla*.

Pyrola media

'In Scotswood dene three miles west of Newcastle - July 16th 1797'. - Smith (1824: 256).	vc67	**1824**
'We found it near Halifax by the way leading to Kighley and in Northumberland'. - Ray (1670: 256). This was on Ray's 1660 Itinerary - see Raven (1950: 115) for this attribution.	vc63	1660

The first record of this species or of *P. minor* is Ray (1670: 256) 'Pyrola Ger. J.B. nostras vulgaris Park. rotundifolia major C.B. In Septentrionalibus Angliae variis locis. We found it near Halifax by the way leading to Kighley and in Northumberland. It grows plentifuly near Halifax in several places. In Hazelwood-woods near Sir Walter Vavasour's Park in York-shire. Mr Witham shew'd me a sort of Pyrola, with a lesser and more pointed leaf'.
At different times this account has been used for both species, with the Hazelwood record for *P. minor*, and the rest for *P. media*. That may well be correct, and I have followed that for the first evidence of each. But it was not until Smith that species were clearly separated and the position clarified. Lees (1888: 320) gives as his first Yorkshire record 'on the moors south of Heptenstall in the way to Burnley in great plenty for nearly a mile's riding'. - Ray (1690)

Pyrola minor

'At Studley, Yorkshire, and Corra Linn [Falls of Clyde]. *Mr. Winch.* Common in many parts of Durham. *Mr. J. Backhouse.* Wood near Brodie house. *Mr. Brodie.* At the Falls of Clyde and many other places in Scotland. *Professor Hooker'*. - Smith (1824: 257).	vc64, 66, 77, 95	**1824**
'Pyrola *Ger. J.B.* nostras vulgaris *Park.* rotundifolia major *C.B.* ... In Hazelwood-woods near Sir Walter Vavasour's Park in York-shire. Mr Witham shew'd me a sort of Pyrola, with a lesser and more pointed leaf ...'. - Ray (1670: 256). 'on Ray's 1668 Itinerary 'in Sylvis Haselwood [near Tadcaster] Eboracens'. - Lankester (1848: 27) and see Raven (1950: 149).	vc64	1668

See notes above. Raven cites Lees (1888: 322) as agreeing that the Ray reference above is probably for this species, and points out that the note in Ray makes clear that it is not *P. rotundifolia* or

P. media. Note though that Lees (*op.cit.*) gives as his first record 'Pyrola groweth in Lansdale, and Craven … especially in a close called Craggeclose'. - Gerard (1597: 330). Raven too (1947: 302) cites How (1650: 100) who cites Gerard's record! On balance I feel that Ray's 1670 record is the safer.

Pyrola rotundifolia

* 'Pyrola nostras vulgaris … In Yorkshire Lancashire and further North yea even in Scotland in the woods everywhere and seldom in fields'. - Parkinson (1640: 510.1).		**1640**

P. vulgaris … It growes in a bogge by Roscree in the King's county. Mr Heaton'. - How (1650: 100). In common with other species first found by Richard Heaton but not published till 1650 by How, this is reckoned by Walsh (1978) to have been discovered by 1641.

Pyrus communis s.l.

* 'Pyrum … wyld Pere tre … well knowen'. - Turner (1562: 108).		**1562**

No early herbarium specimens have been traced other than a very poor specimen in Buddle's (d. 1715) collection in Hb. Sloane 126: 18, **BM**. There is no possibility of identifying this to a segregate.

Pyrus communis s.s.

Druce (1908, 1928a) separating it from var. b *achras*.		**1908**

Pyrus cordata

as *P. communis* var. *Briggsii* 'hedge between Thornbury and Common Wood [Egg Buckland]'. - Syme (1871).	vc3	**1871**
Marren (1999) has '1865, *Fl. Plym.*', but there is no such date in the *Flora* text. However Brown, in Syme (1892) cites a specimen collected by Briggs in May 1865, in Watson's herbarium [possibly at **K**?].	vc3	1865

Pyrus pyraster

Druce (1908, 1928a) separating it as *P. achras* from *P. communis*.		**1908**
'Once, and once only have I met with a wild Pear-tree in flower, and this was in Crow's-nest Wood, St. John's'. - Lees (1867), cited as this species [from Crowneast Wood] in Maskew (2014).	vc37	1867

The separation of this species from *P. communis s.s.* is still contentious.

Quercus agg.

* 'It was told me by a learned man, a frende of myne, that in the year of our lorde MDLVII that there was a great plentye of galles found upon oke leves in the North countre of England, and namely about Hallyfax [Halifax]'. - Turner (1562: 109).	vc64 vc64	**1562** 1557

Quercus' is in Turner (1548: F. v. verso) 'in english an Oke', and presumably the first record is centuries before that.

Quercus petraea

'Quercus latifolia mas, quae brevi pediculo est *C.B.* In Bagley-Wood and divers other places, observed by Mr Bobart who gave us the first intelligence of it'. - Ray (1724: 440).   vc22   **1724**
   vc22   1719

This species was omitted by Clarke. Bagely Wood is south of the Thames, opposite Oxford. Bobart died in 1719.

Quercus robur

'Quercus latifolia *Park.* vulgaris *Ger. The common Oak-tree*'. - Ray (1696: 286).   **1696**

Although there can be lttle doubt that earlier writers were describing this species, it is in Ray (*op.cit.*) that the two are contrasted for the first time in English literature. Note though that Oswald & Preston (2011: 271) identify the oak in Ray (1660: 129) as this species despite Ray citing the polynomials for both species. However *Q. petraea* is rare in Cambs.

Radiola linoides

* 'Millegrana minima, Lob. polygonum polyspermum, quorundam'. Between Sandwich and Canterbury. - Johnson (1632: 23).   vc15   **1632**

'I found this in Kent on a Heath not farre from Chistehurst [Chislehurst], being in company with Mr Bowles and divers others, in July 1630'. - Johnson (1633: 569).   vc16   1630

The Chislehurst record came from Goodyer, see Edgington (2007). It is slightly odd that Johnson didn't cite his own 1632 record.

Ranunculus acris

* 'Ranunculus surrectis cauliculis. In pastures and medowes almost every where'. - Gerard (1597: 804).   vc21   **1597**

Both Stearn (1965) and Rydén *et al.* (1999) identify 'Ranunculus … Crowfote … Kyngcuppe aut a Golland'. - Turner (1538: C.), with either *R. acris* or *R. bulbosus*.

Ranunculus aquatilis s.l.

'Ranunculum quartum … with a white flower … swimmeth above the water in poules'. - Turner (1562: 114). 'Millefolium sive Maratriphyllum flore & semine Ranunculi Aquatici, Hepaticæ facie. Crowfoote, or water Milfoile. They may be found in lakes and standing waters, or in waters that run slowly, I have not founde such plentie of it in any one place as in the water ditches adjoining to Saint George his fielde [Southwark] neere London'. - Gerard (1597: 679).   1562

This heading covers all ten species of the subgenus Batrachium, the white-flowered aquatic species marked † in the *Ranunculus* accounts following. Although various forms were described, with the exception of *R. hederaceus/omiophyllus* it was not until the very late 18th to mid nineteenth century that today's concepts were described in a satisfactory fashion. The dates of first evidence are not completely satisfactory either. In an ideal world the best evidence would have been a herbarium specimen, expertly determined. This is possible in two species, *R. baudotii* and *R. omiophyllus*, but for the others I have had to use local evidence and probabilities, as interpreted by experts such as Babington, Druce and Preston.

Gerard describes both this species and *Hottonia* on this page; it is not certain which he is referring to in this location.

Ranunculus aquatilis s.s. †

* First properly clarified as *R. heterophyllus* 'I have obtained this plant from Cambridgeshire, Leicestershire, Chichester, the River Lea near Hertford, Battersea in Surrey and Pangbourn in Berkshire'. - Babington (1855: 393).	vc13, 17, 20, 22, 29, 55	**1855**
'Ranunculus aquatilis *Dod. Ger.*'. - Ray (1660: 130). Both Babington (1860b) and Perring *et al.* (1964) cite this as the first Cambridge record, as do Oswald & Preston (2011).	vc29	1660

The Ray record and its subsequent confirmation for Cambridge seem much more satisfactory than other attributions. Clarke adds that '*R. aquatilis*' - E.B. 101 (1793) represents this species and that Babington (1851: 5) depicts this as *R. aquatilis ± heterophyllus*. Druce cites 'R. aquaticus' - Johnson (1633: 829) and 'Hampstead' - Ray (1724: 249.3). Nelson (1959) favours Turner (1562: 114). There is a record cited in Skidmore *et al.* (n.d.) from Rossington (vc63), T.Tofield, 1779 (date of death), which is a possibility.

Ranunculus arvensis

* 'Ranunculus arvorum … groweth commonly in fallow fieldes where corne hath beene lately sowen'. - Gerard (1597: 805).	**1597**
'in Anglia' - annotations made by T. Penny on C. Gesner's plant illustrations dated as by 1565 in Foley (2006b)	1565

Ranunculus auricomus

* 'Ranunculus auricomus'. - Gerard (1597: 807), but without any location in Britain. 'Growes in meadows and about the sides of woods'. - Johnson (1633: 958).	**1597**

Turner also has a figure of it (1562: 113 verso), but no text. 'Ranunculus auricomus, Golden Crowfoote … common in this Countrie'. - Lyte (1578: 417).

Ranunculus baudotii †

* 'I possess it from Edinburgh, Seaton Carew in the county of Durham, Burnham in Norfolk, Chepstow in Monmouthshire and Gloucestershire, Shirehampton, near Bristol in Gloucestershire, Dunster in Somersetshire'. - Babington (1855: 395).	vc6, 28, 34, 35, 66, 83	**1855**
'Guernsey. Gosselin, J., c.1790'. - see McClintock (1982: 38).	vc113	1790

Ranunculus bulbosus

* 'Ranunculus … Crowfote ... Kyngcuppe aut a Golland'. - Turner (1538: C.). 'R. bulbosus'. - Gerard (1597: 806).   **1538**

Both Stearn (1965) and Rydén *et al.* (1999) identify Turner's description with either *R. acris* or *R. bulbosus*.

Ranunculus circinatus †

* 'Christ-church meadows [Oxford]'. - Sibthorp (1794: 175).	vc23	**1794**
'Ranunculus aquaticus albus, circinatis tenuissimè divisis foliis, floribus ex alis longis pediculis innixis *Phytogr. t55. f2*'. - Plukenet (1696: 311). 'Provenit in aquis coenosis frequentissime ... D. Plukenet' - Ray (1724: 249-250). 'R. aquaticus albus, circinatus folio., Isle of Dogs'. - Druce & Vines (1907: 80).		1696

Moss (1920: 150) too cites Ray (1724: 249) which includes both Plukenet's description and his habitat and it seems reasonable to assume that he was describing this species. The Isle of Dogs record is also given in Kent (1975) as c. 1730 and in Druce as 1724. The record from Druce & Vines is dated as 1747, the date of Dillenius' death. Kent (1950) cites a record from Tothill fields, 1736, Blackstone.

Ranunculus flammula

* 'Flamula ... Sperewurte or spergrasse. It groweth in moyste places'. - Turner (1548: H. iij).   **1548**

Moss (1920: 150) cites Plukenet in Ray (1724: 249). But Plukenet gives no location, and it is not clear if this is a British record.

Ranunculus fluitans †

* 'In rivers and also in stagnant water. I have seen it in a perfectly stagnant ditch at Mildenhall, Suffolk'. - Babington (1839b: 229).	vc29	**1839**
'Ranunculo aquatili ablo affine Millefolium Maratriphyllum fluitans *J.B.* Fennel-leaved Water-Crowfoot ... In the river Tame about Tamworth, Middleton, etc in Warwick-shire; as also in the river Ouse [Isis] about Oxford, plentifully'. - Ray (1670: 259)	vc23, 38	1670

It was Babington who first sorted out this species from others. Cadbury *et al.* (1971) and Druce (1897:13, 1927 & 1932) both give the Ray reference as their first record. Clarke adds 'first record by this name; but it is the *R. fluviatilis* of Sibthorp (1794: 176)' ('rivers … Isis, Cherwell, Windrush'), and only gives the Ray record as a probability.

Ranunculus hederaceus †

* 'Ranunculus hederaceus, *Lugd*'. Between Sandwich and Canterbury. - Johnson (1632: 21).   vc15   **1632**

This species is now rare in Kent, but was much more frequent in the period of Hanbury & Marshall (1899). Another possibility, *R. omiophyllus*, is virtually un-recorded there.

Ranunculus lingua

'R. flammeus maior *Ger*. … In some ditches at Teversham moor, and abundantly in many great ditches in the Fens in the Isle of Ely'. - Ray (1660: 131).   vc29   **1660**

'Ranunculus flammeus major *G.961*'. This is in a ms. list of Cambridge plants from Samuel Corbyn, dated 20 May,1657. See Preston (in press).   vc29   1657

Clarke and Druce cited 'Ranunculus flammeus major, *Tab*.' East of Canterbury. - Johnson (1632: 23). But Rose, in Gilmour (1972: 121), suggests this is more likely to be a robust form of *R. flammula* [which he marks as having seen in the area]. Gerard (1597: 814) has this species, with a figure, and merely says (along with *R. flammula* and one of the aquatic species) 'grows … almost every where', which is not terribly satisfactory, though Edgington (2014) is convinced that Johnson (1638: 22) was describing this species from Tottenham. Parkinson (1640: 1215) merely says 'scarce to meete with', and How (1650) and Merrett (1666: 101) merely list it with no descriptive comment.

Ranunculus omiophyllus †

* '*R. hederaceus* β *grandiflorus*' (Babington (1843: 5), and then as *R. Lenormandi* in Babington (1845: 141).   **1843**

'In the ditches on Lancaster Marsh', from the collection of George Crossfield the elder. - Salmon (1912). Benson Knott, vc69, 1833, det. Cook, **LIV**.   vc60   1780

G. Crossfield died in 1820, but Salmon states that the specimens were collected c.1780.

Ranunculus ophioglossifolius

'St Peter's Marsh, Jersey …6th June, 1838'. - Babington (E.B.S. 2833 & 1839: 1).   vc113   **1839**
    vc113   1838

The first record other than from the Channel Islands is 'Found in 1878 near Hythe, S. Hants'. - Groves (1883). This specimen is in **BM**, dated 8th July 1878.

Ranunculus paludosus

'The specimens sent by Dr. Bull were collected in the early part of May on a small piece of dry uncultivated land sloping to the west, and on clay slate, close to the town of St. Aubin's [Jersey]'. - Trimen (1872b).   vc113   **1872**

Le Sueur (1984) gives an excellent history.

Ranunculus parviflorus

* 'Ranunculus parvus echinatus *Ger. em.* ... In locis humidis. Litle rough-headed creeping Crow-foote'. - Johnson (1634: 63). 'Ranunculus hirsutus annuus flore minimo'. - Ray (1663: 8).   **1634**

Ranunculus peltatus †

* 'I possess this plant from St Pierre in Monmouthshire, (where it was first noted as being a distinct species by the Rev. F. J. A. Hort) Bream in Gloucestershire, and Hoveton in Norfolk'. - Babington (1855: 398).   vc27, 34, 35   **1855**

In a letter from Littleton Brown to Dillenius, dated Jan 14th 1726, he refers to a *Ranunculus* identified by Druce & Vines (1907: lxx) as *R. peltatus* from 'ye [River] Teme & Corve by Ludlow (Salop)'. Druce & Vines (1907: lxxxiv *et seq.*) also cite a letter from Richardson, 26th Dec 1727, giving, *inter alia*, a record of what he [Druce] identifies as this species, as 'a very common plant in the river Aire from Apperly Bridge to Bingley, floating on the water amongst the stones where the water is not deepe'. 'Llyn Coch, Snowdon, Mr Br[ewer]'. - Druce & Vines (1907: 80). Brewer died in 1743.   vc40   1726

Druce's (1932) first record is 'R. aquatilis'. - Turner (1562: 114), but this is too tenuous.

Ranunculus penicillatus †

Baker & Foggitt (1865).   **1865**

'Ranunculo sive Polyanthemo aquatili albo affine Millefolium Maratriphyllum fluitans. In the river between Cambridge and Chesterton'. - Ray (1685: 15). 'Harefield River'. - Blackstone (1737: 85), cited in Trimen & Dyer (1869) & Kent (1975). Mildenhall, vc26, 1837, det. Webster, **E**.   vc29   1685

Oswald & Preston (2011: 475n) endorse the Ray record (as subsp. *pseudofluitans*). Syme (1863: i. 20) had included this as *R. aquatilis*? var.γ *pseudofluitans*, but it was properly described only in 1865 (See Pearsall, 1919). Druce cites 'R. sive polyanthemo, etc., Hounslow Heath, Middlesex'. - Ray (1724: 250. 6). Green (2008) postulates that it was this species recorded near Lismore, County Waterford, in 1746.

Ranunculus repens

* 'Ranunculus pratensis etiamque hortensis ... In pastures and medowes almost everywhere'. - Gerard (1597: 805). **1597**

Ranunculus reptans

* 'At the west end of Loch-Leven in Kinross-shire, Dr. Parsons'. - Lightfoot (1777: 289 and figured on title-page). vc85 **1777**

'At the north east side of Loch Leven plentifully, Aug 7th, 1764'. - John Hope's ms. 'List of Plants growing in the Neighbourhood of Edinburgh', reproduced in Balfour (1907). vc85 1764

Both the species and its hybrid with *R. flammula* (*R.* x *levenensis*) occured here. Preston & Croft (1997) point out that Lightfoot's drawing is of the hybrid, and further, that it seems very persistent, whereas *R. reptans* seems more fugitive. We will never know which Hope or Parsons did find. The next field record of the species is from Loch Leven in 1869 (Preston & Croft, 1997).

Lightfoot, J. 1777. *Flora Scotica.*

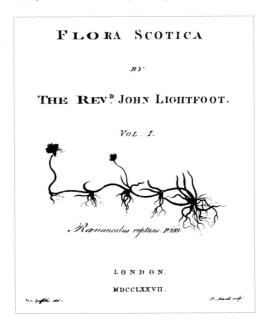

Ranunculus sardous

* 'Ranunculus rectus foliis pallidioribus hirsutus. *J.B.* ... In locis humidis et lutosis ubi per hyemem aquae stagnarunt'. - Ray (1663: 8). 'Below Hamstead in the Meadows betwixt the Town and Heath'. - Merrett (1666: 102). vc29 **1663**

'Around Cambridge', in a letter from Ray to Courthope, dated from Trinity College in July 1661. See Gunther (1934). vc29 1661

Thompson (1974) has '*Ranunculus vectus* …', clearly an error.

Ranunculus sceleratus

* 'Ranunculus ... Secundum genus ... vocatur apium risus and herba sardonia'. - Turner (1538: C.). 'Ranunculus rotundifolius forte Apium risus'. Hampstead Heath. - Johnson (1629: 12).   **1538**

Turner adds 'it is from that demonination that 'the sardonic smile' originates'!

Ranunculus trichophyllus †

* *R. aquatilis* β *trichophyllus* (Godr.). - Babington (1851: 3). As *R. trichophyllus* 'Plentiful in Cambridgeshire, Norfolk, and Suffolk'. - Babington (1855: 390).   **1851**

Edgington (2014) dealing with Johnson's Tottenham plants of 1638, interprets 'Millefolium sive Maratriphyllum flore & semine Ranunculi Aguas. Hepaticæ facie. *Ger.* Fœniculum aquaticum 3. *Tab.* Water Fennell, Crow-foote Milfoile' - Johnson (1634: 51, 1638: 17) as this species. Oswald & Preston (2011: 242n), although stressing the convoluted taxonomy of Batrachian *Ranunculus*, are reasonably confident that both Parkinson (1640: 1257) and Ray (1660: 99), citing the same polynomials as Johnson, were describing this species as currently circumscribed.   vc21   1638

Rose, in Gilmour (1972: 117), identifies 'Ranunculus tricophyllon aquaticus medioluteus, *Col*, sive Millefolium maratriphyllon tertium, &c. *Lob.*', from Johnson (1632: 19) as 'almost certainly' this species.

Ranunculus tripartitus †

'Innes Moor, Roche, E. Cornwall [1876]'. - Tellam (1877). 'From deep pools at Roche, near St. Austell, Cornwall'. - Trimen (1877b).   vc2   **1877**

'In a pond near the brickfield on the right-hand side of the road from Southborough to Tunbridge Wells, 1846, Reeves'. - Hanbury & Marshall (1899).   vc16   1846

For a fuller report see Babington (1878), where he notes that the plant described and illustrated under this name in E.B.S. 2946 is not this species. This was Clarke's first date, and should be disregarded. Babington adds that records previous to those reported by Tellam and Trimen (*op.cit.*) were errors for *R. intermedius*. Presumably these errors include his own in the Third (1851) and later editions of his *Manual*.

Raphanus maritimus

* 'Raphanus maritimus flore luteo, siliquis articulatis, secundum longitudinem eminenter striatis ... Mr. Stonestreet found it under the Cliffs by the Sea-side about half a Mile Westward of the Fisher-houses at Bourn [Eastbourne] in Sussex'. - Ray (1696: 342).   vc14   **1696**

To treat these entries for *Raphanus* as two species is not consistent with Stace (2010), but I have done so since one is a native and the other is an archaeophyte.

Note Wolley-Dod (1937: 50) 'But there is a note [?a gloss] in the British Museum copy of Merrett's *Pinax* (1666) of '*Rapistrum maximum* on the rocks near Bordin in Sussex and west from Hastings near the town'. … [and that] there is little doubt that this species is meant and that the station was Eastbourne (Bourne). It cannot be used as a record since the entry is not dated, though it was probably made by Newton, circ. 1683'.

Raphanus raphanistrum

* '*Raphanus sylvestris* … upon the borders of banks and ditches cast up & in the borders of fields'. - Gerard (1597: 186). '*Rapistrum flore albo*, as common in Corn fields in Surrey as the Yellow. Mr. Brown uti etiam in pluribus Cantii locis'. - Merrett (1666: 103). — **1597**

Reseda lutea

* '*Rheseda Plinii* … In sundrie places of Kent as at Southfleete neere master Swanne's house upon longfielde downes … and at Greenehithe'. - Gerard (1597: 216). — vc16 **1597**

Reseda luteola

* '*Luteola* … Locis Galliae, & Angliae ruderibus, et semitibus occurrat'. - L'Obel (1571: 149). Between Gravesend and Rochester. - Johnson (1629: 2). — **1571**

Grigson (1955) gives a reference from Chaucer, and notes that it was also grown as a crop, as well as imported from France.

Rhamnus cathartica

* '*Rhamnus solutinus* … In Kent in sundrie places as at Farningham … in Southfleete … at Dartford'. - Gerard (1597: 1154). — vc16 **1597**

Rhinanthus angustifolius

* '*Pedicularis major angustifolia ramosissima flore minore luteo labello purpureo* D. Richardson … within a mile of Burrowbridge amongst the Corn … betwixt Wetherby and Catall (both these Places are within ten Miles of York). I also observed it in this Year 1723, amongst the Corn nigh Westnewton [c16 miles S. of Berwick] in Northumberland, upon the Borders of Scotland'. - Ray (1724: 284*) — vc64, 68 **1724**

'*Crista galli montana angustifolia* C.B. Prodr. … a common weed in most of the cornfields'. - Richardson to Sloane, 18th August 1722 (British Library: Sloane ms. 4046, f.284). — vc68 **1722**

In his *Flora of Nidderdale*, Lees (1894: 45) adds more details including 'In Ray's time in then newly-reclaimed cornfields, c1696'. No evidence is given for the date of 1696 (which is also the date of the second edition of Ray's *Synopsis*). Richardson was active by then, so this could be correct.

The history of this species would make an interesting subject in that it appeared in many places in the north of England, consequent in the breaking up of old pastures. It then gradually vanished almost to the point of extinction. Where did it come from?

Rhinanthus minor

* 'Christa galli ... groweth in drie medows and pastures and is to them a great annoiance'. - Gerard (1597: 912).		**1597**
'Ratle grasse in meddowes very vulger'. - Manuscript notes of William Mount, of c.1584, in Gunther (1922: 255 - 263), and identified by him as this species.	vc16	1584

Rhynchospora alba

* 'Gramen junceum leucanthemum ... I never found this but once, and that was in the companie of Mr. Thomas Smith and Mr. James Clarke Apothecaries of London; we riding into Windsore Forest upon the search of rare plants'. - Johnson (1633: 30).	vc22	**1633**

Druce (1897) says the locality might have been in Surrey.

Rhynchospora fusca

* 'Cyperus minor angustifolius palustris capitulis fuscis paleaceis ... In Occidentalibus Angliae'. - Morison (1699: 239). 'I found this plentifully in a bog between Southampton and Limington in August'. - Petiver (1716: ii. 149).		**1699**

Was Petiver Morison's source?

Ribes alpinum

* 'Ribes Alpinus dulcis *J.B.* ... In agro Eboracensi invenit D. Dodsworth'. - Ray (1688: 1486).	vc64	**1688**

Probably vc64; see Lees (1888).

Ribes nigrum

* 'Ribes fructu nigro ... By the rivers side at Abington'. - Ray (1660: 139)	vc29	**1660**
'Ribesium sylvestre, neither in *Ger.* nor *Parkinson*, found in Kent & Bedfordsheire'. This is in a ms. list of Cambridge plants from Samuel Corbyn, dated 20 May, 1657. - see Preston (in press).	vc29	1657

Ribes rubrum

'Rhibes ... is called in some places of Englanda a Rasin tree'. - Turner (1548: iiij). 'Ribes ... By a waters side at Clouer [Clewer, nr. Cheddar] in Somerset shyre in the possession of Maister Horner'. -	**1548**

Turner (1568: 63). 'Ribes sylv. *Matth.* ... in sylvis in Septentrionali comitatus Eboracensis parte circa Greta-bridge inque Dunelmensi Episcopatu [Greta Bridge, is S. of Barnard Castle in North-west Yorkshire] & Westmorlandia. D. Johnson'. - Ray (1677: 254).

Roach (1985) notes that the reference in Turner (1568) is of particular interest as currant seeds have been discovered in the nearby Glastonbury lake village site.

Ribes spicatum

'Near Richmond in Yorkshire and between Piersbridge and Gainford in the county of Durham'. - Robson (1797).	vc65, 66	**1794**
Druce cites 'Brackenbrow ...'. - Ray (1724: 456)', but I cannot see that there. Lees (1888: 243) gives the Ray reference, 'Ribes vulgaris fructu dulci Clus.'. Under this polynomial Ray does have a bracketted record '(Near Settle in Yorkshire; Dr Richardson)', but no mention of Brackenbrow [which is opposite Helk's Wood, near Settle]. However, there seems little doubt that it was this species that he was describing.	vc64	1724

Ribes uva-crispa

'Mr Robson finds it plentifully in woods and hedges around Darlington'. - E.B. 1292.	vc66	**1804**
'I. Lyons,1763'. - Perring *et al.* (1964).	vc29	1763
Doubtless there are earlier records, but early authors, such as Culpeper (1653), only recorded it from gardens. John Martyn added a record of this species in his copy of Ray (1660), probably after his *Methodus* of 1727, and before Lyon's record, but the actual date is not known (A. Leslie, *pers.comm.*).		

Romulea columnae

* '*Trichonema bulbocodium* Ker grows wild on the Warren (a sandy tract) between Dawlish and Exmouth, Devon. My friend, W.C. Trevelyan and myself, found it there on 24th March, 1834'. - Milford (1834).	vc3	**1834**
McClintock (1975: 235) mentions an earlier introduction to English gardens from the Fermain area of Guernsey in 1726, where 'it will be what Knowlton saw on Monday 6 June 1726, in two or three places on the island … but always upon the highest part of the rocks'.	vc113	1726

Guernsey. Sent to Sir J. Banks by Finlay in 1787. See McClintock (1982: 163).

The only other English record, and one from a more typical site than that at Dawlish, is 'Top of the cliffs near the Coastguard Battery, Polruan, E. Cornwall, May 1879, Miss Kemp'. - Davey (1909). This site was only rediscovered by I.J. Bennallick in 2002.

Rorippa amphibia

* 'Raphanus aquaticus ... In ditches, standing waters and rivers ... in the chincks amongst the mortar of a stone wall that bordereth upon the Thames by the Savoy in London'. - Gerard (1597: 186). | vc21 | **1597**

The Gerard location is clarified in Johnson (1633: 240). See also 'Laver ... yellowe water cresse groweth in water sydes and in sprynges & wel heades'. - Turner (1548: D.viij. verso), which Stearn (1965) equates with this species.

Rorippa islandica s.s.

'Specimens were collected from Garsow, N. Ronaldsay by M. Spence in 1908'. - Randall (1974). | vc111 | **1974**

Renvylc, Galway, H16, 1831, R.J. Shuttleworth, det. Jonsell, **BM**. | vcH16 | 1831

Differences from *R. palustris* were clarified by B. Jonsell in Sweden in 1968. Randall (*op.cit.*) alludes to earlier records, such as in Griffith (1895) which might date back to 1813.

Rorippa palustris

* 'Small Jagged Water Radish ... Moist ponds, &c., Lond. [London]'. - Petiver (1713: xlix. 9). 'Tothill Fields'. - Curtis (1783: fasc.5, pt 49). | vc21 | **1713**

Rorippa sylvestris

* 'Eruca aquatica ... I have sometimes found in wet places'. - Johnson (1633: 247). 'Tothill Fields, Westminster'. - Blackstone (1746: 20). | | **1633**

Rosa agrestis

* as *R. rubiginosa* var. *sepium*? 'near Bidford Grange, at the back of the Brick-kiln house, on the Stratford road, July 1818'. - Bree in Purton (1821: 41). | vc38 | **1821**

also Newington, vc21, 1818, J.J. Woods, det. Sandwith, **LINN**. | vc21, 38 | 1818

Clarke adds 'In E.B.S. 2653, and elsewhere, wrongly quoted 'Bridport'.

Rosa arvensis

* 'Rosa Canina sylv. unico fl. & fructu. In the fields near Hackney in the way thence to London'. - Merrett (1666: 105). | vc21 | **1666**

Rosa caesia

as *R. coriifolia*, Wolley-Dod (1908: 97). | | **1908**

Ulverston, vc69, 1824, det. Primavesi, **E**. | vc69 | 1824

Rosa canina s.l.

* 'Cynosbatos ... anglis a wylde heptre or a brere tre'. - Turner (1538: A. iv. verso). **1538**

L'Obel (1571: 446) refers to another rose, scentless, but otherwise like *R. rubiginosa* 'Dunen Rosem. Pomifera. Inter Syluestres non postremæ venustatis alia Angliæ perquàm familiaris, & Esglantinæ cognatior'. Like Ray (1660: 139) he refers to the 'Bedeguard', 'Robin's pincushion' often growing on it.

```
Esglentine ou    as in hortis Italicis.
Esglentier.      Species est altera folijs minoribus Myrtinis, sed latioribus, flore & elegātia Rosæ Mos-
                 chatæ: fructum similem cæteris fert, & æqualem: tota planta multò odoratior, vulgò
Spongia Bede-    Esglentine, Angliæ & Flandriæ hortis, ad odoris gratiam colitur.
guaris.          Amis omnium, sed præcipuè huius villosum, coloriumq;, muscidum excrementū hæ-
                 rere solitum floccis, velleris Sericei, instar, Bedeguard vocatur.
Dunen Rosem.     Inter Syluestres non postremæ venustatis alia Angliæ perquàm familiaris, & Esglanti-
Pomifera.        næ cognatior: fruticat pumilla, bicubito non altior, sed inodora: folio minore, & pere-
Morifera.        leganti: Mespilo Aroniæ non dispari: pomo donata minore multò, colore viuidiore, ru-
Baccifera.       bello subpuniceo, per Augustum maturo: flos Rosæ syluestris est.
```

Pena, P. & L'Obel, M. de. 1571. *Stirpium adversaria nova*.

Rosa micrantha

* 'Mr. W. Borrer has communicated to us the present Rose under the above name. He observes that it is common in hedges and thickets'. - E.B. 2490 (1812). **1812**

Cleadon, vc66, 1810, W. Robertson, det. Graham, **HAMU**.   vc66   1810

Garry (1903-4: 66) notes that the the E.B sheet is annotated 'W. B[orrer], 1811 and 1812. Scent less powerful….'.

Rosa mollis

* as *R. villosa* var. β 'In the way from Edinburgh to Ravelston wood'. - Smith (1800: ii. 539).   vc83   **1800**

as *R. villosa*, probably from the Craven area of Yorkshire - see notes of a 1660 Itinerary by Ray in Raven (1950: 114).   vc64   1660

Lees (1888) also cites the Ray record. Smith in E.B. 2459 quotes 'Rosa sylvestris folio molliter hirsuto, &c. Ray (1724: 478)' as a synonym. The continental *R. villosa* has only been found as an alien in Britain.

Rosa obtusifolia

Wolley-Dod (1908: 67).   **1908**

as *R. rubiginosa* var. *inodora* 'nr. Halesworth'. - Graves & Hooker (1821: t.117). Little Malvern, vc37, 1841 det. Primavesi, **OXF**.   vc25   1821

Wolley-Dod is the first to clearly describe it.

Rosa rubiginosa

* 'Rosa sylvestris odora, the Eglantine, or sweete brier … In the borders of fieldes & woods in most parts of England'. - Gerard (1597: 1087.1). 'Rosa sylvestris odora Eglenteria. Sweet Briarbush: Growes wild in divers hedges by Darfield [Yorkshire], but with more smooth shining leaves than the Garden kind'. - How (1650: 105).

1597

These references might, of course, have been *R. micrantha*, but though neither is common in Yorkshire, *R. rubiginosa* is less rare there.
Turner (1538: A. iv. verso) 'Cynorrhodos is, as far as I am able to discern, a bush (whose leaves smell lovely at the beginning of spring), which I believe people call Swetebrere or Eglentyne'. But this might well have been only in gardens, as he has only describes that habitat in his 1548 work (C.vj. verso).

Rosa sherardii

Davies (1813: 49).

Wolley-Dod (1910: 69) notes that 'there is a specimen from Davies, which is his type, from Anglesea, at South Kensington [**BM**]. vc52 **1813**

Rosa spinosissima

* 'Rosa pimpinella … groweth very plentifully in a field as you go from a village in Essex called Graies [Grays] (upon the brinke of the river Thames) unto Horndon on the hill, insomuch that the fielde is full fraught therewith all over … [and by] a village hard by London called Knights bridge, unto Fulham, a village thereby'. - Gerard (1597: 1088). vc18, 21 **1597**

Rosa stylosa

* as *R. collina* 'Mr. William Borrer … first distinguished this Rose in England … [it] is common in Sussex'. -   E.B. 1895. vc13, 14 **1808**

Rosa tomentosa

'About London, Hudson. At St Faith's, Catton, and other places near Norwich. Common in Shropshire and Wales'. - Smith (1800: ii. 539). vc21 **1800**

Clarke & Druce both cited as their first record 'Rosa sylvestris fructu majore hispido … In sepibus non infrequens a D. Dale observata'. - Ray (1696: 296), which may well be correct and is cited as a synonym in Smith (*op.cit.*). 1696

Rubia peregrina

* 'Rubia … the wilde kinde … In the yle of Wyght,' and' besyde Wynchester in the way to Southampton'. - Turner (1562: 118). 'Master George Bowles found it growing wilde on Saint Vincents vc10 **1562** rocke; and out of the Cliffes of the rockes at Aberdovie in Merioneth shire'. - Johnson (1633: 1120).

Allen (1982 & *in litt.*) suggests convincingly that Turner's record for Hampshire was an error for *R. tinctoria*. But the Isle of Wight locality and the fact that Johnson's other record, from Aberdovey, is also a *bona fide* site for this species, are convincing evidence that he was describing this species.

The records from St Vincent's Rocks are more unsatisfactory. Although *R. peregrina* is a frequent plant there, Johnson cites an account in L'Obel (1576: 464), and uses his illustration. But in L'Obel's account of 'Rubia minima. Altera perpusilla, quam in rupibus Vincentij [St Vincent Rocks] non procul Bristolio Angliae enatam legimus' he describes 'a very small plant, upon creeping stalks some inch and a half high' and the illustration is definitely one of an annual, and one looking more like a *Sherardia*, were it not for the yellow colour of the flowers. Indeed Ray (1724) lists L'Obel's plant in his *Plantarum Dubiarum*, and cites *R. peregrina* as 'Rubia sylvestris aspera, quae sylvestris Dioscoridis *C.B. Pin. 333*'. He refers to the St Vincent Rocks site for this in his 1670 work, but it is not clear whether he saw it himself on his 1662 Itinerary, or whether he is copying Bowles' record.

Rubus arcticus

'Mr Sowerby has been favoured by Richard Cotton Esq. with a dry wild specimen from the high regions of Ben-y-Glo, Blair, in Scotland'. - E.B. 1585 (1806). vc89 **1806**

The reference in Smith (1800: ii. 546) is to 'In the Isle of Mull, Rev. Dr. Walker'. This is not supported by a herbarium specimen, and has never been really accepted. See also Druce (1920b) and Harley (1956).

Rubus caesius

* 'Rubus saxatilis. Stone blacke Berrie tree … in divers fieldes in the Isle of Thanet harde by a village called Birchinton'. - Gerard (1597: 1091). 'Rubus repens fructu caesio'. - Johnson (1633: 1271). vc15 **1597**

Rubus chamaemorus

'Chamaemorus anglicana … thornless bramble … by the eminent Thomas Penny [with a wonderful description of the plant!]…on mount Inglebarrow [Ingleborough] the highest in all England, twelve miles from Lancaster'. - L' Ecluse (1583: 117-119). 'Chamaemorus … Knotberry … upon Ingleborough hils … and Stanemore, between Yorkshire and Westmerland'. - Gerard (1597: 1091). vc64 **1583**

See Raven (1947: 154), commenting on Penny's (c.1530 - 1589) record 'the admirable picture and history … probably obtained in his early years'.

Rubus fruticosus agg.

'Rubus … angli vocant a bramble, aut a Blakbery busshe. Fructus eius vocatur aut blacke beryes, aut blacke byers, aut bumble berys'. - Turner (1538: C.).

1538

Rubus idaeus

* 'The Raspis Bush … it groweth not wilde that I know of [except] I have found it among the bushes of a cawsey neere unto a village called Wisterson [Wistaston] where I went to schoole two miles from the Nantwich in Cheshire'. - Gerard (1597: 1091).

vc58 **1597**

Rubus saxatilis

* 'R. alpinus saxatilis, … In Yorkshire'. - Parkinson (1640: 1015). 'Soon-a-man-meene; In English, the juyce of a faire Woman: In a Woode neer Eddenderry … Mr Heaton'. - How (1650: 116).

vc64 **1640**

See Nelson (1959: 117) with comments on Turner's *R. fruticosus* (1562: 118) - ' … the lesser kynde hath sometyme rede berries … but a few sedes … much pleasanter to eat then the greater berries be'.

In common with other species first found by Richard Heaton but not published till 1650 by How, this is reckoned by Walsh (1978) to have been discovered by 1641. Edenderry is in Co. Offaly (vcH18).

Rumex acetosa

* 'Lapathon … acetosum … quam vulgus appellat Sorell, aut Sourdoc'. - Turner (1538: B.ij. verso).

1538

Rumex acetosella

* 'We have two kindes of wilde Dockes … and so many kindes have we also of Oxalis or Sorell, for the one hath a rounder leafe and the other sharper with sharpe thinges resembling abrode arrow head'. - Turner (1562: 133 [after 121]). 'Oxalis tenuifolia … tanto luxu sabulosis … Brabantiae, Picardiae, Nortmaniae, & Angliae importuna'. - L'Obel (1571: 120).

1562

Rumex alpinus

'This species, I am recently informed by Mr Maughan, has been found truly wild by Mr G. Don, in the Ochill-hills, Clackmannanshire'. - Hooker (1821: ii. 208).

vc85 **1821**

Moss (1914: 133) equates it with 'Hippolapathium rotundifolium' - Gerard (1597: 313.4), but it seems far more likely to be 'Hippolapathium sativum, Patience or Munkes Rubarb … groweth in gardens' - Gerard (1597: 313.3). There are specimens marked 'R. obtusifolium? alpinum?, within ⅛ mile of Moffat, on the roadside, and at Portsoy, Banffshire', in Balfour's (1907) account of

Hope's herbarium list of 1768.. All of these dates are very late for an archaeophyte that must have been used for centuries. Indeed, sccds have been recovered from silted-up fifteenth century drains at Paisley Abbey (Dickson & Dickson 2000).

Rumex aquaticus

'in 1935 Mr R.M. Mackechnie sent me specimens from Stirlingshire (vc 86) …'. - Lousley (1939).   vc86   **1939**
    vc86   1935

Rumex conglomeratus

'Lapathum acutum minimum. In medowes and by rivers sides'. - Gerard (1597: 311).   vc16   **1597**

Clarke had 'Lapathum acutum vulgare'. Near Rochester. Johnson (1629: 2), but I feel that Gerard's citation, which is followed by Druce, is right. See below, *R. crispus*.

Rumex crispus

* 'A varietie [of Lapathum acutum] with crisped or curled leaves'. - Johnson (1633: 387).   **1633**

But Rose, in Gilmour (1972: 51), thinks 'Lapathum acutum vulgare'. Gravesend to Rochester. - Johnson (1629: 2) is likely to be this species.

Rumex hydrolapathum

* 'Great Water Docke … in ditches & watercourses very common throughout Englande'. - Gerard (1597: 312).   **1597**

Chapman *et al*. (1995) identify as this Turner's (1562: 133) ' … they call one Oxilapathon, and it groweth in ponds and ditches'. There is a good picture in Turner and the source seems possible. See also Turner (1538: B. ij. verso) 'Lapathon … oxilapathon est a water doc'. Nelson (1959) has Turner's plant as *R. obtusifolius*, which seems unlikely.

Rumex longifolius

* 'Moist places near Ayr, Mr. Goldie'. - Hooker (1830: 168).   vc75   **1830**

Rumex maritimus

* 'Lapathum folio acuto, flore aureo, *Bauh*.… In uliginosis. Golden Docke' - Johnson (1641: 24). 'Circa Crowland [S. Lincs.]; ad Trentam fluvium prope Swarston [Swarkestone] pontem in Derbia, &c'. - Ray (1670: 188).   **1641**

Although most authorities accept the dates given for this species and *R. palustris*, Petiver (1713: t.2. f.7 & 8) was the first to satisfactorily describe both species. However it was not until Smith (1800: 393) that the separation into two species was consistently adopted.

Rumex obtusifolius

* 'Lapathum sylvestre fol. minus acuto'. - Johnson (1633: 388. 3).

1633

But see Turner (1562: 121) where several 'kindes of Docke' are described, one having a ' leafe much rounder' than others. See also Turner (1538: B. i. verso) 'Lapathon', which both Stearn (1965) and Rydén *et al.* (1999) identify as this species.

Rumex palustris

'Hydrolapathum minus. Small water dock'. - Gerard (1597: 312). 'Hydrolapathum minus'. Erith Marshes. - Johnson (1629: 10).

1597

See *R. maritimus*. Gerard's drawing looks very convincing, but Hanbury & Marshall (1899) cite Johnson. Rose, in Gilmour (1972: 63), points out that '*R. palustris* was not then distinguished from *R. maritimus*; the former occurs here now, and the latter was recorded in 1877, so either is possible'. But in Trimen & Dyer (1869) Plukenet's (1700: 112) 'Lapathum longo angustoque folio Anthoxantho plurimum accedens, &c ... lately found in Westminster, or Tothill Fields as is commonly called, by Mr. Isaac Rand', is given as the first British record.

Rumex pulcher

* 'Lapathum pulchrum Bononiense *J.B.* quod non inepte vocari potest, The Fidle-Dock propter figuram foliorum inferiorum, Common in St. George's Fields [Southwark], and other places about London'. - Merrett (1666: 69).

vc17   **1666**

Rumex rupestris

'I found my plant on the 26th of July last at Wembury, about five or six miles to the east of Plymouth, growing in plenty on a strong shore, associated with *R. crispus* and *R. conglomeratus*'. - Briggs (1875) - *see image p.349*.

vc3   **1875**

' ... known since 1862, when Prof. Babington mentioned it with some reserve in the fifth edition of his *Manual*, as found in Jersey ... the date of the observation of the plant in Jersey by Mr. Newbould and Prof. Babington was 1842'. - Trimen (1876).

vc113   1842

The first record other than from the Channel Islands appears to be 'Three Cliffs Bay [Glamorgan], G.Bentham, 1859'. - Wade *et al.* (1994).

Rumex sanguineus

'Lapathum sativum sanguineum. Bloudwoort ... sometimes red in every part thereof ...' - Johnson (1633: 390).

1633

Clarke and Druce both cite 'Lapathum sanguineum ... As a pot herb planted in gardens yet found wild also'. - Parkinson (1640: 1227). Moss (1914), however, cites Johnson.

Briggs, T.R.A. 1875. *Rumex rupestris*, Le Gall, a British species. *J. Bot.* 13: 294-295.

See species account: Rumex rupestris, p.348

> **SHORT NOTES.**
>
> RUMEX RUPESTRIS, *Le Gall*, A BRITISH SPECIES.—This past summer I have been giving some time to the study of a Dock, which occurs in plenty on several parts of the coast near Plymouth, differing in some respects from all the generally recognised British species, but agreeing so well with *Rumex rupestris*, Le Gall, as described in some of the Continental Floras, that I have no doubt it is this. In Babington's "Manual" (ed. 7 and some previous ones) there may be seen a remark as to the probability of *R. rupestris* being a Jersey species. I have been favoured by Mr. Baker with a Continental specimen of *R. rupestris*, obtained from a plant cultivated by M. Gay in the Luxemburg Garden, 1834, from seed obtained at Cap de Carteret, Manche, in 1831. A comparison of it with the Plymouth Dock has left no doubt in my mind of the identity of the two. I first found my plant on the 26th of July last at Wembury, about five or six miles to the east of Plymouth, growing in plenty on a strong shore, associated with *R. crispus* and *R. conglomeratus*; between which two I was at first disposed to think it might be a hybrid. Since then I have found it two or three miles further east, on the shore of Bigbury Bay, growing on low damp rocks; and also on the Cornish coast, near Port Wrinkle; and at Downderry, in the parishes of Sheviocke and St. Germans. I believe it to be entirely confined to the open coast, and not to be one of those species that follow the salt water up the sides of estuaries and creeks. It bears considerable resemblance to *R. conglomeratus*, but the branches are much less leafy, stiffer, and straighter, forming a compact panicle, so that its outline is more like that of *R. nemorosus*. The cauline leaves

Ruppia cirrhosa

as *R. maritima* 'Salt marshes. Guernsey'. - Babington (1843).　　vc113　　**1843**

Druce (1932) cites a specimen in Herb Du Bois 'W. Stonestreet [died 1716]', but there is no Du Bois record in Druce (1928). As *R. maritima*, 'In the ditches in the marsh going from Brean down to Uphill, July 1773'. - White (1912: 70), describing a journey by Banks and Lightfoot.　　　　　　　　　　　　　　　　　　　　　　　　1716

Confusingly, at least to our concepts, Babington (and many later writers) describe this species as *R. maritima*, and our *R. maritima* as *R. rostellata*.

Clarke and Druce both cite Hooker (1830: 77) but it seems that, under *R. maritima*, he was only groping for two species. Interestingly, Druce (1932) adds that Stonestreet cites Ray (1686: 190.10) 'Potamogiton maritimum pusillum alterum ... in fossis palustribus via à Camaloduno [Colchester] ad Goldhanger'. Ray does indeed list two *Ruppia* species and Gibson (1862) appears to decide that at least one was this species.

Ruppia maritima

* 'Potamogeiton pusillum gramineo folio maritimum ... about Maldon in Essex I first observed it'. - Ray (1670: 251).　　vc18　　**1670**

Ruscus aculeatus

* 'Ruscus ... buchers brome or Petigrue ... groweth in Kent wilde by hedge sydes'. - Turner (1548: F. viij).　　vc15, 16　　**1548**

But see Turner (1538, C) 'Ruscus ... angli Butchers broome and Petygrew'. Since the species was of household use he may, of course, be describing a garden or imported plant.

Sagina apetala s.l.

* 'Saxifraga Anglica annua Alsine folio … it grows plentifully in the walks of Balliol College gardens, and on the fallow fields around Headington and Cowley, and many other places'. - Plot (1677: 146).	vc23	**1677**
'1634, from an unstated locality'. Goodyer, ms.11, f. 14, see Gunther (1922: 186).		1634

Sagina apetala s.s.

as *A. ciliata* 'Found near Thetford in Suffolk by the Rev. W.W. Newbould on June 6th, 1847'. - Babington (1848: 153).	vc26 vc26	**1848** 1847

Sagina filicaulis

Babington (1851). Whilst his previous (1848) paper described the 'new' species, this edition was the first to give both species of the aggregate.		**1851**
as *S. apetala* subsp. *erecta* 'Blackwell, 1805, Backhouse jnr'. - Graham (1988).	vc66	1805
Butcher & Strudwick (1930), as a rare or overlooked plant, were the first to provide an illustration.		

Sagina maritima

* 'We originally received this plant from Mr. R. Brown, who gathered it in 1799, at Bally-castle in Ireland, near the Giant's Causeway. Mr. G. Don sent the same from the summit of Ben Nevis in 1803'. - E.B. 2195.	vc97, H39	**1810**
' … Insula Mona [Anglesey] (loco Llandwyn dicto) nascens et repta Aug. 1726'. - Druce & Vines (1907: 150). This is presumably Dillenius' own record from his Welsh trip of that year.	vc52	1726
Mabberley (1985: 36) gives Brown the credit for being the first to realise that it was a new species. He (Brown) first described it in his unpubished MS in 1797, with specimens dated 1795 - 1797. Specimens from Don (1794) and R. Brown (1795) are in **BM**.		
But note that if this was the plant seen by Thos Lawson, before 1690 'On Whinny field-bank by Cullercoats near Tinmouth in Northumberland', which has traditionally assigned to *S. subulata* (see Whittaker, 1986, Swan, 1993), his record would pre-date all the other records cited.		

Sagina nivalis

* 'Mr. Boswell Syme has shown to me a specimen of this arctic plant, which was picked on Ben Lawers, several years ago, by Professor Balfour'. - Watson (1863).	vc88	**1863**
' … it was apparently first detected by Mr Boswell Syme, amongst specimens gathered on Ben Lawers by Prof. Balfour, on August 25, 1847'. - Babington (1864c).	vc88	1847

Druce cited a record from Glas Moel from Mr J. Backhouse in 1847 or 1848, but Babington (*op.cit.*) identified that as *S. saginoides*.

Sagina nodosa

* 'Saxifraga palustris alsinefolia ... This groweth plentifully on the boggy ground below the red Well of Wellingborough in Northamptonshire. This hath not been described that I finde. I observed it ... August 12, 1626. John Goodyer'. - Johnson (1633: 568). 'About Bath and divers other places'. - Johnson (1634: 19).	vc32	**1633**
Gunther (1922: 53) has 11th August 1625.	vc32	1625

Sagina procumbens

* 'Saxifraga Anglicana'. Near Rochester. - Johnson (1629: 2).	vc16	**1629**
Rose, in Gilmour (1972: 52), points out, quite reasonably, that this might refer to *S. apetala* or *S. procumbens*.		

Sagina saginoides

'On Ben Lawers, where it was discovered by Mr. J. Mackay in 1794'. - Smith (1800: ii. 504).	vc88	**1800**
'I first discovered this plant on rocks on the East Side of Malgyrdy in July 1789 in company of Mr Jas. Brown jun. nursery man Perth'. This is a note on a herbarium sheet from the Museum Inn, Farnham, Dorset: see Edmondson (2009).	vc88	1789
In E.B 2105 is added 'Mr. G. Don appears to have found it previously on Mal-ghyrdy'. Mal-ghyrdy is assumed to be Meall Ghaordaidh. This seems to be correct as the Edmondson paper also cites a specimen from **YRK** with almost the identical text, where Don 'was evidently the collector'.		

Sagina subulata

* 'Saxifraga graminea pusilla foliis brevioribus crassioribus et succulentioribus D. Lawson. On Whinny field-bank by Cullercoats near Tinmouth in Northumberland'. - Ray (1688a: 20, 1690: 146).	vc67	**1688**
Whittaker (1986) identifies this record as referring to *S. maritima* and indeed Baker & Tate (1868), Raven (1950) and Swan (1993) agree. But Hudson (1762: 64) and Smith (1824; 339) both cite the Ray polynomial for this species, omitting his locality.		

Sagittaria sagittifolia

* 'Pistana Magonis, sive Plinii, Sagittaria aquatica ... In Anglia prope Oxonium pone moenia ... etiam Londini ad arcis Regiae vallum [Tower of London] & in Tamesis crepidinibus paludosis'. - L'Obel (1571: 126).	vc21, 23	**1571**

Salicornia agg.

* 'Kali ... very plentuous in many places of England ... Saltwurt ... Glas wede'. - Turner (1568: 37).		**1568**
'in Anglia' - annotations made by T. Penny on C. Gesner's plant illustrations dated as by 1565 in Foley (2006b)		1565

Whilst there is no doubt that, among others, Ray, Smith (1824: 2-3), Babington and Woods (1851), were aware of several different species, I feel that it was not until Moss (1912) and particularly, Ball & Tutin (1959), that our current understanding evolved.

Salicornia dolichostachya

'Found at North Bull, Dublin'. - Moss (1912).	vcH21	**1912**
Hunstanton, vc28, 1895, A. Hosking, det. Sell, **CGE**.	vc28	1895

There is an earlier record in CGE, from Wisbeach, 1831, Henslow. This was determined as this species by Wilmott, but as ?*fragilis* by Sell. But note also the specimen figured in E.B. 2475 (1813), 'sent from Yarmouth [vc. 27] by Mr Turner', which *Fl. Europaea* equates with either this species or *S. fragilis*. The account in E.B. adds that 'Mr W. Borrer has long remarked this as a new *Salicornia*', presumably from Sussex..

Salicornia emerici

as *S. nitens*, Ball & Tutin (1959).		**1959**
North Hayling, 1957, P.W. Ball.	vc11	1957

There is an earlier record from Shingle Street, Hollesley, vc25, F.W. Simpson, 1950, which is within the range of the species, and quite possible, but has not, to my knowledge, been determined.

Salicornia europaea

as *S. herbacea*, Moss (1912)		**1912**
Frindsbury [Rochester], 7/9/1825, H.D. Henslow, **CGE**.	vc16	1825

Salicornia fragilis

Ball & Tutin (1959).		**1959**
Maldon, vc18, 1827, L. Blomefield, det. Sell, **CGE**.	vc18	1827

Salicornia obscura

Ball & Tutin (1959).		**1959**
North Hayling, vc11, 1957, P.W. Ball, **BM**.	vc11	1957

Salicornia pusilla

Ball & Tutin (1959).		**1959**
Woods (1851)		1851

Although Woods (1851) had described this as a new form, and Moss (1912) had followed this, the key point of the single flower remained unelaborated until Ball & Tutin's paper. Woods (1851), postulating six species or forms, separated this species from the *S. procumbens* aggregate, and attempted to fit into his synonymy those described by earlier writers.

Salicornia ramosissima

'*S. ramosissima* … this name is applied to a form gathered last October in Haling Island'. - Woods (1851).   vc11   **1851**

'Kali ramosius, erectum, foliis brevibus, Cupressiforme. In a Salt Marsh on ye east side of Poole, Dorset. Found by Mr Stonestreet'. - Druce (1928: 482).   vc9   1716

Druce's citation includes both *S. appressa* and *S. ramosissima*. Stonestreet died in 1716.

Salix alba

* 'Salix, Common willow. Willowes grow in divers places of England'. - Gerard (1597: 1203).   **1597**

See also Turner (1548: F. viij, under 'Salix'), where two or three species of Willow are described or referred to. Stearn (1965) refers 'Particularis salix is the greate Wylowe tree' to this species.

Salix arbuscula

as *S. myrsinites* 'Upon the Highland mountains as upon Ben Achulader in Glenurchy [Beinn Achaladair in Glen Orchy], Malghyrdy [Meall Ghaordie], in Breadalbane, and on Craig-vore [?Creag Mhor SW of Beinn Heasgarnich]'. - Lightfoot (1777: 599).   vc88, 98   **1777**

as *S. vacciniifolia* 'found [by Walker] in July 1762 on the east side of the high mountain of White-Coom-Edge, in the head of Annandale'. - Taylor (1959).   vc72   1762

Meikle (1984) points out that Lightfoot's records, cited as the first record for *S. myrsinites* in both Clarke and Druce, are all of this species (a fact already realised by Smith, in E.B.1366).

Salix aurita

* 'Salix folio rotundo minore … in sepibus prope Chisselhurst … Dr. Dillenius'. - Ray (1724: 450).   vc16   **1724**

Salix caprea

* 'Salix Caprea rotundifolia. The Goat round leafed Willow. Willowes grow in divers places of England' - Gerard (1597: 1204.3).   **1597**

Salix cinerea

'Salix folio ex rotundo acuminato auriculata. Common Sallow'. - Ray (1660: 145). Ray, who clearly separated this from *S. caprea*, includes a very detailed description which leave one in little doubt as to the identification. See Oswald & Preston (2011: 285).   vc29   **1660**

Moss (1914) cited Ray (1724: 449.16) as the first record, but it is obvious that Ray was describing the same tree in 1660. Clarke preferred 'Specimens from Cumberland and from Fream Wood are in Mr. Lightfoot's herbarium,' - Smith (1804: 1063). Druce & Vines (1907: 29) cited three early specimens, all dating from before 1747: 'A tree on ye road near Lee (Kent). In a hedge near Beaumaris (Anglesey). In the meadow a little beyond the Harrow'.

Salix fragilis

* 'Salix folio lata splendente fragilis … (unde vulgo Cantabridgiensi The Crack Willow dicitur)'. - Ray (1660: 143).   vc29   **1660**

This is now considered a hybrid between *A. alba* and *S. euxina*'.

Salix herbacea

'Salix humilis saxatilis, sive Chamaeitea repens montana rotundifolia. Round leaved Mountaine Willow'. - Johnson (1641: 32).   vc49   **1641**

Seen by Thomas Johnson and party, on Snowdon, Aug 1639. See Raven (1947: 289).   vc49   1639

Clarke, Druce and Moss (1914) all prefer 'Salix pumila folio rotundo *J.B*. ... On Ingleboroughhill on the highest Rock next to the Beacon; and on a hill called Whern-side over against Ingleborough-hill on the other side the subterraneous River. Th. Willisel'. - Ray (1670: 273). However the Johnson record seems more than likely to be this species.

Salix lanata

* 'On the rocks in the Highlands of Scotland. Found by Mr. T. Drummond on rocks among the Clova Mountains, *Mr W. Robertson*'. - Smith (1828: 205).   vc90   **1828**

'First found by Don in 1812, in Glen Callater'. See a letter to Mr D. Booth, dated Nov. 1812, reproduced in Druce (1904: 67).   vc92   1812

David Don too (under E.B.S. 2666, *Carex Vahlii*), says his father found this spcies in Glen Callater'. - Druce (1904).

Salix lapponum

* 'On the Highland mountains as on Creg-chaillech [Creag na Caillich] and Mal-ghyrdy [Meall Ghaordie] in Breadalbane, &c. Mr. Stuart'. - Lightfoot (1777: 604).   vc88   **1777**

'It was first observed in July, 1762, among the rocks on the north side of the mountain, which lies south from the foot of Loch-Skene [Skeen] in the parish of Moffat'. - Walker (1812: 447). vc72 1762

Walker's record is given in more detail in Taylor (1959), as 400ft below *S. vacciniifolia* [*arbuscula*] on White-Coom-Edge, in the head of Annandale. There is an entry for this species 'Mr John Stewart', with no other details, in Hope, 1768, in Balfour (1907).

Salix myrsinifolia

as *S. nigricans* 'in Wrongay Fen [?Wormegay], Norfolk and in osier grounds in other places, not uncommon'. - Smith (1802: 120). vc28 **1802**

See Meikle (1984) for an illuminating discussion on the critical nature of the species and on the possibilities of this record. If the Norfolk record is not accepted then the first satisfactory account is, as *S. Andersoniana,* 'we are obliged to Mr Anderson ... [who reported it] found in various parts of Scotland'. - E.B. 2343 (1812). Smith's account adds that the Rev. Walker used to call it *S. dalbensis*, which implies an earlier collection, but I have found no trace of this name in Walker (1812).

Salix myrsinites

In a tour through the Highlands in 1789 … *S. retusa*, *Linn. Spec. Plant. 1528*, in rupibus siccis, Ben Lawers'. - Dickson (1794). vc88 **1794**

There is an entry for this species 'Mr John Stewart', with no other details, in Hope, 1768, in Balfour (1907). vc88 1768

Meikle (1984) points out that Lightfoot's records (1777: ii. 599), cited as the first record for this species in both Clarke and Druce, are all of *S. arbuscula* (a fact already realised by Smith, in E.B.1366). But from Stuart's [Stewart] observations on its occurrence on Schiehallion, in a letter to Sir Robert Menzies, 25th Feb. 1777 (Banks correspondence, ADD. MS. 33977. 65-72, **BM**) it seems that his (Stuart's) identification was correct (see also Mitchell 1986).

Salix pentandra

* 'Salix folio laureo, sive lato glabro odorato folio, nondum descripta. Willow Bay'. - Johnson (1641: 32). 'At Wolverhampton'. - How (1650: 108). **1641**

Seen by Thomas Johnson and party, near 'Vulcani municipium' [Wolverhampton], July 1639. See Raven (1947: 288). vc39 1639

Salix phylicifolia

* 'At Finlarig [N. of Killin], Bredalbane, Rev. Mr. Stuart'. - Smith (1802: 123). vc88 **1802**

Smith cites a specimen sent to Mr Lightfoot by Mr Stuart, of Luss.   vc88   1788

Under folder 5 of *S. arbuscula* 'Ben-challum [Ben Challum] Oct. 19th [Stuart]', in the catalogue of Lightfoot's Herbarium. - Bowden (1989). The date of the collection and annotation are not known but Lightfoot died in 1788, and this is possibly the specimen that Smith cites.

Salix purpurea

'Salix humilior foliis angustis, subcaeruleis, ut plurimum sibi invicem oppositis ... By the horse-way side to Cherry-Hinton'. - Ray (1660: 144).   vc29   **1660**

Clarke & Druce both cite 'S. folio longo non auriculato vimine rubro ... In the osiar-holts by the river Cams side'. - Ray (1660: 146). Oswald & Preston (2011: 286n) give their reasons for rejecting this and preferring the entry that I have cited.
Another early record is 'Thorpe Meadows near Norwich, Mr. Crowe'. - Withering (1787: ii. 1102).

Salix repens

* 'Salix humilis I found the dwarfe Willowes growing neere to a bog or marish ground at the further end of Hampsteed Heath'. - Gerard (1597: 1205.6).   vc21   **1597**

Salix reticulata

* 'Upon many of the Highland mountains, in a talky soil, such as upon Creg-chaillech, Mal-ghyrdy and Mal-grea mountains, in Breadalbane, &c.'. - Lightfoot (1777: 601).   vc88   **1777**

Mitchell (1986) argues that if the first record was from Creag na Caillich then it could equally have been James Stuart with James Robertson in 1771 or with Lightfoot in 1772 .   vc88   1772

Lightfoot's locations seem to equate to Creag na Caillich, Meall Ghaordaidh and possibly Meall Greigh at the eastern end of the Lawers range. Ray's records (1724: 449.13) quoted in Smith (1824) refer to *S. herbacea*.

Salix triandra

'Salix folio splendente auriculato flexilis'. - Ray (1660: 144). Ray includes a very detailed description which leaves one in little doubt as to the identification.   vc29   **1660**

Both Babington (1860b) and Oswald & Preston (2011: 284) follow the attribution in Ray (1660). Clarke had 'Salix humilis, corticem abjiciens. Near the small brook that runs into the River at Darking in Surrey ... Th. WilliseII'. - Ray (1670: 272), and it is slightly odd that Ray used a completely different polynomial only 10 years later.

Salix viminalis

'Salix folio longissimo. The Osiar'. - Ray (1660: 146)   vc29   **1660**

Both Clarke and Druce cite 'Viminalis ... Osyer tree ... that baskettes are made of'. - Turner (1548: C. i). Clarke adds 'Salix vulgaris longis & angustis foliis ' Hampstead. - Johnson (1632: 35). Edgington (2014) cites Johnson (1638: 23) 'Salix humilis angustifolia, Οἶσος, *Theophr.* Viminalis Dod. amerina, *Plin.* The Oysier'. All are possibilities, but Moss (1914) gives Ray, who wonders if earlier writers were mistaken, as his first record.

Salsola kali

* 'These herbes [*Kali and Salicornia*] grow in saltish groundes, by the sea side, or coaste, in Zealand, and England' - Lyte (1578: 116).   **1578**

Salvia pratensis

'Horminum pratense foliis serratia *C.B.P.* ... In agri Cantiani Vivario Cobhamensi [Cobham]'. - Plukenet (1696: 185).   vc16   **1696**

Salvia verbenaca

* Under Orminum 'horminum sylvestre ... a nostris etia arbitror dici Clare aut Wylde clare'. - Turner (1538: B. iiij).   **1538**

Two other early localised records are 'at Greenwich near the race-course of the Palace, 1579'. - L' Ecluse (1601: xxxi), and 'In the fieldes of Holburne neere unto Graies Inne'. - Gerard (1597: 628).

Sambucus ebulus

* 'Ebulus ... Walwurt or Danewurt ... groweth abrode in Cambryge fieldes in great plentie'. - Turner (1548: C. viij).   vc29   **1548**

See Turner (1538: A. iv) 'Chameacte … ab angli Danwort aut walwort vocatur', which implies familiarity with the plant as a wild species.

Sambucus nigra

* 'Sambucus ... ab anglis an Elder tree, aut a bour tre vocatur'. - Turner (1538: C).   **1538**

Samolus valerandi

* 'Anagallidem aquaticum tertiam, *Lob.*'. From a saltmarsh near Dartford. - Johnson (1629: 9).   vc16   **1629**

'Anagallis aquatica tercia. … by a mill at Emsworth, 1618'. - Goodyer ms.11, f.51, see Gunther (1922: 111).   vc11   1618

Rose, in Gilmour (1972: 63), suggests this is *Veronica catenata* (and does not mention *Samolus*). Yet on p.115, with the same citation, in Johnson (1632: 17) he equates this to *Samolus*!

Sanguisorba officinalis

* 'Bipennella italica ... Burnet ... groweth much about Syon and Shene and in many other places of England'. - Turner (1548: H. j. verso).

vc17, 21   **1548**

Britten (1887) referred this to *S. minor*, but all other commentators follow Clarke.

Sanicula europaea

* 'Sanicula ... groweth communely in woddes'. - Turner (1548: H. iiij). 'Sylvarum copiosissima & notissima Anglis & Germanis'. - L'Obel (1571: 297).

1548

Saponaria officinalis

* 'Saponaria ... groweth wilde of itselfe neere to rivers and running brooks in sunnie places'. - Gerard (1597: 360). Between Erith and Gravesend. - Johnson (1629: 3).

1597

Ray, J. 1688a. *Fasciculus stirpium Britannicarum, post editum Plantarum Angliae Catalogum observatarum.*

See species account: *Sarcocornia perennis*, p.359

Sarcocornia perennis

* 'Kali geniculatum majus, siva nova Species Kali perennis a D. Sloane observatum est prope Insulam Shepey'. - Ray (1688: 1857, 1688a: 13) - *see image p.358*. vc15 **1688**

Clarke (1909) adds 'This is the "Perennial Kali" found near King's Ferry in Sheppy, referred to by Sloane in letter to Ray dated Aug. 10, 1686. See Lankester (1848: 186)'. vc15 1686

Saussurea alpina

* 'Carduus mollis flo. caeruleo, an Cardus mollis foliis Lapathi, *Clus. Ger. emac*.? Thistle Gentle. This growes on the Rockes on the highest part of Snowdon'. - Johnson (1641: 18). vc49 **1641**

Seen by Thomas Johnson and party, on Snowdon, Aug 1639. See Raven (1947: 289). vc49 1639

Saxifraga aizoides

* 'Sedum Alpinum luteum minus. On the sides of Ingleborough-hill ... I found it also in the like places about Shap in Westmorland'. - Ray (1670: 279). vc64, 69 **1670**

'Sedi alpini parva species - on one small spot of ground about Shap in Westmorland'. - Ray's letter to Mr Willughby, 14th Sept 1661, see Lankester (1848: 3). vc69 1661

See also Lankester (1846: 163) from on Ray's 1661 Itinerary 'August .. about Shap .. Sedi quaedam species flore luteo'. However Lees (1888: 248) gives the first record as Merrett (1666: 111) 'Sedum sive illecebra fol. oblongis. On the North side of Ingleborough hill, neer a bog by the side of an underground river'. Lees adds that Ray (1724: 270) assigned this to *Sedum villosum*, but states that was an error. Pulteney, presumably following Ray, also equates this with *S. villosum*.

Saxifraga cernua

* 'Amongst the rocks on the summit of Ben Lawers'. - Dickson (1794). vc88 **1794**

'Discovered by Mr Townson on rocks on the summit of Ben Lawers, 1790'. - Hooker (1821: 130). vc88 1790

Dickson's visit was in 1792, and his paper read to the Linnean Soc. in February 1793. Robert Brown collected the plant in August 1793 (Mabberley 1985).

Saxifraga cespitosa

* 'On alpine rocks above lake Idwell in Carnarvonshire, rare, J. W. Griffith, Esq., in Herb. Soc. Linn'. - Smith (1800: ii. 455). vc49 **1800**

'in 1778 [found by Griffith] on alpine rocks above L. Idwell [L. Idwal]'. - Carter (1955a) and Jones (1996). vc49 1778

In Henderson & Dickson (1994: 20) 'this is the most probable identity for the "S...c..." recorded by Robertson on Ben Avon in 1771. The plant still occurs there, but there is no specimen listed for Hope's herbarium'. Elsewhere they write of the competition to find new plants, and the poaching of others' records, explaining the use of codes for secrecy - see notes under the entries for *Phyllodoce* and *Schoenus ferrugineus*.

Saxifraga granulata

* 'Saxifrage' ' ... whyte floures growe in the toppes, the rote is full of litle knoppes lyke pearles ... in diverse places of England'. - Turner (1568: 67).		**1568**

Saxifraga hirculus

* 'Geum angustifolium autumnale flore luteo guttato ... Found by Dr. Kingstone on Knotsford-moor, Cheshire and there shew'd to Dr. Richardson, by him now growing in plenty'. - Ray (1724: 355).	vc58	**1724**
Clarke (1909) adds 'This was found at Knutsford at least as early as 1720. Sherard, writing to Richardson on March 28th, 1721, says "I had a letter lately from Dr. Fowlkes [Robert Foulkes, of Llanbedr], who ... says *Saxifraga angustifolia fl. luteis punctatis*, Breyn. Cent., was found near Knutsbridge [*sic*] Mills in Cheshire". - *Richardson Corresp.*168', see Turner (1835: 168).	vc58	1720
Richardson saw it himself in 1721 (Richardson to Sloane, 18th August 1722, British Library: Sloane ms. 4046, f. 284). De Tabley (1899: lxxxv) gives 1717 as the date of discovery, but with no further details. Lusby & Wright (1996) give 'sometime between 1696 and 1720'.		

Saxifraga hirsuta

* 'On the mountains of Keri [Kerry] Sanicula guttata grows in abundance'. - Lhwyd (1711a).	vcH1	**1711**
Lhwyd's visit to S.W Ireland was in 1700, and the letter, which forms the article referred to, is dated August 25th, 1700.	vcH1	1700

Saxifraga hypnoides

* 'Sedum Alpinum laciniatis Ajugae foliis ... On the Mountaines of Lancashiere with us as Mr. Hosket [Hesketh] told us'. - Parkinson (1640: 739).	vc64	**1640**
Hesketh died in 1613. Also seen by Thomas Johnson and party, on Snowdon, Aug 1639. See Raven (1947: 289).	vc64	1613
Hesketh's record is more likely to be in Yorkshire, since this species has always been very rare indeed in vc60 and early writers often referred to the W Yorks. hills as in Lancashire.		

Saxifraga nivalis

* 'Sedum serratum, sive Umbilicus Veneris alter *Matth*. The Princes Feather'. - Johnson (1641: 33). 'On Snowdon hill'. - Mcrrett (1666: 111).	vc49	**1641**
Seen by Thomas Johnson and party, on Snowdon, Aug 1639. See Raven (1947: 289).	vc49	1639

Saxifraga oppositifolia

* 'Sedum Alpinum ericoides purpureum vel caeruleum ... In rupibus ad latus Septentrionale montis Ingleborough'. - Ray (1677: 269).	vc64	**1677**
'Sedum ericoides J.B. in summitatis montis Ingleborough' in Letter from Ray to Mr Lister, September 1668. - Lankester (1848: 26) and also Raven (1950: 148) and Roos (2015).	vc64	1668

Saxifraga rivularis

* 'In monte Ben Nevis Scotiae hanc plantam primus in Britannia invenit anno 1790, D. Robertus Townson. *J.E. Smith*'. - Linnaeus (1792: 143).	vc97 vc97	**1792** 1790

Saxifraga rosacea

* as *S. petraea* 'We are indebted to ... J. Wynne Griffith, Esq., for this beautiful addition to our flora. He found it on the rocks of Cwm ldwell, above Llyn ldwell, near Twll dû'. - Withering (1796: iii. 890).	vc49	**1796**
See E.B. 455 (as *S. palmata*).		

Saxifraga spathularis

* 'Cotyledon, sive Sedum serratum Latifolium Montanum guttato flore Parkinsoni & Raii, vulgarly called by the gardners London Pride ... Grows plentifully here with us in Ireland on a mountain called the Mangerton in Kerry'. - Molyneux (1697: 510).	vcH2	**1697**

Under *S. umbrosa*, Colgan & Scully (1898), Praeger (1909a) and Walsh (1978) all suggest that it was this plant referred to in How (1650), but with no details or locality, and that How may have had the information from Richard Heaton. But none of those authors suggest under which name How included it, though it might be his 'Sedum serratum, sive Umbilicus Veneris, alter. *Matth*. Cotyledon altera, 4. *Clus. Hist.* The Princes Feather'. - How (1650: 111) - but see *S. nivalis*.

Lhwyd (Lhwyd to Richardson, 21st October 1700, British Library: Sloane ms. 4063, f.48.) seems to call the plant 'Sedum serratum foliis pediculis etc'.

Saxifraga stellaris

* 'Cotyledon hirsuta sive sedum petraeum hirsutum. Hairy Kidney-wort … Upon the moyst Rockes at Snowdon'. - Johnson (1641: 19).   vc49   **1641**

Seen by Thomas Johnson and party, on Snowdon, Aug 1639. See Raven (1947: 289).   vc49   1639

Saxifraga tridactylites

* 'Paronychia rutaceo folio … Upon bricke and stone wals, upon olde tiled houses, which are growen to have much mosse on them … upon the bricke wall in Chancerie Lane belonging to the Earle of Southampton in the suburbes of London'. - Gerard (1597: 500).   vc21   **1597**

The Gerard entry is certainly for this species, but the location for Chancery Lane might apply to this species or to *Erophila verna*. Kent (1975) identifies it as this species.

Scabiosa columbaria

* 'Scabiosa tenuifolia'. Between Gravesend and Rochester. - Johnson (1629: 3).   vc16   **1629**

But see figure and description of 'S. minor, sive columbaria … growe in pastures, medowes … almost everywhere'. - Gerard (1597: 582.2). L'Obel (1571: 232), favoured in Pulteney, has 'on the edges of fields in northern regions …'. This latter seems less likely.

Scandix pecten-veneris

* 'Scandix … Pinke nedle or storkes bill. And I have judged it to be an herbe ye groweth in ye corne with a fayre whyte floure'. - Turner (1562: 130).   **1562**

Schedonorus arundinaceus

* 'Gramen arundinaceum aquaticum panicula avenacea … a D. Sam. Doody … ad ripas Thamesis fluvii inter Londinum et Chelseiam observatum ad nos siccum transmissum est'. - Ray (1688: 1909; 1688a: 9).   vc21   **1688**

Schedonorus giganteus

* 'Gramen avenaceum glabrum panicula e spicis rarioribus strigosis composita, aristis tenuissimis … Fulhamiae prope Londinum in aggere inter locum excensionis e cymbis & Episcopale palatium a D. Doody observatum & ad nos transmissum est'. - Ray (1688: 1909; 1688a: 9).   vc21   **1688**

Schedonorus pratensis

* 'Gramen paniculatum elatius, paniculis seu spicis muticis, squamosis ... in pratis'. - Ray (1670: 153).   **1670**

Scheuchzeria palustris

* 'Discovered by the Rev. [James] Dalton in June, 1787, growing abundantly, along with *Lysimachia thrysiflora*, in Lakeby Car, near Borough-bridge, Yorkshire'. - E.B. 1801.   vc62   **1807**

There is a note in *Wats. BEC.* (1917, 3: 35) citing specimens from 1787 (and 1790 & 1807) in Herb. Yorks. Phil. Soc.   vc62   1787

Schoenoplectus lacustris

* 'Juncus aquaticus maximus ... In standing pooles and by rivers sides'. - Gerard (1597: 31). 'Harefield' (Middlesex). - Blackstone (1737: 46).   **1597**

Pulteney cites L'Obel (1571: 44) as his first record. Here, under 'Aquaticus maximus', L'Obel has 'Extremely well known everywhere, even in northern regions, principally for [its] spongy pith for weaving rush-mats for spreading [on floors?]'. Whether Britain, or N. Europe was the 'northern region' is not known.

Schoenoplectus pungens

'Juncus acutus maritimus caule triquetro, rigido, mucrone pungente *Pluk. Alm. 200. T 40*. ... In insula Jerseia, D. Sherard'. - Ray (1724: 429).   vc113   **1724**

Sherard's visit was in 1689, but it is odd that this record was not published until 1724.   vc113   1689

The English, presumed alien site, at Southport, Lancs, was reported in 1928 (Druce 1929). But in fact the plant was collected by Travis in 1909, and not identified until 1928 - see Smith (2005).

Schoenoplectus tabernaemontani

* 'Juncus aquaticus medius *C.B. Park*. sylvaticus Tabernaemontani *J.B.* ... In the Sea-ditches at Bricklesey [Brightlingsea] and Mersey-Island. Mr. Dale'. - Ray (1696: 273).   vc19   **1696**

But see 'Juncus sive Scirpus medius, sylaticus *Tab*. The lesser Bull-rush'. - Johnson (1634: 44; 1638: 13, from Tottenham) and 'Juncus laevis vulgaris ... this rush is in all things like the former [*S. lacustris*] but much lesser and shorter'. - Parkinson (1640: 1191). Edgington (2014) makes a persuasive case for the identification as this species, pointing out that Ray (*op. cit*.) uses the same polynomial. Oswald & Preston (2011: 226n) are more cautious, but possibly only concerning the record in Cambridgeshire.

Schoenoplectus triqueter

* 'Juncus caule Triangulati, At the Horse-ferry at Westminster, Hunc mihi primus ostendit Dr. Dale insignis Britannicus'. - Merrett (1666: 67).   vc21   **1666**

'by ye horsferry [Westminster]'. - How, manuscript addition to his copy of *Phytologia Britannica*, cited in Gunther (1922: 286). How died in 1656.   vc21   1656

Kent (1975) erroneously gives the How record as 1650.

Schoenus ferrugineus

* 'Collected beside Loch Tummel in July last by Mr. [James] Brebner'. - White (1885).   vc88   **1885**
   vc88   1884

White wrote in July 1885 that he had received the specimens in July last, that is in 1884.

There is an intriguing record from James Robertson (cited in Henderson & Dickson (1994: 186, n.60) where he records, on Aug. 27th 1771, 'On the hills between Megarnie [Meggernie Castle, in upper Glen Lyon] and Ranach [Rannoch] I found the S Rush, which I have never met with before and which no Botanist has numbered among the British plants'. Robertson used code for his rarest finds (see entries for *Phyllodoce* and *Sax. caespitosa*), but the nearest known record for this species was at L. Tummel, 15 miles to the east.

Schoenus nigricans

* 'Juncus palustris panicula glomerata ex rubro nigricante ... Every where in the watery places of Hinton and Teversham Moors'. - Ray (1660: 82).   vc29   **1660**

Scilla autumnalis

* 'Hyacinthus autumnalis minor ... groweth wilde in many places of England. I gathered divers rootes for my garden from the foote of a high banke by the Thames side at the hither end of Chelsey before you come at the Kings Barge-house'. - Parkinson (1629: 132).   vc21   **1629**

For the location see the note under *Butomus umbellatus*.

Scilla verna

* 'Hyacinthus stellarius vernus pumilus, *Lob*. At the Kings-end neere Dublin, Mr. Heaton'. - How (1650: 60).   vcH21   **1650**

In common with other species first found by Richard Heaton but not published till 1650 by How, this is reckoned by Walsh (1978) to have been discovered by 1641.   vcH21   1641

Scirpoides holoschoenus

* 'Juncus acutus maritimus capitulis rotundis *C.B.* ... Nuper etiam in Anglia detexit in comitatu Somerseti D. Stephens'. - Ray (1688: 1303). 'Braunton burroughs, Mr Stevens'. - Ray (1688a: 13 & 1696: 272).   vc4   **1688**

There are old records of this plant from near Minehead (see Murray, 1896), but Smith (1824: i. 58) gives Stephens' record as from Braunton Burrows in Devon. Martin & Fraser (1939) give the first record for Devon as Ray in Gibson (1695) where he does indeed say 'found by Mr Stephens in Braunton-buroughs in this county [Devon]'.

Scirpus sylvaticus

* 'Cyperus gramineus, sive miliaceus, *Lob*'. Hampstead Heath. - Johnson (1632: 36).   vc21   **1632**

'on ye west parte of Gloster Hall by Oxford, July 6, 1622'. - Goodyer ms.11, f. 7v, see Gunther (1922: 176).   vc23   1622

Scleranthus annuus

'Polygonum selinoides, *Ger*. sive Knawel Germanorum; an Vermiculata nova planta montana, *Col?*'. Around Birchington. - Johnson (1632: 13).   vc15   **1632**

'Polygonum germanis. 20 Julii 1618 ... in a barren field of rye belowe Tichfield Bay'. - Goodyer ms.11, f. 56v, see Gunther (1922: 110).   vc11   1618

There has been a long-standing argument over the record of L'Obel's (1571: 183) of 'Anglica saxifraga ... acclivem cretaceum, & aridum montem, arte militari aggrestum inter Chipnam & Malburú, Angliae, Bristoliense à Londino via'. Some take it to be this species, others *Asperula cynanchica*. I have chosen the latter, but the matter is unresolvable.

Clarke and Druce both cite 'Saxifraga Anglicana, Alsines minimum genus Daleschampii polygonum selinoides Gerardi'. Between Gravesend and Rochester. - Johnson (1629: 2), but this reference had already been cited for *Sagina procumbens*! Rose, in Gilmour (1972: 52), ignores this but accepts (1972: 112) the 1632 record.

Scleranthus perennis

* 'Polygonum Germanicum incanum flore majore ... in the sandy grounds and balks of corn-fields about Elden in Suffolk plentifully'. - Ray (1677: 239).   vc26   **1677**

Scorzonera humilis

'In some plenty, in a moist grassy field, bordered by heathland, in Dorset, Mrs and Noel Sandwith'. - Druce (1916a).   vc9   **1916**
    vc9   1915

There is a slight puzzle over the date of discovery. Druce gives 1914, as does Bowen (2000). Yet there is nothing in the B.E.C. report (which is *Plant notes for 1915*) to suggest that it was in fact collected the year before.

Scorzoneroides autumnalis

* 'Hieracium minus praemorsa radice … pratis montosis Angliae admodum familiare'. - L'Obel (1571: 88). Near Rochester. - Johnson (1629: 2).   **1571**

Scrophularia auriculata

* 'The common water betony groweth commonly about water sydes'. - Turner (1551: L. ij. under 'Clymenum').   **1551**

Turner goes on to describe another 'betony', growing in woods and hedges, but in Germany, which sounds like *S. nodosa*, with a picture.

Scrophularia nodosa

* 'Scrophularia major … In greatest abundance in a wood as you go from London to Harnesey [Hornsey] and also in Stowe Woode and Shotouer neere Oxenforde'. - Gerard (1597: 580).   vc18, 23   **1597**

See the note on Turner (1551) above.
Pulteney cites Lyte (1578: 43.1 & 2), who gives two species, '[both] grow very plentifuly in this countrey, in the borders of fieldes, and under hedges, and about lakes and ditches'. But of course we do not know whether he was referring to this country or to Europe.

Scrophularia scorodonia

* 'Scrophularia Scorodoniae folio *H.R. Bl. & Hist.Ox. II. 482*. By the Rivulets Sides betwixt the Port and S. Hilary in the Isle of Jersey; Dr Sherard, (and since by Mr. Edward Lhwyd near the Sea-shore about St. Ives in Cornwal'. - Ray (1724: 283*).   vc1, 113   **1724**

Sherard was in Jersey in 1689.   vc113   1689

The first mainland record is ' Dr Sherard's Scrophularia Scorodoniae folio; Pensans and St Ives'. - Lhwyd (1711). Lhwyd's letter, which forms the article referred to, is dated September 22nd 1700.

Scrophularia umbrosa

* as *S. ehrharti* 'Edinburgh, Mr. W. H. Campbell. Cramond Woods, West Lothian, Dr. A. Hunter'. - Stevens (1840).   vc83, 84   **1840**

as *S. aquatica* 'On the sides of the Whiteadder below Claribad mill [c.5 miles W. of Berwick], plentiful; Mr R. Dunlop'. - Johnston (1833). Leighton (1840: 300), in attempting to separate from *S. auriculata* and from the descriptions in Linnaeus, describes a specimen gathered in 1835, at Sutton, near Shrewsbury.   vc81   1833

Scutellaria galericulata

* 'Lysimachia Galericulata caeruleo - purpurea. Rivularum & fluminum ripas & depressos agrorum margines Londinensis sequitur'. - L'Obel (1576: 186). 'This I found in a waterie lane leading from the Lord Treasurers house called Thibals unto the backside of his slaughter house'. - Gerard (1597: 387).   vc21   **1576**

Scutellaria minor

* 'Gratiola. Hedge Hyssope ... I found it growing upon the bog or marrish ground at the further end of Hampsteed heath, and upon the same heath towards London, neere unto the head of the springs that were digged for water to be conveied to London 1590, attempted by that carefull citizen Sir John Hart, Knight, Lord Maior of the Citie of London: at which time my selfe was in his Lordships company, and viewing for my pleasure the same goodly springs, I found the said plant, not heretofore remembred'. - Gerard (1597: 466.2 not the figure).   vc21   **1597**

Sedum acre

* 'Sedum ... minus puto esse herbam quam vulgus appellat Thryft aut Stoncrop'. - Turner (1538: C. verso).   **1538**

Sedum album

* 'Sedum minus flo. albo ... In locis saxosis et asperis. White-flowred Prick-madame'. - Johnson (1634: 67). 'Very plentifully on many of the thatch'd houses in Chatteresse in the Isle of Ely'. - Ray (1660: 153).   **1634**

Sedum anglicum

* 'Sedum minimum flo. mixto ex albo & rubro. On the West of Ingleborough hill in the mud of the hollow topped stones'. - Merrett (1666: 111). 'In sterilioribus Suffolciae itinere à Yarmouth ad Dunwich plurimum observavimus. Nec minus abundat in rupibus Westmorlandicis & Lancastrensibus prope lacum Winandermere'. - Ray (1670: 280).   vc64   **1666**

Sedum forsterianum

* 'Sedum Divi Vincentii [Bristol] N. D. Mr. Goodyer'. - Merrett (1666: 111). — vc34 — **1666**

Gunther (1922: 77) refers to this in describing a visit by Goodyer to St Vincent Rocks in 1638 - apparently his only visit. — vc34 — 1638

Ray visited the site in 1662, on his third itinerary, see 'Thursday, June 19th … on St Vincent's rock … Sedum medium'. - Lankester (1846: 180) Chapman *et al.* (1995: 563, n.270) wonder if Turner (1562: 131) is including this species in his descriptions of kinds of *Sempervivum*, but this can only be conjecture.

Sedum rosea

'Rhodea radix … in Anglia in monte Engleborreno [Ingleborough] dicto … D. Thomas Pennaeus Anglus'. - Camcrarius (1588: 139). — vc64 — **1588**

The record comes from Foley (2009). Clarke and Druce both cited Gerard (1597: 426) 'Rhodia radix … Upon sundry mountains in the north part of England especially in a place called Ingleborough Fels'.

Sedum telephium

* 'Crassula sive Faba inversa … Plentifully in … Englande'. - Gerard (1597: 416). — **1597**

But see Turner (1538, C. verso) 'Telephium … vocant vulgus Orpyne appellat'. Orpine has been grown for centuries as a garden plant.

Sedum villosum

* 'Sedum parvum palustre flo. incarnato … On the moist springs about Ingleborough-hill and elsewhere, as v.g. Hartside-hill near Gamblesby in the way to Often Cumber [Hartside, Cumberland]'. - Ray (1677: 270). — vc64, 70 — **1677**

This was on Ray's trip to the north of England in 1671, with T. Willisel. See Raven (1950: 154). — vc64 — 1671

Pulteney cites Merrett (1666: 111) 'Sedum sive illecebra fol. oblonga … on the north side of Ingleborough hill, near a bog by the side of an underground river' as this species, but see *Saxifraga aizoides*.

Selaginella selaginoides

'Muscus polyspermus, On the North of Ingleborough near the under-ground River'. - Merrett (1666: 79). — vc64 — **1666**

On Ray's first Itinerary 'Sept 5th [1658] … an hill called Caderidris … another small club moss with white seeds'. - Lankester (1846: 130). — vc48 — 1658

Selinum carvifolia

* 'near Broughton Woods, N. Lincolnshire, July 1880'. - Fowler (1881).

See also Lees (1882) for many more details of the find. Brown (in Syme, 1892) is highly sceptical of its claims to nativity.

	vc54	**1881**
	vc54	1880

Senecio aquaticus

* 'Jacobaea latifolia *Ger. emac.* ... in moist & watery places'. - Ray (1660: 80).

vc29   **1660**

Senecio cambrensis

'... a large radiate grounsel was received from Mr H.E. Green, who had seen similar plants, growing by a roadside in Flintshire [at Ffrith], since 1948'. - Rosser (1955).

| vc51 | **1955** |
| vc51 | 1948 |

Possibly it originated earlier as there is a 1925 specimen in **OXF** from Denbigh (det. Rosser) but this may be a fertile tetraploid hybrid (Ingram & Noltie, 1995). They conclude that '... there is little doubt that the species in Wales originated between 1925 and 1948'.

Senecio eboracensis

'First recorded in 1979, near York railway station'. - Lowe & Abbott (2003).

| vc64 | **2003** |
| vc64 | 1979 |

Apparently now extinct.

Senecio erucifolius

'Jacobaea minor foliis magis dissectis'. Margate to Nash. - Johnson (1632: 10).

vc15   **1632**

Clarke cited 'Jacobaea Senecionis folio incano perenne ... In aggeribus sepium & dumetis'. - Ray (1677: 170), but Rose, in Gilmour (1972: 107), cites the 1632 record. Rose (*op.cit.* 50) wonders if the record from Johnson (1629: A. 2) 'Jacobea ... minor', from Gravesend to Rochester, is this species. See also Hanbury and Marshall (1899: 203) where a still earlier record, from Gerard (1597: 219) of 'Jacobea marina' is suggested. This reads 'groweth neere the sea side in sundrie places. I have seene it in the fielde by Margate by Queakes house and Brychenton in the Isle of Thanet [and also at two places in the Isle of Sheppey]'. But Johnson (1633: 281) states that he has been to those places and couldn't find it, and wonders if it is 'hardly wild'.

Senecio jacobaea

* 'Jacobea ... Lande Ragwoort groweth every where in untilled pastures and fields, which are somewhat moist especially'. - Gerard (1597: 219).

1597

Lyte (1578: 691) pictures this as 'groweth almost every where'.

Senecio paludosus

* 'Conyza palustris ... We found it in many places about the Fens, as by a great ditch side near Stretham ferry, &c'. - Ray (1660: 37).	vc29	**1660**
Raven (1947: 296) cites How's record (1650: 115) of 'Solidago Saracenica ... On the five mile Banke neere Whitlesea and between Dudson and Guarthlow. Dr Bowle', pointing out that he has confused this with 'the famous Fenland species *S. paludosus* for which Whittlesea is a good locality'.	vc29	1650

The How record is, as Raven says, a muddle. The report from Dudson to Guarthlow is of *S. sarracenicus*, and is from Shropshire, as Dudston and Gwarthlow in Chirbury parish are in the SW corner of the county near to Montgomery. That must be Dr. Bowles' record, but we do not know who reported the Cambridge record.

Other early records are 'Conyza aquatica laciniata. 19 July 1656 ... growes in great plentie in the fenns in Norfolk near Downam markett neare Linn'. - Goodyer ms.9, f. 186a, see Gunther (1922: 193).

Ray, J. 1660. *Catalogus Plantarum circa Cantabrigiam nascentium.*

Senecio sylvaticus

'*Senecio hirsutus viscidus major odoratus. J.B. III*'. - Ray (1660: 154). vc29 **1660**

Ray's record has traditionally been interpreted as referring to *S. viscosus* but see Oswald & Preston (2011: 293 n.726) where, though a final decision is impossible, it seems much more likely that Ray's plant was *S. sylvaticus*.

'*Erigeron tomentosum alterum*, Lob.'. Thanet. - Johnson (1632: 10 & 1633: 278), is a possibility, but Rose, in Gilmour (1972: 106), is unsure. Another early record, cited by Clarke, is 'Cotton Groundsel. Hamsted'. - Petiver (1713: xvii. 6).

Senecio viscosus

'In arenosis passim'. - Hudson (1762: 316). **1762**

'*Senecio hirsutus viscidus major odoratus R.S. 3. 178.2.* Assington. 14th July, 1749'. - Boulger & Britten (1918). vc26 1749

Hudson seems to be the first to satisfactorily separate this from *S. sylvaticus*. But note '*S. hirsutus viscidus major odoratus*. On all the Fen banks almost in the Isle of Ely'. - Ray (1660: 154). Ray's record has traditionally been interpreted as referring to *S. viscosus* but see Oswald & Preston (2011: 293 n.726) where, though a final decision is impossible, it seems more likely that Ray's plant was *S. sylvaticus*. There must be earlier records than 1749 in herbaria.

Senecio vulgaris

* '*Senecio ... angli vocant Grunswell*'. - Turner (1538: C. verso). **1538**

Serapias parviflora

'In the spring of 1989, two plants … were discovered at a site in vc2, E. Cornwall'. - Cobbing (1989). 'East Cornwall [Mt Edgecombe] in 1989'. - Murphy (1990). vc2 **1989**

Serratula tinctoria

* '*Serratula. ... in nemorosis ... Pedemontis, Northmaniae & Angliae*'. - L'Obel (1571: 231). 'they grow in Hampsteede Woode: likewise I have seene it in great abundance in the wood adjoining to Islington, within halfe a mile from the farther ende of the towne, and in sundrie places of Essex and Suffolke'. - Gerard (1597: 577). **1571**

'in Anglia' - annotations made by T. Penny on C. Gesner's plant illustrations dated as by 1565 in Foley (2006b). 1565

Seseli libanotis

* 'Apium petraeum seu montanum album, *J.B.* On Gogmagog Hills in Cambridge-shire'. - Ray (1690: 70).   vc29   **1690**

In his own copy of Ray (1677) Ray had added 'Pimpinella saxifraga hircina media Herbariis nostratibus dici solita seu Daucus selinoides Cordi. qua in monte Gogmagog Cantabrigiensi provenit, nobis potius oppium petraum sive montanum album videtur, a J. Bauhino descriptum. Qua nota per oblivionem in *Hist*: nostra omissa est'. - See Crompton (2007).   vc29   1677

In fact Oswald & Preston (2011: 261) point out that Ray appears to allude to this species in his Cambridge Catalogue (1660: 119), where, under 'Pimpinella saxifraga minor', he mentions that he had seen a larger species, but had not been attentive enough in describing it.

Sesleria caerulea

* 'Gramen spicatum montanum asperum. E rupium fissuris in monte Ingleborough exit denso cespite'. - Ray (1670: 155). ' Ab amico optimo D. Fitz-Roberts accepi, qui alicubi in Cumberlandia collegit'. - Ray (1696: 325).   vc64   **1670**

See Foley (2008) for further details on FitzRoberts' record.

Sherardia arvensis

* 'Alyscon Plinii is a rare herbe whiche I coulde never see but once in Englande and that was a litle from Syon'. - Turner (1548: A. vij. verso). 'The floures are blu with purple ... amongst the corne beside Sion and ones in a corne fielde in Dorsetshire ... and diverse tymes in the hilles about Welles in Summersetshyre'. - Turner (1568: i. 35).   vc21   **1548**

Sibbaldia procumbens

* 'Fragariae sylvestri affinis planta flore luteo ... Transmissa fuit ad Hortum Medicum a regione *Iernensi* ubi in sylvis sponte provenit'. - Sibbald (1684: ii. 25).   **1684**

The location or meaning of 'Iernensi' has not been traced. It seems to be a word for Hibernia, that is Scotland in general.

Sibthorpia europaea

* 'Alsine spuria pusilla repens foliis Saxifragae aurea ... cum Campanula Cymbalariae foliis [*Wahlenbergia*] in Cornubia & Devonia frequens'. - Ray (1677: 17).   **1677**

'Near St Ives'. This was on 1st July 1662, on Ray's third Itinerary . See Lankester (1846: 188) and Raven (1950: 128).   vc1   1662

Silaum silaus

* 'In Englande there is a wilde kinde of Daucus with longe smal leaves which groweth commonlye in ranke medowes that our Countremen call Saxifrage'. - Turner (1568: 67).   **1568**

But 'Saxifragia … the englishe mens Saxifragia, which they call Saxifrage, hath leaves lyke smal perseley, & it groweth in middowes'. - Turner (1548: H. iiij. verso) seems probable.

Silene acaulis

* 'Caryophyllus montanus minimus sive C. pumilio Alpinus … Mosse-pinkes'. - Johnson (1641: 18). 'Lychnis alpina minima … Nuperrime in Cambriae Septentrionalis altissimo monte Snowdon dicto, a D. Lloyd detecta & observata est'. - Ray (1688: 1004).   **1641**

Seen by Thomas Johnson and party, on Snowdon, Aug 1639. See Raven (1947: 289).   vc49   1639

Silene conica

* 'Lychnis sylvestris angustifolia caliculis turgidis striatis *C.B. Pin. 205*. ... A little to the North of Sandown Castle, plentifully; Mr. J. Sherard, in company with Mr. Rand'. - Ray (1724: 341).   vc15   **1724**

'A new *Lychnis* found at Dover by Mr Sherard in 1715'. - Druce (1928: 468).   vc15   1715

Silene dioica

* 'Lychnis sylvestris rubello flore … Common in many places'. - Gerard (1597: 382).   **1597**

Silene flos-cuculi

* 'Armoraria pratensis … in uliginosis pratis Angliae, Belgiae, & Normaniae'. - L'Obel (1571: 189).   **1571**

Silene gallica

* 'Lychnis segetum parva viscosa flore albo. Mouse-eare Campion'. - Johnson (1634: 49). 'This Mr. Dent Apothecary in Cambridge, found among corn near the Devils-ditch in Cambridge-shire'. - Ray (1670: 202).   **1634**

Dent's specimen is photographed in Walters (1981).

Silene latifolia

'Lychnis sylvestris alba … wilde white Campion'. - Gerard (1597: 384). 'Lychnis sil. flo: albo'. Stoke to Cliffe. - Johnson (1629: 8).   **1597**

Clarke and Druce both cited the Johnson record as their first date, but the Gerard reference, cited by Moss (1920: 70) seems perfectly adequate.

Silene noctiflora

* 'Lychnis noctiflora, *C.B. Park*. ... Found by Mr. Dale among corn'. - Ray (1688a: 15). 'Found by Mr Dale among Corn, between Newmarket and Wood-ditton [Cambs]'. - Ray (1696: 201).   vc29   **1688**

Rose, in Gilmour (1972: 118), suggests that 'Lychnis silvestris parva'. West of Sandwich. - Johnson (1632: 20) is this species, and this record, and the previous three [in Johnson's list] suggested sandy arable land. Parkinson (1640: 632.7), describes it under 'Lychnis noctiflora ... Morpheus sweete wild Campion', but gives no indication that he knows it as a British plant. Morison (1680: 538) similarly describes it, but again, with no indication that he has seen it here.

Silene nutans

* 'Lychnis sylvestris alba nona Clusii *Ger. emac*. ... On the Walls of Notingham-castle, found by T. Willisell'. - Ray (1670: 202).   vc56   **1670**

' … whether you have heeded the *Polemonium Petraeum Gesneri* which he [Thos. Willisell] brought us from Nottingham Castle walls'. - Ray, in letter to Mr Lister, 10th Dec 1669, see Lankester (1848: 48).   vc56   1669

Druce cites 'Behen flore albo elegantiori, three miles from Dover, in the way to Rye, on the beach, all along betwixt Hide [Hythe] and Rumney'. - Merrett (1666:14). But this record is not given in Hanbury & Marshall (1899) nor Smith (1824: ii. 296).

Silene otites

* 'Otites *Tabern*, sive sesamoides parv. Muscipula salamantica minor, *Spanish Catchfly*. Found on Newmarket Heath, Mr. Sare'. - How (1650: 86).   vc26   **1650**

Ray says only in Suffolk, not Cambs, in a letter to Mr Hans Sloane, 1687, see Lankester (1848: 194). Oswald & Preston (2011n) confirm this. Ray (1660: 154) also points out that How's claim that this was the 'salamantica minor' is mistaken - he only found the larger species.

Silene suecica

* 'On rocks near the summit of Clova in Angusshire, but very rare; first observed by Mr. [G] Don in 1795'. - Smith (1811).   vc90   **1811**
                                                                                                                          vc90   1795

There are no records of this in Don (1804-1806).

Silene uniflora

* 'Lychnis marina, Anglica ... in aggeribus maritimis Vectis Insulae [Isle of Wight] Angliae'. - L'Obel (1571: 143). 'Lychnis marina Anglica … groweth by the sea side in Lancashire at a place called   vc10   **1571**

Lytham, five miles from Wygan'. - Gerard (1597: 382). ' ... By Hurst Castle neare the Isle of Wight'. - Parkinson (1640: 640).

There is no record of L'Obel's visiting the Isle of Wight, and the record does not seem to be mentioned in any of the floras of there or Hampshire. Yet the record, and the illustration, seem quite acceptable.

Silene viscaria

* 'Lychnis sylvestris viscosa rubra angustifolia *C.B. Park.* ... Upon the Rocks in Edinburgh-Park. Tho. WiIliseI'. - Ray (1670: 202).   vc83   **1670**

Lusby & Wright (1996) give Willisel's discovery as 1668.   vc83   1668

Moss (1920: 68) cites Johnson's (1633: 601) 'Muscipula angustifolia' as this species, and although Johnson notes that these, along with other catchflies, 'growe wild in the fields in the West parts of England, among the corne', he adds 'we have them in our London gardens'. There can be no evidence that he was referring to this plant in the wild.

Silene vulgaris

* 'Behen album Spatling poppy ... almost in every pasture'. - Gerard (1597: 551). 'Ocimoides, Been [?Behen] album Monspelliensium, sive Papaver spumeum'. Between Gravesend and Rochester. - Johnson (1629: 1).   **1597**

Silybum marianum

'Carduus Mariae. Our Ladies -Thistle ... groweth upon waste and common places by high waies, and by dung-hils almost everie where'. - Gerard (1597: 989).   **1597**

The Gerard reference was overlooked in Preston *et al.* (2004) where 'Carduus mariae, Leucographis *Plinii*'. Sandwich to Canterbury. - Johnson (1632: 21) is given. See Rose, in Gilmour (1972: 119). Turner (1662: 146) gives 'Carduus marie, and in English milk thistle or maries thystel', but this may well be as a pot herb. Raven (1947: 202) cites a reference in Lyte (1578: 530) quoting Cooper's Dictionarie of 1563, where an English name for this species was given - Saint Marie Thistel. But whether that was a wild or cultivated plant we do not know. However Pulteney cites Lyte (1578: 524), where he does include another thistle 'Spina alba, Our Ladyes Thistel ... [with] white, greene leaves, speckled with many white spots ... groweth in this countrie, almost in every garden of pot herbes, and is also founde in rough untoyed places'. This looks much more convincing as a first record.

Simethis mattiazzii

' … has been found by Mr Thaddeus O'mahony, growing in a perfectly wild situation on hills near Derrynane Abbey'. - Harvey (1848).　　vcH1　　**1848**

The record cited by Harvey is a delightful piece of pleading for a record from 'a perfectly wild situation', where it 'has never been cultivated'. But it was found the year before by Miss Charlotte Wilkins [Wilson] 'amongst heath and furze in a lonely spot [Poole Heath] more than two miles from Bourne' [Bournemouth, Hants]. - Anon (1847). The Bournemouth site has always been assumed to be alien, probably brought in with *Pinus maritima*. (note also Allen, D.E., *in litt*., - the Bournemouth discoverer was Wilkins, not Wilson). For full details see Pearman & Edgington (2016).

Sinapis alba

* 'Sinapi agreste Apii, sive Laveris folio *Lob*.'. Road from Gillingham to Sheppy. - Johnson (1629: 5).　　vc15　　**1629**

Rose, in Gilmour (1972: 56), interprets this as 'probably *S. arvensis*, but possibly *S. alba*'. However on p.105, he identifies 'Sinapi alterum siliqua falcata, sive sinapi alterum sativum, *Lob*.' ex Johnson (1632: 4) as probably *S. alba*. Chapman *et al*. (1995: 301) assign one of Turner's Mustardes (1562: 137) to this species 'It that hath the whyte sede is muche shorter then the kindes that have the brown sede. It that groweth in the gardin groweth unto a great hyght … it that groweth in the corne in Somersetshyre, a litle from Glassenberrye, is muche shorter than the gardine mustarde is, but nothynge behynde it in biting and sharpnes'. But see *Brassica nigra*.

Sinapis arvensis

* 'Lampsana Plinii ... wylde Cole, and in other places Carlocke, it groweth communely amonge the corne'. - Turner (1548: D. viij).　　**1548**

Sison amomum

* 'Sison ... Ther groweth a kinde of this besyde Shene … wylde Persley'. - Turner (1548: G. iij. verso).　　vc17　　**1548**

See also Turner (1562: 139). 'of the herb called Sison'.

Sisymbrium officinale

* 'Erysimum Dioscoridis L'Obelii. Banck Cresses ... In stony places among rubbish by path waies, upon earth or mudde wals, and in other untoiled places'. - Gerard (1597: 198).　　**1597**

Sisyrinchium bermudiana

* 'Communicated by the Rev. H.L. Jenner, by whom it had been received as an indigenous Irish plant, collected in a wood near Woodford, Co. Galway' [by Mr. James Lynam in 1845]. - D[oubleday] (1846).   vcH15   **1846**
vcH15   1845

This was a notice of exhibits at the Botanical Society of London.

Sium latifolium

* 'Sium majus latifolium … In moorish and marshie grounds'. - Gerard (1597: 200). 'By Redding'. - How (1650: 114).   **1597**

Collected by Goodyer at Oxford, 5th July 1622, ms.11, f.82v, see Gunther (1922: 177).

Smyrnium olusatrum

* 'Our Alexander groweth … in Ilandes compassed about the se as in a certayn Ilade [island] betwene the far parte of Sommerset shere and Wales'. - Turner (1562: 68).   vc6, 41   **1562**

This could be either Steep Holm (in vc6) or Flat Holm (in vc41). Turner (1538: B iij verso) has 'Olus atrum … is by some considered to be Alexander', which might well refer to knowledge of this in England.

Solanum dulcamara

* 'Amara dulcis … Bitter swete … groweth about ditches and watery places and hedges, and rinneth after the maner of a vyne alonge'. - Turner (1568: 2).   **1568**

Solanum nigrum

* 'Solanum Hortense … commeth up in many places, and not only in gardens … But also neere common high waies, the borders of fieldes, by old wals and ruinous places'. - Gerard (1597: 268).   **1597**

See Turner (1538: C. verso) 'Solanum … vulgus Morell aut Nyghtshad, which implies familiarity with the plant as a wild species.

Solidago virgaurea

* 'Aurea Virga … Germaniae, Franciae, Angliae … Septentrionalibus nemorosis et saltuosis opacis'. - L'Obel (1571: 125). 'In Hampstead Wood,' &c. - Gerard (1597: 349).   **1571**

Gerard gives three more sites.

Sonchus arvensis

* 'Hieracium maius'. Near Rochester. - Johnson (1629: 2). vc16 **1629**

Rose, in Gilmour (1972: 51), suggests that a *Hieracium* species seems more likely. On the same page, however, he interprets 'Sonchi Laevis species 2 aut 3s' as *S. oleraceus* and possibly also *S. arvensis*, and later, p.109, attributes as *S. arvensis* 'Sonchus arborescens, *Tab. Ger.*' in Thanet, ex Johnson (1632: 14).
Clarke adds that 'Turner's' Greate hawkewede ... in the medowe a lytle from Shene', (1562 :14) may have been this. Chapman *et al.* (1995) suggest that this is *Picris hieracioides*.

Sonchus asper

* Figured and described as a distinct species in E.B.S. 2765. - Borrer (1833a). **1833**

'Sonchus asper laciniatus & non laciniatus *C.B.*'. - Ray (1660: 158). vc29 1660

The E.B. account gives 'Gerard (1597: 229)', and notes that the figure there is transposed with that of *S. oleraceus*. Clarke wonders if Turner's 'Rough Sowthistel ... a wild one and hath more pricks upon it ... common in everye countre' (1568: i.135) is this (though he erroneously cites the 1551 edition). He also cites Johnson (1633: 292, and a figure). Edgington (2014) identifies Johnson (1638: 25), and therefore Johnson (1634: 70) 'Sonchus asper vulgi, *Lob*. Prickly Sow-thistle' as this species. But Ray's seems the most reliable early record, and this is followed by Babington (1860b) and Oswald & Preston (2011).

Sonchus oleraceus s.l.

* 'Cicerbita ... a nostris Sowthystell'. - Turner (1538: A. iiij). **1538**

Sonchus oleraceus s.s.

Figured and described as a distinct species in E.B.S. 2766. - Borrer (1833b). **1833**

'Sonchus laevis'. - Johnson (1633: 292.3). 1633

Gerard (1597: 229) muddled the *Sonchus* species, and Johnson (*op. cit.*) substiuted new drawings.

Sonchus palustris

* 'Sonchus tricubitalis fol. cuspidato. In the medows betwixt Woolwich and Greenwich by the banks of Thames'. - Merrett (1666: 115). 'Th. Willisellus invenit ad ripas Tamesis fluvii non longe a Grenvico'. - Ray (1677: 278). vc16 **1666**

Sorbus admonitor

'large tree above scree, Watersmeet, N. Devon'. - Rich & Proctor (2009).   vc4   **2007**

'First collected by W.T. Dyer [Thistelton-Dyer] in 1865 from the East Lyn Valley (ex Martin & Fraser, 1939)'. - Rich *et al.* (2010: 180).   vc4   1865

Sorbus anglica

Marshall (1916) citing a 1914 paper from Hedlund; see Rich *et al.* (2010: 87).   **1916**

'Craig Breidden, 1836, Leighton, **CGE**'. - Rich *et al.* (2010: 87).   vc47   1836

See also Salmon (1930), as *S. mougeoti* var. *anglica*.

Sorbus aria agg.

* 'Aria Theophrasti effigie Alni ... In Angliae frigidioribus sylvosis frequentem videas'. - L'Obel (1571: 435).   **1571**

Pena, P. & L'Obel, M. de. 1571. *Stirpium adversaria nova.*

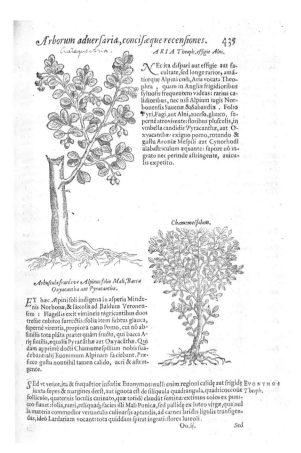

Sorbus arranensis

Wilmott (1934).	vc100	**1934**
'Unusual cut-leaved whitebeams have been known in Arran since at least the 1830s'. '1845, J.H. Balfour, **CGE**, **LIV**. - Rich *et al.* (2010: 72).	vc100	1839

Sorbus aucuparia

* 'Sorbus ... called in Northumberlande a rowne tre or a wicken tre, in the South partes of England, a quick beame tre'. - Turner (1562: 143).	vc67	**1562**

Sorbus bristoliensis

'Avon Gorge at Bristol ; Leigh Woods and Clifton Down (Somerset and Gloucestershire)'. - Wilmott (1934).	vc6, 34	**1934**
'Nightingale Valley, 10 June 1852, Miss M.M. Atwood. **K**'. - Lovatt (2007).	vc6	1852

Sorbus cambrensis

'face of railway cutting cliff, Cwm Clydach'. - Rich & Proctor (2009).	vc42	**2009**
'Brynmawr, Q. Kay. **UCSA**'. - Rich *et al.* (2010: 138).	vc42	1966

Sorbus cheddarensis

'Cheddar Gorge'. - Houston *et al.* (2009).	vc6	**2009**
'Holotype: Cheddar Gorge, 19 September 2007'. - Rich *et al.* (2010: 123).	vc6	2007

First recognised as being a distinct taxon in 2005, but not descrbed till 2009.

Sorbus cuneifolia

'Creigiau Eglwyseg (Cefn Fedw), Llangollen'. - Rich & Proctor (2009).	vc50	**2009**
'from Castell Dinas Bran (as *S. hybrida*) in Hudson, 1798'. - Rich *et al.* (2010: 91).	vc50	1798

Sorbus devoniensis

Warburg (1957), describing a collection of his from Hoo Meavy, S. Devon, in 1934.	vc3	**1957**
As *Crataegus aria* - Polwhele (1797)'. - See Rich *et al.* (2010)		1797

The earliest herbarium specimen is 1848 from Haldon, vc3, **BIRM**. - Cann (2012)

Sorbus domestica

'In May 1983 Marc Hampton observed ... on inaccessible ledges … [on a] sea-cliff in southern Glamorgan'. - Hampton & Kay (1995).	vc41 vc41	**1995** 1983

The first record of this species, presumably of a planted tree, was from Wyre Forest in Worcestershire in 1677. 'Last year I found a rarity growing wild in a forest in this county of Worcestershire …'. - Pitt (1678) and see also Paton (1967) for a summary. But there is a much earlier record from c829, in the *Historia Brittonum*, commonly attributed to Nennius the Monk (Wigginton 1999, and see also Rich *et al.* (2010).

The source of the entry in Ray (1696: 295) 'Sorbus ... The true Service or Sorb. It has been observed to grow wild in many places in the mountainous part of Cornwall, by that ingenious young gentlemen Walter [Moyle], Esquire, in company with Mr Stephens of that county' has never been traced, nor any upland site refound, and has been ignored (though a tree was found just above sea level opposite Padstow in E. Cornwall in 2013). Plot (1686) mentions a tree (S. pyriformis) in Stafford, but this has never been satisactorily confirmed (see Hawksford & Hopkins, 2011).

Sorbus eminens

Warburg (1957), describing a collection of his from Offa's Dyke, Tidenham, W. Glos. in 1935.	vc34	**1957**
Rich *et al.* (2010: 104) refer to early records of 'good material collected by White and others from Cheddar' but there is a record from Leigh Woods, 1837, det Sell, **E**.	vc6	1837

Sorbus eminentiformis

'coppiced tree in woodland, Seven Sisters, Great Dowland'. - Rich & Proctor (2009).	vc36	**2009**
'Great Doward, 1856, W.H. Purchas, **OXF**'. - Rich *et al.* (2010: 108).	vc36	1856

Sorbus eminentoides

'Cheddar Gorge'. - Houston *et al.* (2009).	vc6	**2009**
'Cheddar Gorge, 2006'. - Rich *et al.* (2010: 111).	vc6	2006

Sorbus hibernica

Warburg (1957), describing a collection of his from Ballynahinch, Recess, W. Galway, in 1938.	vcH16	**1957**
'Knockdrin, H23, 1885, **DBN**'. - Rich *et al.* (2010: 148).	vcH23	1885

Sorbus lancastriensis

Warburg (1957), describing a collection of his from Humphrey Head, Westmorland, in 1937, in **BM**.	vc69 vc69	**1957** 1937

Rich (*opp. cit.*) also states that the records for *S. aria* in Ray (1688) 'in parco Witherslacensi [Witherslack] … Consick [Cunswick] Scar, Silverdale, Arnside' almost certainly refer to this species.

Sorbus leighensis

'Quarry 4, Leigh Woods'. - Rich (2009).	vc6	**2009**
'Known since the 1980s'. - Rich *et al.* (2010: 126).	vc6	1989

Sorbus leptophylla

Warburg (1957), describing a collection by A.J. Wilmott from Mynydd Llangattwg, Brecon, in 1933, in **BM**.	vc42 vc42	**1957** 1933

Sorbus leyana

'above Dan-y-graig near Merthyr Tydfil, Breconshire'. - Wilmott (1934).	vc42	**1934**
'Limestone cliff near Cefn Coed, Breconshire, 6th June 1899'. - Ley (1901), although note that Rich *et al.* (2010) give 1896. However there is a reference (Britten, 1897b) citing a letter from Prof. Koehne referring to an earlier specimen sent to him by Ley, from a limestone cliff near Merthyr Tydvil [*sic*], on May 28th, 1896'.	vc42	1896

Sorbus margaretae

'Desolate, vc4'. - Rich & Proctor (2009).	vc4	**2009**
'Desolate, 1865, BM'. - Rich *et al.* (2010: 168).	vc4	1865

Sorbus minima

Ley (1895).	vc42	**1895**
'Craig y Cilau, A. Ley, 1893'. - Rich *et al.* (2010: 78).	vc42	1893

Sorbus parviloba

'Ship Rock, Coldwell Rocks, Symonds Yat'. - Rich *et al.* (2009)	vc34	**2009**
'Symonds Yat, 1999'. - Rich *et al.* (2010: 113).	vc34	1999

Sorbus porrigentiformis

Warburg (1957), describing a collection by A.J. Wilmott from Offa's Dyke, Tidenham, W. Glos. in 1933, in **BM**.	vc34	**1957**
This refers to *S. porrigentiformis* s.s.	vc34	1933

Sorbus pseudofennica

Warburg (1957), describing a collection of his from Glen Catacol, Arran, in 1937.	vc100	**1957**
'N. end of Arran, J. Mackay, 1797'. - Rich *et al.* (2010: 63).	vc100	1797
This was figured, as *Pyrus pinnatifida*, in E.B. 2331 (1811) 'sent by the late Mr. J. Mackay, which he gathered in rocky parts of several mountains at the north end of the isle of Arran'.		

Sorbus pseudomeinichii

'Glen Catacol, main burn, east bank'. - Robertson (2006).   vc100   **2006**

'Glen Catacol'. - Rich *et al.* (2010: 54).   vc100   1949

Sorbus rupicola

Syme (1864: 3. 244).   **1864**

as *S. aria* 'Holy Island' [off Arran], James Robertson to Hunt in 1768. - Dickson (1982). 'Before 1821, Rev. H. Davies on the Great Orme, **BM**.' - Rich *et al.* (2010: 159).   vc100   1768

There seems little doubt that this is what was reported to Hope - there are recent records of this species from the island. There is a much earlier record of what might be this species 'Aria Theofrasti. *Ger*. Silverdale. T. Lawson'. - Ray (1688a: 26).

McClintock (1966: 56) notes that Sir Thomas Molyneux recorded a Whitebeam, *Sorbus rupicola* in 1726. This would have come from Threlkeld (1727) where he records 'Sorbus torminalis & Crategus Theophrasti, mespilus apii folio sylvestris non Spinosa'. Whether this was this species or *S. hibernica* we shall never know.

Sorbus rupicoloides

'Cheddar Gorge'. - Houston *et al.* (2009).   vc6   **2009**

'first discovered in 2006 by L. Houston'. - Rich *et al.* (2010: 163).   vc6   2006

Sorbus saxicola

'Cliffs on west side of Symonds Yat'. - Rich *et al.* (2009).   vc34   **2009**

'Symonds Yat, 1999'. - Rich *et al.* (2010: 129).   vc34   1999

Sorbus scannelliana

'Ross Island, Killarney, Kerry (vc. H2)'. - Rich & Proctor (2009).   vcH2   **2009**

   vcH2   1988

Sorbus stenophylla

'Tarren yr Esgob'. - Rich & Proctor (2009).   vc42   **2009**

'Tarren yr Esgob and Darren Lwyd, A. Ley, 1874. **MANCH**'. - Rich *et al.* (2010: 141).   vc42   1874

Sorbus stirtoniana

'Craig Breidden'. - Rich & Proctor (2009).   vc47   **2009**

'Craig Breidden, 1955, **CGE**'. - Rich *et al.* (2010: 156).   vc47   1955

Sorbus subcuneata

'Greenaleigh near Minehead, Somerset'. - Wilmott (1934).   vc5   **1934**

'Watersmeet, Babington, C.C. **CGE**'. - Rich *et al.* (2010: 187).   vc4   1850

Sorbus torminalis

* 'Sorbus torminalis ... There be many smal trees thereof in a little woode a mile beyond Islington from London: in Kent ... about Southfleete and Gravesend'. - Gerard (1597: 1287).	vc16, 21	**1597**
'In English 'Service'; in southern England it is common, particularly in woods' - an annotation made by T. Penny on C. Gesner's plant illustrations. Foley (2006b) dated these as by 1565, but did not include this species. For further details see Zoller & Steinmann (1987-91).		1565
Turner (1548: G. iv) has 'Sorbus … the fourth kynde in Plinie is called sorbus torminalis, in englishe a service tree'.		

Sorbus vexans

Warburg (1957), describing a collection of his from between Lynmouth and Watersmeet, N. Devon, in 1935.	vc4	**1957**
'Culbone Woods, Babington, 1849'. **CGE**'. - Rich *et al.* (2010: 165).	vc5	1849

Sorbus whiteana

'Quarry 3, Leigh Woods'. - Rich & Houston (2006).	vc6	**2006**
'Clifton Down, White, J.W., 1920. **K**'. - Rich *et al.* (2010: 131).	vc34	1920

Sorbus wilmottiana

Warburg (1967).		**1957**
Clifton Down, White, J.W. 1922, **BM**.	vc34	1922

Sparganium angustifolium

* as *S. natans* 'In lakes. Island of N. Uist; and Galloway; Scotland. Snowdon'. - Babington (1851: 338).	vc49, 74, 110	**1851**
'A loch at Fethaland, 1808, J. Fleming, **GLAM**'. - Scott & Palmer (1987).	vc112	1808
Druce (1932) writes 'noted in a letter of J. Fleming, dated 1814, as having been seen by George Don'. In an earlier paper, Druce (1899) cites a letter from J. Fleming to Sir James Smith, saying 'that it extremely probable that Don knew both plants [this and *S. natans*] and recorded it from Skye, Ben Lawers and the head of Mar Forest'. Fleming adds that 'it is probably referred to by Mr Neill in his 'Tour to Orkney and Shetland (1806)'.		
Early records of this species are almost invariably referred to as *S. natans*, which is confusing as that species was, until recently, called *S. minimum*. Thus, for instance, the records in Lightfoot (1777: 540) of *S. natans*, probably belong to this species.		

Sparganium emersum

* 'Sparganium latifolium. Burre-reed… plentifully grow in Lincolnshire, and such like places, in the ditches of Saint George his fields, and in the ditch right against the place of execution at the end of Southwarke neere London, called Saint Thomas Waterings'. - Gerard (1597: 41). 'Sparganium non ramosum sive latifolium *Ger*'. - How (1650: 117).   vc17, 53, 54   **1597**

St Thomas Waterings was on the Old Kent Road, where pilgrims to Canterbury crossed a little brook.
Another early record is 'Sparganium alterum sive minus. In Kantio inter Rochestriam et Maidenstone'. - L'Obel (1655: 63). L'Obel died in 1616.

Sparganium erectum

* 'Sparganium ... groweth most commonly in waters and fennes ... Thys herbe is comon in England ... it maye be called bede sedge or knop sedge'. - Turner (1562: 143b).   **1562**

Sparganium natans

* 'Sparganium minimum, *P.1205. sine icone*. On the East side of Scrooby nigh a great Wood where the foot way is cast up Notinghamshire'. - Merrett (1666: 115).   vc56   **1666**

How (1650: 117) gives the polynomial, without any locality or other details. Scrooby is west of Gainsborough and east of Sheffield.

Spartina anglica

Hubbard (1957).   **1957**

'Murray, R.P., 1887, Redbridge, Southampton, vc11, **BM**, det. Horlor, M.D., 1973'. - Brewis *et al.* (1996).   vc11   1887

See also Gray *et al.* (1990) for what they then thought was the first collection of the fertile amphidiploid, Lymington, 1892.

Spartina maritima

* 'Gr. Sparteum capite bifido vel gemino, At Crixey [Creeksea] ferry in Essex'. - Merrett (1666: 58).   vc18   **1666**

This is interpreted as this species in almost all sources, and Hubbard (1965) points out that there are specimens from Merrett's collection in Hb. Sloane 33 & 34, **BM**. Gibson (1862) has 1703, Buddle ms. Flora. See also 'Spartum nostras parvum'. Sheerness. - Johnson (1629: 7), which Hanbury & Marshall (1899) identify as this species, though Rose, in Gilmour (1972: 58), has 'probably this, but possibly *Agropyron pungens* (*Elytrigia atherica*).

Spergula arvensis

* 'Saginae Spergula sive Spurry Belgarum et Anglorum'. - L'Obel (1571: 357). Between Gravesend and Rochester. - Johnson (1629: 2). **1571**

Spergularia marina/media s.l.

British *Lepigona*, More (1860). **1860**

'Spergula marina *Daleschamp*. Anthylloides *Thalii*'. Nr Chatham. - Johnson (1629: 4). vc15 1629

More's article, a translation of a continental work, clearly set out the differences between the two species, with a key.

Spergularia marina

'Spergula marina, *Ludg.* Forte Anthylloides, *Thal.* ad salinas Saxonicas & Alsine maritima Neopolitana, *Col.*'. Westgate Bay, Thanet. - Johnson (1632: 18). vc15 **1632**

Rose (in Gilmour 1972: 113), has 'probably *S. marina*', as more likely on a beach than *S. media*. Clarke cited (under *Lepigonum marinum*) 'Spergula marina Daleschamp'. Near Chatham, Kent. - Johnson (1629: 4). There is an obvious problem here - Clarke does not separate the species. Rose, in Gilmour (1972), is equivocal, and also assigns (1972: 59) 'Rubia marina' nr. Queenborough Castle, Johnson (1629: 7) to either of these species.

Spergularia media

'Spergula marina *Daleschamp*. Anthylloides *Thalii*'. Nr Chatham. - Johnson (1629: 4). 'Spergula foemin foliaceo nigro, circulo membranaceo albo cincto *Hort. Blaes*. On the shore everywhere [in Jersey]. I have found it near Southampton'. - Sherard in Ray (1690 app.: 239). vc15 **1629**

Druce cites this as his first record, whereas Clarke did not separate the two species. Rose (in Gilmour 1972: 55) is unsure; see the note under *S. marina*. Ray (1724: 351) seems to satisfactorily separate the two, but Babington (1843) was still doubtful and all through the nineteenth century various synonyms were used.

Note, however, that Le Sueur (1984) is sure that Sherard's Jersey record from 1689 refers to *S. marina*, as *S. media* is not found on the island.

The *Hort. Blaes* reference is to Morison's 1669 edition of *Hortus Regius Blesensis Auctus Cum Notulis Durationis*.

Spergularia rubra

* 'Spergula flore rubro; an Alsine Spergulae facie minor *Bauh*'. Between Sandwich and Canterbury. - Johnson (1632: 20). vc15 **1632**

The record came from Goodyer, see Edgington (2007).

Spergularia rupicola

as *Arenaria marina* var. *hirsuta* 'From Newlyn Cliff, near Penzance, Cornwall'. - Gibson (1842). Perhaps better 'Guernsey (J.T. Syme) and 'Isle of Wight' (A.G. More) - More (1859). vc1 **1842**

Guernsey. Gosselin, J., c.1790'. - see McClintock (1982: 61). vc113 1790

Spiranthes aestivalis

'In Jersey'. - Babington (1837). This was in a bare list of additions to the flora of the island. 'In the island of Jersey … on the 25th of July 1837. … It grows in a wet sandy spot upon the banks of St Ouen's Pond, 1837'. - C.C. Babington in E.B.S. 2817. vc113 vc113 **1837** 1837

Chatters (2009: 84) refers to a composite herbarium sheet at **BM** with one specimen, apparently collected near Lyndhurst by Dr Bossey, with a number, possibly the date '1813' adjacent. Apart than that slightly unsatisfactory reference the first record other than from the Channel Islands is 'Mr. Janson exhibited specimens of the *Neottia aestivalis* discovered in August last by himself and Mr. Branch near Lyndhurst, Hampshire, being the first time it had been observed in England'. - Meeting on Nov. 17th, 1840 (*Proc. Linn. Soc. i. 80*).

Spiranthes romanzoffiana

* as *Neottia gemmifera* 'Near Castletown opposite to Bearhaven on the northern side of Bantry Bay, County of Cork, in small quantities. … Mr. Drummond communicated to me in August, 1810'. - Smith (1828: 36). vcH3 vcH3 **1828** 1810

Colgan & Scully (1898) note that Drummond first reported this, as a 'nondescript *Neottia*', in 1820, in the Munster Farmer's Magazine. This record was from 'on the strand of the mainland opposite the western redoubt, Bere Island, Bantry Bay, August 1810'.

Spiranthes spiralis

* 'Satyrion … groweth besyde Syon, it bryngeth furth whyte floures in the end of harveste, and it is called Lady traces'. - Turner (1548: F. viij. verso). vc21 **1548**

Gerard (1597: 168) gives detailed sites from Islington, Barn Elms and Stepney, all from the London area, where is is now long extinct.

Spirodela polyrhiza

* 'Lenticula palustris major, *Commel. Cat. Pl. Holl.* … In fossis et aquis purioribus passim occurrit'. - Ray (1724: 129 with a figure). **1724**

Stachys alpina

'On the 30th of June this year … to a wooded hill near Wotton-under-Edge …'. - Bucknall (1897). vc33 **1897**

Stachys arvensis

* 'Sideritis Alsines trixaginis foliis *Bauh.*'. Between Margate and Sandwich. - Johnson (1632: 19). vc15 **1632**

'I first found it August 1626 ... not far from Greenhithe in Kent'. - Johnson (1633: 699). vc16 1626

> (23)
>
> *pia ubique in vliginosis, & salsis.*
> *Beta sylvestris spontanea maritima, Lob.*
>
> Tota die in hisce inveniendis consumpta, post solis occasum defessi domum nos conferimus. Dein cibo, somnoque refecti, sequente mane Margata post tergam relicta, Sandwich versus teudimus, & antequam ad Maris littus venimus, collectæ fuerunt plantæ sequentes.
>
> *Euphrasia vulgaris.*
> *Sideritis alsines trixaginis foliis, Bauh.*
> *Œnanthe Angustifolia, Lob.*
> *Cicutaria palustris, Lob. Phellandrium, Plinii. Dod.*
> *Sagittaria aquatica, Plinii. Major, Matth. Dod. Phleos mas latifolia, Lugd.*
> *Hydrolapathum minus, Lob.*
> *Anagallis aquatica major foliis acutioribus, floribus albidis.*
> *Polygonum fœmina semine vidua, Lob.*
> *Equisetum, 1. Matth. Hippuris major. Dod.*
> *Arundo vulgaris vellatoria, Lob. palustris, Matth.*
> *Hydropiper, persicaria acris.*
> *Persicaria mitis maculosa.*
> *Iacea Floribus albis.*
> *Gram. spica Tritici mutici, Bauh.*
> *Cichoreum sylvestre flore cæruleo.*
> *Tragopogon vulgare luteo flore.*
>
> *Cyno-*

As Sideritis alsines from Johnson, T. 1632. *Descriptio Itineris Plantarum Investigationis.*

Stachys germanica

* 'Stachys *Fuch*. ... My kinde friend Mr. Buckner an Apothecary of London the last yeare, being 1632, found [this] growing wilde in Oxfordshire in the field joyning to Witney Parke a mile from the Towne'. - Johnson (1633: 696). vc23 **1633**
  vc23 1632

Stachys palustris

* 'Panax Coloni ... almost everie where especially in Kent about Southfleete neere to Gravesend and likewise in the medows by Lambeth neere London'. - Gerard (1597: 852). vc16, 17 **1597**

Stachys sylvatica

* 'Galeopsis vera *Dios*. Urtica Heraculea *Taber*.'. Stoke to Cliffe. - Johnson (1629: 8). vc16 **1629**

see also 'Galeopsis … called in englishe red Archaungel. It is lyke Archaungel, but it hath a purple floure, and lesse leaves and shorter. It groweth in hedges'. - Turner (1548: D. ij. verso). Clarke and others identify this as *Lamium purpureum*. Lyte (1578: 130.2) has a good description, but with no specific reference to Britain.

Stellaria alsine

'Alsine fontana … in the brinks and borders of wels, fountaines and shallow springs'. - Gerard (1597: 490.9). **1597**

Clarke and Druce chose 'Alsine longifolia uliginosis proveniens locis'. - Ray (1660: 8) as their first record. But Moss (1920: 63) cites Gerard, as does Smith (1824: ii. 304), the latter with a note 'good'.

Stellaria graminea

'Gramen leucanthemum alterum'. - Gerard (1597: 43). 'Holosteum Ruellii, gramen Leucanthemum minus'. Between Hampstead and Kentish Town. - Johnson (1629: 12). **1597**

Clarke and Druce both cite the Johnson record, but Pulteney and Moss (1920: 63) prefer the Gerard, and it must be said that the illustration and description there look convincing.

Stellaria holostea

* 'Stychewort groweth only in hedge sides and in woddes and shadowy places'. - Turner (1562: 13, under 'Grasse'). **1562**

Stellaria media

* 'Alsine … herba ilIa est quam nostrates mulieres vocant Chykwede aut Chykenwede, Qui alunt aviculas caveis inclusas … hac solent illas (si quando cibos fastidiant) recreare'. - Turner (1538: A. ij). **1538**

See Raven's note (1947: 106) where he claims that Turner's illustrations purporting to be of this species are probably not.

Stellaria neglecta

as *S. media* β *umbrosa*, Forster (1842: 29, suppt). vc14 **1842**

'Hedges near Worcester. J. Stokes (Stokes [ie Withering] 1787)'. - Maskew (2014). Barton, vc29, 1831, Downes, det. Warburg, **CGE**. '1837, Hendle (*sic*) Wood, Maresfield, ex hb. J. Woods, W. Borrer in hb. Bab. as *S. grandiflora*Ten.'. - Wolley-Dod (1937: 73). vc37 1787

Maskew (*op.cit*.) argues that Stokes stated that his plants had 10 stamens, and thus were this species.

Stellaria nemorum

* 'Alsine montana folio Smilacis instar, flore laciniato *H. Ox. II. 550*. Found by Dr. Richardson in Bingley Parish'. - Ray (1724: 347). vc63 **1724**

'Alsine latifolia montana flore laciniato *C.B.P.* which I have found nigh this place [Bingley, his home] in abundance'. - Richardson to Sloane, 3rd March 1721, British Library: Sloane ms. 4046, f. 70.	vc63	1721
A record that seems to be of this species, and would predate any of these is 'Alsine nemorosa maxima montana, common on the shady banks of the R. Were, as near the New-bridge at Durham, and several other places'. - Gibson (1722: 962), and repeated in Graham (1988). The 1722 edition of Gibson was a reprint of the 1695 edition, and if that is the case then the record might well date from before 1695.		

Stellaria pallida

as *Alsine pallida*, Babington (1864b). as *S. media* var. β *pallida*, Babington (1867).	vc31	**1864**
'Bluntisham, Hunts., gathered in 1846 by Mr. Newbould'. - Babington (1864b).	vc31	1846
See also Anon [Irvine, A.] (1860) as *S. boraeana*, Jord (*S. apetala* Auct.) discussing an Isle of Wight plant.		

Stellaria palustris

* 'Caryophyllus holosteus arvensis medius. D. Stonestreet invenit in Insula Eliensi [Isle of Ely] & D. Sherrard prope Oxoniam'. - Doody in Ray (1690: 245).	vc23, 29	**1690**

Stratiotes aloides

* 'Militaris aizoides, Fresh water Souldier ... I found this growing plentifully in the ditches about Rotsey a small village in Holdernesse, and my friend Mr. William Broad observed it in the fennes in Lincolnshire'. - Johnson (1633: 825). 'In the new Ditches of the Dutch workes of Hatfeild, within three or four years after they were made'. - How (1650: 75).	vc53, 61	**1633**
Johnson's visit to Lincolnshire was in 1626 (see Raven 1947: 274).	vc53	1626
This is mentioned in Gerard (1597: 677), but with no indication of a British site.		

Suaeda maritima

* 'Kali minus ... Ad septentrionem Angliae familiaris'. - L'Obel (1571: 170).		**1571**

Suaeda vera

'Vermicularis frutex minor *Ger.* ... ego in isthmo illa, insulam Portlandicum & litus Dorcestriae interjacente, abunde provenientem multis abhinc annis observavimus. Eundem antea ostendit mihi Cl. Vir D. Tho. Brown Norvicensis medicus in litore	vc9, 28	**1670**

Norfolciae collectum'. - Ray (1670: 313). Ray (1690: 37), with almost identical wording.

The Dorset record must have been on Ray's 1667 Itinerary. Although Ray states that Browne's record was earlier, we do not know when, though see the note below.   vc9   1667

Ray's (1670) record is accepted by Smith (1824), Petch & Swann (1968), Bowen (2000) and others, but Raven (1950; 116) identifies this plant with *S. maritima*. (Clarke & Druce both cited 'Blitum fruticosum maritimum ... In Sinus Bristoliensis Anglici Oceani Insulis vocatis Homs'. - L'Obel (1571: 162)). But *S. vera* has never been found in the Bristol Channel, other than as a casual. See also White (1912: 503), who dismisses the record. It is also odd because the L'Obel entry they cite is actually for 'Arborescens graecum, Illyricum, & Anglicum, sive maritimum' and Pulteney cites, as this species, but 'with a bad figure', the <u>next</u> page of L'Obel (1571: 163) 'An Chamaepithys maior *Diosc*. ... maritimus littoribus ... Angliae' which seems to be a more likely reference. All very confusing.

Johnson (1633: 523.4) gives 'Vermicularis frutex minor, the lesser shrubby Sengreen … the Isle of Purbecke in England: and on Raven-spurne in Holdernesse, as I my selfe have seene'. This seems a possible record for Dorset (though it has never been recorded for Yorkshire. But where did the Dorset record come from? In conection with this the distinguished Dorset botanist Bowles Barrett (1905) makes a very interesting claim which I have been unable to verify. 'It [*S. vera*] was discovered by the eminent Sir Thomas Browne [1605-1682] ... in all probability about the year 1630 on [his] way to or from France'. Mansell-Pleydell cites the same in the first edition of his *Flora* (1874) but not in the second (1897). It might be a mis-reading of Ray's text, but it would be nice to resolve this.

See also Raven (1947: 103) where in describing Turner's 1550 visit to Dorset, states that he visited Poole and noted (1562: 148) 'the Stechas that Dioscorides writeth of is very plentiful in the toun of Poule and in diverse places of the West countrey where it is called Cassidonia or Spanish lavendar'. Raven wonders if this is *Lavendula vera*, but I wonder if it might be *S. vera*.

Subularia aquatica

* 'Graminifolia aquatica thlaspeos capitulis rotundis septa medio siliculam dirimente ... ā D. Sherard inventore ex Hibernia in Angliam nuperrime nobis transmissa est'. - Plukenet (1692: 188. f. 5).   1692

Ray (1696: 281) adds that it was found 'Lough Neagh ... D. Sherard'. Lough Neagh is bordered by five vice-counties.

Succisa pratensis

* 'The devils bite ... Morsus diaboli, & succisa ... groweth abroade in untilled places as in meaddowes and plaine feldes'. - Turner (1568: 43). **1568**

Symphytum officinale

* 'Symphytum herbarij vocant consolidam majorem, vulgus Comfrey'. - Turner (1538: C. verso); and see also (1562: 148). **1538**

Pulteney (1790: 72) and others mention that John Falconer had communicated English plants to Amatus Lusitanus whilst in Ferrara during the period 1540 - 1547, and the DNB mentions that one of these was this species, as 'Symphyto petraeo'.

Symphytum tuberosum

* 'It has been oberved in several places in Scotland. Mr. Yalden found it growing sparingly opposite the new well at the Water of Leith, but more plentifully in Dr. Robinson's walks at North Marchiston [Edinburgh], where it seems to be a native'. - Lightfoot (1777: 1092). vc83 **1777**

'By the water of Leeth [Edinburgh] opposite to the new well, May 18th 1765'. - John Hope's ms. 'List of Plants growing in the Neighbourhood of Edinburgh', reproduced in Balfour (1907). There is also a specimen in his herbarium, dated 1768. vc83 **1765**

D.E. Allen (*in litt.*) 'I happen to have researched Thomas Yalden (1750 -1777) and have established that his entire Scottish fieldwork was in 1773 - 1775'.

Tamus communis

'The black Brionye hath leaves lyke Ivy, but lyker to the leaves of Smilax, but greater and so are the stalkes ... It groweth in the hedges that go aboute the closse that is nexte unto them'. - Turner (1562: 167b). 'Bryonia nigra ... in hedges and bushes almost everywhere'. - Gerard (1597: 721). **1562**

Turner's illustration looks like a *Solanum* (Pulteney (1799) has that as *Clematis vitalba*), but the text surely describes *Tamus*, and Chapman *et al.* (1995: 601) identify it with this species. They also interpret the last word 'them' as 'Shene', but there is no justification at all for this. It is even more clearly 'them' in the 1668 edition.

Tanacetum parthenium

'Matricaria, Feverfew ... groweth in hedges, gardens and about old wals, it joyeth to grow amongst rubbish'. - Gerard (1597: 526). **1597**

The Gerard record was overlooked in Preston *et al.* (2004), where Johnson, 1638 (from his Tottenham list), ex Kent (1975) is given as the first record. Pulteney cites Lyte (1578: 19) 'Parthenium, Feverfew … groweth well in dry places, by olde walles'. But in fact Turner (1548) mentions the English name, Feverfew, twice. It was obviously known, but was almost certainly cultivated, and the Gerard record is the first clearly in the wild in England.

Tanacetum vulgare

* 'Tanacetum … groweth wilde in fields as well as in gardens'. - Gerard (1597: 525).   **1597**

Pulteney cites Lyte (1578: 17) 'Tanacetum majus … groweth about high wayes and is very common in this countrie' for this species too, but whether this is is a record for Britain is not known. Earle (1880) cites a thirteenth century use of 'Tanesetum, taneseie, helde'.

Taraxacum agg.

* 'Intubus … Dan de lyon … groweth everywhere'. - Turner (1548: D. vj. verso).   **1548**

See also Turner (1538: B. ij) 'Intubum … non desunt qui putent Dandelyon'. Here Turner is struggling to make an English plant fit a classical description. He probably is alluding to *Taraxacum*, but it seems safer to go with the 1548 reference.

Taxus baccata

* 'Taxus an Uhe tree unde hodie apud nos fiunt arcus'. - Turner (1538: C. verso). 'Comune Ughe, groweth in diverse partes of Yorke shyre'. - Turner (1548: G. vj).   **1538**

Teesdalia nudicaulis

* 'Nasturtium petraeum … Mr. Bowles found [this] growing in Shropshire in the fields about Birch in the parish of Elsmere'. - Johnson (1633: 250).   vc40   **1633**

See Raven (1947: 295) dating Bowles' trip to 1632.   vc40   1632

Tephroseris integrifolia

* 'Jacobaea montana lanuginosa angusti folia non laciniata … On Gogmagog hills and Newmarket heath'. - Ray (1660: 80).   vc29   **1660**

'Jacobaea, sive Senecion minimum. Crescit in Agro Cantabrigiensi in parvis collibus non procul Stapleford'. - How ms.1650–6, iIllustrated with a watercolour, held in Magdalen College, Oxford, see Gunther (1922: 289).   vc29   1656

One of a series of ms annotations made by How in his copy of his *Phytologia* before his death in 1656. Druce (1912b) assigns to *Senecio paludosus* 'Jacobea angustifolia Pannonica *non* laciniata', from a ms. list of Cambridge plants from Samuel Corbyn, dated 20 May,1657. But Oswald & Preston (2011: 16) and Preston (in press) identify this as *Tephroseris integrifolia*, with which I concur. The Welsh subsp. *maritima* was discovered 'on cliffs near Holyhead', by Rev. H. Davies (Smith 1800: ii. 896).

Tephroseris palustris

'Conyza foliis laciniatis *Ger. emac*. helenitis foliis laciniatis *Lob. ico. Park*. … In the Fen ditches about March and Chatteresse in the Isle of Ely'. - Ray (1660: 37).   vc29   **1660**

Samuel Corbyn records this as growing in his garden in Cambridge in 1656 (see Preston, in press) as 'Conyza Helenites foliis laciniatis'. It must be very likely that this came from a wild source.

How (1650: 80) cited a record 'Conyza foliis laciniatis ... A stones cast from the East end of Shirley Poole neere Rushie moore belonging to Mr. Darcy Washington. In Yorkshire, Hoary Fleabane, Mr. Heaton'. This was repeated by many others, and accepted by Clarke & Druce. But Lees (1888: 293), comments that though the district so precisely described is still very marshy, no one has seen the plant since. However Coles (2011) notes that T. Tolfield of Doncaster discovered this plant within a few miles of Heaton's site, sometime before his death in 1779, so perhaps Lees was being too harsh. It might be significant that Goodyer's plant (ms.9, f.186a, see Gunther, 1922: 193), with the very similar polynomial 'Conzya aquatica laciniata. 19 July 1656 ... growes in great plentie in the fenns in Norfolk near Downam markett neare Linn' has been identified as *Senecio paludosus*. That has not been recorded for Yorkshire, but the habitat is right.

Ray's 1660 work was published in February of that year, and there is a further possible record from later in the same year. Greenwood (2012) points out that Ray (1670: 79) gives a further record apart from those in Cambridgeshire from 'ditches about Pillin-mosse in Lancashire', which he thinks could have been made in 1660, following a tour of northern England and the Isle of Man. This site has never been found again, but Greenwood adds that Ray would probably have returned from the Isle of Man via the Wyre estuary and Garstang, which would have taken taken him past Pilling - and, of course, he knew the plant from Cambridgeshire.

Teucrium botrys

* 'On Saturday last, the 17th instant [August], when in company with William Bennett, I found several fine plants of *Teucrium botrys*, in a wild stony locality, far from any house or garden, at the back of Box Hill, in Surrey'. - Ingall (1844).   vc17   **1844**

Teucrium chamaedrys

'a small form looking quite wild, downs east of Cuckmere Haven, 1945, A.W. Graveson'. - Wallace (1951). vc14 **1951**

'In pretty good quantity in thick old turf on a high down about three miles N.W. of Lewes, Sussex, looking very wild'. - Mr Justice Talbot, July 1925, **OXF**. 'Near Offham, in grass, 1924-1937'. - Wolley-Dod (1937). vc14 1924

Offham is at the eastern end of the South Downs, just NW of Lewes. This is the first date for a possibly wild site. All the other known records are of aliens, of which the first is 'Walls of Winchilsea Castle, J. Sherard'. - Herb. Du Bois. (Druce 1932). Sherard died in 1738.

Teucrium scordium

* 'Scordium ... I heare saye that it groweth besyde Oxforde ... water Germander'. - Turner (1548: G. j. verso). 'In Oxfordshyre and in Cambridgeshyre in good plenty'. - Turner (1562: 131). vc23 **1548**

Teucrium scorodonia

* 'Scordion alterum Plinii, latioribus foliis, Mentastro simile … sive salvia agrestis. Wood Sage … in sylvis Germaniae & Angliae'. - L'Obel (1571: 210). **1571**

Rydén *et al.* (1999) identify as this species Turner (1538: B) 'Eupatorium … ut alii falso scribunt, wylde sauge, sed herba est quam omnes hodie vocant Agrimony', but Stearn (1965) and others identify it with *Agrimonia*.

Thalictrum alpinum

* 'Thalictrum minimum montanum atro rubens foliis splendentibus. In udis scopulis & ad rivulos in Alpibus Arvoniae [Carnarvonshire]. D. Lloyd'. - Ray (1690: 62). vc49 **1690**

Thalictrum flavum

* 'Thalictrum majus … Alongst the ditch sides leading from Kentish streete unto Saint Thomas Watrings the place of execution, on the right hande; they growe also upon the Thames bankes leading from Blacke Wall to Woolwich neere London'. - Gerard (1597: 1067). vc16, 17 **1597**

St Thomas Waterings was on the Old Kent Road, where pilgrims to Canterbury crossed a little brook.

Thalictrum minus

* 'Thalictrum minus *Ger* … About Newmarket … and also among the corn between that pit and Cambridge road: also about Bartlow and Linton in the chalky grounds'. - Ray (1660: 162). vc29 **1660**

This plant is in Gerard (1597: 1067), with its distribution as for *T. flavum*, but the given habitat is not convincing.

Thelypteris palustris

'Dryopteris Penae & L'Obelii … Many yeares past I [Goodyer] found this same in a very wet moore or bog.…called Whitrow Moore…a mile from Petersfield in Hampshire; and this 6 Julii 1633, I digged up there many plants …'. - Johnson (1633: 1136).   vc11   **1633**

See Goodyer ms.11, f. 140, in Gunther (1922: 183).

Thesium humifusum

* 'Linaria adulterina ... Mr. Goodyer found it growing wilde on the side of a chalkie hill in an inclosure on the right hand of the way as you go from Droxford to Poppie hill in Hampshire'. - Johnson (1633: 555).   vc11   **1633**

as 'Anthyllis montana. 5 julii 1620'. - Goodyer ms.11, f. 84, see Gunther (1922: 117).   vc11   1620

Thlaspi arvense

* 'Thlaspi … plentuously besyde Syon … dysh mustard or triacle Mustard'. - Turner (1548: G. vj. & 1662: 152). 'In the corne fields between Croydon and Gods stone in Surrey, at South-fleete in Kent, Harnsey unto Waltham crosse [Essex], and many other places'. - Gerard (1597: 206).   vc21   **1548**

Thymus polytrichus

* 'Serpyllum … Wylde Tyme … groweth … in sandy fieldes and bare groundes'. - Turner (1548: G. ij).   **1548**

Thymus pulegioides

* as *T. chamaedrys* 'Devil's Ditch in Cambridgeshire; Box Hill, Surrey; and How Chapel, Herefordshire'. - Babington (1853a: 431).   vc17, 29, 36   **1853**

as *T. chamaedrys* in Druce & Vines (1907: 75). 'Aug. 24 [1852]. Devil's Ditch … on the lower inner slope of the ditch between the railroad and Stetchworth, we found the true *T. chamaedrys*'. - Babington (1897:164).   1747

Druce has 'c.1720', but I can find no justification for this. The record from Druce & Vines is dated as 1747, the date of Dillenius' death. See also Jorden (1853) where he claims to have noticed the differences over 'the last 10 years'. For the difficulties in attributing early records of this genus see, for instance, Oswald & Preston (2011: 477n).

Thymus serpyllum

'divided by Jalas in 1947'. - Warburg (1949) but with full details in Pigott (1951, 1954).   **1949**

'Sir J.C. Cullum, 1773'. - Trist (1979).   vc26   1773

There is abiding confusion between the current concept of this species and the use of the name for the common wild thyme, *T. polytrichus*. Sanford (2010) and his predecessors in Suffolk, cite Babington (1853a) as being the first to separate, but he was distinguishing between *T. polytrichus* (as *T. serpyllum*) and *T. pulegioides* (as *T. chamaedrys*), as he did in a later paper (Babington (1863b)). However earlier recorders evidently knew they were seeing something different. Jalas' paper is in *Acta Bot. Fennica 39: 3 - 85*. Pigott's (1951) paper contains a reference to an earlier note in the *Phytologist*, which contained the intriguing observation that sheep will eat one species of thyme but not another. One of the attendees at Pigott's talk suggested that more attention should be paid to the biochemistry of plants.

Thyselium palustre

* as *Selinum palustre* 'In paludibus prope Doncaster, D. Tofield'. - Hudson (1778: 115).   vc63   **1778**

Lees (1888: 260) cites 'Tolfield (1762 in Huds.)', but this seems to be a slip.

Tilia cordata

* 'Tilia ... Lind tre ... groweth very plentuously in Essekes in a parke within two mile from Colichester, in the possession of one maister Bogges'. - Turner (1562: 153 verso). 'Neere Colchester, and in many places alongst the highway leading from London to Heningham, in the Countie of Essex'. - Gerard (1597: 1299).   vc19   **1562**

Both Turner and Gerard seem to intend *T.* x *europaea* by their description, but according to Ray (1688: 1694), were in error in so doing. 'Turnerum & Gerardum errasse existimo cum in Essexia Angliae hoc genus copiose provenire aiunt, nam quamvis ipse Essexiae in cola sum, neque inibi neque alibi in Anglia Tiliam foeminam vulgarem platyphyllon sponte nascentem vidi. Quae frequens in sepibus & sylvis apud nos invenitur Tilia est minore folio J.B. & aliorum'.

Tilia platyphyllos

* 'Tilia ulmifolia semine Hexagono, At Whitstable in Surrey [?*sic*] and near Darkin [Dorking]'. - Merrett (1666: 118).   vc17   **1666**

Salmon (1931) suggests that West Humble in Surrey was intended; this seems very likely, as Westhumble (Wystumble in 1248) is just north of Dorking.

Of course the Merrett record is that of a planted specimen. In a letter to Mr Lister, 15th Nov 1669, see Lankester (1848: 43), Ray writes 'the major [*T. major* = *europaea*], I have not as yet seen anywhere with us spontaneous'. An early record of a presumed native site might be 'in a position precluding all doubt that it was native ... at Craig Cille, nr. Crickhowell, Brecon, 23rd Aug 1909, at 1300ft'. - Ley (1909). There must be earlier records, but botanists have never been very good at recording trees. Abraham & Rose (2000) give a record from W. Borrer (d.1862) from Chanctonbury Hill, Washington, W. Sussex.

Tofieldia pusilla

* 'Asphodelus palustris Scoticus minimus. Found about two miles north of Barwick near a small rivulet'. - Ray (1677: 30).	vc68	**1677**
July 23rd, 1671, Berwick'. - Lankester (1846: 150-151 note), and see Raven (1950: 154).	vc68	1671

Braithwaite (2004) argues convincingly for Ray's location to be near Loughend in North Northumberland.
See also Horsman (1995) referring to Gibson (1722: col 962) 'Pseudo-Asphodelus palustris Scoticus minimus *Raii*. On a fell in this county about a mile East of Birdale in Westmoreland'.

Tordylium maximum

'Tordylium sem. minus hirto & limbo quasi Iaevi seu parum granulato ... in marginibus fossarum & aggeribus sepium, sponte passim circa Blaesas, Londinum, & alibi provenit'. - Morison (1672: 40).	vc21	**1672**
'Betwixt St. James and Chelsea'. - notes written (c.1670) by Christopher Merrett in a copy of his *Pinax* (in the **BM** Library). See Kent (1975).	vc21	1670

The location of Blaesas is unknown - there are at least 30 references to it in Morison's work, and it is most likely to be in the south of France.

Torilis arvensis

* 'Caucalis minor semipedalis, *G.1023*. huius tantum meminit sub finem descriptionis quintae, Amongst wheat plentifully neer Peters field [Hants], Mr. Goodyer, who called it Caucalis pumila segetum'. - Merrett (1666: 24).	vc11	**1666**
'Caucalis minor flore rubente *Park. 921 G. 1022*'. This is in a ms. list of Cambridge plants from Samuel Corbyn, dated 20 May,1657. - see Preston (in press).	vc29	1657

The Goodyer field record would probably predate all of these, if we could find a reference, but I can find nothing in Gunther for a date for this in Goodyer's papers, other than the list that Merrett used (1922: 195). However Preston (*in litt.*) points out that that where Ray (1660: 30) refers under 'Caucalis minor flosculis rudentibus *Ger. Emac.*', to it occurring 'alongside hedges and even amongst standing corn', he must be referring to both *T. japonica* as well as this species. This view is reinforced by his marking with a 'C' [for Cambridge] the entry for *T. arvensis* in his *Fl. Anglica* (1670: 61).

Torilis japonica

* 'Caucalis semine aspero flosculis subrubentibus, *Bauh.*'. Near Margate. - Johnson (1632: 9).	vc15	**1632**
See also 'Caucalis sylvestris'. - Johnson (1629: 2). Gravesend to Rochester. Clarke queries this, but Rose, in Gilmour (1972: 51), identifies as 'probably this species'.		

Torilis nodosa

* 'Caucalis nodoso echinato semine, *Bauhini*'. Chalkedale near Dartford. - Johnson (1629: 9). 'Upon the bankes about S. James and Piccadilla'. - Johnson (1633: 1023).	vc16	**1629**

Tragopogon pratensis

* 'Barba Hirci ... Tragopogon , . . groweth in the fieldes aboute London plentuously ... gotes bearde'. - Turner (1548: B. v).	vc21	**1548**

Trichomanes speciosum (gametophyte)

' … discovered by D.R. Farrar in the English Lake distict in 1989'. - Rumsey *et al.* (1990).	vc69, 70 vc69, 70	**1990** 1989

Trichomanes speciosum (sporophyte)

'Filix humilis repens, folis pellucidis & splendentibus … found by Dr Richardson at Belbank, scarce half a Mile from Bingley [Yorkshire], at the head of a remarkable Spring ...'. - Ray (1724: 127 with a figure) - *see image p.400*.	vc63	**1724**
'a specimen or two of a very beautyfull Capillary which I lately found on some shady moyst rocks not far of [*sic*] in my searches for mosses which I take to be new'. - Richardson to Sloane, 3rd March 1721, British Library: Sloane ms. 4046, f. 70. Another early specimen, from the same site, is in Uvedale's herbarium in the Sloane collection in **BM**; see Dandy (1958a: 225). Uvedale died in 1722.	vc63	1721
See also Lees (1888: 500) for a very full account. For Ireland 'first found in fructification on moist shady rocks near the waterfall between Mangerton and Turk mountains, Killarney, 1804'. - Mackay (1806). Edgington (2013) gives an intruiging and plausible account of a find by E. Lhwyd in Snowdonia in 1688, which was described by Ray (1696: 47) as 'Adiantum trichoides', that might well have been this species.		

Ray, J. 1724. *Synopsis Methodica Stirpium Britannicarum*, ed 3.

See species account : Trichomanes speciosum (sporophyte), p.399

Trichophorum alpinum

* 'Found by Mr. Brown and Mr. Don in a moss about three miles east of Forfar. A specimen of this was presented to the Linnean Society some time ago by Mr. Teesdale'. - Dickson (1794: 290). vc90 **1794**

'Moss of Restenet, Forfarshire, first found in Aug. 1791, in company with Mr. George Don'. - R. Brown in **BM**. vc90 1791

Teesdale's specimen was presented in 1792 (minute book in *Trans. Linn. Soc. 2: 356*).
There is a specimen (as *Eriophorum alpinum*) of this in Hope, 1768, in Balfour (1907) marked 'I am uncertain from whence it came or where it was found'. Mabberley (1985: 20n) suggests this may have been from a foreign source.

Trichophorum cespitosum s.l.

* 'Gr. Sparteum capitulis Equiseti, Beyond the Windmill at Adington in Surry where Peat is dig'd'. - Merrett (1666: 58). 'Juncus parvus montanus cum parvis capitulis luteis *J.B.* ... in pratis udis & locis palustribus saepius occurrit circa Middleton & alibi in agro Warwicensi'. - Ray (1670: 181). vc17 **1666**

Trichophorum cespitosum s.s.

as *Trichophorum cespitosum* subsp. *cespitosum*, Swan (1999), giving four sites in S. Northumberland.	vc67	**1999**
'Twyford Vownog, nr. West Felton, vc40, 1840, Leighton, det. Swan, **BM** & **E**'. - Swan (1999).	vc40	1840

The presence of two species or subspecies had long been suspected, see, for instance, Druce (1928) & Clapham *et al.* (1952). But Swan (*op.cit*) demonstrated the existance of the hybrid between the two species, and re-determined as this hybrid the records from the only two locations given for subsp. *caepitosum* by Clapham *et al*, from Ingleborough and Ben Lawers.

Trichophorum germanicum

as *Trichophorum cespitosum* subsp. *germanicumum*, Swan (1999).		**1999**
'Gr. Sparteum capitulis Equiseti, Beyond the Windmill at Adington in Surry where Peat is dig'd'. - Merrett (1666: 58). 'Juncus parvus montanus cum parvis capitulis luteis J.B. ... in pratis udis & locis palustribus saepius occurrit circa Middleton & alibi in agro Warwicensi'. - Ray (1670: 181).	vc17	1666

This is the common species, and the first record for the aggregate will be the first record for this as it is outside the range of *T. cespitosum*.

Trientalis europaea

* 'Pyrola Alsines flora major ... In betul[et]is Scotiae nata[m] D. Cargillus, ex Scotia misit'. - Bauhin (1620: 100). 'It growes in woods and the shadowie places of Mountaines both in Wales and Scotland'. - Johnson (1641: 31).	**1620**
'in 1603 he [James Cargill of Aberdeen] sent the first recorded specimen of *Trientalis europaea* ... to Bauhin who reported them in his *Prodromus*'. - Raven (1947: 236).	1603

Presumably Johnson is mistaken over its presence in Wales, as there are no subsequent records from there.

Trifolium arvense

* 'Lagopus ... groweth much amog the corne ... rough Trifoly or harefote'. - Turner (1548: D. vij. verso).	**1548**

Trifolium bocconei

* Edinburgh Catalogue (1841).	vc1	**1841**
'Found by Messrs. Borrer and Babington at Cadgewith, Cornwall, but in small quantity'. - Hore (1842). Only in Hore (1845) does he date the record 'in July, 1839'.	vc1	1839

E.B S. 2868 gives the record as Aug. 1840.

Trifolium campestre

* 'Trifolium luteum majus lupulinum, sive lupulus sylvaticus, *ThaI.*'. Thanet. - Johnson (1632: 4).   vc15   **1632**

Trifolium dubium

* 'Trifolium lupulinum alterum minus'. - Ray (1660: 166).   vc29   **1660**

Rose, in Gilmour (1972: 50 & 105), identifies the records 'Medica semine racemosa' in Johnson (1629: 1) and 'Trifolium luteum minimum' (1632: 4) to *T. dubium*. Both Clarke and Druce ignore Johnson (1633: 1186) 'Trifolium luteum minimum', which Smith (1825: 310) equates with *T. dubium*. Note that Johnson does not separate *T. micranthum*.

Trifolium fragiferum

* 'Trifolium fragiferum Clusii'. Cliffe. - Johnson (1629: 8).   vc16   **1629**

See also Johnson (1633: 1208.4) 'Dartford salt marsh, in those below Purfleet'.

Trifolium glomeratum

* 'Trifolium cum glomerulis ad caulium nodos rotundis ... Prope Saxmundham in Suffolcia'. - Ray (1670: 305).   vc25   **1670**

Trifolium incarnatum

* 'I found this plant in two places, about a quarter of a mile distant from each other, in 1839 near the Lizard lighthouse'. - Hore (1842).   vc1   **1842**
    vc1   1839

This is the native subsp. *molinerii*. However in Hore (1845) he gives 1838, as does Byfield (1986).

Trifolium medium

* 'Trifolium majus flore purpureo ... Lately in an enclosed ground near the river Cam, not farre from Newnham by the foot way to Grantcester'. - Ray (1660: 168).   vc29   **1660**

Trifolium micranthum

* 'Trifolium lupulinum minimum *H. Ox II. 142.* ... Locis arenosis subhumidis circa Putney, Blackheath & alibi'. - Ray (1724: 331).   vc16, 21   **1724**

'Buddle, about 1710, who carefully distinguished it from small *T. minus [dubium]*'. - Trimen & Dyer (1869: 82), who cite '[In Tuttle fields Westminster: *Budd. MSS* and Herb. vol. cxix. fol. 39, **BM**]'.   vc21   1710

Trifolium occidentale

'In Sept 1957 I became aware [of this new species] ... (Lizard Peninsula) ...'. - Coombe (1961).   vc1   **1961**

'Guernsey, Gosselin, J., c.1790'. - see McClintock, (1982: 74).   vc113   1790

Trifolium ochroleucon

* 'Trifolium pratense hirsutum majus flore albo-sulphureo … About Cherry-Hinton in many pasture closes and elsewhere'. - Ray (1660: 168).   vc29   **1660**

Trifolium ornithopodioides

* 'Trifolium siliquis Ornithopodii nostras. Mr. Newton, in our Company, found it on Sandy Banks by the Sea side at Tolesbury [Tollesbury, on the Blackwater Estuary] in Essex'. - Ray (1690: 136).   vc19   **1690**

The traditional first record, cited by Clarke & Druce, was 'Trifolium Ornithopodium siliquis … Found by Tho. Willisel among the Corn half a mile on this side Tadcaster … and also near Oxford'. - Ray (1677: 291), which were also cited in Ray (1690: 135). But see Druce (1897: 133) and Lees (1888: 193) where both of these records are doubted. 'Among the corn' is a strange habitat.

Trifolium pratense

* 'Trifolium … common Trifoly or clauer that groweth in myddoes, somtyme wyth a whyte floure and somtymes wyth a purple, which is called trifolium pratense in Latin'. - Turner (1562: 158).   **1562**

Trifolium repens

* 'Trifolium pratense album'. Thanet. - Johnson (1632: 5).   vc15   **1632**

Nelson (1959) and Chapman *et al.* (1995) both assume that Turner (1562: 158) is describing both *T. pratense* and this species. It does not seem clear that he is.

Trifolium scabrum

* 'Trifolium flosculis albis, in glomerulis oblongis asperis, cauliculis proxime adnatis … At Newmarket where the Sesamoides salamanticum [*Silene otites*] grows, and divers other places'. - Ray (1696: 194).   vc26   **1696**

See also Ray (1690: 134-5), where it is probable that the same plant is recorded, but with the wrong name.

Trifolium squamosum

* 'T. stellatum glabrum … I first observed it in Dartford salt marsh the tenth of June 1633'. - Johnson (1633: 1208).   vc16   **1633**

Trifolium striatum

* 'Trifolium dilute purpureum glomerulis florum oblongis sine pediculis caulibus adnatis … In all the closes you pass through going from Cambridge to Chesterton Church'. - Ray (1660: 169).   vc29   **1660**

In Ray's account he wonders if this species is How's (1650: 114) 'T. nodiflorum vel juxta folia floridum nondum descriptum, [In pratis. Knotted Trefoile]', or Bauhin's (1651: 2. 378) 'T. cujus caules ex geniculis glomerulos oblongos proferunt'. Certainly the Bauhin illustration looks like this species, but with no indication that it was found in Britain, and How gives no details of location or finder.

Trifolium strictum

'West of Jersey'. -. Woods (1836: 279).   vc113   **1836**

The first record other than from the Channel Islands is '9th July 1847 … on the same bank [called Cairn William, or Caerthillian, between the Lizard Head and Kynance Cove] I also discovered in great abundance, but past flowering, *Trifolium strictum*'. - Johns (1847). Johns' account includes the famous lines about him covering six clovers with his hat. Druce gives the Jersey record as 1826 rather than 1836.

Trifolium subterraneum

* 'Trifolium pumilum supinum flosculis longis albis, nondum descriptum. In montosis. White Dwarfe Trefoile'. - Johnson (1634: 73). 'At Gamlingay not far from the windmill; in the road between Eltham to Deptford in Kent; in the road between Burntwood and Brook-street in Essex, &c. abundantly'. - Ray (1670: 306).   **1634**

Gilmour (1972: 152) notes that this, with exactly the same wording, is added, supposedly in Johnson's handwriting, to the copy of the *Descriptio* that apparently belonged successively to Johnson, How and Goodyer. This would presumably date the record to 1632 and locate it to Kent.   vc15   1632

This was the plant described by Ray in a letter to Courthope in 1662, see Gunther (1934), and published in Ray (1663).

Trifolium suffocatum

* 'The President presented a specimen of *T. suffocatum* found wild on the sands about Yarmouth by Mr. Lilly Wigg', 1792. - Minute book for May 7th 1793 (Trans. Linn. Soc. ii. 357). 'Aldborough [*sic*], Suffolk [1794]'. - Groves (1905).   vc27   **1794**
    vc27   1792

Groves cited letters from George Crabbe, the poet, written in the summer of 1794, where he was relating details of a plant that he had discovered which he thought was new to Britain.

Triglochin maritima

* 'Gramen marinum spicatum. … In marish and waterie places neere to the sea'. - Gerard (1597: 18.1).   **1597**

Triglochin palustris

* 'Gramen aquaticum spicatum ... In mirie and muddie grounds'. - Gerard (1597: 12).

1597

Trinia glauca

* 'Peucedanum ... I hear say that it groweth also in England, and I found a root of it at saynt Vincentis rock a litle from Bristow'. - Turner (1562: 83b). 'Peucedani facie pusilla planta ... Bristoiae in Anglia ad rupem Vincentii ... magna copia'. - L'Obel (1571: 331).

vc34   **1562**

Tripleurospermum inodorum

'Chamaemelum inodorum annuum humilius, foliis obscure virententibus *H. Ox. III. 36.15.* ... Observed by Mr Rand along the Way to Chelsea'. - Ray (1724: 186.6).

vc21   **1724**

Kent (1975) has as his first record 'Morley, c1677' [the ms. for which is in the Sloane Collection at the British Library].

vc21   1677

Kay (1994) dates the first certain record to Ray (1690), but the only reference traced is in the 1724 edition. Morison (1699: 36) has the species but with no indication the record is from Britain. The first record using this name is in Babington (1843) citing E.B. 676 (1800).

Plate 979, 1802.
Smith J. and Sowerby, J.
English Botany, vol 14.

See species account :
Tripleurospermum
maritimum s.s., p.406

Tripleurospermum maritimum s.l.

* 'There be three kinds of wilde Cammomils ... one stinking and two other not stinking ... They growe in Corne fieldes, neare unto pathwaies, and in the borders of fieldes'. - Gerard (1597: 617).	vc15	**1597**

Tripleurospermum maritimum s.s.

first described as a species in Babington (1843) citing '*Pyrethrum maritimum*, from Manorbier, Pembroke. E.B. 979, 1802' - *see image p.405*.	vc45	**1843**
'Cockbush ad litt. maris Junio'. - Druce & Vines (1907: 66). Marked by Druce as 'first record for Britain and the type specimen'.	vc13	1747
Cockbush Common is near West Wittering. The record from Druce & Vines is dated as 1747, the date of Dillenius' death.		

Trisetum flavescens

* 'Gramen avenaceum panicula flavescente, locustis parvis ... In pratis & pascuis'. - Ray (1670: 141).	vc29	**1670**
This is marked by Ray as a Cambridge plant, but it is not in his 1660 or 1663 works.		

Trollius europaeus

'Trollius flos … Eum etiam superiore anno in Anglia inter primulas flore rubro [*Primula farinosa*]'. - L'Ecluse (1583: 372).		**1583**
Philip Oswald (*in litt.*) points me to an annotation made by T. Penny on C.Gesner's plant illustrations (dated as 'by 1565' in Foley (2006b)) which he translates as 'We have in England along the course of the River Lune another plant entirely similar to this [*Caltha palustris*] except that the flowers are larger, many-petalled [and] with a very pleasant scent. And because the flowers are many-petalled and dense it is called lockergolling'. This is very convincing on a number of grounds.	vc65	1565
L'Ecluse's trip was in 1581, see Raven (1947: 169). He inspected Penny's herbarium and took from him several pictures and descriptions. Foley (2009) too cites the 1583 record. L'Ecluse (1601: 237) cites it again 'Trollius flos … ex septentrionalibus Angliae montibus'.		

Tuberaria guttata

'Cistus flore pallido punicante … nere Grosnez Castle, Jersey, W. Sherard'. - Ray (1690: 238).	vc113	**1690**
Sherard's visit was in 1689.	vc113	1689
The first record other than from the Channel Islands was that of Samuel Brewer in 1726, near Holyhead (vc52) - Clarke, citing his MS 'Botanical Journey' in Bot. Dept. **BM**. But in fact Brewer's diary covers 1726 and 1727. I can find no mention of the discovery		

in Dillenius' account of the visit to Anglesey in 1726, and yet in Brewer's 1727 diary he refers to receiving better specimens of the plants collected 'last year' (see Hyde (1931: 24)). Druce & Vines (1907: 105.4) also cite a specimen from Brewer, presumably from the same source. Raven (1947: 156) and Foley (2006b: 98) mention a drawing by Thomas Penny of this species 'T.P.Angli ex picta' (dated by Foley as before 1565), but this is probably from a continental source.

Turritis glabra

* 'Turritis; Towers Mustard … In the west part of England upon dunghills and such places … at Pyms by a village called Edmonton, neere London, by the citie wals of West-chester in the corne fields and where flax did growe about Cambridge'. - Gerard (1597: 213). vc21, 29, 58 **1597**

Westchester is Chester - see *Cochlearia officinalis*.

Tussilago farfara

* 'Tussilago … Bulfote or horsehofe … groweth by water sydes and in marishe groundes'. - Turner (1548: G. vj. verso). **1548**

Typha angustifolia

'Peculiare quoddam genus Typhe veluti medium observabam anno M.D. LXXXI. copiose nascens palustri quadam fossa apud Tyburn facellum … [a certain unusual kind of Typha growing copiously in a certain marshy ditch by the chapel at Tyburn, not far from that place in which are buried those whose neck has been broken by a noose because of their misdeeds, at the first mile from the city of London, where it looks West]. - L'Ecluse (1601: ccxv). vc21 **1601**
vc21 1581

L'Ecluse goes on to contrast this species with 'Typhae vulgaris folia'. Kent (1975 and Edgington (2011) both accept this record. But Hanbury & Marshall (1899: 356) refer to Gerard (1597: 42) 'smaller kinde' of Reed-mace' observed by him in the 'Ile of Shepey' as this species. Harvey's (1981) *Service Index to Gerard* does not include this.
Pulteney cites a reference in L'Obel (1571: 41) as a record of this ' … Monente peritissimo & admodum accurato plantarum expensore Thoma Penio Anglo, qui hanc vulsit superioribus annis & nobis impertiuit', but I see this more as a reference to the man, who was in France at the time.
Clarke's first date was 'Typha palustris media J.B. … vidi in rivulo quodam juxta aedes Nobiliss. Comitis Warwicensis Leez-house dictas in Essexia'. - Ray (1670: 308)'.

Typha latifolia

* 'Typha groweth in fennes and water sydes among the reedes … cattes tayle, or a Reedmace'. - Turner (1548: G: vij). **1548**

Ulex europaeus

* 'Genista, Spartium spinosum majus ... Brike Browme ... In Anglia, ubi frequentissima in sterilibus et ericetis'. - L'Obel (1571: 409). 1571

An earlier record might be Turner (1538, B. iv) 'Paliurus ... quem aliqui a Whyn, alii a Furre nominant'. In addition Turner probably referred to this as 'a fur; whyche in manye places of Englande is called a Whyne'. - (1551, D. j), where it is contrasted with 'pety whine'. Both Nelson (1959) and Stearn (1965) cite the 1538 reference for this species.

Ulex gallii

* 'Three years ago ... my attention was aroused ... specimen from Dorsetshire'. - Planchon (1849). vc9 1849

'We have two sorts [of the furze-bush] ... one a dwarf-furse of a small prickle ... which we call Cornish furze, never growing three feet high, flowering in the autumn; the other five, six and eight feet high, we call French furze ... blossoms in the spring'. - Borlase (1758: 224). as *U. nana* 'on the Goonhilly Downs [The Lizard]'. - Jones (1820). 'Mull of Galloway, 1834, **E**'. 'Aug 26th 1836 ... A *Ulex* which appears to be new [on St Vincent's Rocks], Mr Forbes said it resembled *U. provincalis* of the south of France'. - Babington (1897: 57). vc1 1758

Planchon's paper is followed directly by two more, from Forbes and Babington, with other sites. White (1912) does not mention the reference in Babington's Journal.

Davey (1909: xxx) cites Carew, in his Survey of Cornwall (1602) as recording both *U. europaeus* and *U. gallii* in Cornwall (and gives 1912: 110) Carew's reference to the latter as *U. nanus*). I cannot find any specific reference in Carew, though he does write (p68) 'furze - the shrubby sort is called tame, the better grown French'. However, Hopkins (1983) too cites Carew, adding that *U. europaeus* is the 'better grown French' and *U. gallii* equates to 'tame'.

Carter (1986) cites a George Owen of Henllys, in his manuscript 'The description of Pembrookshire' (1603), referring to Gorse 'this last kind blossometh with the heath in the latter end of harvest against winter, whereas the former accompanieth the broom and bloweth in May during the summer'. One way or the other, it is quite plain that botanists have suspected two species for centuries. There is a further very early record that should be noted. Horsman (2011) examines the notebook of Christopher Hunter, kept in the Dean & Chapter Library at Durham. He identifies as *U. minor* the 'Genista spinosa minor *Park*. On the top of ?Brushston Bank [Brusselton] as you go from Aukland .... I saw it there in flower on ye 2d of Octob. 1699'. *U. minor* is a very unlikely record from Durham, but *U. gallii* is possible from the flowering date.

Ulex minor

* 'Genistella Anglica spinosa alpina, sive Chamaespartum supinum. Non procul a Castro South Sea Castlc in Comitatu South-hampton. Creeping Dwarfe furze'. - Johnson (1641: 21).   vc11   **1641**

Parkinson (1640: 1003) has 'Genista spinosa minor … growing in our owne land', which might be this species.

Ulmus

'Ulmus ... an Elm tree ... it groweth comonly in all countrees'. - Turner (1562: 169b).   **1562**

See Armstrong & Sell (1996) for a full account of the historical records, and also Melville's papers in *Journal of Botany* (1938; 1939a, b, c: 1940).

Ulmus glabra

'Ulmus folio glabro … Witch elme … between Rumford and Stubbers [North Ockendon] in the yeere 1620'. - Goodyer in Johnson (1633: 1478.4).   vc18   **1633**
vc18   1620

See Armstrong & Sell (1996) for a full discussion. Clarke cited 'Ulmus ... an Elm tree ... it groweth comonly in all countrees'. - Turner (1562: 169b), and Druce has 'Ulmus latifolia … neere South fleete'. - Gerard (1597: 1297). Neither of these seem precise enough.

Ulmus minor

'Ulmus minor folio angusto scabro … in the hedges by the high way … betweene Christ Church and Limmington in the New Forrest in Hampshire, about the middle of September 1624'. - Goodyer in Johnson (1633: 1478.2).   vc11   **1633**
vc11   1624

This is *U. minor sensu* Stace. See also Armstrong & Sell (1996). Apparently (D.E. Allen *in litt.*) Melville (1938) identified the (original?) tree concerned as one clonal extract of *U. angustifolia* (and called it var. *goodyeri*) and discusses (1939c) the concept of *U. minor* Mill.

Ulmus plotii

Melville (1940), citing a tree from Banbury as the type specimen, and referring to Druce's (1911c) first description of this as *U. plotti* Druce *sp.nov.*   vc23   **1940**

Cransley, vc32, 1878, Druce, **BM**.   vc32   1878

Melville (*op.cit.*) cites all of Druce's papers on this tree, but concludes that neither Druce's descriptions nor his reference back to Plot (1677) are sound. Earlier field records might well exist. Druce (1920a) cites a specimen from Wales in the Hortus Siccus of Edward Morgan, who died in 1689, in **OXF**, but I have not had this re-determined.

Ulmus procera

'Ulmus vulgatissima folio lato scabro, the Common Elme tree'. - Johnson (1633: 1478.1).   **1633**

See Armstrong & Sell (1996) for a full discussion.

Umbilicus rupestris

* 'Umbilicus Veneris ... in welles and divers places of Summerset shyre ... wall penny grasse'. - Turner (1562: 169).   vc5, 6   **1562**

Is Turner's reference to 'wells', the habitat, or 'Wells', the place? Chapman *et al.* (1995) adopt the latter, but the former seems more likely.

Urtica dioica

* 'Our comon nettel of Englande'. - Turner (1562: 170).   **1562**

This seems very late for such a widespead species, but in Turner (1548: H. i) he specifically states 'The true netel groweth not in England out of gardines'.

Urtica urens

* 'Urtica minor … also cometh up in untilled places'. - Gerard (1597: 571). 'Urtica minor acrior'. Thanet. - Johnson (1632: 4).   **1597**

Lyte (1578: 128) pictures this as 'groweth in all places, as by hedges, quick settes, and walles'.

Utricularia australis

* 'We have been shown a specimen of this plant in **BM**, collected by the late Edward Forster in a gravel pit in Henhault Forest, Essex'. - Seemann (1867).   vc18   **1867**

as *U. major* 'in standing water in ye Horse road about a mile before I came to Melin Wen near Newboro in Anglesey'. - Druce & Vines (1907: 147). Presumably this is Dillenius' own record from his 1726 journey   vc52   1726

The record from Anglesey has been assigned to this segregate as *U. vulgaris s.s* is not claimed from that county.
See also Webb (1876) for further notes on this species and on the Henhault record.

Utricularia bremii

'New Forest, discovered in 1990s'. - Stace (2010).   vc11   **2010**

'Andy [Smith] had first found the population [at Denny Lodge] in the early 1990s'. - Rand (2011).   vc11   1995

See also Hall (1939), setting out other possible records, including specimens collected in 1833 from Moss of Inshoch, vc96 (specimen in **BM** from 1898) and Loch Spynie, vc95 (specimen in **LINN**). He decided that there was no firm evidence for their occurrence in Britain.

Utricularia intermedia s.l.

as *U. media* 'observed by Dr Scott, during successive seasons, growing in a peat drain near Scottiborough, county of Fermanagh'. - Mackay (1806: 129). 'Found by the late Dr. Scott near Dublin, and by Miss Hutchins in the West of Ireland'. - E.B. 2489.	vcH33	**1806**
'In a marsh near Mr McKinnon's at the foot of Ben-na-callich, Sky, Sept. 1778', det. (as *U. ochroleuca*) A. Bennett, 2/11/1912, **E**. 'Almost certainly *U. ochroleuca*', F. Rumsey (*in litt.* 2017).	vc104	1778

Colgan & Scully (1898) suggest that Dr Scott found the plant in Fermanagh about 1804. The Dublin record has never been accepted.

Utricularia intermedia s.s.

Bennett (1910b).	**1910**

It is not at all easy to trace a first record in the literature; the Bennett reference has been cited as that was the first referring to two varieties of *U. intermedia s.l.* However a more detailed discussion, well worth noting, is in Clarke & Gurney (1921). It has been impossible to cite an earlier reliable herbarium specimen.

Utricularia minor

* 'Millefolium palustre galericulatum minus flora minore'. - Ray (1677: 200). 'Found by Mr. Dent on Teversham Moor in Cambridgeshire'. - Ray (1685: 13).	**1677**

In Ray (1677) this appears merely as a name, with no other details.

Utricularia ochroleuca

as *U. intermedia* var. *ochroleuca* 'In 1903 I reported it for Scotland, based on specimens gathered at Broadford, Skye, by Mr Symers M. Macvicar, July 1895'. - Bennett (1910b).	vc104	**1910**
'In a marsh near Mr McKinnon's at the foot of Ben-na-callich, Sky, Sept. 1778', det. (as *U. ochroleuca*) A. Bennett, 2/11/1912, **E**. 'Almost certainly *U. ochroleuca*', F. Rumsey (*in litt.* 2017). Druce (1911b) 'Kylemore, Galway, 1875, Loch Mallachie, Easterness, 1882, both det. Gluck'.	vc104	1778

Bennett's 1903 reference is to a note of his in *Annals of Scottish N.H* (12: 123), but I feel that the reference there was still only a tentative identification. Druce (*ibid.*) also cites another specimen of his from near Aviemore, 1869, but does not say whether that was determined or not.

Utricularia stygia

Stace (1991). **1991**

L. Maddy, N. Uist, vc110, 1894, W.A. Shoolbred, det. Rich & Smith, **BM**. vc110 1894

This was first described by Thor (1987, 1988), but not incorporated into the British flora until Stace (*op. cit.*).

Utricularia vulgaris s.l.

'Millefolium palustre galericulatum … they may be found in lakes and standing waters, or in waters that run slowly, I have not founde such plentie of it in any one place as in the water ditches adoiyning to Saint George his field [Southwark] neere London'. - Gerard (1597: 679). 'About Oxford'. - How (1650: 76). vc17 1597

The illustration was added in Johnson (1633: 828.5). The Southwark location given in Gerard covers this species, *Hottonia palustris* and *Myriophyllum spicatum* but Salmon (1931) only cites it for this species.

Vaccinium microcarpum

as *V. oxycoccus* var. *microcarpum* 'from Cheshire and Sutherlandshire'. - Moss (1912). vc108 **1912**

'Near L. Brandy, G. Don, 1806'. - Ingram & Noltie (1981). Ben ??Bedl, July 1845', det. Warburg, **OXF**. 'Glen Callater', A. Croall, 1855, det. E.F. Warburg, **BM**. vc90 1806

Druce (1914a) cited Moss, adding more details, but see Lousley (1936) for a fuller treatment. Ingram & Noltie make a convincing case for the Don record. I was not able to clearly read the location on the **OXF** sheet, and have not been able to locate 'Ben ??Bedl' (?Ledi, vc87).
No evidence has been found for the Cheshire record, which is totally outside the presumed range.

Vaccinium myrtillus

* 'Vaccinia nigra aut rubra … In Anglia, Belgio Gallia & Germania … fructum esitavimus'. - L'Obel (1571: 417). 'In certayne woods of … Englande'. - Lyte (1578: 670). **1571**

<1379. ' … Pety Juniper … in plenty on a heath on the borders of Dorset and Wiltshire, near Shaftesbury'. - Henry Daniel, see Harvey (1981), and also Allen (2001).

Vaccinium oxycoccos

* 'Upon bogs ... And fennie places especially in Cheshire and Staffordshire'. - Gerard (1597: 1367). vc39, 58 **1597**

Clark & Druce both cite 'Vacinia palustria. Marrish Whorts ... Fenberries'. - Lyte (1578: 671), but that reference seems to only cite Holland.

Vaccinium uliginosum

* 'Vaccinia nigra fructu majore *Park*. ... At Osten [Orton] in Cumberland [*sic*], a village in the mid-way between Hexham and Pereth [Penrith] in the moorish pastures. Th. Willisell'. - Ray (1670: 309).   vc69   **1670**

Vaccinium vitis-idaea

* 'Vaccinia rubra ... In Westmerland at a place called Crossby Ravenswaith ... [and] in Lancashire also upon Pendle Hills'. - Gerard (1597: 1230). 'In the wilde moores of Northumberland'. - Johnson (1641: 35).   vc59, 69   **1597**

Valeriana dioica

* 'Valeriana minor ... in moist places hard to river sides'. - Gerard (1597: 917).   **1597**

Valeriana officinalis

* 'Phu ... growing about water sydes, and in the moyst plasshes and in morish groundes, and it is called in englishe wylde Valerian'. - Turner (1548: F. iij).   **1548**

Valerianella carinata

* 'Gathered by Mr. E. Forster at Ongar in Essex'. - Woods (1835b: 427).   vc18   **1835**

'Guernsey. Gosselin, J., c.1790'. - see McClintock (1982: 136).   vc113   1790

Valerianella dentata

* 'Found in Cornwall by Mr. E. Forster, jun., in 1799'. - Smith (1804: 1385).   vc1   **1804**

'Valerianellae vulgaris species major, serotina, *Mor. Praelud. 319* ... from Mr Dale [between London-Coney and St Albans, Herts]'. - Druce & Vines (1907: 69). 'Valerianella vulgaris, seu Lactucae agninae species major serotina *Moris. Praelud.* Found among corn at Chiselhurst in Kent. Herb. Du Bois'. - Druce (1928: 474).   vc20   1724

Ray (1724: 201) gives three species of *Valerianella* and Druce after his entry on *V. rimosa*, notes the Herb. Dill. plant is *dentata*. This must have been gathered by 1724. However Druce (1927) gives 1746, ex Dillenius ms., and this is the date used in Preston *et al*. (2004). Druce implies that the Chiselhurst reference is in Ray (1724) but it is not. I cannot find a satisfactory paragraph in Ray, and Smith does not cite him.

Hanbury & Marshall (1899) cite an earlier record from Jacob (1777) 'Lactuca agnina seu Valerianells foliis serratis. In Fields among Corn, and on old Walls, common'. This may well be this species, but there can be no certainty.

Valerianella eriocarpa

'Portland, waste rocky ground under Pennysylvania Castle'. - Mansel-Pleydell (1895: 145).	vc9 vc9	**1895** 1874

This is the first record as a presumed native; in this second edition he refers to the first record as 1874, which is the date of the first edition, but there is no record of this species in that. See Pearman & Edwards (2002) for a full treatment of the status and distribution. The first earliest record, but as an alien, is 'Between Henley Castle and Barnard Green, Worcestershire, collected by Mr. E. Lees ... in 1845'. - Syme (1865: iv. 244).

Valerianella locusta

* 'Phu minimum ... saepe nobis visa & enata in Anglia'. - L'Obel (1571: 319)'. 'Lactuca agnina ... wilde in the corne fieldes'. Gerard (1597: 243).		**1571**

Valerianella rimosa

* 'Valerinellae vulgaris species major, serotina *Mor. Praelud. 319* ... In the Corn fields between Ore and the Foot Ferry to Shepey Isle in Kent. Also in the third or fourth Field on the right Hand of the Road going from London-Coney [Colney] towards St. Albans in Hertfordshire; Mr Dale'. - Ray (1724: 201).	vc15, 20	**1724**

See Druce & Vines (1907: 69) who identify the Hertfordshire plant as *V. dentata*. However, specimens from Ore are in the herbarium of Joseph Andrews 'given to him by the discoverer, Mr Dale and now in **BM**', and are this species. - Boulger (1918).

Verbascum blattaria

* 'Blattaria flore luteo, Yellow Moth Mullein ... Blackheath, and near De[p]tford'. - Parkinson (1640: 64.3).	vc16	**1640**
from Tottenham 'Blattaria major flore luteo, sive Blattaria Plinii, *Lob.*'. - Johnson (1638: 4).	vc21	1638

See the note on *V. virgatum*. Johnson's 1638 record is not in Kent (1975), though we know that he had seen Johnson's 1638 manuscript.

Verbascum lychnitis

* 'Tapsus barbatus flore albo ... Blacke Heath next to London ... Eltham neere unto Dartford in Kent ... Hiegate neere London'. - Gerard (1597: 629).	vc16, 21	**1597**

The location in Gerard covers both this and 'Great Mullein' (*V. thapsus*).

Verbascum nigrum

* 'Verbascum nigrum latifolium luteum … pratis Brabanticis, atque in Francia & Anglia … sponte provenit'. - L'Obel (1571: 242). 'In many places in Kent'. - Parkinson (1640: 61).

1571

Verbascum pulverulentum

* 'Verbascum pulverulentum flore luteo parvo *J.B.* … circa moenia Norvici [Norwich] urbis'. - Ray (1670: 312).

vc27   1670

Verbascum thapsus

* 'Verbascum … a vulgo Mullen aut Longwort appellatur'. - Turner (1538: C. verso). 'Tapsus barbatus … Upon the end of Black Heath next to London as also about the Queens house at Eltham neere unto Dartford in Kent'. - Gerard (1597: 630).

1538

Verbascum virgatum

* 'First shewn me by my late worthy friend Mr. Waldron Hill of Worcester … near that town'. - Withering (1787: 228).

vc37   1787

In Hanbury & Marshall (1899) Johnson's 'Blattaria vulgaris flo. luteo' (1629: 10) is referred with some doubt to this species. Rose, in Gilmour (1972: 64), notes that Johnson did not distinguish between this species and *V. blatteria*.

Verbena officinalis

* 'Verbenaca, in english Vervine … groweth in many places of England'. - Turner (1548: G. vij).

1548

See also Turner (1538: C. verso) 'Verbena recta scilicet, que vulgo dicitur Veruyne', which implies familiarity with the plant as a wild species. Grigson (1958) cites the Anglo-Saxon use of this species in the '*Lacnunga*'.

Veronica agrestis

* 'Alsine foliis Trissaginis … being gotten into a garden ground it is hard to be destroied but naturally commeth up from yeere to yeere as a noisome weede'. - Gerard (1597: 491).

1597

Veronica alpina

* In montibus prope Garway Moor et in Ben Nevis' on his trip 'last year' [ie 1789]. - Dickson (1790: 29 & 1794: 288).

vc96, 97   1790

' … found in the Highlands of Scotland by Mr. Dickson in 1786, and not before in this island'. - Smith (1791).

vc96   1786

Garway Moor is probably Garvamore (NN5294), on the upper R. Spey. Dickson recorded both this species and *Phleum alpinum* from this site, and here are recent records of the latter from Carn Dubh, three miles to the south-west.

There is a specimen of '*V. alpina*' in Hope, 1768 (see Balfour 1907), 'from Dr de la Roche and Messrs. Fabricius, 1767 Ben Nevis', but Smith (1824: i. 20), citing Lightfoot (1777) too, makes this name a synonym of *V. serpyllifolia* subsp. *humifusa*, for which that would be the first record. Note that Fletcher (1959) accepts, I feel erroneously, that record as being of this species.

Veronica anagallis-aquatica s.l.

* 'Anagallis aquat. angustioribus foliis flo. albidis'. Erith. - Johnson (1629: 10).	vc16	**1629**

Both Clarke and Druce ignore Gerard (1597: 496) 'Anagallis aquatica, by rivers sides, small running brookes, and waterie ditches', which Ray (1724) and Smith (1824: i. 21) equate to this species. This is certainly possible, but not conclusive. Rose, in Gilmour (1972: 63), suggests that Johnson's record might be of *V. scutellata* and that 'Anagallis aquat. Latioribus foliis.flo. caeruleis'. Erith Marshes - Johnson (1629: 10) is this species, but this is an attribution that he (1972: 113) also makes for 'Anagallis aquatica major foliis acutioribus, floribus albidis'. - Johnson (1632: 19).

Veronica anagallis-aquatica s.s.

Druce (1911a), Britton (1928). See also *V. catenata*.		**1911**
'near Peterburrow House by ye Thames side', an annotation in Petiver's hand in H.S.152: 41, **BM**.	vc21	1718

Petiver died in 1718. Kent (1950) dates Petiver's record to 1695, but I cannot see any evidence to support this. Kent also notes that Petiver's site is now occupied by Millbank.

Veronica arvensis

* 'Ad insulam Sheppey … Alsine foliis veronicae'. - Johnson (1629: 5).	vc15	**1629**

This is in Gerard (1597: 489), as 'alsine foliis veronicae', but with no specific mention as a British plant, other than a general habitat, though Pulteney gives this as his first record.

Veronica beccabunga

* 'Cepaea … Brooklem ... groweth in water sydes and by brookes & sprynges'. - Turner (1548: C. i).	**1548**

Grigson (1955) notes that 'Lemmington in Northumberland (*Lemocton* in 1201) was the homestead where the *hleomoc*, or Brooklime grew, and that a plant with this name comes into Anglo-Saxon remedies in the *Lacunga* [a collection of miscellaneous Anglo-Saxon medical texts]'.

Veronica catenata

'Our British plant appears (*teste* Prof. Hugo Glück) to be *Veronica aquatica*, S.F. Gray'. - Druce (1911a). 'Southport, Lancashire, Hampstead Norris, Berkshire, and near Galway'. - Druce (1912a). **1911**

'Guernsey, Grande-mare. Gosselin, J., c.1790'. - see McClintock (1982: 119). vc113 1790

The Glück work referred to is vol 3 of *Biologische und morphologische Untersuchungen uber Wasser- und Sumpfgewachse* (1911). Druce's account (*op.cit.*) is not at all clear; see Britton (1928) for a better exposition of the two species.

Veronica chamaedrys

* 'Teucrium pratense, Ang. Wilde Germander … in nemoribus & pratis'. - L'Obel (1571: 209). 'In many places about London'. - Gerard (1597: 531). **1571**

Veronica fruticans

* as *V. saxatilis* 'in rupibus, Ben Lawers [in 1789]'. - Dickson (1790: 29 & 1794: 288). vc88 **1790**
vc88 1789

Hudson (1778: 4) gives a record of *V. fruticulosa* 'in montibus, in monte Ben Nevis dicto in Scotia', and *V. saxatilis* and *V. fruticulosa* were maintained as separate species until Babington (1867). But Ben Nevis is outside the range of V. *fruiticans*.

Veronica hederifolia

* 'Alsine hederacea … In gardens among pot herbes … and in the fieldes after the corne is reaped'. - Gerard (1597: 493). **1597**

Veronica montana

* 'Chamaedrys spuria foliis pediculis oblongis insidentibus'. - Ray (1663: 4). 'Alyssum mont. Columnae … In Ericeto Hamstediano'. - Merrett (1666: 6). vc29 **1663**

Veronica officinalis

'Veronica vera et major … groweth upon bankes, borders of fields, and grassie mole-hils, in sandy grounds, and in woods, almost everywhere'. - Gerard (1597: 502). 'Veronica mas *Dod. Fuch.* vera et major *Lob.* … Speedwell Fluellen'. - Johnson (1634: 76). **1597**

Clarke & Druce both cite 'Veronica groweth in many places of England, and it is called in englishe Fluellyng'. - Turner (1548: H. v). This does not seem to be convincing enough to cite as a first record.

Veronica scutellata

* 'Anagallis aquatica 4ta *Lob.* angustifolia *Bauh.* in uliginosis. Narrow leafed Brooklime'. - Johnson (1641: 15). 'Found on Teversham moor near Fulborn, and in a close, compassed about with a great ditch and hedge in the open field within a quarter of a mile of Barnwall' - Ray (1660: 11).

1641

Both Clarke and Druce ignore Johnson (1633: 621) 'Anagallis aquatica quarta', which Smith (1824: i. 21) equates to this species, but Johnson does have his doubts, especially over the illustration that he provides. Ray (*op.cit.*) too expresses his doubts over the identification with L'Obel's plant, but concludes by saying 'perhaps it may be the same'.

Veronica serpyllifolia

* 'Betonica Pauli ... groweth in Englande in a parke besyde London ... Paules betony or wodde Peny ryal'. - Turner (1548: B. v. verso).

vc21   **1548**

Johnson, T. 1641. *Mercurius Botanicus pars altera.*

Veronica spicata

* 'Veronica recta mas, *Lob. Ger.* ... Great Speed-well or Fluellin. Found at Saint Vincente Rocke by Master Goodyer'. - Johnson (1641: 36). 'Ad latera montis cujusdam Craig Wreidhin [Craig Breidden] dicti in comitatu montis Gomerici Wallire D. Lloyd'. - Ray (1690: 119).

vc34   **1641**

This is recorded as an annotation, of plants found his native county, by Sir John Salusbury, of Denbighshire, in his copy of Gerard (1597) dated to 1606 - 1608. See Gunther (1922: 244). However this record is not mentioned in Carter's (1960) account. Found by Goodyer on St Vincent Rocks Bristol in 1638. See Gunther (1922: 76). vc50 1608

Veronica triphyllos

* 'Alsine foliis hederaceis Rutae modo divisis, *Lob*. Sent by Tho. Willisell, and found by him at Rowton in Norfolk, betwixt the town and the highway, twelve miles before you come to Norwich; and at Mewell in Suffolk [? Methwold in Norfolk], between the two Wind-mills, and the Warren-lodge in wheat-ground, on the right hand of Lynne-road; and in gravel-pits, two miles beyond Barton-mills, on the ridge of a hill where a small Cart-way crosseth the road to Lynne; and in the grass thereabout plentifully'. - Ray (1670: 340). vc28 1670

Ray corrected his citation to 'alsine folio profunde secto, flore purpureo aut violaceo, *J.B*'. - Letter to Mr Lister, July 17th 1670, see Lankester (1848: 61).

Veronica verna

* 'Found by Sir John Cullum near Bury in Suffolk'. - Rose (1775: app. 444). vc26 1775

Simpson (1882) has Sir J. Cullum, 1771, as does Hind (1889). Trist (1979) has 1773. vc26 1773

Cullum's papers are in the West Suffolk Record Office in Bury St Edmunds (M. Sanford *pers.comm*.) and have not been checked, but 1773 seems the more likely date.

Viburnum lantana

* 'Viurna vulgi Gallorum & Ruellii ... In Italiae, Angliae, & Galliae Burgundae ... senticosis & sylvosis passim'. - L'Obel (1571: 436). 'In the chalkie groundes of Kent, about Cobham Southfleete and Gravesend, and al the tract to Canterburie'. - Gerard (1597: 1305). 1571

Viburnum opulus

* 'Montana ambucus aquatica ... In Angliae ... pratensibus udis convalliumque'. - L'Obel (1571: 444). 1571

Vicia bithynica

* 'Ad sepes prope Doncaster in agro Eboracensi, D. Tofield. in insula Purbeck [Dorset]'. - Hudson (1778: 320). vc9 1778

The Doncaster record is excluded by Lees (1888: 202) and not mapped by Preston *et al*. (2002). Yet Skidmore *et al*. (n.d.) say that 'the original specimen was in Tofield's herbarium when in the possession of William Younge, where it was seen by Smith in 1800'. If that record is correct then it was probably as a casual. The Dorset record is accepted as a native.

Vicia cracca

'Vicia multiflora sive spicata, Tufted Vetches, woods and moist grounds with us, among hedges and bushes, roote creepeth underground farre about'. - Parkinson (1640: 1072.5) **1640**

Clarke and Druce both cite 'Aracus sive Cracca major *Lob*.'. Hampstead. - Johnson (1629: 12), but Rose, in Gilmour (1972: 67), identifies this as *V. sativa*. However he accepts the record from Johnson (1632: 23) 'Vicia maxima dumetorum, *Bauh*. Cracca major, *Tab*. Woods east of Canterbury', which Clarke and Druce have cited for *V. sepium*, and I have followed.

Vicia hirsuta

* 'Vicia sylvestris Strangle Tare … ramping and climbing among corne ... the herbe is better knowne then desired'. - Gerard (1597: 1053), 'Aracus, sive Cracca minima *Lob*'. Between Sandwich and Canterbury. - Johnson (1632: 23). **1597**

Aracus, sive Cracca major & minor, *Lob*.'. Hampstead to Kentish Town.- Johnson (1629: 12) is accepted as referring to this and *V. sativa* by Rose, in Gilmour (1972), and Edgington (2011). Pulteney cites Lyte (1578: 484.2) where he contrasts what seems to be this species, with two seeds, 'grey-speckled … amongst Rye & Otes'.

Vicia lathyroides

* 'Vicia minima praecox Parisiensium *H.R. Par*. Found by Mr. J. Sherard and Mr. Rand on the Chalky banks near Green-hithe in Kent'. - Ray (1724: 321). vc16 **1724**

'Vicia pratensis verna seu praecox Solomensis semine cubico, seu Hexaedron referente moris. ... Blount's key near Newcastle'. - Ray in Camden (1695). vc66 1695

The Blount's Quay record is cited by Graham (1988) as from '(John Ray) Camden 1722'. The 1722 edition was a reprint of the 1695 edition, and if that is the case then the record must date from before 1695. Blount's Quay has not been traced, but there is a Blount Street on the east side of Newcastle.

Vicia lutea

'On Glastenbury Tor-hill; circa Weymouth in agro Dorsetsiensi'. - Hudson (1778: 319). vc6, 9 **1778**

Clark (1900) adds that there are specimens in **BM** collected at Glastonbury Tor, 1739 (Herb. Rand), and at Weymouth by Lightfoot in 1774. 'The late Mr. Humphrey of Norwich found this plant many years ago on the beach at Orford, Suffolk ... Our vc6 1739 specimens were gathered at Weymouth, by the Rev. Mr. Baker and A. B. Lambert, Esq., in August, 1795'. - E.B. 481.

De Tabley (1899: 89) relates an occurrence of this in Cheshire in 1862, suggesting it might well be the survivor of a ballast alien dating back to 1621!

Vicia orobus

* 'Orobus sylvaticus nostras … At Bigglesby [Gamblesby] in the way to Pereth [Penrith] in Cumberland from Hexham, in the hedges and pastures: as also below Brecknock-hills, in the way to Caerdyff; and in Merionith-shire, near a little village going down the hill from Denbigh, not far from Bala'. - Ray (1670: 339).   vc48, 70   **1670**

Vicia parviflora

* as *V. gracilis* 'On Barrow Hill, Bath'. - Babington (1839c: 74).   vc6   **1839**

as *V. gracilis* 'In a lane near Cobham leading to Ifield from the corner of Cobham Park, 1836, Quekett in Herb. Borrer'. - Hanbury & Marshall (1899). 'W.A. Bromfield, near Coppid Hall, Ryde, July 16. 1839'. (Note on original drawing for E.B.S. 2904, see Garry (1903: 53)).   vc16   1836

Probably the *E. tetrasperma*, var. 2, of Withering (1787: 781), found by Mr Woodward 'on a remarkably dry gravel' near Cambridge, was *V. gracilis*'. - Crompton (2007), citing Babington (1860b: 62).

Vicia sativa

'Aracus, sive Cracca maior & minor, *Lob*.'. Hampstead. - Johnson (1629: 12).   vc21   **1629**

Clarke & Druce both interpreted this as *V. cracca*, but Kent (1975) and Edgington (2011) prefer this species. Druce cited 'Cracca major. Sandwich, Canterbury'. - Johnson (1632: 20). Rose, in Gilmour (1972: 119), has some doubts this record (see *V. cracca*) but suggests that 'Aracus, *Tab*. Galegae silvestris, *Dod*.' Ash nr Canterbury. - Johnson (1632: 21) might be either *V. sativa* or *V. sepium*. There are earlier records, such as in Turner (1562: 162), cited by Pulteney, that are certanly a *Vicia*, but without any certainty that it is this species.

Vicia sepium

* 'Vicia maxima dumetorum, *Bauh*. Cracca major, *Tab*.'. East of Canterbury. - Johnson (1632: 23).   vc15   **1632**

Johnson (1633: 1227.2) has this too, as does Ray (1696: 304) under 'V. sepium perennis *J.B.* maxima dumetorum, *C.B. Ger. emac. Park*. Bush-Vetch'. But Rose, in Gilmour (1972: 121), identifies Johnson's 1632 record as *V.cracca* (see that entry). See the note on *V. sativa* above.

Vicia sylvatica

* 'Vicia maxima sylvatica, nondum descripta … Great wood Vetch, or Fetch. In a wood nigh Bathe'. - Johnson (1634: 76).   vc6   **1634**

Vicia tetrasperma

* 'Viciae sive Craccae minimae species cum siliquis glabris *J.B.* … In the corn as you goe from Hoginton [Oakington] to Huntington roade; and in the woods at St. George Hatley'. - Ray (1660: 175).   vc29   **1660**

'4th June 1657 in Goodyer's descriptions, ms. 11, f. 145'. - See Gunther (1922: 193), with no location, but where Gunther adds '?First evidence for Hants'.   1657

Vinca minor

* 'Vinca pervinca … Periwincle … Groweth plentuously in Englande in gardynes and Wylde also in the west cuntre'. - Turner (1551: K. vj. verso).   **1551**

Raven (1947: 103) implies that Turner saw this plant on his trip to Purbeck and area in 1550.   vc9   1550

Turner (1538: A. iv. & 1548: C. iij. verso) refers to 'Clematis daphnoides … anglis Perwyncle dicitur', which implies familiarity with the plant, but with no certainty that he was referring to a wild record.

Viola arvensis

* 'Common in cornfields'. - E.B.S. 2712, Forster (1831) and separated from *V. tricolor*. He adds that Johnson's (1633: 864.4) altered figure 'is an indifferent representation, though the description under 864.3 evidently belongs to it.   **1831**

'Viola tricolor sylvestris … the flowers [of which] are of a bleake and pale colour farre inferior in beautie to that of the garden'. - Gerard (1597: 704.3).   1597

Not satisfactorily distinguished from *V. tricolor* until Forster. Clarke wondered if Gerard's wild Pansy was this species. It seems that Gerard's next species (704.2) 'V. tricolor petraea, Stony Hearts-ease' could equally have been this. Ray (1660: 177) too was familiar with this species (see Oswald & Preston (2011: 314n), and if the Gerard record is unsatisfactory, this would be the first evidence.

Viola canina

* 'Violae caninae varietatem, si non speciem diversam observavit D. Du-Bois … In pascuis circa Mitcham'. - Ray (1724: 364, with a figure).   vc17   **1724**

Merrett (1666: 125) has 'Viola canina fl. albo, in Hamstead wood on that side the Chestnut walk where the two wayes meet'. Pulteney and Smith (1824) identify that record as this species, but this is not followed by Kent (1975).

Viola hirta

* 'Viola fol. Trachelii serotina hirsuta radice lignosa. In Charlton Wood and in the lane leading to Sittingbourn, and in the way to Lewsham in a great Gravel pit'. - Merrett (1666: 125).   vc16   **1666**

Viola kitaibeliana

'Jersey, April 1871 ... a dwarf pansy, which agrees well with ... *V. nemensensis* ... var. *nana*'. - Trimen (1871b).   vc113   **1871**

Recognised as a separate species by Baker (1901). Note that Le Sueur (1984: 99) says 'it is more than likely that Piquet's 1851 record of Field Pansy being common on the coast referred in part to this species'. The first record other than from the Channel Islands is '... discovered on Tresco by J. Ralfs in July 1873, and named by him as *V. curtisii* var. *mackaii*'. - Lousley (1971). He cites Curnow (1876) as the first published record 'near the mill, Tresco'.

Viola lactea

* as *V. canina var*. 3 leaves spear-shaped flowers pale 'Found by Mr. Stackhouse at Pendarves [S. of Camborne] in Cornwall'. - Withering (1796: ii. 262). 'Near Tunbridge Wells', Kent. - T. F. Forster, E.B. 445 (1798).   vc1   **1796**

Viola lutea

'Jacea sive herba Trinitatis elegantissima, flore lutea amplissimo. Knapweed with a most ample yellow flower: neer Eldenhole [Eldon Hole] in the Peake, and about Buck-stones in Darbyshire'. - How (1650: 61). 'Viola montana lutea grandiflora *C.B.* ... I have found this in many places among the mountains in Wales, and in the North'. - Ray (1670: 318).   vc57   **1650**

How's source is unknown. Clarke & Druce (and Raven 1947: 212) both cite 'The yellow Violet ...found by Master Thomas Hesketh ... growing upon the hils in Lancashire neere unto a village called Latham'. - Gerard (1597: 701) as this species. This is presumably Lytham, nr Kirkham in vc60 (see Savidge *et al.* (1963: 188)), but this is completely out of the native range of this species, and is almost certainly *V. tricolor*. Eric Greenwood, formerly Vice-county recorder for vc60, concurs.

Viola odorata

* 'Viola purpurea'. Between Gravesend and Rochester. - Johnson (1629: 3). 'Viola martia alba odoratissima, From Cornwall, Dr. Gunthorp'. - Merrett (1666: 125).   vc16   **1629**

Viola palustris

* 'Viola rubra striata Eboracensis. Master Stonehouse a reverend Minister of Darfield in Yorkeshiere assured me he found a kind of wild Violet near unto his habitation whose leaves were rounder and thinner then of others and the flowers reddish with sadder veines therein'. - Parkinson (1640: 755.5). vc63 **1640**

Seen by Thomas Johnson and party, on Snowdon, Aug 1639. See Raven (1947: 289). vc49 1639

Raven is presumably referring to 'Viola Martia palustris,' found by Johnson on Snowdon in Aug. 1639. - See Johnson (1641: 8 & 36)

Viola persicifolia

* 'Mr. John Nicholson of Lincoln has found a very remarkable state of *V. lactea* in the neighbourhood of that city (at Boultham Lane, on both sides of the road)'. - Hooker (1839). vc53 **1839**

'A specimen without collector's name from Otmoor in Herb Oxon, May 28th, 1821'. - Druce (1927). vc23 1821

Gibbons (1975) states that Nicholson found it in Lincolnshire in 1836, and that the actual discoverer, in 1833, was Miss C. Cautley.

Viola reichenbachiana

* 'Mr H.C. Watson … from Surrey, and Mr A.G. More from the Isle of Wight'. - Lees (1861). The account contains a proper differentiation between this and *V. riviniana* and notes that Jordan first described the two species in 1857. vc10, 17 **1861**

'Vc34, 1840, Hb. Sandys'. - Riddlesdell *et al.* (1948). 'Nr Twickenham, 1841, **K**'. - Kent (1975). vc34 1840

Lees (1888) also has a ?1840 record, from Cantley, nr. Doncaster.

Viola riviniana

* 'Viola canina, caerulea inodora, sylvestris, serotina, *Lob*'. In woods near to Faversham. - Johnson (1632: 28). vc15 **1632**

But Gerard (1597: 701) may have referred to this species when he wrote:'Of [wilde field Violet] I have found another sort growing wild neere unto Blackeheath by Greenewich, at Eltham parke, with flowers of a bright reddish purple colour'. Druce alone claims this record. However Trimen & Dyer (1869) cite Merrett (1666: 125) 'Viola canina fl. albo. In Hamstead Wood on that side of the Chestnut Walk, where the two wayes meet' as *V. sylvatica* [*V. riviniana* sl.]. Again, this is not followed by later commentators.

Viola rupestris

* ' ... for several years past Messrs. James Backhouse, father and son, have noticed a small and remarkable-looking Violet, growing upon ... the Sugar Limestone at the upper end of Teesdale on the north side of the river. In 1861 the younger of those gentlemen ... transplanted some into his garden'. - Babington (1863c).

vc66 **1863**
vc66 1861

Viola tricolor s.l.

* 'Trinitatis herba ... called in english two faces in a hoode or panses ... groweth ofte amonge the corne'. - Turner (1548: H. v).

1548

Viola tricolor s.s.

E.B.S. 2712, Forster (1831).

1831

'Viola tricolor', in Hb. Sloane 124: 27, **BM**. This is a good specimen with the petals much longer than the sepals.

1715

The Hb. Sloane specimen is in the Herbarium of Buddle, who died in 1715. Forster separated the two taxa - *V. arvensis* and this species - and provided a key.

Viscum album

* 'Viscum, angli vocant Mysceltyne, aut Myselto. ... tot saecula anglis ignotum fuisse plurimum demiror quum in pyris et malis sylvestribus nusq. non proveniat'. - Turner (1538: C. i). 'Plentye of righte oke miscel' sent to 'Hugh Morgan ... oute of Essex'. - Turner (1562: 165).

1538

Vulpia bromoides

* 'Gramen paniculatum bromoides minus paniculis aristatis, unam partem spectantibus. In marginibus herbosis, & ad sepes'. - Ray (1670: 154).

vc29 **1670**

This is marked by Ray (*op.cit.*) as a Cambridge plant, but it is not in his 1660 or 1663 works.

Vulpia ciliata

* as '*F. uniglumis* β 'On the Dover, Ryde, 1839'. - Bromfield (1856).

vc10 **1856**
vc10 1839

This was clarified further and raised to species level in More (1861).

Vulpia fasciculata

'Festuca sterilis humilima spica unam partem spectante ... Found by Mr. Dale in Mersey-Isle near Colchester'. - Petiver (1716: 101).

vc19 **1716**

For another early record see Druce & Vines (1907: 121), under *F. uniglumis*, who note 'Cockbush and Brekelsham coast [Sussex]'. Riddelsdell (1905) gives a discovery in Glamorgan [in 1773] as 'a new grass [which] was described and named by Solander in 1789.

Ray (1724: 413) gives Petiver's record and also one from Jersey, but Smith (1824: 114) notes that the latter is, in fact, *Anisantha diandra*, which would make that a first record for that alien species.

Vulpia myuros

* 'Gramen murorum spica longissima ... Found by Mr. Goodyer upon the wals of the ancient city of Winchester'. - Johnson (1633: 30). vc11 **1633**

'Gramen ἀλεκτρυονρος, 10 Feb. 1622, Goodyer ms.11, f. 54'. - see Gunther (1922: 171). vc11 **1622**

Gunther adds a note that 'The correctness of the determination of this grass as *Festuca myuros* L. by Druce is doubted by D. Stapf. [presumably Dr. O. Stapf]. However all other authorities have accepted the record, with Smith (1824: 143) noting that Johnson's illustration (which certainly does not look like this species) is bad.

Vulpia unilateris

'*Festuca maritima* L. ... has been gathered in Lincolnshire'. - Bennett (1903). 'Carlby, Lincs, A. Woodruffe-Peacock, 9th June 1903'. - Hubbard (1936). vc53 **1903**

For a full history, and for corrections to Bennett's account, see Stace (1961), who notes that the original discover was a Miss S.C. Stow.

Wahlenbergia hederacea

* 'Campanula Cymbalariae foliis ... First discovered to grow in England by Master George Bowles Anno 1632 who found it in Montgomerieshire on the dry bankes in the high-way as one rideth from Dolgeogg a Worshipfull Gentleman's house called Mr. Francis Herbert unto a market toune called Mahuntleth [Machynlleth] and in all the way from thence to the sea side'. - Johnson (1633: 452). vc47 **1633**
 vc47 1632

Wolffia arrhiza

* 'In a pond near Staines Common, Middlesex'. - Trimen (1866). vc21 **1866**

See also 'Walthamstow, 1866, M. Moggeridge, **BM**'. This record is also included in Trimen & Dyer (1869).

But see Gray (1866) writing that 'About fifty years ago Mr Bennett and myself had some specimens of *Lemna arrhiza*, brought to us as having been discovered in the neighbourhood of London, I believe Putney Common. It was collected by M. Gérard, an old Frenchman, who had been head gardener at Versailles'. See also Salmon (1931). There was a Mr J.J. Bennett (1801-1876), who was Assistant Keeper in the Dept of Botany at the BM from 1827.

Woodsia alpina

'*Acrostichum alpinum*, from the mountains of Wales'. - Bolton (1790: 76. t. 42).	vc49	**1790**
Buddle's herbarium in Hb. Sloane (54: 11), **BM**. This would also have been gathered by Lhwyd in 1688 or 1689.	vc49	1689

Bolton was the first to satisfactorily separate these two *Woodsia* species. However Ray (1690: 27) had described this species as 'Filix alpina, Pedicularis rubrae foliis subtus villosis D. Lhwyd. pumila. Stone-fern with red-rattle leaves, hairy underneath ... Clogwyn y Garnedh ...Dr Richardson'. Although earlier writers have argued that the description better fits *W. ilvensis*, all Lhwyd's specimens labelled as this are *W. alpina* (F.J. Rumsey *in litt.*). However Edgington (2013) relates that Lhwyd found two different species in 1688 or 1689, the other 'Polypodium Ilvense [which was *W. alpina*!] ... A small Fern resembling ... Polypody of Elba', on the summit rocks overlooking St Peris Church (see Ray (1690: 27)).

Woodsia ilvensis

'*Acrostichum ilvense* ... first found to grow near the top of Snowdon by Dr Lhwyd, who discovered it growing horizontally from the chinks of the rocks on Clogwyn y Garnedh'. - Bolton (1785: 14).	vc49	**1785**
See Lhywd in Buddle's Herbarium (in Hb Sloane 54: 13) 'Ignota filices species', from Creigiau' Eryran, almost certainly gathered in the area in 1688 or 1689 (F.J. Rumsey *in litt.*)	vc49	1689

Druce & Vines (1907: lxxxiv) cite a letter from Richardson to Dillenius, 26th December 1727, where he states Lhwyd showed him the fern in 1695. But they then go on, almost certainly wrongly, to equate the record to *W. hyperborea* (= *alpina*).

Zannichellia palustris

* 'Potamogeito affinis gramen aquaticum'. - Ray (1660: 125).	vc29	**1660**

Zostera marina

* 'Alga membranacea ceranoides *Bauh.* in maritimis *wrack* duae species'. - How (1650: 4). 'Alga, Grass Wrack ... in maritimis & folio angustiori, from the Severn Sea'. - Merrett (1666: 3).		**1650**

Zostera noltei

* 'Poole Harbour, Dorset, Mr. Borrer'. - E.B.S. 2931.	vc9	**1847**
Jarrow Slake, 1799, W. Weighell, det. P. Sell, **CGE**.	vc66	1799

Graham (1988) included the Jarrow Slake record under *Z. marina*.

Bibliography

Key texts in **bold**

Abbott, C. 1798. *Flora Bedfordiensis*. Bedford.

Abraham, F. & Rose, F. 2000. Large-leaved Limes on the South Downs. *British Wildlife* 12: 86-90.

Adams, W.H.D. 1873. *Nelson's hand-book to the Isle of Wight; its history, topography, and antiquities*. London: T. Nelson & Sons.

Addyman, M. 2016. *The Natural History of the North in Tudor England*. Morpeth.

Akeroyd, J.R. & Beckett, G. 1995. *Petrorhagia prolifera* (L.) P.W. Ball & Heywood (Caryophyllaceae), an overlooked native species in eastern England. *Watsonia* 20: 405-407.

Allen, D.E. 1982. A probable sixteenth-century record of *Rubia tinctoria* L. *Watsonia* 14: 178.

Allen, D.E. 1986. *The Botanists: a history of the Botanical Society of the British Isles through a hundred and fifty years*. London: St Paul's Bibliographies.

Allen, D.E. 1986a. The discoveries of Druce. In Noltie, H.J. ed. *The long tradition – the botanical exploration of the British Isles*. B.S.B.I. conference report no 20. Kilbarchan: Scottish Natural History Society.

Allen, D.E. 1993. Natural history in Britain in the eighteenth century. *Archives of Natural History* 20: 343-347. (also in Allen, D.E. 2001. *Naturalists and Society*. Aldershot: Ashgate Publishing Limited).

Allen, D.E. 1996. The struggle for specialist Journals: natural history in the British periodicals market in the first half of the nineteenth century. *Archives of Natural History* 23: 107-123. (also in Allen, D.E. 2001. *Naturalists and Society*. Aldershot: Ashgate Publishing Limited).

Allen, D.E. 1999. C. C. Babington, Cambridge botany and the taxonomy of British flowering plants. *Nature in Cambridgeshire* 41: 2-11. (also in Allen, D.E. 2001. *Naturalists and Society*. Aldershot: Ashgate Publishing Limited).

Allen, D.E. 2001. Localised records from the fourteenth century. *BSBI News* 88: 23.

Allen, D.E. 2011. Botanical hotspots in Britain and Ireland: who revealed them and when. *New Journal of Botany* 1: 58-61.

Allen, D.E. & Hatfield, G. 2004. *Medicinal Plants in folk tradition: an ethnobotany of Britain & Ireland*. Portland, Oregon & Cambridge: Timber Press.

Alston, A.H.G. 1949. *Equisetum ramosissimum* as a British plant. *Watsonia* 1: 149 153.

Amherst, A. 1895. *A History of Gardening in England*. London: Bernard Quaritch Ltd.

Amplett, J. & Rea, C. 1909. *The Botany of Worcestershire*. Birmingham: Cornish Brothers Ltd.

Andrews, C.R.P. 1900. Two grasses new to the Channel Islands. *J. Bot.* 38: 33-37.

Anon. [Pamplin, W.]. 1841. New British Plant. *Gardeners' Chronicle* 1: 671 (issue for Oct. 9th, 1841).

Anon. 1847. *Gardeners' Chronicle*, Saturday July 17th, 1847: p. 467.

Anon. [Irvine, A.]. 1860. Chickweeds. *Phytologist N.S.* 4: 172-174.

Anon. 1875. *Gentiana pneumonanthe* in Bucks. *J. Bot.* 13: 295.

Arber, A. 1912. *Herbals: their origin and evolution*. Cambridge: Cambridge University Press.

Ardagh, J. 1929. Bibliographical notes XCIV. "Bernardgreyn". – *J. Bot.* 67: 307-308.

Armstrong, J.V. & Sell, P.D. 1996. A revision of the British elms (*Ulmus* L., Ulmaceae): the historical background. *Bot. Jour. Linn. Soc.* 120: 39-50.

Arsene, L. 1930. *Limonium lychnidifolium*, in report for 1929. *B.E.C.* 9: 231.

Ashfield, C.J. 1862. *Pulmonaria officinalis. Phytologist (N.S.)* 6: 351.

Auquier, P. 1973. Une fetuque nouvelle de Bretagne: *F. huonii. Candollea,* 28: 15-19.

Babington, A.M. 1897. *Memorials, Journal and Botanical Correspondence of Charles Cardale Babington*. Cambridge: Macmillan and Bowes. [the only note of authorship is 'A.M.B' at the end of the Preface.]

Babington, C.C. 1834. *Flora Bathoniensis*. Bath: E. Collings.

Babington, C.C. 1836. On several new or imperfectly understood British & European plants. *Trans. Linn. Soc.* 17: 451 464.

Babington, C.C. 1837. A Notice, with the results, of a Botanical Expedition to Guernsey and Jersey, in July and August 1837. *Mag. Zoology and Botany* 2: 397-399.

Babington, C.C. 1838. On the botany of the Channel Islands. *Annals of N.H.* 2: 348-350.

Babington, C.C. 1839a. *Primitiae Florae Sarnicae*. London: Longman & Co.

Babington, C.C. 1839b. On *Ranunculus aquatilis* of Smith. *Annals of N.H.* 3: 225-230.

Babington, C.C. 1839c. *Supplement to the Flora Bathoniensis*. Bath.

Babington, C.C. 1840. On the British species of *Fumaria. Trans. Bot. Soc. Edin.* 1: 31-38.

Babington, C.C. 1842. On the authority upon which several Plants have been introduced into the 'Catalogue of British Plants', published by the Botanical Society of Edinburgh. *Phytologist* 1: 309-311.

Babington, C.C. 1843. *Manual of British botany* (ed.1 1843, ed.2 1847, ed.3 1851, ed.5 1862, ed. 6 1867, ed.7 1874b). London: Jan van Voorst.

Babington, C.C. 1845. *Ranunculus lenormandii* F.W. Schultz. *Annals of N.H.* 16: 141.

Babington, C.C. 1846. *Agrimonia odorata*, Aiton. *Annals of N.H.* 16: 210.

Babington, C.C. 1848. *Sagina ciliata* (Fries). *Annals of N.H. 2nd series* 1: 153-154.

Babington, C.C. 1850a. A notice of *Potamogeton trichoides* of Chamisso as a native of Britain. *Botanical Gazette* 2: 285-288.

Babington, C.C. 1850b. *Cicendia candollei*, Griseb. *Botanical Gazette* 2: 327.

Babington, C.C. 1853a. Remarks upon some British Plants. *Annals of N.H. 2nd series* 11: 265-273, 360-368, 427-433.

Babington, C.C. 1853b. On British Plants. *Bot. Soc. Edin.* 4: 165-169.

Babington, C.C. 1855. On the Batrachian *Ranunculi* of Britain. *Annals of N.H. 2nd series* 16: 385-404.

Babington, C.C. 1856. On some species of *Epilobium*. *Annals of N. H. 2nd series* 17: 236-247.

Babington, C.C. 1857. *Gladiolus imbricatus*, Linn. *Annals of N.H. 2nd series* 20: 158.

Babington, C.C. 1860a. On the *Fumaria capreolata* of Britain. *Proc. Linn. Soc.* 4: 157-167.

Babington, C.C. 1860b. *Flora of Cambridgeshire*. London: John Van Voorst.

Babington, C.C. 1860c. Note concerning *Statice Dodartii* and *S. occidentalis*. *Annals of N.H. 3rd series.* 5: 401-404.

Babington, C.C. 1862. On the Discovery of *Carex ericetorum*, Poll., as a Native of Britain. *Bot. J. Linn. Soc.* 6: 30-31.

Babington, C.C. 1863a. On British species of *Isoetes*. *J. Bot.* 1: 1-5.

Babington, C.C. 1863b. *Thymus serpyllum*. *Trans. Bot. Soc. Edin.* 4: 178-183.

Babington, C.C. 1863c. *Viola arenaria*, De Cand., as a British plant. *J. Bot.* 1: 325-326.

Babington, C.C. 1864a. On *Hypericum undulatum*, Schousb. *J. Bot.* 2: 97-100.

Babington, C.C. 1864b. On *Alsine pallida* Dum. *J. Bot.* 2: 202-204.

Babington, C.C. 1864c. On *Sagina nivalis*, Lindbl. *J. Bot.* 2: 340-342.

Babington, C.C. 1874a. *Carex ornithopoda*, Willd. in England. *J. Bot.* 12: 371.

Babington, C.C. 1878. On *Ranunculus tripartitus*, DeCand. *J. Bot.* 16: 38-39.

Baily, Miss. 1833. *The Irish Flora*. Dublin: Hodges and Smith.

Baker, E.G. 1901. Some British Violets. *J. Bot.* 39: 9-12.

Baker, E.G. & Salmon, C.E. 1920. Some segregates of *Erodium cicutarium* l'Herit. *J. Bot.* 58: 121-127.

Baker, J.G. 1858. '*Draba verna*'. *Phytologist N.S.* 2: 501-50.

Baker, J.G & Foggitt, W. 1865. *R. pseudo-fluitans*, Newbould. *In Curator's report, Thirsk B.E.C.*

Baker, J.G & Tate, G.R. 1868. A new Flora of Northumberland and Durham. *Nat. hist. Trans. Newcastle* 2: 1-316.

[Balfour, I.B.] 1907. Eighteenth Century Records of British Plants. *Notes from the Royal Botanic Garden, Edinburgh* 4: 123-190. [this includes John Hope's ms. 'List of Plants growing in the Neighbourhood of Edinburgh'].

Balfour, J.H., Babington, C.C. & Campbell, W.H. 1841. See *Edinburgh Catalogue*.

Ball, J. 1844. On some British species of *Oenanthe*. *Trans. Bot. Soc. Edin.* 2: 105-108.

Ball, P.W. & Tutin, T.G. 1959. Notes on annual species of *Salicornia* in Britain. *Watsonia* 4: 192-205.

Banckes's Herbal. 1525. See Larkey & Pyles.

Barling, D.M. 1959. Biological studies in *Poa angustifolia*. *Watsonia* 4: 147-168.

Barrett, W.B. (1905). Notes on the Flora of the Chesil Bank and the Fleet. *Dorset NHAS.* 26: 251-265.

Bates, S. & Spurgin, K. 1994. *Stars in the Grass.* Truro: Dyllansow Truran. [Life of F.H. Davey]

Bauhin. C. 1620. Πποδρομος *theatri botanici*. Francofurti ad Moenum [Frankfurt].

Bauhin, J. & Cherler, J.H. 1650–1651. (as I. Bauhinus & I.H. Cherlerus) *Historia plantarum universalis ... quam recensuit et auxit Dominicus Chabraeus. Volume 1* (1650); *Volumes 2–3* (1651). Ebroduni [Yverdon].

Bebber, D.P, Harris, S.A., Gaston, K.J. & Scotland, R.W. 2007. Ethnobotany and the first printed records of British flowering plants. *Global Ecology and Biogeography* 16: 103-108.

Beeby, W.H. 1872. *Juncus capitatus* Weigel, in England. *J. Bot.* 10: 337.

Beeby, W.H. 1887. *Juncus alpinus* probably a Scotch plant. *Scot. Nat.* 3 (ns): 92.

Bennett, A. 1883a. A new British Plant. - *Najas major* All. *J. Bot.* 21: 246.

Bennett, A. 1883b. On *Najas marina* L. as a British Plant. *J. Bot.* 21: 353-354.

Bennett, A. 1884. *Carex trinervis* Degland, in England. *J. Bot.* 22: 125.

Bennett, A. 1885a. New British and Irish Carices. *J. Bot.* 23: 50.

Bennett, A. 1885b. *Calamagrotis strigosa* Hartm. In Britain. *J. Bot.* 23: 253.

Bennett, A. 1893. *Alisma Plantago*, L., var. *lanceolatum*, With., in Report for 1892–1893. *Wats. Bot. Exch. Club* 9: 16.

Bennett, A. 1903. *Festuca maritima* L. in Britain. *J. Bot.* 41: 314.

Bennett, A. 1908. *Potamogeton pensylvanicus* in England. *Naturalist, Hull* 1908: 10-11.

Bennett, A. 1910a. *Medicago sylvestris*, *M. falcata*, *Carex ericetorum* and *Psamma baltica* in England. *Trans Norf. Norw. Nat. Soc.* 9: 16-25.

Bennett, A. 1910b. Notes on the British species of *Utricularia*. *Trans. Bot. Soc. Edin.* 24: 59-63.

Bennett, A. 1914. *Hydrilla verticillata* Casp. In England. *J. Bot.* 52: 257-258.

Bentham, G. 1858. *Handbook of the British Flora.* London: Lovell Reeve. [11 eds. up to 1924].

Bicheno, J.E. 1818. Observations on the Linnean Genus *Juncus*, with the Characters of those Species, which have been found growing wild in Great Britain. *Trans. Linn. Soc.* 12: 291-337.

Birkinshaw, P.R. & Sanford, M.N., 1996. *Pulmonaria obscura* Dumort. (Boraginaceae) in Suffolk. *Watsonia* 21: 169-178.

Blackstock, N & Ashton, P.A. 2001. A re-assessment of the putative *Carex flava* agg. (Cyperaceae) hybrids at Malham Tarn (v.c.64): a morphometric analysis. *Watsonia* 23: 505-516.

Blackstone, J. 1737. *Fasciculus Plantarum circa Harefield sponte nascentium*. London.

Blackstone, J. 1746. *Specimen Botanicum quo Plantarum plurium variorum Angliae indigenarum loci natales illustrantur*. London.

Blakelock. R.A. 1952. *Kew Bulletin for 1951*: 325.

Blakelock, R.A. 1953. *Artemesia norvegica* Fries in Scotland. *Kew Bulletin 1952*: 173.

Bolton, J. 1785. *Filices Britannicae*. Leeds: John Binns.

Borlase, W. 1758. *The Natural History of Cornwall*. Oxford.

Borrer, W. 1830. *Myosotis collina*. In Smith & Sowerby supp. Vol. 1 (ed. by Hooker, W.J.). London.

Borrer, W. 1833a. *Sonchus asper*. In Smith & Sowerby supp. Vol. 2 (ed. by Hooker, W.J.). London.

Borrer, W. 1833b. *Sonchus oleraceus*. In Smith & Sowerby supp. Vol. 2 (ed. by Hooker, W.J.). London.

Borrer, W. 1844. Note on the discovery of *Leersia oryzoides* in Sussex. *Phytologist* 1: 1140.

Botanical Society of Edinburgh 1837. First Report.

Boulger, G.S. 1883. Samuel Dale. *J. Bot.* 21: 193-197, 225-231.

Boulger, G.S. 1891. Robert Uvedale. *J. Bot.* 29: 9-18.

Boulger, G.S. 1900. History of Essex botany. *Essex Naturalist* 11: 57-68, 169-184, 229-236.

Boulger, G.S. 1918. Joseph Andrews and his herbarium. *J. Bot.* 56: 323-331.

Boulger, G.S. & Britten, J. 1918. Joseph Andrews and his herbarium. *J. Bot.* 56: 257-261 *etc*.

Bowden, J.K. 1989. *John Lightfoot, his work and travels*. Kew & Pittsburgh.

Bowen, H.J.M. 2000. *The Flora of Dorset*. Newbury: Pisces Press.

Braithwaite, M.E. 2004. *Tofieldia pusilla* (Michx.) Pers. at the Scottish border. *Watsonia* 25: 207-208.

Bree, W.T. 1831. *Aspidium dilatatum* var. *recurvum*. *Loudon's Magazine of Natural History* 4: 162.

Bree, W.T. 1843. Description of *Aspidium recurvum*. *Phytologist* 1: 773-774.

Bree, W.T. in Purton, T. 1821. *A botanical Description of British Plants in the Midland Counties. Appendix*. London.

Brewis, A., Bowman, R. P. & Rose, F. 1996. *The Flora of Hampshire*. Colchester: Harley Books.

Brichan, J.B. 1842. *Potamogeton praelongus*. *Phytologist* 1: 236-237.

Bridson, G.D.R., Phillips, V.C. & Harvey, A.P. 1980. *Natural History Manuscript Resources in the British Isles*. London: Mansell Publishing.

Briggs, D. & Gorringe, E. 2002. The struggle to produce a Flora of the British Isles (1933-1952). *Watsonia* 24: 1-15.

Briggs, T.R.A. 1864. *Hypericum undulatum*, Schousb., a recent addition to the British Flora. *J. Bot.* 2: 45-46.

Briggs, T.R.A. 1875. *Rumex rupestris*, Le Gall, a British species. *J. Bot.* 13: 294-295.

Briggs, T.R.A. 1880. *The Flora of Plymouth*. London: Jan van Voorst.

Britten, J. 1879. *Gentiana pneumonanthe* in Berks. *J. Bot.* 17: 44.

Britten, J. 1887. *William Turner, The Names of Herbes*. London: English Dialect Society.

Britten, J. 1892. The Deptford Pink. *J. Bot.* 30: 177-178.

Britten, J. 1897. John Whitehead. *J. Bot.* 35: 89-92.

Britten, J. ed. 1897b. Hybrid forms of *Pyrus*. *J. Bot.* 35: 99-100.

Britten, J. 1909. A seventeenth century English botanist. *J. Bot.* 47: 99-104.

Britten, J. 1911. William Ambrose Clarke (1841-1911). *J. Bot.* 49: 167-169.

Britten, J. 1915. *Melampyrum arvense*. *J. Bot.* 53: 91.

Britten, J. 1917. In memory of Daniel Oliver. *J. Bot.* 55: 89-95.

Britten, J. & Boulger, G.S. 1893. *A biographical index of deceased British and Irish botanists*. London: West, Newman & Co. [supplements 1899, 1904, 1908 and revised, Rendle, 1931.]

Britten, J. & Holland, R. 1886. *A Dictionary of English Plant-names*. London: Trübner, for the English Dialect Society.

Britton, C.E. 1922. British *Centaureas* of the *nigra* group. *B.E.C.* 6: 406-417.

Britton, C.E. 1928. *Veronica anagallis* L. and *V. aquatica* Bernh. *B.E.C.* 8: 548-550.

Bromfield, W.A. 1841. *Prunus cerasus*. In Smith & Sowerby supp. Vol. 3 (ed. by Hooker, W.J.). London.

Bromfield, W.A. 1843. Notice of a new British *Calamintha*, discovered in the Isle of Wight. *Phytologist 1: 768 – 770*.

Bromfield, W.A. 1847. Remarks on *Polygala depressa* of Wenderoth. *Phytologist* 2: 966-968.

Bromfield, W.A. 1849. *Orobanche picridis*. *Phytologist* 3: 604.

Bromfield, W.A. 1856. *Flora Vectensis*. London: William Pamplin.

Brunker, J.P. 1950. *Flora of the County Wicklow*. Dundalk: Dundalgan Press (W. Tempest) Ltd.

Bucknall, C. 1897. *Stachys alpina* L. in Britain. *J. Bot.* 25: 380-381.

Bunting, M.J., Briggs, D. & Block, M. 1995. The Cambridge British Flora (1914-1920). *Watsonia* 20: 195-204.

Burtt, B.L. 1950. *Koenigia islandica* in Britain. *Kew Bulletin 1950*: 266.

Butcher, R.W. 1922. *Tillaea aquatica* L. in Plant notes, etc., for 1921. *B.E.C.* 6: 281-282.

Butcher, R.W. 1961. *A New Illustrated British flora*. London: Leonard Hill (Books) Limited.

Butcher, R.W. & Strudwick, F.E. 1930. *Further illustrations of British plants*. Ashford: L. Reeve & Co. Ltd.

Byfield, A.J. 1986. The Lizard Flora: a history of discovery. In Noltie, H.J. ed. *The long tradition – the botanical exploration of the British Isles*. B.S.B.I. conference report no 20. Kilbarchan: Scottish Natural History Society.

Cadbury, D. A., Hawkes, J. G. & Readett, R. C. 1971. *A computer-mapped flora: a study of the county of Warwickshire*. London: Academic Press.

Caius, J. 1570. *Britanni de Rariorum Animalium atque Stirpium Historia*. London.

Camerarius, J. (the Younger). 1588. *Hortus medicus et philosophicus*. Francofurti ad Moenum [Frankfurt].

Campbell, M.S. 1937. Three weeks botanising in the Outer Hebrides. *B.E.C.* 11: 304-318.

Campbell, M.S. 1938. Further botanising in the Outer Hebrides. *B.E.C.* 11: 529-560.

Camus, J.M. 1991. *Changes in the list of British Pteridophytes, in The History of British Pteridology, 1891 – 1991*. London: British Pteridological Society.

Cann, D. 2012. *Sorbus devoniensis* at Little Haldon, Devon. *BSBI News* 121: 42.

Carter, P.W. 1955a. Some account of the botanical exploration of Caernarvonshire – Part 1. *Trans. Caerarvonshire Hist. Soc. Historical Society* 9: 1-32.

Carter, P.W. 1955b. Some account of the history of botanical exploration in Glamorgan. *Reps. & Trans. Cardiff Naturalists' Soc. 1952-3.* 82: 5-31.

Carter, P.W. 1960. Some account of the botanical exploration of Denbighshire. *Denbighshire Historical Society* 9: 1-32.

Carter, P.W. 1986. Some account of the history of botanical exploration in Pembrokeshire [part 1]. *Nature in Wales* 5: 33–44.

Chapman, G.T.L. & Tweddle, M.N. eds. 1989. *William Turner: A New Herbal. Part I. 1551*. Ashington: Mid Northumberland Arts Group and Carcanet Press.

Chapman, G.T.L., McCombie, F. & Wesencraft, A. eds. 1995. *William Turner: A New Herbal. Part II, 1562 & Part III, 1568*. Cambridge: Cambridge University Press.

Chapman, M.A. & Stace, C.A. 2001. Tor-grass is not *Brachypodium pinnatum*! *BSBI News* 87: 74.

Chater, A.O. 1984. An unpublished botanical notebook of Edward Llwyd. *BSBI Welsh Bulletin* 40: 4-15.

Chatters, C. 2009. *Flowers of the Forest*. Old Basing: Wildguides.

Clapham, A.R. 1946. Check-list of British vascular plants. *J. Ecol.* 33: 308-347.

Clapham, A.R., Tutin, T.G. & Moore, D.M. 1987. *Flora of the British Isles* 3rd ed. Cambridge: Cambridge University Press.

Clapham, A.R., Tutin, T.G. & Warburg, E.F. 1952. *Flora of the British Isles*. Cambridge: Cambridge University Press.

Clapham, A.R., Tutin, T.G. & Warburg, E.F. 1962. *Flora of the British Isles* 2nd ed.. Cambridge: Cambridge University Press.

Clarke, W.A. 1892. First Records of British Flowering Plants. *J. Bot.* 30: 19-25.

Clarke, W.A. 1900. *First Records of British Flowering Plants*, 2nd ed. London: West, Newman & Co.

Clarke, W.A. 1909. First Records of British Flowering Plants [a supplement]. *J. Bot.* 47: 413-416.

Clarke, W.G. & Gurney, R. 1921. Notes on the Genus *Utricularia* and its distribution in Norfolk. *Trans. Norf. & Norw. Nat. Soc.* 11: 128-161.

Clement, E.J. 1985. Selfheals (*Prunella* spp.) in Britain. *BSBI News* 41: 20.

Clokie, H.M. 1964. *An account of the Herbaria of the Department of Botany in the University of Oxford*. Oxford: Oxford University Press.

Clusius – see L'Ecluse.

Cobbing, P. 1989. *Serapias parviflora* Parl. *BSBI News* 52: 11.

Coleman, W.H. 1844. Observations on a new species of *Oenanthe*. *Trans. Bot. Soc. Edin.* 2: 91-93.

Coles, G.L.D. 2011. *The story of South Yorkshire botany*. York: Yorkshire Naturalists' Union.

Coles, W. 1657. *Adam in Eden: or, natures paradise*. London: Nathaniel Brooke.

Colgan, N. & Scully, R.W. 1898. *Contributions towards a Cybele Hibernica*. Ed. 2. Dublin: Edward Ponsonby.

Compton, S.G. & Key, R.S. 2000. The Biological Flora of the British Isles, *Coincya wrightii* (O.E. Schulz) Stace. *J. Ecol.* 88: 535-547.

Coombe, D.E. 1961. *Trifolium occidentale*, a new species related to *T. repens*. *Watsonia* 5: 68-87.

Cope, T.A. 2009. *The Wild flora of Kew Gardens*. Kew: Royal Botanic Gardens.

Cope, T. & Gray, A. 2009. *Grasses of the British Isles*. London: Botanical Society of the British Isles.

Cope, T.A. & Stace, C.A. 1978. The *Juncus bufonius* L. aggregate in western Europe. *Watsonia* 12: 113-128.

Crabbe, J.A., Jermy, A.C. & Walker, S. 1970. The distribution of *Dryopteris assimilis* in Britain. *Watsonia* 8: 3-15.

Crackles, F.E. 1990. *Flora of the East Riding of Yorkshire*. Hull: Hull University Press.

Crompton, G. 2007. Catalogue of Cambridgeshire flora records since 1538. [see http://www.cambridgeshireflora.com]

Crotch, W.R. 1855. *Epipogium aphyllum*. Phytologist, N.S. 1: 118.

Crouch, H.A. 2011. *Bolboschoenus laticarpus. Somerset rare plants group newsletter* 12: 10-11.

Culpeper, N. 1653. *The English physician enlarged*. London.

Curnow, J. 1876. A botanical trip to the Scilly Islands. *Science Gossip* 12: 162.

Curtis, W. 1777- 98. *Flora Londinensis*. London.[for dates of each fascicle see Stevenson (1961)]

Curtis, W. 1782. *A catalogue of certain plants growing wild, chiefly in the Environs of Settle, in Yorkshire, observed by W. Curtis, in a Six Weeks Botanical Excursion from London, made at the Request of J.C. Lettsom, MD, FRS &c, in the months of July and August, 1782.* (forming Appendix in vol. i. *Flora Londinensis*, 1777-1787).

Curtis, W. 1790. *Practical Observations on the British Grasses*, 2nd ed. London.

D[oubleday], G.E. 1846. Botanical Society of London. 1846. Notice of exhibits at the meeting on March 6th 1846. *Phytologist* 2: 499-501.

Dale, S. 1730. *History and Antiquities of Harwich and Dovercourt*. London.

Dandy, J.E. 1958a. *The Sloane Herbarium*. London: Trustees of the British Museum.

Dandy, J.E. 1958b. *List of British Vascular Plants*. London: British Museum (Natural History).

Dandy, J.E. n.d. *The British species of* Potamogeton: annotated typescript of an unpublished monograph held in the archives of the Natural History Museum, London. [a key to the species, with a detailed synonymy and list of specimens for each taxon, compiled initially with Sir G. Taylor, and still unfinished at Dandy's death in 1976].

Dandy, J.E. & Taylor, G. 1938a. Studies of British Potamogetons. - I. *J. Bot.* 76: 89-92.

Dandy, J.E. & Taylor, G. 1938b. Studies in British Potamogetons. - III. *Potamogeton rutilus* in Britain. *J. Bot.* 76: 239-241.

Dandy, J.E. & Taylor, G. 1939. Studies of British Potamogetons. - IV. The identity of *Potamogeton drucei*. *J. Bot.* 77: 56-62.

Dandy, J.E. & Taylor, G. 1940. Studies of British Potamogetons. - XIII. *Potamogeton berchtoldii* in Great Britain. *J. Bot.* 78: 49-66.

Davey, F.H. 1907. *Euphrasia vigursii. J. Bot.* 45: 217-220.

Davey, F.H. 1909. *Flora of Cornwall*. Penryn: F. Chegwidden.

David, R.W. 1977. The distribution of *Carex montana* L. in Britain. *Watsonia* 11: 377-378.

David, R.W. 1981a. The distribution of *Carex ericetorum* Poll. in Britain. *Watsonia* 13: 225-226.

David, R.W. 1981b. The distribution of *Carex punctata* Gaud. Poll. in Britain, Ireland and the Isle of Man. *Watsonia* 13: 318-321.

David, R.W. & Kelcey, J.G. 1985. *Carex muricata* L. aggregate. *J. Ecol.* 73: 1021-1039.

Davies, H. 1810. A Determination of Three British Species of *Juncus*, with jointed Leaves. *Trans. Linn. Soc.* 10: 10-14.

Davies, H. 1813. *Welsh Botanology*. London.

De Tabley, Lord. 1899. *The Flora of Cheshire*. London: Longmans, Green, and Co.

De Winter, 2015. The Dutch rush: history and myth of the *Equisetum* trade. *Fern Gazette* 20: 23-45.

Deakin, R. 1845. *Florigraphia Britannica*, vol.2. London: R. Groombridge.

Dean, M., Ashton, P. A., Hutcheon, K., Jermy, A. C. & Cayouette, J. 2008. Description, ecology and establishment of *Carex salina*, Wahlenb. (Saltmarsh Sedge) - a new British species. *Watsonia* 27: 51-57.

Dean, M., Hutcheon, K., Jermy, A. C., Cayouette, J. & Ashton, P. A. 2005. *Carex salina* – a new species of sedge for Britain. *BSBI News* 99: 17-19.

Delforge, P. 2000. Nouvelles contributions taxonomiques et nomenclaturales aux Orchidées d'Europe. *Les Naturalistes belges* 81: 396-398.

Delforge, P. & Gevaudan, A. 2002. Contribution taxonomique et nomenclaturale au groupe d' *Epipactis leptochila*. *Les Naturalistes belges* 83: 26.

Derham, W. 1760. *Select remains of the learned John Ray*. London : George Scott

Desmond, R. 1977. *Dictionary of British and Irish botanists and horticulturalists*. London: Taylor and Francis Ltd.

Desmond, R. 1994. *Dictionary of British and Irish botanists and horticulturalists* 2nd ed. London: Taylor and Francis Ltd.

Dick, R. 1854. Notice of the Discovery of *Hierochloe borealis*, near Thurso. *Annals of N.H. N.S.* 14: 314-315.

Dickenson, S. 1798. *A catalogue of plants ascertained to be indigenous in the county of Stafford*, in Shaw, S. *The History and Antiquities of Staffordshire*. London: J. Nichols.

Dickie, G. 1860. *The Botanist's Guide to the counties of Aberdeen, Banff, and Kincardine*. Aberdeen: A. Brown & Co.

Dickson, C & Dickson, J. 2000. *Plants & People in Ancient Scotland*. Stroud: Tempus Publishing Ltd.

Dickson, J. 1790. *Plantarum cryptogamicarum Britanniae*. Fasc.2. Londoni.

Dickson, J. 1794. An account of some plants newly discovered in Scotland. *Trans. Linn. Soc.* 2: 286-291 [read Feb.1793].

Dickson, J.H. 1982. The earliest record of a Whitebeam on Arran. *Glasgow Naturalist* 20: 261-262.

Dodoens, R. 1568. *Florum et coronariarum odorataramque nonnullarum herbarum historia*. Antverpiae.

Don, G. 1804. *Herbarium Britannicum*. Fasciculi 1 & 2. Edinburgh.

Don, G. 1805. *Herbarium Britannicum*. Fasciculi 3 & 4. Edinburgh.

Don, G. 1806. *Herbarium Britannicum*. Fasciculi 5 to 9 [Fasc 9 not issued before 1812 or 1813]. Edinburgh.

Don, G. 1813. *Account of the Native Plants in the County of Forfar and the animals to be found there*, as an appendix to Headrick, *General View of the Agriculture of the County of Angus or Forfarshire*. Edinburgh.

Dony, J.G. 1953. *Flora of Bedfordshire*. Luton: Corporation of Luton Museum and Art Gallery.

Dony, J.G. 1967. *Flora of Hertfordshire*. Hitchen: Hitchen Museum.

Druce, G.C. 1882. On *Lycopodium complanatum* as a British Plant. *J. Bot.* 20: 321-323.

Druce, G.C. 1883. The Botanical Work of George Don of Forfar. *Scottish Naturalist* 6: 126-129, 176-178, 217-223, 258-269.

Druce, G.C. 1895. A new *Bromus*. – *Bromus interruptus*. *J. Bot.* 33: 344.

Druce, G.C. 1896a. Review of Wettstein, von R.V. (1896). Monographie der Gattung Euphrasia. *J. Bot.* 34: 369-370.

Druce, G. C. 1896b. On a new species of grass, *Bromus interruptus*, in Britain. *Jour. Linn. Soc.* 32: 426-430.

Druce, G.C. 1897. *The Flora of Berkshire*. Oxford: The Clarendon Press.

Druce, G.C. 1899. An early Scottish locality for *Sparganium affine*, Schizl. *Annals of Scottish N.H.* 31: 186-187.

Druce, G.C. 1900. *Galium sylvestre*, Poll., var. *nitidulum*. In Report for 1898. *B.E.C.* 1: 577.

Druce, G.C. 1904. The life and work of George Don. *Notes Royal Botanic Garden Edinburgh* 3: 53-290. (parts XII - XIV). [now available as a print-on-demand volume]

Druce, G.C. 1905. *Koeleria splendens* as a British plant. *J. Bot.* 43: 313-317.

Druce, G.C. 1908. *List of British Plants*. Oxford: The Clarendon Press.

Druce, G.C. 1908a. *Luzula pallescens*, in Report for 1907. *B.E.C.* 2: 312.

Druce, G.C. 1909a. *Orobanche reticulata*, Wallr., var. *procera* (Koch). *B.E.C.* 2: 334-337.

Druce, G.C. 1909b. *Schoenus ferrugineus* Huds. = *Scirpus pauciflorus* Lightf. *J. Bot.* 47: 108-109.

Druce, G.C. 1911a. (*Veronica anagallis-aquatica* b. *anagalliformis*), in Plant notes for 1910, etc. *B.E.C.* 2: 505.

Druce, G.C. 1911b. (*Utricularia*), in Plant notes for 1910, etc. *B.E.C.* 2: 511-520.

Druce, G.C. 1911c. *Ulmus plottii* Druce *sp. nov. J. Northamptonshire. Nat. Hist. Soc.* 16: 88.

Druce, G.C. 1912a. *Veronica anagallis-aquatica*, in Plant notes for 1911, etc. *B.E.C.* 3: 26-27.

Druce, G.C. 1912b. Samuel Corbyn's Catalogue of Cambridge Plants. *J. Bot.* 50: 76-79.

Druce, G.C. 1913. *Poa irrigata* Lindman. *B.E.C.* 3: 181-184.

Druce, G.C. 1914a. *Oxycoccus quadripetalus* Gilib., var. *microcarpus*. In Plant Notes for 1913, etc. *B.E.C.* 3: 327-328.

Druce, G.C. 1914b. Notes on the Marsh Orchids. *B.E.C.* 3: 339-341.

Druce, G.C. 1915a. *Hydrilla verticillata*. In Plant notes, etc., for 1914. *B.E.C.* 4: 22-24.

Druce, G.C. 1915b. *Orchis maculata* L. and *O. fuchsii*. *B.E.C.* 4: 99-108.

Druce, G.C. 1916a. *Scorzonera humilis* L. In Plant notes for 1915. *B.E.C.* 4: 202.

Druce, G.C. 1916b. *Lycopodium complanatum* L. In Plant notes for 1915. *B.E.C.* 4: 219-222.

Druce, G.C. 1917. *Limonium lychnidifolium* Kuntze, var *corymbosum* Salm. in New County and other records. *B.E.C.* 4: 494.

Druce, G.C. 1917a. James Dickson. *Jour. Northamptonshire N.H.S.* 19: 116.

Druce, G.C. 1918. Notes on the British Orchids. *B.E.C.* 5: 149-180.

Druce, G.C. 1919. *Centaurium scilloides* Druce. In Plant notes for 1918. *B.E.C.* 5: 290-295.

Druce, G.C. 1920a. Edward Morgan's *Hortus Siccus*. *B.E.C.* 5: 722-724.

Druce, G.C. 1920b. The extinct and dubious plants of Britain. *B.E.C.* 5: 731-799.

Druce, G.C. 1923. *Carex microglochin*. In Plant notes, etc., for 1923. B.E.C. 7: 68-71.

Druce, G.C. 1925. *Galium debile*, in Plant notes, etc., for 1924. *B.E.C.* 7: 438-439.

Druce, G.C. 1926. *The Flora of Buckinghamshire*. Arbroath: T. Buncle & Co.

Druce, G.C. 1927. *The Flora of Oxfordshire*, 2nd ed. Oxford: Clarendon Press.

Druce, G.C. 1928. British plants contained in the Du Bois Herbarium at Oxford, 1690 – 1723. *B.E.C.* 8: 463-493.

Druce, G.C. 1928a. *British Plant List*. Arbroath: T. Buncle & Co.

Druce, G.C. 1929. *Scirpus americanus*, in New county and other records, 1928. *B.E.C.* 8: 762.

Druce, G.C. 1930. The British *Erophila*. *B.E.C.* 9: 177-198.

Druce, G.C. 1930a. *Flora of Northamptonshire*. Arbroath: T. Bunkle & Co.

Druce, G.C. 1932. *The Comital Flora of the British Isles*. Arbroath: Buncle & Co.

Druce, G.C. 1933. Local Floras. *B.E.C.* 10: 399-424.

Druce, G.C. & Leach, T.H. 1915. *Arabis alpina* L. in Report for 1914. *B.E.C.* 4: 115 - 116.

Druce, G.C. & Vines, S. H. 1907. *The Dillenian Herbaria*. Oxford: Clarendon Press.

Dyer, A. F., Parks, J. C. & Lindsay, S. 2000. Historical review of the uncertain taxonomic status of *Cystopteris dickieana* R. Sim (Dickie's bladder fem). *Edinburgh J. Bot.* 57(1): 71-81.

E.B. & E.B.S. See Smith & Sowerby.

Earle, J. 1880. *English Plant Names form the Tenth to the Fifteenth Century*. Oxford: Clarendon Press.

Edgington, J.A. 2003. The strange case of the Gentian of 1812. *BSBI News* 93: 27-29.

Edgington, J.A. 2007. A plant list of 1633: annotations in a copy of Thomas Johnson's *Iter plantarum. Archives of natural history* 34(2): 272-292.

Edgington, J.A. 2010. First British record of *Nardus stricta*. *Watsonia* 23: 123-127.

Edgington, J.A. 2011. Early London botanists: '… in the fields aboute London, plentuously …'. *London Naturalist* 90: 21-45.

Edgington, J.A. 2013. *Who found our ferns?* London: British Pteridological Society, Special Publication No 12.

Edgington, J.A. 2014. The Tottenham plant list of Thomas Johnson, 1638. *London Naturalist* 92: 21-41.

Edinburgh Catalogue. 1841. [Balfour, J.H., Babington, C.C. & Campbell, W.H]. *A Catalogue of British Plants*. Part 1. 2nd ed.]. Edinburgh: Maclachlan Stewart & Co).

Edmondson, J. 2009. A Georgian Herbarium rediscovered. *Naturalist* 134: 79-84.

Edmondston, T. 1839. List of plants observed in the Island of Unst, Shetland, during the summer of 1837. (in Hooker, W. D., *Notes on Norway*, ed. 2: 111-117)

Edwards, Z.J. 1862. *The Ferns of the Axe and its tributaries*. London: Hamilton, Adams, and Co.

Egerton, F.N. 2010. A History of the Ecological Sciences, Part 36: Hewett Watson, Plant Geographer and Evolutionist. *Bulletin of the Ecological Society of America* 91: 294-312.

Ekwall, E. 1922. *The Place-names of Lancashire*. Manchester: Manchester University Press.

Ellis, R.G. 1974. *Plant hunting in Wales*. Cardiff: National Museum of Wales.

Ellis, R.G. 1983. *Flowering Plants of Wales*. Cardiff: National Museum of Wales.

Fitzherbert, A. 1523. *The Boke of Husbandrie*. London.

Fletcher, H.R. 1959. Exploration of the Scottish flora. *Trans. Bot. Soc. Edin.* 38: 30-47.

Foley, M.J.Y. 2005. Some early English plant records of Thomas Penny (c1530 - 1588). *BSBI News* 99: 71-72.

Foley, M.J.Y. 2006a. Christopher Merrett's *Pinax rerum naturalium britannicarum* (1666): annotations to what is believed to be the author's personal copy. *Archives of Natural History* 33(2): 191-201.

Foley, M.J.Y. 2006b. Thomas Penny, the sixteenth century English botanist: some of his European plant records. *Saussurea* 36: 88-101.

Foley, M.J.Y. 2008. John Fitz-Roberts: A little-known seventeenth century botanist. *Watsonia* 27: 131-141.

Foley, M.J.Y. 2009. Some localised early plant records from North-west England: then and now. *Watsonia* 27: 355-364.

Foley, M. & Clarke, S. 2005. *Orchids of the British Isles*. Cheltenham: Griffen Press.

Forster, E. 1831. *Viola arvensis*, in Smith & Sowerby supp. Vol. 2 (ed. by Hooker, W.J.). London.

Forster, T.F. 1816. *Flora Tonbridgensis*. London.

Forster, T.I.M. 1842. *Flora Tonbridgensis*. Ed. 2. London.

Fowler, W. 1881. *Selinum carvifolia*, in the general locality list. *Bot. Rec. Club. Report for the years 1880*.

Francis, G.W. 1837. *An analysis of British Ferns and their allies*. London: Simkin Marshall & Co.

Fraser-Jenkins, C.R. 1980. *Dryopteris affinis*: a new treatment for a complex species in the European pteridophyte flora. *Wildenowia* 10: 107-115.

French, C.N., Murphy, R.J. & Atkinson, M.J.C. 1999. *Flora of Cornwall*. Camborne: Wheal Seton Press.

Fryer, A. 1898. *The Potamogetons (Pond Weeds) of the British Isles*, pp 25-36, tt. 13-24. London: L. Reeve & Co. Limited.

Fryer, A. 1899. *Potamogeton Drucei* Fryer. *J. Bot.* 37: 524.

Garry, F.N.A. 1903-4. Notes on the drawings for 'English Botany'. *J. Bot. 1903-4.* supplements.

Gerard[e], J. 1597. *The Herball, or general history of Plantes*. London.

Gibbons, E.J. 1975. *The Flora of Lincolnshire*. Lincoln: Lincolnshire Naturalists' Union.

Gibbons, E.J. & Lousley, J.E. 1958. An inland *Armeria* overlooked in Britain. *Watsonia* 4: 125-135.

Gibson, E. 1695. *Camden's Britannia, newly translated into English: with large additions and improvements*. London.

Gibson, E. 1722. *Camden's Britannia, newly translated into English: with large additions and improvements*. 2nd ed. London. .

Gibson, G.D. & Taylor, I. 2005. Biological Flora of the British Isles: *Festuca longifolia*. *J. Ecol.* 93: 214 - 226.

Gibson, G.S. 1848. Notice of the discovery of *Filago Jussiaei* near Saffron Walden. *Phytologist* 3: 216.

Gibson, G.S. 1862. *The Flora of Essex*. London: William Pamplin.

Gibson, S. 1842. Notes on *Arenaria rubra, marina*, and *media*. *Phytologist* 1: 217-218.

Gilmour, J.S.L. ed. 1972. *Thomas Johnson. Botanical Journeys in Kent & Hampstead. (a facsimile reprint with Introduction and Translation of his Iter Plantarum 1629 [and] Descriptio Itineris Plantarum 1632)*. Pittsburgh: Hunt Botanical Library.

Godfery, M.J. 1919. *Epipactis viridiflora* Reich. *J. Bot.* 57: 37-42.

Godfery, M.J. 1923. *Orchis fuchsii* Druce. *J. Bot.* 61: 306-309.

Goodenough, S. 1792. Observations on the British species of *Carex*. *Trans. Linn. Soc.* ii: 126-211.

Goodenough, S. 1797. Additional observations on the British species of *Carex*. *Trans. Linn. Soc.* iii: 76-79.

Goodway, K.M. 1955. The forms of *Galium pumilum* in Britain (Exhibit). *Proc. BSBI* 1: 383.

Goodway, K.M. 1957. *The species problem in* Galium pumilum. *(Exhibit)*. In Lousley, J.E. ed. *Progress In The Study Of The British Flora*. London: BSBI.

Gorham, G.C. 1830. *Memoirs of John Martyn and Thomas Martyn*. London: Hatchard & Son

Goss, H. 1899. *Orchis cruenta* in Cumberland. *J. Bot.* 37: 37.

Gosse, P.H. 1874. *Land and Sea*. London: J. Nisbet.

Gourlie, W. 1855. Letter to Linnean Society, 22nd March, 1855. *Proc. Linn. Soc.* 2: 374.

Graham, G.G. 1988. *The Flora and Vegetation of County Durham*. Durham: Durham Flora Committee and the Durham County Conservation Trust.

Graves, G. & Hooker, W.J. 1819-1828. - *Flora Londinensis, a new edition enlarged* (vol. 4.1, 1819; vol. 4.2, 1821; vol. 5, 1828). London: George Graves.

Gray, A., Benham, P.E.M. & Raybould, A.F. (1990). *Spartina anglica* - the evolutionary and ecological background. In: A.J. Gray & P.E.M. Benham (eds) *Spartina anglica* - a research review, pp.5-10. ITE research publication number 2, HMSO, London.

Green, J.R. 1904. *A history of botany in the United Kingdom*. London: J.M. Dent & Sons Ltd.

Green, P.R. 2008. *Flora of County Waterford*. Dublin: National Botanic Gardens of Ireland.

Greenwood, E.F. 2012. *Flora of North Lancashire*. Lancaster: Palatine Books.

Greenwood, E.F. 2015. *Hunting plants. The story of those who discovered the flowering plants & ferns of North Lancashire*. Lancaster: Scotforth Books.

Grey, J.E. 1866. A new British station for *Wolffia arrhiza*. *J. Bot.* 4: 263-264.

Griffith, J.E. n.d. [1895]. *The Flora of Anglesey and Caernarvonshire*. Bangor.

Grigson, G. 1955. *The Englishman's Flora*. London: Phoenix House Ltd.

Grigson, G. 1974. *A Dictionary of English Plant Names*. London: Allen Lane.

Grose, D. 1957. *The Flora of Wiltshire*. Devizes: Wiltshire Archaeological and Natural History Society.

Groves, H. 1883. *Ranunculus ophioglossifolius* in England. *J. Bot.* 21: 51-52.

Groves, J. 1905. Crabbe as a botanist. *Proc. Suffolk Inst. Arch. and NH.* 12: 223-232.

Gunther, R.T. 1922. *Early British Botanists and their gardens*. Oxford: Oxford University Press.

Gunther, R.T. ed. 1928. *Further correspondence of John Ray*. London: Ray Society.

Gunther, R.T. 1934. Letters from John Ray to Peter Courthope. *J. Bot.* 72: 217-223.

Gunther, R.T. 1945. *Early Science in Oxford, Vol XIV. Life and Letters of Edward Lhwyd*. Oxford: Oxford University Press.

Hackney, P. 1992. *Stewart & Corry's Flora of the North-East of Ireland*. 3rd ed. Belfast: Institute of Irish Studies, The Queen's University.

Hall, P.M. 1939. The British species of *Utricularia*. *B.E.C.* 12: 100-117.

Hall, T.B. 1839. *The Flora of Liverpool*. London: Whitaker & Co.

Halliday, G. 1990. *Crepis praemorsa* (L.) Tausch, new to Western Europe. *Watsonia* 18: 85-87.

Halliday, G. 1997. *The Flora of Cumbria*. Lancaster: Lancaster University.

Halliday, R. 1999. A Suffolk miracle? The Sea Pea Harvest of 1555. *Suffolk Review N.S.* 33: 37-46.

Hambrough, A. 1854. Notice of the Occurrence of *Arum italicum* at Steephill, Isle of Wight. *Phytologist* 5: 194-195.

Hampton, M. & Kay, Q.O.N. 1995. *Sorbus domestica* L., new to Wales and the British Isles. *Watsonia* 20: 379-384.

Hanbury, F.J. & Marshall, E.S. 1899. *The Flora of Kent*. London: privately published.

Hardy, G. 2011. "To get those plants … likely to prove interesting at Edinburgh": Robert Brown of Perth and James McNab's North American tour of 1834. *Sibbaldia* 9: 191-222.

Harley, R.M. 1956. *Rubus arcticus* L. in Britain. *Watsonia* 3: 237-238.

Hart, H. C. 1887. *Arabis alpina* in Skye. *J. Bot.* 25: 247.

Harvey, J. (compiler). 1981. *A Service Index of Latin Binomials to Gerard's Herbal, as revised by Thomas Johnson, 1633*. Privately printed.

Harvey, J. 1981. *Medieval Gardens*. London: B.T. Batsford Ltd.

Harvey, J.H. 1985. The first English garden book: Mayster Jon Gardener's treatise and its background. *Garden History* 13: 83-101.

Harvey, J.H. 1987. Henry Daniel: A Scientific Gardener of the Fourteenth Century. *Garden History* 15: 81-93.

Harvey, J.H. 1998. Published writings on garden history and related topics. *Garden History* 26: 102-105.

Harvey, W.H. 1848. Account of a new British Saxifrage. *London J. Bot.* 7: 569-571.

Hazlitt, W.C., ed. 1866. *Scoggin's jests … gathered by Andrew Boord*. [Reprint of the original 1626 edition with introduction and notes.] London: Willis and Sotheran.

Hawksford, J.E. & Hopkins, I.J. 2011. *The Flora of Staffordshire*. Stafford: Staffordshire Wildlife Trust.

Hayward, I.M. & Druce, G.C. 1919. *The Adventive Flora of Tweedside*. Arbroath: T. Buncle & Co.

Hayward, W.R. 1872 *et seq*. *The Botanist's Pocket-Book*. London: Bell & Daldy.

Henderson, D.M. & Dickson, J.H. 1994. *A Naturalist in the Highlands: James Robertson, his Life and Travels in Scotland, 1767-1771*. Edinburgh: Scottish Academic Press.

Henrey, B. 1975. *British Botanical and Horticultural Literature before 1800*. 3 vols. Oxford: Oxford University Press.

Henrey, B. ed Chater, A.O. 1986. *No ordinary gardener: Thomas Knowlton, 1691–1781*. London: British Museum (Natural History).

Henslow, J.S. 1832. *Fumaria vaillantii* a British Plant. *Magazine of Natural History* 5: 88.

Heslop Harrison, J.W. 1917. The genera *Orchis* & *Gymnadenia* in Durham. *Vasculum* 9: 86-89.

Heslop Harrison, J.W. 1948. Potamogetons in the Scottish Western Isles. *Bot. Soc. Edin.* 35: 1-25.

Hiern, W.P. 1901. *Limosella aquatica* L. var. *tenuifolia* Hook. f. *J. Bot.* 39: 336-339.

Hiern, W.P. 1909. *Euphrasia minima. J. Bot.* 47: 165-172.

Hind, W.M. 1889. *The Flora of Suffolk*. London: Gurney & Jackson.

Hooker, J.D. 1870 *et seq. The Student's Flora of the British Islands*. London: Macmillan and Co.

Hooker, W.J. 1821. *Flora Scotica*. London.

Hooker, W.J. 1830. *The British Flora*. (ed. 1, 1830; ed. 2, 1831; ed. 3, 1835c; ed. 4., 1838; ed. 5, 1842). London: Longman, Orme, Brown, Green, & Longmans.

Hooker, W.J. 1835a. Botanical Information. *Companion to the Botanical Magazine.* 1: 119, 1: 307.

Hooker, W.J. 1835b. Another Heath found in Ireland. Botanical Information. *Companion to the Botanical Magazine.* 1: 158.

Hooker, W.J. 1836. Botanical Information. *Companion to the Botanical Magazine.* 2: 191-192.

Hooker, W.J. 1839. *Viola lactea*. Annals of N. H. 2: 383.

Hooker, W.J. 1854. *Epipogium gmelini* Rich., a British plant. *Hooker's Journal of Botany* 6: 318.

Hope, J. 1769. A Letter from John Hope, M.D., F.R.S. Professor of Physic and Botany in the University of Edinburgh, to William Watson, M.D., F.R.S. on a Rare Plant Found in the Isle of Skye. *Phil. Trans. Royal Soc.* 59: 241-246.

Hopkins, J.J. 1983. *Studies of the historical ecology, vegetation and flora of the Lizard district, Cornwall, with particular reference to heathland*. PhD. Thesis for University of Bristol. Unpublished.

Hore, W.S. 1842. Lists of plants found in Devonshire & Cornwall, not mentioned by Jones in the *Flora Devoniensis*, with remarks on the rarer species. *Phytologist* 1: 160-163.

Hore, W.S. 1845. A day's Botanizing on the Lizard. *Phytologist* 2: 235 – 239.

Horsman, F. 1995. 'Ralph Johnson's Notebook'. *Archives of Natural History* 22(2): 147-167.

Horsman, F. 2011. The Earliest Botanists in Teesdale. *Teesdale Record Society Journal.* 3rd series 19: 25-35.

Horwood, A.R. & Noel, C.W.F. 1933. *The flora of Leicestershire and Rutland*. Oxford: Oxford University Press.

Houston, L., Robertson, A., Jones, K., Smith, S.C.C., Hiscock, S. & Rich, T.C.G. 2009. An account of the Whitebeams (*Sorbus* L., Rosaceae) of Cheddar Gorge, England, with descriptions of three new species. *Watsonia* 27: 283-300.

[How(e)], W. 1650. *Phytologia Britannica natales exhibens lndigenarum Stirpium sponte emergentium*. Londini.

Howard, H.W. & Lyon, A.G. 1950, The identification and distribution of the British Watercress species. *Watsonia* 1: 228-233.

Howard, H.W. & Manton, I. 1946. Autopolyploid and Allopolyploid Watercress with the description of a new species. *Annals of Botany N.S.* 10: 1-13.

Howarth, W.O. 1925. On the Occurrence and Distribution of *Festuca ovina* L., *sensu ampliss*. in Britain. *Bot. J. Linn. Soc.* 47: 29-39.

Hroudová, Z., Zákravský, P., Ducháček, M. & Marhold, K. 2007. Taxonomy, distribution and ecology of *Bolboschoenus* in Europe. *Annales Botanici Fennici* 44: 81-102.

Hubbard, C.E. 1936. Some uncommon British grasses. *Proc. Linn. Soc.* 148: 108-113.

Hubbard, C.E. 1954. *Grasses*. London: Penguin Books.

Hubbard, C.E. 1957. Report to BES Symposium on *Spartina*. *J. Ecol.* 45: 613-616.

Hubbard, C.E. 1965. The earliest record of *Spartina maritima* in Britain. *Proc. BSBI* 6: 119.

Hudson, W. 1762. *Flora Anglica*. London.

Hudson, W. 1778. *Flora Anglica*, ed 2. London.

Huon, A. 1970. Les fétuques de L'Ouest de la France. *Botanica Rhedonica, sér. A,* 9: 186 & 259-260.

Ingall, T. 1844. *Teucrium botrys* found in Surrey. *Phytologist* 1: 1086.

Ingram, R. & Noltie, H.J. 1981. *The Flora of Angus*. Dundee: Dundee Museums and Art Galleries.

Ingram, R. & Noltie, H.J. 1995. Biological Flora of the British Isles. No. 186. *Senecio cambrensis*. *J. Ecol.* 83: 537-546.

Ingrouille, M.J. 1985. The *Limonium auriculae-ursifolium* (Pourret) Druce group (Plumbaginaceae) in the Channel Isles. *Watsonia* 15: 221-229.

Ingrouille M.J. & Stace, C.A. 1986. The *Limonium binervosum* aggregate (Plumbaginaceae) in the British Isles. *Bot. J. Linn. Soc.* 92: 177-217.

Irvine, A. 1859. Wandsworth Plants. *Phytologist N.S.* 3: 330-350.

Jackson, A.B. 1913. *Maianthemum bifolium* Schmidt in England. *J. Bot.* 51: 202-208.

Jackson, B.D. 1881. *Guide to the Literature of Botany; being a Classified Selection of Botanical Works*. London: Longmans Green & Co.

Jackson, B.D. 1899. Legré, L. La Botanique en Provence au XVI Siècle; Pierre Pena et Mathias de Lobel (Book review). *J. Bot.* 37: 88-92. [examining the role of Pena with L'Obel]

James, T.J., Jimenez-Mejias, P. & Porter, M.S. 2012. The occurrence in Britain of *Carex cespitosa*, a Eurasian sedge rare in western Europe. *New Journal of Botany* 2: 20-25.

Jeffers, R.H. 1967. *The Friends of John Gerard*. Falls Village, Conncticut: The Herb Grower Press.

Jeffers, R.H. 1969. *The Friends of John Gerard. Biographical Appendix*. Falls Village, Connecticut: The Herb Grower Press.

Jenner, E. 1845. *A Flora of Tunbridge Wells*. Tunbridge Wells: J. Colbran.

Jermy, A.C. 1989. The history of *Diphasiastrum issleri* (Lycopodiaceae) in Britain and a review of its taxonomic status. *Fern Gazette* 13(5): 257-265.

Jermy, A. C., Arnold, H. R., Farrell, L. & Perring, F. H. 1978. *Atlas of ferns of the British Isles*. London: Botanical Society of the British Isles & British Pteridological Society.

Jermyn, S.T. 1974. *Flora of Essex*. Colchester: Essex Naturalists' Trust Ltd.

Johns, C.A. 1847. Observations on the plants of the Land's End. *Phytologist* 2: 906-908.

Johnson, T. 1629. *Iter Plantarum lnvestigationis ... in Agrum Cantianum [and] Ericetum Hamstedianum*. Londini. [Iter].

Johnson, T. 1632. *Descriptio Itineris Plantarum lnvestigationis ... in Agrum Cantianum A.D. 1632 et Enumeratio Plantarum in Ericeto Hampstediano locisq. vicinis crescentium*. Londini. [Descriptio].

Johnson, T. 1633. *The herball or generall historie of plantes. Gathered by John Gerarde ... Very much enlarged and amended by Thomas Johnson*. London.

Johnson, T. 1634. *Mercurius Botanicus: sive, Plantarum gratia suscepti itineris anno M.DC.XXXIV description*. Londini.

Johnson, T. 1638. Catalogus plantarum juxta Tottenham lectarum anno. dom. 1638. Manuscript held at Royal Botanic Gardens, Kew. [See Edgington (2014)].

Johnson, T. 1641. *Mercurius Botanicus pars altera*. Londini.

Johnston, G. 1833. List of Plants discovered within the District, since the publication of Dr Johnston's *Flora of Berwick-upon-Tweed*. *History of the Berwickshire Naturalists Club* 1: 29-32.

Jones, D. 1996. *The botanists and guides of Snowdonia*. Llanrwst: Gwasg Carreg Gwalch.

Jones, J.P. 1820. *A Botanical Tour through various parts of the counties of Devon and Cornwall*. Exeter.

Jones, J.P. & Kingston, J.F. 1829. *Flora Devoniensis*. London: Longman, Rees, Orme, Brown & Green.

Jorden, G. 1853. *Thymus Serpyllum* and *T. chamaedrys*. *Phytologist* 4: 1142.

K'Eogh, J. 1735. *Botanica Universalis Hibernica*. Cork.

Kay, Q. O. N. 1971. Biological Flora of the British Isles. No. 122. *Anthemis cotula* L. *J. Ecol.* 59: 623-636.

Kay, Q. O. N. 1997. A review of the taxonomy, biology, geographical distribution and European conservation status of *Asparagus prostratus* Dumort. (*A. officinalis* subsp. *prostratus*), Sea Asparagus. Unpublished report to CCW.

Kemp, R.F.O. 1968. *Lloydia serotina* (L.) Reichenb. - New locality in Radnorshire, vc 43. *Proc. BSBI.* 7: 391-392.

Kenneth, A.G., Lowe, M.R. & Tennant, D.J. 1988. *Dactylorhiza lapponica* (Laest. ex Hartman) Soó in Scotland. *Watsonia* 17: 37-41.

Kent, D.H. 1949. John Blackstone, apothecary and botanist (1712-53). *Watsonia* 1: 141-148.

Kent, D.H. 1950. Tothill Fields, Westminster: a lost botanical area. *London Naturalist* 29: 3-6.

Kent, D.H. 1967. *Index to Botanical Monographs.* London: Academic Press.

Kent, D.H. 1975. *The Historical Flora of Middlesex.* London: Ray Society.

Kent, D.H. 1992. *List of Vascular Plants of the British Isles.* London: Botanical Society of the British Isles.

Kent, D.H. 2000. *Flora of Middlesex, a supplement.* London: Ray Society.

Kent, D.H. 2001. The history of *Fritillaria meleagris* in Britain to 1900. *The London Naturalist,* no. 80: 29-41.

Kent, D.H. & Allen, D.E. 1984. *British and Irish Herbaria.* London: Botanical Society of the British Isles.

Kew, H.W. & Powell, H.E. 1932. *Thomas Johnson: Botanist and Royalist.* London: Longmans, Green & Co.

Keynes, G. 1951. *John Ray, a bibliography.* London: Faber and Faber.

Keynes, G. 1976. *John Ray, 1627-1705: a bibliography, 1660-1970.* Amsterdam: Gerald Th. van Houston.

Killick, J, Perry, R. & Woodell, S. 1998. *The Flora of Oxfordshire.* Newbury: Pisces Press.

Knipe, P.R. 1988. *Gentianella ciliata* (L.) Borkh. in Buckinghamshire. *Watsonia* 17: 94-95.

L'Écluse, C. de. 1576. (as C. Clusius) *Rariorum aliquot stirpium per Hispanias observatarum historia libris duobus expressa.* Antwerp.

L'Ecluse [Clusius], C. 1583. *Rariorum aliquot stirpium, per Pannoniam, Austriam et vicinas quasdam provincias observatarum historia.* Antwerp.

L'Ecluse [Clusius], C. 1601. *Rariorum Plantarum Historia.* Antwerp.

L'Obel 1571, see Pena, P. & L'Obel, M. de. 1571.

L'Obel, M. de. 1576. (as M. de Lobel) *Plantarum seu stirpium historia. Cui adnexum est Adversariorum volumen.* Antverpiae [Antwerp]. [*Stirpium Historia*].

L'Obel, M. de. 1581. *Kruydtboeck oft Beschrÿvinghe van allerleye ghewassen, kruyderen, hesteren, ende gheboomten.* Antwerp.

L'Obel, M. de. 1605. *Adversariorum altera pars.* Londini. (in Pena, P. & L'Obel, M. de. 1605. *Dilucidae simplicium medicamenorum explicationes, Stirpium adversaria.*

L'Obel, M. de. 1655. *Stirpium illustrationes. Plurimas elaborantes inauditas plantas, subreptitiis Joh: Parkinsoni rapsodiis (ex codice MS insalutato) sparsim gravatae.* Londini.

Lankester, E. ed. 1846. *Memorials of John Ray (consisting of his life by Dr. Derham with his Itineries etc)*. London: Ray Society.

Lankester, E. ed. 1848. *The Correspondence of John Ray (consisting of selections from the philosophical letters published by Dr Derham)*. London. Ray Society.

Lansdown, R.V. & Bruinsma, J. 1999. *Callitriche palustris* L. new for Britain and Ireland. *BSBI News* 82: 18-19.

Larkey, S.V. & Pyles, T. 1941. *An Herbal [1525] edited and transcribed into Modern English with an Introduction*. New York: New York Botanical Garden.

Le Sueur, F. 1984. *Flora of Jersey*. Jersey: Société Jersiaise.

Leach, S.J. & Pearman, D.A. 2003. An assessment of the status of *Gaudinia fragilis* (L.) P. Beauv. (Poaceae) in the British Isles. *Watsonia* 24: 469-487.

Lees, E. 1867. *The Botany of Worcestershire*. Worcester: Worcestershire Naturalists Club.

Lees, F.A. 1861. in Curator's Report for 1861. *Thirsk B.E.C.* 1861: 7- 8.

Lees, F.A. 1882. On a new British Umbellifer. *J. Bot.* 20: 129-133.

Lees, F.A. 1888. *The Flora of West Yorkshire*. London: Lovell Reeve & Co.

Lees, F.A. 1894. *The Flora of Nidderdale*, in Speight, H. *Nidderdale and the garden of the Nidd*. London.

Lees, F.A. 1941. *A supplement to the Yorkshire Floras*. London: A Brown & Sons, Limited.

Leighton, W.A. 1841. *A Flora of Shropshire*. London: John van Voorst.

Leslie, A.C. 1987. *Flora of Surrey. Supplement and Checklist*. Guildford: privately published.

Ley, A. 1895. A new form of *Pyrus*. *J. Bot.* 33: 84.

Ley, A. 1901. *Pyrus scandica*? In Report for 1899. *B.E.C.* 1: 605.

Ley, A. 1909. *Tilia platyphyllos* Scop. in Wales. *J. Bot.* 47: 432.

Lhwyd, E. 1711a. Some farther observations relating to the Antiquities and Natural History of Ireland. *Phil. Trans. Royal Soc.* 27: 524-526.

Lhwyd, E. 1711b. An extract of a letter from the late Mr. Edw. Lhwyd to Dr Tancred Robinson; giving an account of some uncommon plants growing about Pensans and St Ives in Cornwall. *Phil. Trans. Royal Soc.* 27: 527.

Lightfoot, J. 1777. *Flora Scotica*. London.

Lindley, J. 1829. *Synopsis of the British Flora*. London.

Lindley, J. 1835. *The genera and species of Orchidaceous plants*. London: Ridgways, Piccadilly.

Lindman, C.A.M. 1928. On *Poa subcaerulea* Sm. and its restoration. *B.E.C.* 3: 179-180.

Linnaeus, C. 1792. *Flora Lapponica*, 2nd ed. London.

Linnean Society. 1795. Extracts from the Minute Book, June 2nd, 1795. *Trans. Linn. Soc.* 3: 333.

Linton, E.F. 1895. *Alchemilla vulgaris* and its segregates. *J. Bot.* 33: 110-112.

Linton, W.R. 1903. *Flora of Derbyshire*. London: Bemrose & Sons Ltd.

Lockton, A.J. & Whild, S.J. 2005. *Rare plants of Shropshire*, 3rd ed. Montford Bridge: Shropshire Botanical Society.

Lockton, A.J. & Whild, S.J. 2015. *The Flora and Vegetation of Shropshire*. Montford Bridge: Shropshire Botanical Society.

Lomax, E.A. 1873. *Echium plantagineum* in England. *J. Bot.* 11: 105-106.

London Catalogue of British Plants. 1844 *et seq*. [Watson, H.C. & Dennes, G.E.]. London: William Pamplin.

Louis, A. 1980. *Mathieu de L'Obel, 1538 – 1616*. Ghent: Story-Scientia.

Lousley, J.E. 1936. *Oxycoccus* Hill. In Notes on some interesting British plants- II. *J. Bot.* 74: 197-200.

Lousley, J.E. 1939. *Rumex aquaticus* L. as a British Plant. *J. Bot.* 77: 149-152.

Lousley, J.E. 1950. Distributor's report for 1948. *B.S.B.I. Year Book 1950*.

Lousley, J.E. 1953. *Flora of the British Isles*. A.R. Clapham, T.G. Tutin, and E.F. Warburg. Book review. *Watsonia* 2: 417-425.

Lousley, J.E. 1957. *Alisma gramineum* in Britain. *Proc BSBI* 2: 346-353.

Lousley, J.E. 1971. *The Flora of the Isles of Scilly*. Newton Abbot: David & Charles.

Lovatt, C. 2007. 'Only one tree I believe'. *Bull. Bristol Nats. Soc.* 466: 15-18.

Lowe, A.J. & Abbott, R.J. 2003. A new British species, *Senecio eboracensis*. *Watsonia* 24: 375-388.

Lowe, E.J. 1891. *British Ferns*. London: Swan Sonnenschein & Co.

Lowe, M.R. 2003. *Dactylorhiza majalis* in Scotland. *Eurorchis* 15: 77-86.

Lusby, P. & Wright, J. 1996. *Scottish Wild Plants: their history ecology and conservation*. Edinburgh: The Stationery Office Ltd.

Lyons, I. 1763. *Fasciculus Plantarum circa Cantabrigiam nascentium*. London.

Lyte, B. & Cope, T. 1999. Plants in peril, 25. *Bromus interruptus*. *Curtis's Botanical Magazine* 16: 296-300.

Lyte, H. 1578. *A niewe herball, or Historie of plantes*. London. [a translation into English of Dodoens' work of the same title].

Mabberley, D.J. 1985. *Jupiter Botanicus. Robert Brown of the British Museum*. Braunschweig: J. Cramer & London: British Museum (Natural History).

Mabey, R. 1996. *Flora Britannica*. London: Sinclair-Stevenson.

Mackay J.T. 1806. A Systematic Catalogue of Rare Plants found in Ireland. *Trans. Royal Dublin Soc.* 5: 127-184.

Mackay J.T. 1825. Catalogue of the indigenous plants found in Ireland. *Trans. Royal Irish Academy* 14.

Mackay, J.T. 1830. Notice of a new indigenous heath, found in Connemara. *Trans. Roy. Ir. Acad.* 16: 127-128.

Mackay, J.T. 1836. *Flora Hibernica*. Dublin: William Curry jun and Company.

Mansel – Pleydell, J.C. 1874. *The Flora of Dorsetshire*. Dorchester.

Mansel – Pleydell, J.C. 1895. *The Flora of Dorsetshire*, 2nd ed. Dorchester.

Manton, I., 1947. Polyploidy in *Polypodium vulgare. Nature* 159: 136.

Manton, I., 1950. *Problems of cytology and evolution in the Pteridophyta*. Cambridge: Cambridge University Press.

Margetts, L.J. & David, R.W. 1981. *A Review of the Cornish Flora 1980*. Redruth: Institute for Cornish Studies.

Marren, P.R. 1983. The History of Dickie's Fern in Kincardineshire. *Trans. Bot. Soc. Edin.* 44: 157-164.

Marren, P.R. 1999. *Britain's rare flowers*. London: T. & A.D. Poyser.

Marshall, E.S. 1894. On an apparently undescribed *Cochlearia* from Scotland. *J. Bot.* 32: 289-292.

Marshall, E.S. 1912. *Utricularia ochroleuca* R. Hartman. *J. Bot.* 50: 132-133.

Marshall, E.S. 1916. Notes on *Sorbus. J. Bot.* 54: 10-14.

Marshall, E.S. & Shoolbred, W. A. 1897. *Carex chordorrhiza* Ehrhart in Britain. *J. Bot.* 35: 450.

Martin, W. Keble & Fraser, G.T. 1939. *Flora of Devon*. Arbroath: T. Buncle & Co. Ltd.

Martyn, J. 1732. *Tournefort's History of plants growing about Paris … Translated into English, with many additions*. 2 vols. London.

Martyn, J. 1763. *Plantae Cantabrigienses*. London.

Martyn, T. 1794. *Flora Rustica* Vol 4. London: F.P. Nodder.

Maskew, R. 2014. *The Flora of Worcestershire*. Tenbury Wells: privately published.

McAllister, H.A. & Rutherford, A. 1990. *Hedera helix* L. and *H. hibernica* (Kirchner) Bean (Araliaceae) in the British Isles. *Watsonia* 18: 7-15.

McClintock, D. 1966. *Companion to flowers*. London: G. Bell and Sons.

McClintock, D. 1972. *Gaudinia fragilis* (L.) Beau. *Watsonia* 9: 143-146.

McClintock, D. 1975. *The Wildflowers of Guernsey*. London: Collins.

McClintock, D. 1982. *Guernsey's Earliest Flora*. London: Ray Society.

McNaughton, I.H. & Harper, J.L. 1964. *Papaver lecoqii* Lamotte (Biological Flora). *J. Ecol.* 52: 784-786.

McVeigh, A, Carey, J.E. & Rich T.C.G. 2005. Chiltern Gentian, *Gentianella germanica* (Willd.) Börner (Gentianaceae) in Britain: distribution and current status. *Watsonia* 25: 339-367.

Meikle, R. D. 1984. *Willows and Poplars of Great Britain and Ireland*. Botanical Society of the British Isles Handbook No. 4. London: Botanical Society of the British Isles.

Melderis, A. 1980. *Elymus* L., in Tutin T.G. *et al.*, eds., *Flora Europaea* 5: 196. Cambridge: Cambridge University Press.

Melville, R. 1938. Contributions to the study of British Elms - I. What is Goodyer's Elm? *J. Bot.* 78: 185-192.

Melville, R. 1939a. Contributions to the study of British Elms - II. The East Anglian Elm. *J. Bot.* 79: 138-145.

Melville, R. 1939b. Ambiguous Elm names – I. *Ulmus sativa*. Mill. *J. Bot.* 79: 244-248.

Melville, R. 1939c. Ambiguous Elm names – II. *Ulmus minor* Mill. *J. Bot.* 79: 266-270.

Melville, R. 1940. Contributions to the study of British Elms - III. The Plot Elm, *Ulmus plotii* Druce. *J. Bot.* 78: 181-192.

Merrett, C. 1666. *Pinax Rerum Naturalium Britannicarum*. London.

Michell, P.E. & Smith, P.H. 2012. Distribution, ecology and conservation of *Epipactis dunensis* in the sand-dunes of the Sefton Coast, Merseyside. *BSBI News* 120: 6 – 16.

Milford, J. 1834. *Trichonema bulbocodium* Ker. *Loudon's Mag. of N.H.* 7: 272.

Miller, P. 1752. *The Gardeners dictionary*, 6th ed. London.

Miller, P. 1768. *The Gardeners dictionary*, 8th ed. London.

Millett, L. & M. 1853. Wild flowers and ferns of the Isles of Scilly observed in June and July. Trans. *Nat. Hist. and Antiq. Soc. Penzance* 2: 75-78.

Milne-Redhead, E. 1947. *Cerastium brachypetalum* Pers. in Britain. *Naturalist* 822: 95-96.

Mitchell, J. 1986. The Reverend John Stuart D.D. (1743 - 1821) and his contribution to the discovery of Britain's Mountain Flowers. *Glasgow Naturalist* 21: 119-125.

Mitchell, J. 1992. Further notes on the Reverend John Stuart's contribution to the discovery of Britain's Mountain Flowers. *Glasgow Naturalist* 22: 103-105.

Mitchell, M.E. 1974. The sources of Threlkeld's *Synopsis Stirpium Hibernicarum*. *Proc. Royal Irish Acad.* 74B: 1-6.

Mitchell, M.E. 1975. Irish Botany in the seventeenth century. *Proc. Royal Irish Acad.* 75B: 275-284.

Mitchell, M.E. 2000. The Irish floras: a checklist of non-serial publications. *Glasra* 4: 47-57.

Mitten, W. 1845. Notice of the Discovery and description of *Carex montana*, (L.). *Phytologist* 2: 289-290.

Mitten, W. 1848. Notice of a species of *Fumaria* new to Britain. *Lond. Journ. Bot.* 7: 556-557.

Molyneux, T. 1697. A Discourse Concerning the Large Horns Frequently Found under Ground in Ireland, Concluding from Them That the Great American Deer, Call'd a Moose, Was Formerly Common in That Island: With Remarks on Some Other Things Natural to That Country. *Phil. Trans. Royal Soc.* 19: 489-512.

Moore, D. 1864. *Neotinea intacta* Reichb., a recent addition to the British Flora. *J. Bot.* 2: 228-229.

Moore, D. 1865. Discovery of *Inula salicana*, De Cand., in Ireland. *J. Bot.* 3: 333-335.

Moore, D. & More, A.G. 1866. *Contributions towards a Cybele Hibernica*. Dublin.

Moore, J. J. 1966. *Minuartia recurva* (All.) Schinz and Thell. new to the British Isles. *Irish Naturalists' Journal* 15: 130-132.

Moore, T. 1845. On the *Glyceria fluitans* and *G. plicata*. Annals of N.H. 2nd series 16:. 230-232.

Moore, T. 1849. Mr Dickie's *Cystopteris*. *Botanical Gazette* 1: 310-312.

Moore, T. 1859. *The Ferns of Great Britain and Ireland*. (Nature-printed by Bradbury, H.). London: Bradbury & Evans.

More, A.G. 1859. In Curator's Report for 1859. *Thirsk B.E.C.* 1859: 8-9.

More, A.G. 1860. British Lepigona. *Phytologist (N.S.)* 4: 193-197.

More, A.G. 1861. On the Occurrence of *Festuca ambigua*, Le Gall, in the Isle of Wight. *Proc. Linn. Soc.* 5: 189-192.

More, A.G. 1862. On the discovery of *Gladiolus illyricus* (Koch) in the Isle of Wight. *Proc. Linn. Soc.* 5-6: 177-182.

More, A.G. 1870. On *Callitriche obtusangula*, Le Gall, as a British Plant. *J. Bot.* 8: 342-343.

Morison, R. 1672. *Plantarum Umbelliferanum distributio nova*. Oxonii.

Morison, R. 1680. *Plantarum Historiae Universalis Oxoniensis. Pars secunda*. Oxonii

Morison, R. 1699. *Plantarum Historiae Universalis Oxoniensis. Tomus tertius*. Oxonii.

Morton, A.G. 1986. *John Hope 1725-1786*. Edinburgh: Edinburgh Botanic Garden (Sibbald Trust).

Moss, C.E. 1912. Remarks on the characters and the nomenclature of some critical plants noticed on the excursion, in The International Phytogeographical Excursion in the British Isles. XII. *New Phytologist* 11: 398-414.

Moss, C.E. 1914. *The Cambridge British Flora. Vol. 2*. Cambridge: Cambridge University Press.

Moss, C.E. 1920. *The Cambridge British Flora. Vol. 3*. Cambridge: Cambridge University Press.

Mott, F.T. 1878. *Prunella vulgaris*, white variety. *Midland Naturalist* 1: 136.

Murphy, R.J. ed. 1990. *Serapias parviflora*, in Progress report. *Botanical Cornwall* 4: 1.

Murphy, R.J. & Rumsey, F.J. 2005. *Cystopteris diaphana* (Bory) Blasdell (Woodsiaceae) – an overlooked native new to the British Isles. *Watsonia* 25: 255-263.

Murray, A. 1836. *The Northern Flora. Part 1*. Edinburgh: Adam & Charles Black.

Murray, C.W. & Birks, H.J.B. 2005. *The Botanist in Skye and Adjacent Islands*. Prabost and Bergen: privately published.

Murray, R.P. 1896. *The Flora of Somerset*. Taunton: Barnicott and Pearce.

Nash, R. & Ross, H.C.G. 1980. *Robert Templeton, Naturalist and Artist (1802–1892)*. Belfast: Ulster Museum.

Nash, R.T. 1781. *Collections for the History of Worcestershire: a catalogue of some rare plants in Worcestershire.* 1: lxxx –xc. London: John Nichols.

Nelmes, E. 1939. Notes on British Carices. - IV. *J. Bot.* 77: 259-266.

Nelmes, E. 1947. Two critical groups of British Sedges. *B.E.C.* 13: 95-105.

Nelson, E.C. 1977. The discovery in 1810 and subsequent history of *Phyllodoce caerulea* (L.) Bab. in Scotland. *Western Naturalist* 6: 45-72.

Nelson, E. C. 1979a. Historical records of the Irish Ericaceae, with particular reference to the discovery and naming of *Erica mackaiana*. *J. Soc. Biblphy Nat Hist.* 9(3): 289-299.

Nelson, E. C. 1979b. Records of the Irish flora published before 1726. *Bull. Irish Biogeog. Soc.* 3: 51-74.

Nelson, E.C. 1998. 'A willing Cicerone': Professor Robert Scott (ca. 1757–1808) of Trinity College, Dublin, Fermanagh's first botanist. *Glasra* 3: 115 143.

Nelson, G.A. 1959. William Turner's contribution to the first records of British plants. *Proc. Leeds Phil. and Lit. Soc., Scientific Section* 8(4): 109-138.

Newman, E. 1843a. *Cystopteris montana* a British Fern. *Phytologist* 1: 671-672.

Newman, E. 1843b. A history of the British *Equiseta*. *Phytologist* 1: 722-731.

Newman, E. 1844. *History of British Ferns and allied genera.* London: Van Voorst.

Newman, E. 1851. Proposed addition of three new Species and three new Genera to our list of British Ferns. *Phytologist* 4: 368-371.

Newman, E. 1853a. *Gymnogramma leptophylla* in the Channel Islands. *Phytologist* 4: 914.

Newman, E. 1853b. *Pseudathyrium alpestre*, and an allied Species. *Phytologist* 4: 974-975.

Newman, E. 1854. *Ophioglossum lusitanicum*. *Phytologist* 5: 80.

Noltie, H.J. ed. 1986. *The long tradition – the botanical exploration of the British Isles.* B.S.B.I conference report no 20. Kilbarchan: Scottish Natural History Society.

Noltie, H.J. 2011. *John Hope (1725-1786)*. Edinburgh: Royal Botanic Garden.

Norman, ?A.M. 1862. Notes on the species to which the Linnean *Polygonum aviculare* has been divided by Continental botanists. *Trans. Tyneside Nat. Field Club* 5: 140-142.

O'Mahony, T. 2009. *Wildflowers of Cork City and County.* Cork: The Collins Press.

Oliver, D. 1850. Discovery of *Naias flexilis* in Ireland. *Bot. Gaz.* 2: 278.

Oliver, D. 1852. Botanical notes of a week in Ireland during the present month [August, 1852]. *Phytologist* 4: 676-679.

Oliver, F.W. (ed.). 1913. *Makers of British Botany.* Cambridge: Cambridge University Press.

Oswald, P.H. 1995, in Trueman, I, Morton, A. & Wainwright, M. *The Flora of Montgomeryshire.* Welshpool.

Oswald, P.H. 2000. Historical records of *Lactuca serriola* L. and *L. virosa* L. in Britain, with special reference to Cambridgeshire (v.c.29). *Watsonia* 23: 149-159.

Oswald, P.H. & Preston, C.D. 1998. A mare's nest of horsetails: John Ray's treatment of "*Equisetum*" in his Cambridge Catalogue (1660). *Nature in Cambridgeshire* 40: 2-18.

Oswald, P.H. & Preston, C.D. 2011. *John Ray's Cambridge Catalogue (1660)*. London: Ray Society.

Park, J.J. 1814. *Topography and Natural History of Hampstead*. London.

Parkinson, A. 2007. *Nature's Alchemist: John Parkinson, herbalist to Charles 1*. London: Frances Lincoln Limited.

Parkinson, J. 1629. *Paradisi in Sole; Paradisus Terrestris*. London.

Parkinson, J. 1640. *Theatrum Botanicum . . . an Herball*. London.

Paton, A. 1967. True Service Trees of Worcestershire. *Proc. BSBI* 7: 9-13.

Pearman, D.A. 2013. Late-discovered petaloid monocotyledons: separating the native and alien flora. *New Journal of Botany* 3: 24-32.

Pearman, D. & Edgington, J. 2016. *Simethis planifolia* Kunth (Kerry Lily) in Britain and Ireland. *BSBI News* 132: 24-26.

Pearman, D.A. & Edwards, B. 2002. *Valerianella eriocarpa* Desv. in Dorset. *Watsonia* 24: 81-89.

Pearsall, W.H. 1914. *Hydrilla verticillata*. The Lancashire and Cheshire Naturalist No. 78 (No. 66 New Series) Sept. 1914 7: 212-213.

Pearsall, W.H. 1919. The British Batrachia. *B.E.C.* 5: suppl. for 1918.

Pearsall, W.H. 1931. *Carex flava*. In Plant notes. *B.E.C.* 9: 529.

Pearsall, W.H. 1935. The British species of *Callitriche*. *B.E.C.* 10: 861-871.

Pena, P. & L'Obel, M. de. 1571. *Stirpium adversaria nova*. Londini.

Pennant, T. 1774. *A tour in Scotland and voyage to the Hebrides; MDCCLXXII. Part 1*. Chester: John Monk.

Penneck, R. 1838. Notices of two plants new to the Cornish Flora. *Jour. Royal Inst. of Cornwall* 42-43.

Perring, F. H., Sell, P. D., Walters, S. M. & Whitehouse, H. L. K. 1964. *A Flora of Cambridgeshire*. Cambridge: Cambridge University Press.

Petch, C.P. & Swann, E.L. 1968. *Flora of Norfolk*. Norwich: Jarrold and Sons Limited.

Petiver, J. 1695. *Musei Petiveriana, Centauria Prima*. Londini.

Petiver, J. 1703. *Gazophylacii naturae et artis*. Londini.

Petiver, J. 1710. An Account of Divers Rare Plants, Lately Observed in Several Curious Gardens about London, and Particularly the Company of Apothecaries Physick Garden at Chelsey. *Phil. Trans. Royal Soc.* 27: 375-394.

Petiver, J. 1713. *Herbarii Britannici clariss. D. Raii Catalogus. Tab. 1-50*. London.

Petiver, J. 1715a. *Botanicum Hortense* IV. Giving an Account of Divers Rare Plants, Observed the Last Summer A. D. 1714. in Several Curious Gardens about London, and Particularly the Society of Apothecaries Physick-Garden at Chelsea. *Phil. Trans. Royal Soc.* 29: 269-284.

Petiver, J. 1715b. *Herbarii Britannici clariss. D. Raii Catalogus. Tab. 51-72*. London.

Petiver, J. 1716. *Graminum, muscorum, fungorum, submarinorum, &c. Britannicorum concordia, etc.* London.

Philipson, W.R. 1937. A revision of the British species of the genus *Agrostis* L. *Jour. Linn. Soc.* 51: 73-151.

Pigott, C.D. 1951. *The geographical distribution of the British species of* Thymus, *and a brief illustration of the application of distributional studies in ecology*. In Lousley, J.E. ed. *The study of the distribution of British Plants*. BSBI Conference Report no.2.

Pigott, C.D. 1954. Species delimination in British *Thymus. New Phytologist* 53: 470-495.

Pitt, E. 1678. Concerning the *Sorbus Pyriformis. Phil. Trans. Royal Soc.* 12: 978-979.

Planchon, J.E. 1849. Observations on the genus *Ulex*, with a description of a new species common to Brittany and the South-east England. *Bot. Gaz.* 1: 281-290.

Plot, R. 1677. *The Natural History of Oxfordshire*. Oxford.

Plot, R. 1686. *The Natural History of Staffordshire*. Oxford.

Plukenet, L. 1692. *Phytographia, Pars Tertia*. Londini.

Plukenet, L. 1696. *Almagestum botanicum sive Phytographiae Pluc'natianae Omomasticum*. Londini.

Plukenet, L. 1700. *Almagesti botanici mantissa*. Londini.

Plukenet, L., 1705. *Amaltheum botanicum*. London.

Polwhele, R. 1797. *History of Devonshire*, Vol.1. Exeter: Cadell, Dilly & Murray.

Pope, C.R. 2003. The first British record for Fringed Gentian (*Gentianella ciliata*). *BSBI News* 92: 14.

Pope, C., Snow, L. & Allen, D. 2003. *The Isle of Wight Flora*. Wimborne: The Dovecote Press.

Praeger, R.L. 1901. Irish Topographical Botany. *Proc. Roy. Irish Acad. Third Series* 7: 1-410.

Praeger, R.L. 1909a. *A Tourist's Flora of the West of Ireland*. Dublin: Hodges, Figgis, and Co., Ltd.

Praeger, R.L. 1909b. *Lastrea remota* in Ireland. *Irish Naturalist.* 18: 151-153.

Praeger, R.L. 1919. *Asplenium adiantum-nigrum var. acutum. Irish Naturalist* 28: 13-19.

Praeger, R.L. 1934. *The botanist in Ireland*. Dublin: Hodges, Figgis, and Co., Ltd.

Praeger, R.L. 1939. A further contribution to the flora of Ireland. *Proc. Roy. Irish Academy* 45B: 231-254.

Praeger, R.L. 1949. *Some Irish naturalists: a biographical note-book*. Dundalk: Dundalgan Press.

Preston, C.D. 2010. The first British records of *Potamogeton compressus* L. and *P. friesii* Rupr. *Watsonia* 28: 82-84.

Preston, C.D. 2016. Francis Willughby and the botany of the expeditions - evidence from plant specimens in the Middleton collection, in Birkhead, T.R. (editor), *Virtuoso by nature: The scientific worlds of Francis Willughby FRS (1635-1672)*. Brill, Leiden.

Preston, C.D. in press. Wild and cultivated plants in Cambridge, 1656–1657: a re-examination of Samuel Corbyn's lists.

Preston, C.D. & Croft, J.M. 1997. *Aquatic plants in Britain and Ireland*. Colchester: Harley Books.

Preston, C.D., Pearman, D.A. & Dines, T.D. 2002. *New Atlas of the British and Irish Flora*. Oxford: Oxford University Press.

Preston, C.D., Pearman, D.A. & Hall, A.R. 2004. Archaeophytes in Britain. Bot. J. Linn. Soc. 145: 257-294.

Pryor, A.R. 1876. Notes on some Hertfordshire Carices. *J. Bot.* 14: 365-372.

Pryor, A.R. 1887. *Flora of Hertfordshire*. London: Gurney & Jackson.

Pugh, J.P. 1953. *Dryopteris borreri* Newm. in the British Isles. *Watsonia* 3: 57-65.

Pugsley, H.W. 1902. The British capreolate fumitories. *J. Bot.* 40: 129-181.

Pugsley, H.W. 1904. A new *Fumaria*. *J. Bot.* 42: 217-220.

Pugsley, H.W. 1912. The Genus *Fumaria* in Britain. *J. Bot.* 50: Supp. 1-76.

Pugsley, H.W. 1913. British *Fumaria* records. *J. Bot.* 51: 50-51.

Pugsley, H.W. 1919a. A Revision of the Genera *Fumaria* and *Rupicanos*. *Bot. J. Linn. Soc.* 44: 233-352.

Pugsley, H.W. 1919b. Notes on British *Euphrasias* - I. *J. Bot.* 57: 169-175.

Pugsley, H.W. 1924a. A New *Statice* in Britain. *J. Bot.* 62: 129-134.

Pugsley, H.W. 1924b. *Gentiana uliginosa* Willd. in Britain. *J. Bot.* 62: 193-196.

Pugsley, H.W. 1929. New British species of *Euphrasia*. *J. Bot.* 67: 224-225.

Pugsley, H.W. 1930. A revision of the British *Euphrasiae*. *Jour. Linn. Soc.* 48: 467-542.

Pugsley, H.W. 1931. A further new *Limonium* in Britain. *J. Bot.* 69: 44-47.

Pugsley, H.W. 1935. On some Marsh Orchids. *Bot. J. Linn. Soc.* 49: 553-592.

Pugsley, H.W. 1936a. *Gentiana amarella* L. in Britain. *J. Bot.* 74: 163-170.

Pugsley, H.W. 1936b. New British Marsh Orchids. *Proc. Linn. Soc.* 148: 121-125.

Pugsley, H.W. 1936c. The *Brassica* of Lundy Island. *J. Bot.* 74: 323-326.

Pugsley, H.W. 1940a. Notes on British *Euphrasias* - VI. *J. Bot.* 78: 89-92.

Pugsley, H.W. 1940b. Notes on *Orobanche* L. *J. Bot.* 78: 105-116.

Pugsley, H.W. 1945. The Eyebrights of Rhum. *The Naturalist* no. 813: 41-44.

Pulteney, R. 1749. *Opusculum botanicum locos plantarum natales circa Loughborough et in agris adjacentibus …* Linnean Society of London ms. collections.

Pulteney, R. 1790. *Sketches of the Progress of Botany in England*. London.

Pulteney, R. 1799. *Catalogue of the Birds, Shells, and some of the more rare plants of Dorsetshire*. London: printed by J. Nicholls for the use of the compiler and his friends. [a revised version by T. Rackett, with additions from others and with a memoir of Pulteney, was published in Hutchins, J. ed. Gough, R. 1813. *The History and Antiquities of the County of Dorset*. Vol 3. London].

Pulteney, R. n.d. (c.1789). *Flora Anglica abbreviata*. Ms., held at **BM**.

Pulteney, R. n.d. (c.1790). *Catalogue of English Plants with the names of first describers, or discoverers, annexed*. Ms., held at **BM**.

Purchas, W.H. & Ley, A. 1889. *A Flora of Herefordshire*. Hereford.

Purton, T. 1821. *A botanical Description of British Plants in the Midland Counties. Appendix*. London: Longman, Hurst, Rees, Orme & Brown.

Ralph, T.S. 1847. *Botanica Thomae Johnsoni*. London: W. Pamplin.

Rand, M. 2011. *Utricularia bremii* Heer ex Koell. in the New Forest. *BSBI News* 116: 8 - 9.

Randall, R.E. 1974. *Rorippa islandica* (Oeder) Borbás *sensu stricto* in the British Isles. *Watsonia* 10: 80-82.

Raven, C.E. 1942. *John Ray, Naturalist*. Cambridge: Cambridge University Press.

Raven, C.E. 1947. *English Naturalists from Neckham to Ray*. Cambridge: Cambridge University Press.

Raven, C.E. 1948. Thomas Lawson's Note-book. *Proc. Linn. Soc.* 160: 3-12.

Raven, C.E. 1950. *John Ray, Naturalist*. ed.2. Cambridge: Cambridge University Press. [reissued, with an introduction by S.M. Walters, 1986]

Raven, C.E. 1957. *The early development of a knowledge of the British flora*. In Lousley, J.E. ed. *Progress in the study of the British Flora*. London: Botanical Society of the British Isles.

Raven, J., 1950. Notes on the flora of the Scilly Isles and the Lizard Head. *Watsonia* 1: 356-358.

Raven, P.H. 1963. *Circaea* in the British Isles. *Watsonia* 5: 262-272.

Ray, J. 1660. *Catalogus Plantarum circa Cantabrigiam nascentium*. Londini.

Ray, J. 1663. *Catalogus Plantarum circa Cantabrigiam nascentium, app.1*.

Ray, J. 1670. *Catalogus Plantarum Angliae et Insularum adjacentium*. Londini.

Ray, J. 1677. *Catalogus Plantarum Angliae et Insularum adjacentium, ed. 2*. Londini.

Ray, J. 1685. *Catalogus Plantarum circa Cantabrigiam nascentiu: continens addenda et emendanda*, ed. 2. Cambridge.

Ray, J. 1686. *Historia Plantarum, vol 1*. Londini.

Ray, J. 1688. *Historia Plantarum, vol 2*. Londini.

Ray, J. 1688a. *Fasciculus stirpium Britannicarum, post editum Plantarum Angliae Catalogum observatarum*. Londini.

Ray, J. 1690. *Synopsis Methodica Stirpium Britannicarum*. Londini.

Ray, J. 1696. *Synopsis Methodica Stirpium Britannicarum*, ed.2. Londini.

Ray, J. 1704. *Historia Plantarum*, vol 3. Londini.

Ray, J. 1724. *Synopsis Methodica Stirpium Britannicarum*, ed 3. [by Dillenius]. Londini.

Rees, A. 1802 – 1820. *The Cyclopædia; or, Universal Dictionary of Arts, Sciences, and Literature.* 39 vols. London.

Relhan, R. 1786. *Florae Cantabrigiensi supplementum*. Cantabrigiae [Cambridge].

Relhan, R. 1802. *Flora Cantabrigiensis*. ed. 2. Cantabrigiae [Cambridge].

Relhan, R. 1820. *Flora Cantabrigiensis*. ed. 3. Cantabrigiae [Cambridge].

Rhode, E.S. 1922. *The Old English Herbals*. London: Longmans, Green and Co.

Ribbons, B.W. 1952. *Homogyne alpina* in Scotland. *Watsonia* 2: 237-238.

Rich, T.C.G. 1996. Is *Gentianella uliginosa* (Willd.) Boerner (Gentianaceae) present in England? *Watsonia* 21: 208-209.

Rich, T.C.G. 2009. Validation of names for new Avon Gorge *Sorbus* (Rosaceae) taxa. *Watsonia* 27: 370.

Rich, T.C.G. & Houston, L. 2006. *Sorbus whiteana* (Rosaceae), a new endemic tree from Britain. *Watsonia* 26: 1-7.

Rich, T.C.G., Houston, L., Charles, C.A. & Tillotson, A.C. 2009. The diversity of *Sorbus* L. (Rosaceae) in the Lower Wye Valley. *Watsonia* 27: 301-313.

Rich, T.C.G., Houston, L., Robertson, A. & Proctor, M.C.F. 2010. *Whitebeams, Rowans and Service Trees of Britain and Ireland*. London: Botanical Society of the British Isles.

Rich, T.C.G. & Jermy, A.C. 1998. *Plant Crib 1998*. London: Botanical Society of the British Isles.

Rich, T.C.G. & Proctor, M.C.F. 2009. Some new British and Irish *Sorbus* L. taxa (Roseaceae). *Watsonia* 27: 207-216.

Richards, A.J. & Porter, A.F. 1982. On the identity of a Northumberland *Epipactis*. *Watsonia* 14: 121-128.

Riddlesdell, H.J. 1905. Lightfoot's visit to Wales in 1773. *J. Bot.* 43: 290-307.

Riddlesdell, H.J. 1917. *Helioscieadium* in Britain. *B.E.C.* 4: 409-412.

Riddelsdell, H. J., Hedley, G. W. & Price. W. R. 1948. *Flora of Gloucestershire*. Cheltenham: Cotteswold Naturalists' Field Club.

Ridley, H.N. 1885. Two new British Plants. *J. Bot.* 23: 289-291.

Rix, E.M. & Woods R.G. 1981. *Gagea bohemica* (Zauschner) J.A. & J.H. Shultes in the British Isles, and a general review of the G. bohemica complex. *Watsonia* 13: 265-270.

Roach, F.A. 1985. *Cultivated fruits of Britain: their origin and history*. Oxford: Basil Blackwell.

Robertson, A. 2006. *Sorbus pseudomeinichii*, a new endemic *Sorbus* (Rosaceae), microspecies from Arran, Scotland. *Watsonia* 26: 9-14.

Robertson, J. 1768. An account and a print, of a new species of *Astragalus*, a plant discovered by James Robertson in 1767. *The Scotch Magazine, July 1768*: 344.

Robinson, F. 1919. *Isoetes hystrix* Durieu in Cornwall. *J. Bot.* 57: 322.

Robinson, F. 2001. James Sutherland's *Hortus Medicus Edinburgensis* (1683). *Garden History* 29: 121-152.

Robinson, J.F. 1902. *The Flora of the East Riding of Yorkshire*. London: A Brown & Sons.

Robson, E. 1797. Description of the *Ribes spicatum*. *Trans. Linn. Soc.* 3: 240-241.

Robson, E. c.1780. Manuscript notes in a copy of Robson, S. *The British Flora*. [see Graham, 1988, above]

Roe, R.G.B. 1981. *The Flora of Somerset*. Taunton: Somerset Archaeological and Natural History Society.

Roger, J.G. 1952. *Diapensia lapponica* L. in Scotland. *Trans. Bot. Soc. Edin.* 36: 34-37.

Rogers, W.M. 1891. Thomas Richard Archer Briggs. *J. Bot. 1891.* 29: 97-106.

Roles, S.J. 1957-1965. *Flora of the British Isles – Illustrations*. Cambridge: Cambridge University Press.

Roos, A.M. ed. 2015. *The Correspondence of Dr. Martin Lister (1639-1712). Volume One: 1662-1677*. Leiden: Brill.

Rose, F. 1991. A new subspecies of *Gymnadenia conopsea* (L.) R. Br. *Watsonia* 18: 319-320.

Rose, H. 1775. *Elements of Botany. Appendix (descriptions of some plants lately discovered in Norfolk and Suffolk)*. London: T. Cadell.

Rosser, E.F. 1955. A new British species of *Senecio*. *Watsonia* 3: 228-232.

Rumsey, F.J. 2007a. An overlooked boreal clubmoss *Lycopodium lagopus* (Laest. ex Hartm.) Zinserl. ex Kusen. (Lycopodiaceae) in Britain. *Watsonia* 26: 477-480.

Rumsey, F.J. 2007b. An early specimen of *Cystopteris diaphana* (Bory) Blasdell supports its native status. *Watsonia* 26: 489-490.

Rumsey, F.J. 2012. *Diphasiastrum tristachyum* (Pursh) Holub – an overlooked extinct British native. *Fern Gazette* 19(2): 55-62.

Rumsey, F.J., Sheffield, E. & Farrar, D.R. 1990. British filmy fern gametophytes. *Pteridologist* 2: 40-42.

Rumsey, F.J. & Spencer, M. 2012. Is *Equisetum ramosissimum* (Equisetaceae: Equisetophyta) native to the British Isles? *Fern Gazette* 19: 37-46.

Rydén, M. 1984. *The English plant names in The Grete Herball (1526). A contribution to the historical study of English plant-name usage*, (Stockholm Studies in English LXI). Uppsala: Almqvist and Wiksell.

Rydén, M., Helander, H. & Olsson, K. 1999. *William Turner, Libellus de re Herbaria Novus 1538*. Uppsala.

Salmon, C.E. 1901. *Limonium lychnidifolium* var. *corymbosum*. *J. Bot.* 39: 193-195.

Salmon, C.E. 1903. Notes on *Limonium*. *J. Bot.* 41: 65-74.

Salmon, C.E. 1912. Early Lancashire and Cheshire records. *J. Bot.* 50: 369-371.

Salmon, C.E. 1914. *Alchemilla acutidens* Buser and other forms of *A. vulgaris* L. *J. Bot.* 52: 281-289.

Salmon, C.E. 1914. *Poa remotiflora* Murb. in Jersey. *J. Bot.* 52: 193-196.

Salmon, C.E. 1926. A new *Myosotis* from Britain. *J. Bot.* 64: 289-295.

Salmon, C.E. 1928. *Alchemilla pubescens* Lam. as a British plant. *J. Bot.* 66: 345-347.

Salmon, C.E. 1930. Notes on *Sorbus*. *J. Bot.* 68: 172-177.

Salmon, C.E. 1931. *Flora of Surrey*. London: G. Bell & Sons Ltd.

Salmon, J.D. 1847. New locality for *Cyperus fuscus*, Linn. *Phytologist* 2: 609.

Sandwith, N.Y. 1935. Report for 1934. *B.E.C.* 10: 992 (under *C. lepidocarpa*).

Sanford, M. & Fisk, R. 2010. *A Flora of Suffolk*. Ipswich: privately published.

Savidge, J. P., Heywood, V. H. & Gordon, V. (eds). 1963. *Travis's Flora of South Lancashire*. Liverpool: Liverpool Botanical Society.

Scannell, M.J.P. & Synott, D.M. 1972. *Census Catalogue of the flora of Ireland*. Dublin: Stationery Office. [2nd ed. 1987].

Scott, E.J.L. 1904. *Index to the Sloane Manuscripts in the British Museum*. London: Trustees of the British Museum.

Scott, M. 2016. *Mountain Flowers*. London: Bloomsbury Natural History.

Scott, W. & Palmer, R.[C.] 1987. *The Flowering Plants and Ferns of the Shetland Islands*. Lerwick: Shetland Times.

Scott, W. 1967. Notes on the Flora of Shetland. *New Shetlander* 82: 16-17.

Scully, R.W. 1916. *Flora of County Kerry*. Dublin: Hodges, Figgis & Co., Ltd

Seemann, B. ed. 1867. *Utricularia neglecta*. Lehm. *J. Bot.* 5: 73.

[Seemann, W. ed.] 1870. *Arenaria ciliata*. L. in Short notes. *J. Bot.* 8: 324.

Sell, P.D. & Murrell, G. 1996→. *Flora of Great Britain and Ireland*. Cambridge: Cambridge University Press.

Shivas, M.G. 1962. The *Polypodium vulgare* complex. *Br. Fern Gaz.* 9: 65-70.

Shivas, M.G. 1969. A cytotaxonomic study of the *Asplenium adiantum-nigrum* complex. *Br. Fern Gaz.* 10: 68-80.

Sibbald, R. 1684. *Scotia Illustrata*. Edinburgh.

Sibbald, R. 1710. *The History, ancient and modern, of the Sheriffdoms of Fife and Kinross*. Cupar.

Sibthorp, J. 1794. *Flora Oxoniensis*. Oxford.

Sim, R. 1848. *Cystopteris Dickieana*. *Gard. Farm. Jour.* 2(17): 308.

Simpson, F. W. 1982. *Simpson's Flora of Suffolk*. East Bergholt: Suffolk Naturalists Society.

Simpson, N.D. 1960. *A Bibliographical Index of the British flora*. Bournemouth: privately published.

Simpson, N.D. & Walters, S.M. 1959. *Juncus bufonius* ssp. *foliosus*. *Proc. BSBI* 3: 335.

Sinker, C. A., Packham, J. R., Trueman, I. C., Oswald, P. H., Perring, F. H. & Prestwood, W. V. 1985. *Ecological Flora of the Shropshire Region*. Shrewsbury: Shropshire Trust for Nature Conservation

Skeat, W. W. 1882. *The Book of Husbandry by Master Fitzherbert*. London Trübner & Co, for the English Dialect Society.

Skidmore, P, Dolby, M.J. & Hooper, M.D. n.d [c.1980]. *Thomas Tofield of Wilsic*. Doncaster: Doncaster Metropolitan Borough Council.

Slack, A.A. 1986. Lightfoot and the Exploration of the Scottish Flora. In Noltie, H.J. ed. *The long tradition – the botanical exploration of the British Isles*. B.S.B.I conference report no 20. Kilbarchan: Scottish Natural History Society.

Smail, H.C.P. 1974. William Borrer of Henfield, botanist and horticulturalist, 1781-1862. *Watsonia* 10: 55-60.

Smiles, S. 1878. *Robert Dick, Baker*. New York: Harper and Brothers.

Smith, C. 1756. *The Ancient and Present State of the County of Kerry*. Dublin.

Smith, G.E. 1828. *Medicago denticulata*, *Orobanche caryophyllacea*, and *Ophrys apifera*. *Loudon's Mag of Nat. Hist.* 1: 398.

Smith, G.E. 1829. *A catalogue of rare or remarkable phaenogamous plants, collected in south Kent*. London: Longman, Rees, Orme, Brown & Green.

Smith, G.E. 1846. *Filago* (*apiculata*: provisional name). *Phytologist* 2: 575-576.

Smith, J.E. 1791. Remarks on the genus *Veronica*. *Trans. Linn. Soc.* 1: 189-195.

Smith, J.E. 1793. Description of *Cerastium cerastoides* a new British plant discovered in Scotland by Mr James Dickson. *Trans. Linn. Soc.* ii: 343-345.

Smith, J.E. 1798. Observations on the British species of *Bromus*, with introductory remarks on the composition of a *Flora Britannica*. *Trans. Linn. Soc.* 4: 276-302.

Smith, J.E. 1800 – 1804. *Flora Britannica* (vol. 1 1800, vol. 2 1800, vol. 3 1804). London.

Smith, J.E. 1800a. Descriptions of five new British species of *Carex*. *Trans. Linn. Soc.* v. 264-273.

Smith, J.E. 1802. Remarks on some British species of *Salix*. *Trans Linn. Soc.* 6: 110-124.

Smith, J.E. 1811. An Account of several Plants, recently discovered in Scotland by George Don, A.L.S., not mentioned in the *Flora Britannica* nor *English Botany*. *Trans. Linn. Soc.* 10: 333-346.

Smith, J.E. 1824 - 1828. *The English Flora* (vol. 1 - 2 1824, vol. 3 1825, vol. 4 1828). London: Longman, Hurst, Rees, Orme, Brown & Green.

Smith, J.E. ed. 1792. *Caroli Linnaei Flora Lapponica*, 2nd ed. Londinii.

Smith, J.E. & Sowerby, J. 1790 - 1814. *English Botany*. London. [dates cited in text are those of the issue of each part].

Smith, J.E. & Sowerby, J. 1831 - 1865. *English Botany supplements*. London. [dates cited in text are those of the plates in each part].

Smith, J.E. & Sowerby, J. 1839. *English Botany*, ed 2. London.

Smith, Lady P. 1832. *Memoir and Correspondence of the late Sir John Edward Smith M.D.* London: Longman, Rees, Orme, Brown, Green & Longman.

Smith, P.H. 2005. *Schoenoplectus pungens* on the Sefton Coast. *BSBI News* 98: 30-33.

Snooke, W.D. 1823. *Flora Vectiana*. London.

Spruce, R. 1844. Note on *Carex paradoxa*. *Phytologist* 1: 842; 1121.

Stace, C.A. 1961. *Nardurus maritimus* (L.) Murb. in Britain. *Proc. BSBI* 4: 248-261.

Stace, C.A. 1991. *New Flora of the British Isles*. Cambridge: Cambridge University Press.

Stace, C.A. 1997. *New Flora of the British Isles* 2nd ed. Cambridge: Cambridge University Press.

Stace, C.A. 2010. *New Flora of the British Isles* 3rd ed. Cambridge: Cambridge University Press.

Stace, C.A., Ellis, R.G., Kent, D.H. & McCosh, D.J. 2003. *Vice-county census catalogue of the vascular plants of Great Britain*. London. Botanical Society of the British Isles.

Stearn, W.T. 1973. *John Ray: Synopsis methodica stirpium Britannicarum editio tertia 1724. Carl Linnaeus: Flora Anglica 1754 & 1759.* Facsimilies with an introduction by W.T. Stearn. London: Ray Society.

Stearn, W.T. ed. 1965. *William Turner: Libellus de re herbaria 1538 The names of herbes 1548: Facsimiles with introductory matter*. London: Ray Society. [actually Britten, J., Jackson B.D. & Stearn, W.T.]

Stephens, C.A. 1840. On *Scrophularia aquatica* of Linnaeus and Ehrhart. *Annals of N.H.* 5: 1-3.

Stephens, H.O. 1847. Notice of the discovery of *Allium sphaerocephalum*, L. on St Vincent Rocks, Bristol. *Phytologist* 2: 929.

Stephenson, T. & Stephenson, T.A. 1918. A new form of *Helleborine viridiflora*. *J. Bot.* 56: 1-4.

Stevenson, A. 1961. *A bibliographic study of William Curtis'* Flora Londinensis *1777 – 98 [1775-], in the Catalogue of Botanical Books in the Collection of Rachel McMasters Miller Hunt*. Pittsburgh: Hunt Foundation.

Stewart, O.M. 1981. A preliminary investigation of *Calamagrostis* Adanson in Scotland (abstract of exhibit at 1980 meeting). *Watsonia* 13: 369-370.

Styles, B.T. 1962. The taxonomy of *Polygonum aviculare* and its allies in Britain. *Watsonia* 5: 177-214.

Sutherland, J. 1683. *Hortus Medicus Edinburgensis*. Edinburgh.

Sutton, C. 1798. A description of five species of *Orobanche*. *Trans. Linn. Soc.* 4: 173-188.

Swan, G. A. 1993. *Flora of Northumberland*. Newcastle upon Tyne: The Natural History Society of Northumbria.

Swan, G.A. 1999. Identification, distribution and a new nothosubspecies of *Trichophorum cespitosum* (L.) Hartman (Cyperaceae) in the British Isles and N.W. Europe. *Watsonia* 22: 209-233.

Swan, G.A. & Walters, S.M. 1988. *Alchemilla gracilis* Opiz. A species new to the British Isles. *Watsonia* 17: 133-138.

Syme, J.B. 1863 – 1892. *English Botany* 3rd ed. (vol. 1, 1863; vol. 2, 1864; vol. 3, 1864; vol. 4, 1865; vol. 5, 1866; vol. 6, 1866; vol. 7, 1867; vol. 8, 1868; vol. 9, 1869; vol. 10, 1870; vol. 11, 1872; vol. 12, 1886, Supp. Vols 1-4, ed. N.E. Brown, 1892). London: G. Bell & Sons.

Syme, J.B. 1871. *Pyrus communis*, in report for the year 1870. *Bot. Ex. Club* 1871: 11.

Symons, J. 1798. *Synopsis plantarum insulis Britannicis indigenarum*. London.

Taschereau, P.M. 1977. *Atriplex praecox* Hulphers: a species new to the British Isles. *Watsonia* 11: 195-198.

[Tattersall, W.M. ed.] 1914. *Hydrilla* in England. In notes, queries and records of field work. *The Lancashire and Cheshire Naturalist. No. 77 (No. 63 New Series) Aug. 1914.* 7: 163.

Taylor, G. 1959. John Walker, 1731-1803, a notable Scottish naturalist. *Trans. Bot. Soc. Edin.* 38: 180-203.

Teesdale, R. 1792. Plantae Eboracenses. *Trans. Linn. Soc.* 2: 103-125.

Teesdale, R. 1798. A Supplement to the *Plantae Eboracenses* printed in the Second Volume of these Transactions. *Trans. Linn. Soc.* 5: 36-95.

Tellam, R.V. 1877. *Ranunculus tripartitus* DC., and *R. intermedius* Knaf. *Bot. Loc. Rec. Club Rep. for 1876*: 175.

Thirsk, J. 1997. *Alternative Agriculture*. Oxford: Oxford University Press.

Thompson, R. 1974. Some newly discovered letters of John Ray. *J. Soc. Biblphy Nat. Hist.* 7(1): 111-123.

Thor, G. 1987. Sumpbläddra, *Utricularia stygia*, en ny svensk art. [*Utricularia stygia* Thor, a new *Utricularia* species in Sweden]. *Svensk Botanisk Tidskrift*, 81: 273-280.

Thor, G. 1988. The genus *Utricularia* in the Nordic countries, with special emphasis on *U. stygia* and *U. ochroleuca*. *Nordic Journal of Botany* 8(3): 213-225.

Threlkeld, C. 1726. *Synopsis stirpium Hibernicarum*. Dublin.

Townsend, F. 1864. Contributions to a Flora of the Scilly Isles. *J. Bot.* 2: 102-120.

Townsend, F. 1879. *Erythraea tenuiflora*, Link. *J. Bot.* 17: 329.

Townsend, F. 1883. *Flora of Hampshire*. London: Lovell Reeve.

Townsend, F. 1891. A new form of *Euphrasia officinalis* L. from Scotland. *J. Bot.* 29: 161-162.

Townsend, F. 1896. *Euphrasia salisburgensis* Funk, native in Ireland. *J. Bot.* 34: 441-44.

Townsend, F. 1897. Monograph of the British species of *Euphrasia*. *J. Bot.* 35: 419-442, 465-468.

Townsend, F. 1904. *Flora of Hampshire,* 2nd ed. Ashford: L. Reeve & Co. Ltd.

Trail, J.W.H. 1923. *Flora of Aberdeen.* Aberdeen: Aberdeen University Press.

Trimen, H. & Dyer, W.T.T. 1869. *The Flora of Middlesex.* London: Robert Hardwicke.

Trimen, H. 1866. *Wolffia arrhiza,* Wimmer, in England. *J. Bot.* 4: 219-223.

Trimen, H. 1870a. *Callitriche truncata,* Guss., as a British plant. *J. Bot.* 8: 154-157.

Trimen, H. 1870b. On *Bromus asper. J. Bot.* 8: 376-379.

Trimen, H. 1871a. *Cyperus fuscus* not a native. *J. Bot.* 9: 148.

Trimen, H. 1871b. Notes in Jersey and Guernsey. *J. Bot.* 8: 198-201.

Trimen, H. 1872a. Another new British *Juncus. J. Bot.* 10: 337.

Trimen, H. 1872b. *Ranunculus chaerophyllos,* L., *auct.* in Jersey. *J. Bot.* 10: 225-228.

Trimen, H. 1873. *Juncus pygmaeus,* Rich., as a British Plant. *J. Bot.* 11: 3-5.

Trimen, H. 1876. *Rumex rupestris* Le Gall as a British Plant. *J. Bot.* 14: 1-5.

Trimen, H. 1877a. *Lavatera sylvestris,* Brot. in the Scilly Isles. *J. Bot.* 15: 16.

Trimen, H. 1877b. *Ranunculus tripartitus,* DC. *J. Bot.* 15: 209.

Trist, P.J.O. 1973. *Festuca glauca* Lam. and its var. *caesia* (Sm.) K. Richt. *Watsonia* 9: 257-262.

Trist, P.J.O. 1995. *Elytrigia repens* (L.) Desv. ex Nevski subsp. *arenosa* (Spenner) A. Löve (Poaceae) in north-western Europe. *Watsonia* 20: 385-390.

Trist, P.J.O. ed. 1979. *An Ecological Flora of Breckland.* Wakefield: EP Publishing.

Trueman, I., Morton, A. & Wainwright, M. 1995. *The Flora of Montgomeryshire.* Welshpool.

Tucker, R. 1870. *Gentiana campestris* L. in Isle of Wight. *J. Bot.* 8: 160.

Turner, D. & Dillwyn, L.W. 1805. *The Botanist's Guide through England and Wales.* London.

Turner, D., comp. 1835. E*xtracts from the Literary and Scientific Correspondence of Richard Richardson.* Yarmouth.

Turner, W. 1538. *Libellus de re Herbaria Novus.* Londini.

Turner, W. 1548. *The Names of Herbes.* [London].

Turner, W. 1551. *A new herball.* London.

Turner, W. 1562. *The seconde parte of Vuilliam Turner's herball.* Collen [Cologne].

Turner, W. 1568. *The thirde parte of Vuilliam Turner's Herball.* Collen.

Tutin, T.G. 1950. *Milium scabrum* Merlet. *Watsonia* 1: 345-348.

Van der Lande, V. 2016. George Anderson FLS (1773-1817): botanist, early fellow and officer of the Linnean Society. *The Linnean* 32: 23-31.

Vines, S. H. & Druce, G.C. 1897. *An Account of the Herbarium of the University of Oxford.* Oxford: Clarendon Press.

Vines, S. H. & Druce, G.C. 1914. *The Morisonian Herbaria.* Oxford: Clarendon Press.

Vines, S. H. & Druce, G.C. 1919. *An Account of the Herbarium of the University of Oxford, part II*. Oxford: Clarendon Press.

Wade, A. E., Kay, Q. O. N., Ellis, R. G. & National Museum of Wales. 1994. *Flora of Glamorgan*. London: HMSO.

Walker, J. 1812. *Essays on Natural History and Rural Economy*. London: Longman, Hurst, Rees, and Orme.

Walker, S. 1961. Cytogenetic studies in the *Dryopteris spinulosa* complex. II. *Am. J. Bot.* 48: 607-614.

Wallace, E.C. 1951. Plant Records. *Watsonia* 2: 36-62.

Walsh, L. 1978. *Richard Heaton of Ballyskenagh, 1601-1666*. Roscrea: Parkmore Press.

Walters, S. M. 1949b. *Aphanes microcarpa* (Boiss. et Reut.) Rothm. in Britain. *Wats* 1: 163-169.

Walters, S.M. 1949a. *Alchemilla vulgaris* in Britain. *Watsonia* 1: 6-18.

Walters, S.M. 1952. *Alchemilla subcrenata* Buser in Britain. *Watsonia* 2: 277-278.

Walters, S.M. 1963. *Eleocharis austriaca* Hayek, a species new to the British Isles. *Watsonia* 5: 329-335.

Walters, S.M. 1981. *The shaping of Cambridge botany*. Cambridge: Cambridge University Press.

Warburg, E.F. 1949. *Thymus*, in Abstracts from Literature, comp. Alston, A.H.G. *Watsonia* 1: 173-174.

Warburg, E.F. 1957. Some new names in the British Flora. *Sorbus*. *Watsonia* 4: 43-46.

Warburg, E.F. 1967. Taxonomic and nomenclatural notes on the British flora. *Sorbus*. *Watsonia* 6: 296.

Watson, H.C. & Dennes, G.E. 1844 *et seq*. See *London Catalogue*.

Watson, H.C. 1835-1837. *The new Botanist's Guide*, 2 vols. London: Longman, Rees, Orme, Brown, Green & Longman.

Watson, H.C. 1847-59, supp. 1860. *Cybele Britannica*. London: Longman & Co.

Watson, H.C. 1850. Explanatory notes on certain British plants for distribution by the Botanical Society of London, in 1850. *Phytologist* 3: 801-811.

Watson, H.C. 1852. *Cybele Britannica*, Vol. 3. London: Longman & Co.

Watson, H.C. 1863. *Sagina nivalis*, Fries, discovered in Scotland. *J. Bot.* 1: 355-366.

Watson, H.C. 1868-1870, supp. 1872. *Compendium to Cybele Britannica*. Thames Ditton: for private circulation.

Watson, H.C. 1873, 1883, with later supplements. *Topographical botany*. London: Bernard Quarich.

Webb, D.A. 1943. *An Irish Flora*. Dundalk: Dundalgan Press.

Webb, D.A. 1986. The hey-day of Irish Botany, 1866-1916. In Noltie, H.J. ed. *The long tradition – the botanical exploration of the British Isles*. B.S.B.I conference report no 20. Kilbarchan: Scottish Natural History Society.

Webb, D. A. & Scannell, M. J. P. 1983. *Flora of Connemara and the Burren*. Dublin: Royal Dublin Society & Cambridge University Press.

Webb, F.M. 1876. On *Utricularia neglecta* Lehmann; and on *U. bremii*, Heer, as a British plant. *J. Bot.* 14: 143-147.

Webster, A.D. 1886. *British Orchids*. Bangor: Nixon & Jarvis.

Webster, M.M. 1978. *Flora of Moray, Nairn & East Inverness*. Aberdeen: Aberdeen University Press.

Wegmüller, S. 1971. A cytotaxonomic study of *Lamiastrum galeobdolon* (L.) Ehrend. & Polatschek in Britain. *Watsonia* 8: 277-288.

Welch, M.A. 1972. Francis Willoughby, F.R.S. (1635–1672). *Journal of the Society for the Bibliography of Natural History* 6: 71-85.

Wheldon, J.A. & Travis, W.G. 1913. *Helleborine viridiflora* in Britain. *J. Bot.* 51: 343-346.

White, F.B. 1885. *Schoenus ferrugineus* L. in Britain. *J. Bot.* 23: 219-220.

White, F.B. 1887. *Juncus alpinus* as a British plant. *Scot. Nat.* 9: 182-183.

White, J.W. 1906. *Prunella laciniata* in Britain. *J. Bot.* 44: 365-366.

White, J.W. 1912. *The Flora of Bristol*. Bristol: John Wright and Sons Ltd.

Whittaker, E.J. 1986. *Thomas Lawson, 1630 -1691*. York: Sessions Book Trust.

Whittaker, E.J. ed. 1981. *A Seventeenth Century Flora of Cumbria – William Nicolson's Catalogue of Plants, 1690*. Durham: The Surtees Society.

Whittle, P. 1831. *Marina; or a historical and descriptive account of Southport, Lytham and Blackpool, situate on the western coast of Lancashire*. Preston: P. & H. Whittle.

Wigginton, M.J. ed. 1999. *British Red Data books. 1. Vascular Plants* 3rd ed. Peterborough: Joint Nature Conservation Committee.

Wilkinson, M.J. & Stace, C.A. 1988. The taxonomic relationships and typification of *Festuca brevipila* Tracey and *F. lemanii* Bastard (Poaceae). *Watsonia* 17: 289-299.

Wilkinson, M.J. & Stace, C.A. 1991. A new taxonomic treatment of the *Festuca ovina* L. aggregate (Poaceae) in the British Isles. *Bot. J. Linn. Soc.* 106: 347-397.

Williams, E. n.d. [c1800]. *Flora of Shropshire*. Ms. in Shropshire County Archives, Shrewsbury.

Williams, J. 1830. *Faunula Grustensis*. Llanrwst.

Wilmore, G.T.D, Lunn, J. & Rodwell, J.S. 2011. *The South Yorkshire Plant Atlas*. York: Yorkshire Naturalists' Union.

Wilmott, A.J. 1918. *Erythraea scilloides* in Pembrokeshire. *J. Bot.* 56: 321-323.

Wilmott, A.J. 1921. *Geranium purpureum* T.F. Forster. *J. Bot.* 59: 93-101.

Wilmott, A.J. 1922. Two Alchemillas new to Britain. *J. Bot.* 60: 163-165.

Wilmott, A.J. 1923. *Myosotis sicula* Gussone in Jersey. *J. Bot.* 61: 212-215.

Wilmott, A.J. 1934. Some interesting British Sorbus. *Proc. Linn. Soc.* 146: 73-79.

Wilmott, A.J. 1936. New British marsh orchids. *Proc. Linn. Soc.* 148: 126-130.

Wilmott, A.J. 1939. Annotationes Systematicae. V. Nomenclature of two British Alchemillas. *J. Bot.* 77: 249-250.

Wilson, A. 1938. *The Flora of Westmorland*. Conway: privately published.

Wilson, W. 1828. Observations on some British plants, particularly with reference to the English Flora of Sir James E. Smith. *Botanical Miscellany* 2: 133-143.

Withering, W. 1776. *A Botanical Arrangement of British plants …, ed 1*. Birmingham.

Withering, W. 1787-1792. *A Botanical Arrangement of British plants …, ed 2*. Birmingham.

Withering, W. 1796. *A Botanical Arrangement of British plants …, ed 3*. Birmingham.

Withering, W. 1801. *A Botanical Arrangement of British plants …, ed 4*. London.

Withering, W. 1830. *A Botanical Arrangement of British plants …, ed 7*. London.

Wollaston, T.V. 1845. Note on the Entomology of Lundy Island. *Zoologist* 3: 897-900.

Wolley-Dod, A.H. 1908. The subsection Eu-caninae of the Genus Rosa. *J. Bot.* 46 (supplement).

Wolley-Dod, A.H. 1910. The British Roses (excluding Eu-canina). *J. Bot.* 48 (supplement).

Wolley-Dod, A.H. 1937. *Flora of Sussex*. Hastings: Kenneth Saville.

Wolsey, G. 1861. Recent discovery of *Isoetes Hystrix*. *Phytologist* 5 (N.S.): 45-46.

Woodley, G. 1822. *A view of the present state of the Scilly Isles*. London.

Woods, J. 1835a. Botanical Excursion in the north of England. *Companion to the Botanical Magazine.* 1: 288-299.

Woods, J. 1835b. Observations on the species of *Fedia*. *Trans. Linn. Soc.* 17: 421-433.

Woods, J. 1836. Account of a Botanical Excursion into Brittany. *Companion to the Botanical Magazine* 2: 263-282.

Woods, J. 1851. On the various forms of *Salicornia*. *Proc. Linn. Soc.* 2: 109-112 & *Botanical Gazette* 3: 29-33.

Wright, F.R. Elliston. 1936. The Lundy *Brassica*, with some additions. *J. Bot.* 74: supp. 1-8.

Wycherley, P.R. 1953. Proliferation of Spikelets in British Grasses. *Watsonia* 3: 41-56.

Yeo, P.F. 1971. Revisional notes on *Euphrasia*. *Bot. J. Linn. Soc.* 64: 353-361.

Yeo, P.F. 1978. A taxonomic revision of *Euphrasia* in Europe. *Bot. J. Linn. Soc.* 77: 223-334.

Young, D.P. 1952. Studies in the British *Epipactis*. *Watsonia* 2: 253-259.

Zoller, H., & Steinmann, M. eds. 1987-1991. *Conradi Gesneri Historia plantarum: Gesamtausgabe*. 2 vols. Dietikon-Zurich: Urs Graf Verlag.

Zoller, H., Steinmann, M. & Schmid, K. eds. 1972 - 1980. *Conrad Gesneri Historia Plantarum. Faksimileausgabe*. 8 vols. Dictikon-Zurich: Urs Graf Verlag. [Facsimilies of watercolors and drawings in the Erlangen University Library with transcriptions and German translation].

Index of English names
compiled by Gwynn Ellis

For clarity these are the names recommended by the BSBI and used in Stace (2010) - for example Fumitories are also found under Ramping-fumitory, Sedges also under Tufted-sedge and Tussock-sedge, and so on. I far prefer those used in Clapham *et al.* (1987), and strongly disagree with the exaggerated and bookish use of hyphens.

Adder's-tongue = Ophioglossum vulgatum
    Least = Ophioglossum lusitanicum
    Small = Ophioglossum azoricum
Agrimony = Agrimonia eupatoria
    Fragrant = Agrimonia procera
Alder = Alnus glutinosa
Alexanders = Smyrnium olusatrum
Allseed = Radiola linoides
    Four-leaved = Polycarpon tetraphyllum
Alpine-sedge, Black = Carex atrata
    Close-headed = Carex norvegica
    Scorched = Carex atrofusca
Anemone, Wood = Anemone nemorosa
Angelica, Wild = Angelica sylvestris
Apple = Malus pumila
    Crab = Malus sylvestris
Archangel, Yellow = Lamiastrum galeobdolon
Arrowgrass, Marsh = Triglochin palustris
    Sea = Triglochin maritima
Arrowhead = Sagittaria sagittifolia
Ash = Fraxinus excelsior
Asparagus, Garden = Asparagus officinalis
    Wild = Asparagus prostratus
Aspen = Populus tremula
Asphodel, Bog = Narthecium ossifragum
    Scottish = Tofieldia pusilla
Aster, Goldilocks = Aster linosyris
    Sea = Aster tripolium
Avens, Mountain = Dryas octopetala
    Water = Geum rivale
    Wood = Geum urbanum
Awlwort = Subularia aquatica
Azalea, Trailing = Loiseleuria procumbens

Balm, Bastard = Melittis melissophyllum
Balsam, Touch-me-not = Impatiens noli-tangere
Baneberry = Actaea spicata
Barberry = Berberis vulgaris
Barley, Meadow = Hordeum secalinum
    Sea = Hordeum marinum
    Wall = Hordeum murinum
    Wood = Hordelymus europaeus
Bartsia, Alpine = Bartsia alpina
    Red = Odontites vernus
    Yellow = Parentucellia viscosa
Basil, Wild = Clinopodium vulgare
Bastard-toadflax = Thesium humifusum
Beak-sedge, Brown = Rhynchospora fusca
    White = Rhynchospora alba
Bearberry = Arctostaphylos uva-ursi
    Alpine = Arctostaphylos alpinus
Beard-grass, Annual = Polypogon monspeliensis
Bedstraw, Fen = Galium uliginosum
    Heath = Galium saxatile
    Hedge = Galium album
    Lady's = Galium verum
    Limestone = Galium sterneri s.s.
    Northern = Galium boreale
    Slender = Galium pumilum
    Wall = Galium parisiense
Beech = Fagus sylvatica
Beet = Beta vulgaris
Bellflower, Clustered = Campanula glomerata
    Giant = Campanula latifolia
    Ivy-leaved = Wahlenbergia hederacea
    Nettle-leaved = Campanula trachelium
    Rampion = Campanula rapunculus
    Spreading = Campanula patula
Bent, Black = Agrostis gigantea
    Bristle = Agrostis curtisii
    Brown = Agrostis vinealis
    Common = Agrostis capillaris
    Creeping = Agrostis stolonifera
    Velvet = Agrostis canina s.s.
Bermuda-grass = Cynodon dactylon
Betony = Betonica officinalis
Bilberry = Vaccinium myrtillus
    Bog = Vaccinium uliginosum

Bindweed, Field = Convolvulus arvensis
    Hedge = Calystegia sepium
    Sea = Calystegia soldanella
Birch = Betula
    Downy = Betula pubescens
    Dwarf = Betula nana
    Silver = Betula pendula
Bird's-foot = Ornithopus perpusillus
    Orange = Ornithopus pinnatus
Bird's-foot-trefoil, Common = Lotus corniculatus
    Greater = Lotus pedunculatus
    Hairy = Lotus subbiflorus
    Narrow-leaved = Lotus tenuis
    Slender = Lotus angustissimus
Bird's-nest, Yellow = Hypopitys monotropa
Bistort, Alpine = Persicaria vivipara
    Amphibious = Persicaria amphibia
    Common = Persicaria bistorta
Bitter-cress, Hairy = Cardamine hirsuta
    Large = Cardamine amara
    Narrow-leaved = Cardamine impatiens
    Wavy = Cardamine flexuosa
Bittersweet = Solanum dulcamara
Bitter-vetch = Lathyrus linifolius
    Wood = Vicia orobus
Black-bindweed = Fallopia convolvulus
Black-grass = Alopecurus myosuroides
Black-poplar = Populus nigra s.l.
Blackthorn = Prunus spinosa
Bladder-fern, Alpine = Cystopteris alpina
    Brittle = Cystopteris fragilis
    Diaphanous = Cystopteris diaphana
    Dickie's = Cystopteris dickieana
    Mountain = Cystopteris montana
Bladder-sedge = Carex vesicaria
Bladderseed = Physospermum cornubiense
Bladderwort = Utricularia australis
    Greater = Utricularia vulgaris s.l.
    Intermediate = Utricularia intermedia s.s.
    Lesser = Utricularia minor
    New Forest = Utricularia bremii
    Nordic = Utricularia stygia
    Pale = Utricularia ochroleuca
Blinks = Montia fontana
Bluebell = Hyacinthoides non-scripta
Blue-eyed-grass = Sisyrinchium bermudiana
Blue-sow-thistle, Alpine = Cicerbita alpina
Bogbean = Menyanthes trifoliata
Bog-myrtle = Myrica gale
Bog-rosemary = Andromeda polifolia

Bog-rush, Black = Schoenus nigricans
    Brown = Schoenus ferrugineus
Bog-sedge = Carex limosa
    Mountain = Carex rariflora
    Tall = Carex magellanica
Box = Buxus sempervirens
Bracken = Pteridium aquilinum
Bramble = Rubus fruticosus agg.
    Arctic = Rubus arcticus
    Stone = Rubus saxatilis
Brome, Barren = Anisantha sterilis
    False = Brachypodium sylvaticum
    Interrupted = Bromus interruptus
    Meadow = Bromus commutatus
    Rye = Bromus secalinus
    Smooth = Bromus racemosus s.l.
    Upright = Bromopsis erecta
Brooklime = Veronica beccabunga
Brookweed = Samolus valerandi
Broom = Cytisus scoparius
Broomrape, Bedstraw = Orobanche caryophyllacea
    Common = Orobanche minor
    Greater = Orobanche rapum-genistae
    Ivy = Orobanche hederae
    Knapweed = Orobanche elatior
    Oxtongue = Orobanche picridis
    Thistle = Orobanche reticulata
    Thyme = Orobanche alba
    Yarrow = Orobanche purpurea
Bryony, Black = Tamus communis
    White = Bryonia dioica
Buckler-fern, Broad = Dryopteris dilatata
    Crested = Dryopteris cristata
    Hay-scented = Dryopteris aemula
    Narrow = Dryopteris carthusiana
    Northern = Dryopteris expansa
    Rigid = Dryopteris submontana
    Scaly = Dryopteris remota
Buckthorn = Rhamnus cathartica
    Alder = Frangula alnus
Bugle = Ajuga reptans
    Pyramidal = Ajuga pyramidalis
Bugloss = Anchusa arvensis
Bulrush = Typha latifolia
    Lesser = Typha angustifolia
Burdock = Arctium agg.
    Greater = Arctium lappa
    Lesser = Arctium minus
    Wood = Arctium nemorosum

Bur-marigold, Nodding = Bidens cernua
    Trifid = Bidens tripartita
Burnet, Great = Sanguisorba officinalis
    Salad = Poterium sanguisorba
Burnet-saxifrage = Pimpinella saxifraga
    Greater = Pimpinella major
Bur-parsley, Small = Caucalis platycarpos
Bur-reed, Branched = Sparganium erectum
    Floating = Sparganium angustifolium
    Least = Sparganium natans
    Unbranched = Sparganium emersum
Butcher's-broom = Ruscus aculeatus
Butterbur = Petasites hybridus
Buttercup, Bulbous = Ranunculus bulbosus
    Celery-leaved = Ranunculus sceleratus
    Corn = Ranunculus arvensis
    Creeping = Ranunculus repens
    Goldilocks = Ranunculus auricomus
    Hairy = Ranunculus sardous
    Jersey = Ranunculus paludosus
    Meadow = Ranunculus acris
    Small-flowered = Ranunculus parviflorus
Butterfly-orchid, Greater = Platanthera chlorantha
    Lesser = Platanthera bifolia
Butterwort, Alpine = Pinguicula alpina
    Common = Pinguicula vulgaris
    Large-flowered = Pinguicula grandiflora
    Pale = Pinguicula lusitanica

Cabbage = Brassica oleracea
    Isle of Man = Coincya monensis
    Lundy = Coincya wrightii
Calamint, Common = Clinopodium ascendens
    Lesser = Clinopodium calamintha
    Wood = Clinopodium menthifolium
Campion, Bladder = Silene vulgaris
    Moss = Silene acaulis
    Red = Silene dioica
    Sea = Silene uniflora
    White = Silene latifolia
Canary-grass, Reed = Phalaris arundinacea
Candytuft, Wild = Iberis amara
Caraway = Carum carvi
    Whorled = Carum verticillatum
Carrot, Moon = Seseli libanotis
    Wild = Daucus carota
Catchfly, Alpine = Silene suecica
    Night-flowering = Silene noctiflora
    Nottingham = Silene nutans
    Sand = Silene conica

    Small-flowered = Silene gallica
    Spanish = Silene otites
    Sticky = Silene viscaria
Cat-mint = Nepeta cataria
Cat's-ear = Hypochaeris radicata
    Smooth = Hypochaeris glabra
    Spotted = Hypochaeris maculata
Cat's-tail, Alpine = Phleum alpinum
    Purple-stem = Phleum phleoides
    Sand = Phleum arenarium
    Smaller = Phleum bertolonii
Celandine, Greater = Chelidonium majus
    Lesser = Ficaria verna
Celery, Wild = Apium graveolens
Centaury, Common = Centaurium erythraea
    Guernsey = Exaculum pusillum
    Lesser = Centaurium pulchellum
    Perennial = Centaurium scilloides
    Seaside = Centaurium littorale
    Slender = Centaurium tenuiflorum
    Yellow = Cicendia filiformis
Chaffweed = Centunculus minimus
Chamomile = Chamaemelum nobile
    Corn = Anthemis arvensis
    Stinking = Anthemis cotula
Charlock = Sinapis arvensis
Cherry, Bird = Prunus padus
    Dwarf = Prunus cerasus
    Wild = Prunus avium s.s.
Chervil, Bur = Anthriscus caucalis
    Rough = Chaerophyllum temulum
Chestnut, Sweet = Castanea sativa
Chickweed, Common = Stellaria media
    Greater = Stellaria neglecta
    Jagged = Holosteum umbellatum
    Lesser = Stellaria pallida
    Upright = Moenchia erecta
    Water = Myosoton aquaticum
Chickweed-wintergreen = Trientalis europaea
Chicory = Cichorium intybus
Chives = Allium schoenoprasum
Cinquefoil, Alpine = Potentilla crantzii
    Creeping = Potentilla reptans
    Hoary = Potentilla argentea
    Marsh = Comarum palustre
    Rock = Potentilla rupestris
    Shrubby = Potentilla fruticosa
    Spring = Potentilla tabernaemontani
Clary, Meadow = Salvia pratensis
    Wild = Salvia verbenaca

Cleavers = Galium aparine
    Corn = Galium tricornutum
Cloudberry = Rubus chamaemorus
Clover, Bird's-foot = Trifolium ornithopodioides
    Clustered = Trifolium glomeratum
    Crimson = Trifolium incarnatum
    Hare's-foot = Trifolium arvense
    Knotted = Trifolium striatum
    Red = Trifolium pratense
    Rough = Trifolium scabrum
    Sea = Trifolium squamosum
    Strawberry = Trifolium fragiferum
    Subterranean = Trifolium subterraneum
    Suffocated = Trifolium suffocatum
    Sulphur = Trifolium ochroleucon
    Twin-headed – Trifolium bocconei
    Upright = Trifolium strictum
    Western = Trifolium occidentale
    White = Trifolium repens
    Zigzag = Trifolium medium
Clubmoss, Alpine = Diphasiastrum alpinum
    Cypress = Diphasiastrum tristachyum
    Fir = Huperzia selago
    Hare's-foot = Lycopodium lagopus
    Interrupted = Lycopodium annotinum
    Issler's = Diphasiastrum complanatum
    Lesser = Selaginella selaginoides
    Marsh = Lycopodiella inundata
    Stag's-horn = Lycopodium clavatum
Club-rush, Bristle = Isolepis setacea
    Common = Schoenoplectus lacustris
    Floating = Eleogiton fluitans
    Grey = Schoenoplectus tabernaemontani
    Pedunculate = Bolboschoenus laticarpus
    Round-headed = Scirpoides holoschoenus
    Sea = Bolboschoenus maritimus
    Sharp = Schoenoplectus pungens
    Slender = Isolepis cernua
    Triangular = Schoenoplectus triqueter
    Wood = Scirpus sylvaticus
Cock's-foot = Dactylis glomerata
Colt's-foot = Tussilago farfara
    Purple = Homogyne alpina
Columbine = Aquilegia vulgaris
Comfrey, Common = Symphytum officinale
    Tuberous = Symphytum tuberosum
Copse-bindweed = Fallopia dumetorum
Coral-necklace = Illecebrum verticillatum
Coralroot = Cardamine bulbifera

Cord-grass, Common = Spartina anglica
    Small = Spartina maritima
Corncockle = Agrostemma githago
Cornel, Dwarf = Cornus suecica
Cornflower = Centaurea cyanus
Cornsalad, Broad-fruited = Valerianella rimosa
    Common = Valerianella locusta
    Hairy-fruited = Valerianella eriocarpa
    Keeled-fruited = Valerianella carinata
    Narrow-fruited = Valerianella dentata
Corydalis, Climbing = Ceratocapnos claviculata
Cotoneaster, Wild = Cotoneaster cambrica
Cottongrass, Broad-leaved
      = Eriophorum latifolium
    Common = Eriophorum angustifolium
    Hare's-tail = Eriophorum vaginatum
    Slender = Eriophorum gracile
Cottonweed = Achillea maritima
Couch, Bearded = Elymus caninus
    Common = Elytrigia repens
    Neglected = Elytrigia campestris
    Sand = Elytrigia juncea
    Sea = Elytrigia atherica
Cowbane = Cicuta virosa
Cowberry = Vaccinium vitis-idaea
Cowslip = Primula veris
Cow-wheat, Common = Melampyrum pratense
    Crested = Melampyrum cristatum
    Field = Melampyrum arvense
    Small = Melampyrum sylvaticum
Crack-willow = Salix fragilis
Cranberry = Vaccinium oxycoccos
    Small = Vaccinium microcarpum
Crane's-bill, Bloody = Geranium sanguineum
    Cut-leaved = Geranium dissectum
    Dove's-foot = Geranium molle
    Long-stalked = Geranium columbinum
    Meadow = Geranium pratense
    Round-leaved = Geranium rotundifolium
    Shining = Geranium lucidum
    Small-flowered = Geranium pusillum
    Wood = Geranium sylvaticum
Creeping-Jenny = Lysimachia nummularia
Cress, Shepherd's = Teesdalia nudicaulis
    Thale = Arabidopsis thaliana
Crocus, Sand = Romulea columnae
Crosswort = Cruciata laevipes
Crowberry = Empetrum nigrum

Crowfoot, Ivy-leaved = Ranunculus hederaceus †
    Round-leaved = Ranunculus omiophyllus †
    Three-lobed = Ranunculus tripartitus †
Cuckooflower = Cardamine pratensis
Cudweed, Broad-leaved = Filago pyramidata
    Common = Filago vulgaris
    Dwarf = Gnaphalium supinum
    Heath = Gnaphalium sylvaticum
    Highland = Gnaphalium norvegicum
    Jersey = Gnaphalium luteoalbum
    Marsh = Gnaphalium uliginosum
    Narrow-leaved = Filago gallica
    Red-tipped = Filago lutescens
    Small = Filago minima
Currant, Black = Ribes nigrum
    Downy = Ribes spicatum
    Mountain = Ribes alpinum
    Red = Ribes rubrum
Cut-grass = Leersia oryzoides
Cyphel = Minuartia sedoides

Daffodil = Narcissus pseudonarcissus
Daisy = Bellis perennis
    Oxeye = Leucanthemum vulgare
Dandelion = Taraxacum agg.
Darnel = Lolium temulentum
Dead-nettle, Cut-leaved = Lamium hybridum
    Henbit = Lamium amplexicaule
    Northern = Lamium confertum
    Red = Lamium purpureum
    White = Lamium album
Deergrass = Trichophorum germanicum
    Cotton = Trichophorum alpinum
    Northern = Trichophorum cespitosum s.s.
Dewberry = Rubus caesius
Diapensia = Diapensia lapponica
Dittander = Lepidium latifolium
Dock, Broad-leaved = Rumex obtusifolius
    Clustered = Rumex conglomeratus
    Curled = Rumex crispus
    Fiddle = Rumex pulcher
    Golden = Rumex maritimus
    Marsh = Rumex palustris
    Northern = Rumex longifolius
    Scottish = Rumex aquaticus
    Shore = Rumex rupestris
    Water = Rumex hydrolapathum
    Wood = Rumex sanguineus
Dodder = Cuscuta epithymum
    Greater = Cuscuta europaea

Dog-rose = Rosa canina s.l.
    Round-leaved = Rosa obtusifolia
Dog's-tail, Crested = Cynosurus cristatus
Dog-violet, Common = Viola riviniana
    Early = Viola reichenbachiana
    Heath = Viola canina
    Pale = Viola lactea
Dogwood = Cornus sanguinea
Downy-rose, Harsh = Rosa tomentosa
    Sherard's = Rosa sherardii
    Soft = Rosa mollis
Dropwort = Filipendula vulgaris
Duckweed, Common = Lemna minor
    Fat = Lemna gibba
    Greater = Spirodela polyrhiza
    Ivy-leaved = Lemna trisulca
    Rootless = Wolffia arrhiza

Eelgrass = Zostera marina
    Dwarf = Zostera noltei
Elder = Sambucus nigra
    Dwarf = Sambucus ebulus
Elecampane = Inula helenium
Elm, English = Ulmus procera
    Plot's = Ulmus plotii
    Small-leaved = Ulmus minor
    Wych = Ulmus glabra
Enchanter's-nightshade = Circaea lutetiana
    Alpine = Circaea alpina
Eryngo, Field = Eryngium campestre
Everlasting, Mountain = Antennaria dioica
Everlasting-pea, Narrow-leaved = Lathyrus sylvestris
Eyebright = Euphrasia agg.

False-brome, Heath = Brachypodium pinnatum s.s.
Fat-hen = Chenopodium album
Fennel = Foeniculum vulgare
    Hog's = Peucedanum officinale
Fen-sedge, Great = Cladium mariscus
Fern, Beech = Phegopteris connectilis
    Jersey = Anogramma leptophylla
    Killarney = Trichomanes speciosum (sporophyte)
    Lemon-scented = Oreopteris limbosperma
    Limestone = Gymnocarpium robertianum
    Maidenhair = Adiantum capillus-veneris
    Marsh = Thelypteris palustris
    Oak = Gymnocarpium dryopteris
    Parsley = Cryptogramma crispa
    Royal = Osmunda regalis

Fern-grass = Catapodium rigidum
    Sea = Catapodium marinum
Fescue, Bearded = Vulpia ciliata
    Blue = Festuca longifolia
    Breton = Festuca armoricana
    Confused = Festuca lemanii
    Dune = Vulpia fasciculata
    Giant = Schedonorus giganteus
    Huon's = Festuca huonii
    Mat-grass = Vulpia unilateris
    Meadow = Schedonorus pratensis
    Rat's-tail = Vulpia myuros
    Red = Festuca rubra s.s.
    Rush-leaved = Festuca arenaria
    Squirreltail = Vulpia bromoides
    Tall = Schedonorus arundinaccus
    Wood = Festuca altissima
Feverfew = Tanacetum parthenium
Field-rose = Rosa arvensis
    Short-styled = Rosa stylosa
Field-speedwell, Green = Veronica agrestis
Figwort, Balm-leaved = Scrophularia scorodonia
    Common = Scrophularia nodosa
    Green = Scrophularia umbrosa
    Water = Scrophularia auriculata
Filmy-fern, Tunbridge
     = Hymenophyllum tunbrigense
    Wilson's = Hymenophyllum wilsonii
Flat-sedge = Blysmus compressus
    Saltmarsh = Blysmus rufus
Flax, Fairy = Linum catharticum
    Pale = Linum bienne
    Perennial = Linum perenne
Fleabane, Alpine = Erigeron borealis
    Blue = Erigeron acris
    Common = Pulicaria dysenterica
    Irish = Inula salicina
    Small = Pulicaria vulgaris
Fleawort, Field = Tephroseris integrifolia
    Marsh = Tephroseris palustris
Flixweed = Descurainia sophia
Flowering-rush = Butomus umbellatus
Fluellen, Round-leaved = Kickxia spuria
    Sharp-leaved = Kickxia elatine
Fool's-water-cress = Apium nodiflorum
Forget-me-not, Alpine = Myosotis alpestris
    Changing = Myosotis discolor
    Creeping = Myosotis secunda
    Early = Myosotis ramosissima
    Field = Myosotis arvensis

    Jersey = Myosotis sicula
    Pale = Myosotis stolonifera
    Tufted = Myosotis laxa
    Water = Myosotis scorpioides
    Wood = Myosotis sylvatica
Foxglove = Digitalis purpurea
Fox-sedge, False = Carex otrubae
    True = Carex vulpina
Foxtail, Alpine = Alopecurus magellanicus
    Bulbous = Alopecurus bulbosus
    Marsh = Alopecurus geniculatus
    Meadow = Alopecurus pratensis
    Orange = Alopecurus aequalis
Fragrant-orchid, Chalk = Gymnadenia conopsea s.s.
    Heath = Gymnadenia borealis
    Marsh = Gymnadenia densiflora
Fritillary = Fritillaria meleagris
Frogbit = Hydrocharis morsus-ranae
Fumitory, Common = Fumaria officinalis s.s.
    Dense-flowered = Fumaria densiflora
    Few-flowered = Fumaria vaillantii
    Fine-leaved = Fumaria parviflora

Galingale = Cyperus longus
    Brown = Cyperus fuscus
Garlic, Field = Allium oleraceum
Gentian, Alpine = Gentiana nivalis
    Autumn = Gentianella amarella
    Chiltern = Gentianella germanica
    Dune = Gentianella uliginosa
    Early = Gentianella anglica
    Field = Gentianella campestris
    Fringed = Gentianopsis ciliata
    Marsh = Gentiana pneumonanthe
    Spring = Gentiana verna
Germander, Cut-leaved = Teucrium botrys
    Wall = Teucrium chamaedrys
    Water = Teucrium scordium
Gladiolus, Wild = Gladiolus illyricus
Glasswort, Common = Salicornia europaea
    Glaucous = Salicornia obscura
    Long-spiked = Salicornia dolichostachya
    One-flowered = Salicornia pusilla
    Perennial = Sarcocornia perennis
    Purple = Salicornia ramosissima
    Shiny = Salicornia emerici
    Yellow = Salicornia fragilis
Globeflower = Trollius europaeus
Goat's-beard = Tragopogon pratensis
Goldenrod = Solidago virgaurea

Golden-samphire = Inula crithmoides
Golden-saxifrage, Alternate-leaved
  = Chrysosplenium alternifolium
  Opposite-leaved
  = Chrysosplenium oppositifolium
Gold-of-pleasure = Camelina sativa
Good-King-Henry
  = Chenopodium bonus-henricus
Gooseberry = Ribes uva-crispa
Goosefoot, Fig-leaved = Chenopodium ficifolium
  Many-seeded = Chenopodium polyspermum
  Maple-leaved = Chenopodium hybridum
  Nettle-leaved = Chenopodium murale
  Oak-leaved = Chenopodium glaucum
  Red = Chenopodium rubrum
  Saltmarsh = Chenopodium chenopodioides
  Stinking = Chenopodium vulvaria
  Upright = Chenopodium urbicum
Gorse = Ulex europaeus
  Dwarf = Ulex minor
  Western = Ulex gallii
Grape-hyacinth = Muscari neglectum
Grass-of-Parnassus = Parnassia palustris
Grass-poly = Lythrum hyssopifolia
Greenweed, Dyer's = Genista tinctoria
  Hairy = Genista pilosa
Gromwell, Common = Lithospermum officinale
  Field = Lithospermum arvense
  Purple = Lithospermum purpureocaeruleum
Ground-elder = Aegopodium podagraria
Ground-ivy = Glechoma hederacea
Ground-pine = Ajuga chamaepitys
Groundsel = Senecio vulgaris
  Heath = Senecio sylvaticus
  Welsh = Senecio cambrensis
  Sticky = Senecio viscosus
Guelder-rose = Viburnum opulus
Gypsywort = Lycopus europaeus

Hair-grass, Bog = Deschampsia setacea
  Crested = Koeleria macrantha
  Early = Aira praecox
  Grey = Corynephorus canescens
  Silver = Aira caryophyllea
  Somerset = Koeleria vallesiana
  Tufted = Deschampsia cespitosa
  Wavy = Deschampsia flexuosa
Hairy-brome = Bromopsis ramosa
  Lesser = Bromopsis benekenii
Hampshire-purslane = Ludwigia palustris

Hard-fern = Blechnum spicant
Hard-grass = Parapholis strigosa
  Curved = Parapholis incurva
Harebell = Campanula rotundifolia
Hare's-ear, Sickle-leaved = Bupleurum falcatum
  Slender = Bupleurum tenuissimum
  Small = Bupleurum baldense
Hart's-tongue = Asplenium scolopendrium
Hartwort = Tordylium maximum
Hawkbit, Autumn = Scorzoneroides autumnalis
  Lesser = Leontodon saxatilis
  Rough = Leontodon hispidus
Hawk's-beard, Leafless = Crepis praemorsa
  Marsh = Crepis paludosa
  Northern = Crepis mollis
  Rough = Crepis biennis
  Smooth = Crepis capillaris
  Stinking = Crepis foetida
Hawkweed = Hieracium agg.
Hawthorn = Crataegus monogyna s.s.
  Midland = Crataegus laevigata
Hazel = Corylus avellana
Heath, Blue = Phyllodoce caerulea
  Cornish = Erica vagans
  Cross-leaved = Erica tetralix
  Dorset = Erica ciliaris
  Irish = Erica erigena
  Mackay's = Erica mackayana
  St Dabeoc's = Daboecia cantabrica
Heather = Calluna vulgaris
  Bell = Erica cinerea
Heath-grass = Danthonia decumbens
Hedge-parsley, Knotted = Torilis nodosa
  Spreading = Torilis arvensis
  Upright = Torilis japonica
Hellebore, Green = Helleborus viridis
  Stinking = Helleborus foetidus
Helleborine, Broad-leaved = Epipactis helleborine
  Dark-red = Epipactis atrorubens
  Dune = Epipactis dunensis
  Green-flowered = Epipactis phyllanthes
  Lindisfarne = Epipactis sancta
  Marsh = Epipactis palustris
  Narrow-leaved = Cephalanthera longifolia
  Red = Cephalanthera rubra
  Violet = Epipactis purpurata
  White = Cephalanthera damasonium
  Narrow-lipped = Epipactis leptochila
Hemlock = Conium maculatum
Hemp-agrimony = Eupatorium cannabinum

Hemp-nettle, Bifid = Galeopsis bifida
    Common = Galeopsis tetrahit s.s.
    Downy = Galeopsis segetum
    Large-flowered = Galeopsis speciosa
    Red = Galeopsis angustifolia
Henbane = Hyoscyamus niger
Herb-paris = Paris quadrifolia
Herb-Robert = Geranium robertianum
Hogweed = Heracleum sphondylium
Holly = Ilex aquifolium
Holly-fern = Polystichum lonchitis
Holy-grass = Hierochloe odorata
Honewort = Trinia glauca
Honeysuckle = Lonicera periclymenum
    Fly = Lonicera xylosteum
Hop = Humulus lupulus
Horehound, Black = Ballota nigra
    White = Marrubium vulgare
Hornbeam = Carpinus betulus
Horned-poppy, Violet = Papaver bivalve
    Yellow = Glaucium flavum
Hornwort, Rigid = Ceratophyllum demersum
    Soft = Ceratophyllum submersum
Horse-radish = Armoracia rusticana
Horsetail, Branched = Equisetum ramosissimum
    Field = Equisetum arvense
    Great = Equisetum telmateia
    Marsh = Equisetum palustre
    Rough = Equisetum hyemale
    Shady = Equisetum pratense
    Variegated = Equisetum variegatum
    Water = Equisetum fluviatile
    Wood = Equisetum sylvaticum
Hound's-tongue = Cynoglossum officinale
    Green = Cynoglossum germanicum
Hutchinsia = Hornungia petraea

Iceland-purslane = Koenigia islandica
Iris, Stinking = Iris foetidissima
    Yellow = Iris pseudacorus
Ivy, Atlantic = Hedera hibernica
    Common = Hedera helix s.s.

Jacob's-ladder = Polemonium caeruleum
Juniper, Common = Juniperus communis

Knapweed, Chalk = Centaurea debeauxii
    Common = Centaurea nigra s.l.
    Greater = Centaurea scabiosa

Knawel, Annual = Scleranthus annuus
    Perennial = Scleranthus perennis
Knotgrass = Polygonum aviculare s.s.
    Cornfield = Polygonum rurivagum
    Equal-leaved = Polygonum arenastrum
    Northern = Polygonum boreale
    Ray's = Polygonum oxyspermum
    Sea = Polygonum maritimum

Lady-fern = Athyrium filix-femina
    Alpine = Athyrium distentifolium
Lady's-mantle, Alpine = Alchemilla alpina
    Clustered = Alchemilla glomerulans
    Hairy = Alchemilla filicaulis
    Large-toothed = Alchemilla subcrenata
    Least = Alchemilla minima
    Pale = Alchemilla xanthochlora
    Rock = Alchemilla wichurae
    Shining = Alchemilla micans
    Silky = Alchemilla glaucescens
    Smooth = Alchemilla glabra
    Starry = Alchemilla acutiloba
    Velvet = Alchemilla monticola
Lady's-slipper = Cypripedium calceolus
Lady's-tresses, Autumn = Spiranthes spiralis
    Creeping = Goodyera repens
    Irish = Spiranthes romanzoffiana
    Summer = Spiranthes aestivalis
Leek, Round-headed = Allium sphaerocephalon
    Sand = Allium scorodoprasum
    Wild = Allium ampeloprasum
Lettuce, Great = Lactuca virosa
    Least = Lactuca saligna
    Prickly = Lactuca serriola
    Wall = Mycelis muralis
Lily, Kerry = Simethis mattiazzii
    May = Maianthemum bifolium
    Snowdon = Gagea serotina
Lily-of-the-valley = Convallaria majalis
Lime, Large-leaved = Tilia platyphyllos
    Small-leaved = Tilia cordata
Liquorice, Wild = Astragalus glycyphyllos
Little-Robin = Geranium purpureum
Lobelia, Heath = Lobelia urens
    Water = Lobelia dortmanna
Loosestrife, Tufted = Lysimachia thyrsiflora
    Yellow = Lysimachia vulgaris
Lords-and-Ladies = Arum maculatum
    Italian = Arum italicum

Lousewort = Pedicularis sylvatica
    Marsh = Pedicularis palustris
Lovage, Scots = Ligusticum scoticum
Lucerne = Medicago sativa
Lungwort, Narrow-leaved = Pulmonaria longifolia
    Unspotted = Pulmonaria obscura
Lyme-grass = Leymus arenarius

Madder, Field = Sherardia arvensis
    Wild = Rubia peregrina
Male-fern = Dryopteris filix-mas
    Borrer's = Dryopteris borreri
    Golden-scaled = Dryopteris affinis s.s.
    Mountain = Dryopteris oreades
    Narrow = Dryopteris cambrensis
Mallow, Common = Malva sylvestris
    Dwarf = Malva neglecta
    Rough = Malva setigera
Maple, Field = Acer campestre
Mare's-tail = Hippuris vulgaris
Marigold, Corn = Glebionis segetum
Marjoram, Wild = Origanum vulgare
Marram = Ammophila arenaria
Marsh-bedstraw, Common = Galium palustre
    Slender = Galium constrictum
Marsh-mallow = Althaea officinalis
Marsh-marigold = Caltha palustris
Marsh-orchid, Early = Dactylorhiza incarnata
    Hebridean = Dactylorhiza ebudensis
    Irish = Dactylorhiza kerryensis
    Narrow-leaved
        = Dactylorhiza traunsteinerioides
    Northern = Dactylorhiza purpurella
    Southern = Dactylorhiza praetermissa
Marshwort, Creeping = Apium repens
    Lesser = Apium inundatum
Masterwort = Imperatoria ostruthium
Mat-grass = Nardus stricta
Mayweed, Scented = Matricaria chamomilla
    Scentless = Tripleurospermum inodorum
    Sea = Tripleurospermum maritimum s.s.
Meadow-grass, Alpine = Poa alpina
    Annual = Poa annua
    Bulbous = Poa bulbosa
    Early = Poa infirma
    Flattened = Poa compressa
    Glaucous = Poa glauca
    Narrow-leaved = Poa angustifolia
    Rough = Poa trivialis
    Smooth = Poa pratensis s.l.
    Spreading = Poa humilis
    Wavy = Poa flexuosa
    Wood = Poa nemoralis
Meadow-rue, Alpine = Thalictrum alpinum
    Common = Thalictrum flavum
    Lesser = Thalictrum minus
Meadowsweet = Filipendula ulmaria
Medick, Black = Medicago lupulina
    Bur = Medicago minima
    Spotted = Medicago arabica
    Toothed = Medicago polymorpha
Medlar = Mespilus germanica
Melick, Mountain = Melica nutans
    Wood = Melica uniflora
Melilot, Tall = Melilotus altissimus
Mercury, Annual = Mercurialis annua
    Dog's = Mercurialis perennis
Mezereon = Daphne mezereum
Mignonette, Wild = Reseda lutea
Milk-parsley = Thyselium palustre
    Cambridge = Selinum carvifolia
Milk-vetch, Alpine = Astragalus alpinus
    Purple = Astragalus danicus
Milkwort, Chalk = Polygala calcarea
    Common = Polygala vulgaris
    Dwarf = Polygala amarella
    Heath = Polygala serpyllifolia
Millet, Early = Milium vernale
    Wood = Milium effusum
Mint, Corn = Mentha arvensis
    Round-leaved = Mentha suaveolens
    Spear = Mentha spicata
    Water = Mentha aquatica
Mistletoe = Viscum album
Moneywort, Cornish = Sibthorpia europaea
Monk's-hood = Aconitum napellus s.l.
Monk's-rhubarb = Rumex alpinus
Moonwort = Botrychium lunaria
Moor-grass, Blue = Sesleria caerulea
    Purple = Molinia caerulea
Moschatel = Adoxa moschatellina
Mouse-ear, Alpine = Cerastium alpinum
    Arctic = Cerastium nigrescens
    Common = Cerastium fontanum
    Dwarf = Cerastium pumilum
    Field = Cerastium arvense
    Grey = Cerastium brachypetalum
    Little = Cerastium semidecandrum
    Sea = Cerastium diffusum
    Starwort = Cerastium cerastoides

  Sticky = Cerastium glomeratum
Mouse-ear-hawkweed = Pilosella officinarum
  Shaggy = Pilosella peleteriana
  Shetland = Pilosella flagellaris
Mousetail = Myosurus minimus
Mudwort = Limosella aquatica
  Welsh = Limosella australis
Mugwort = Artemisia vulgaris
  Norwegian = Artemisia norvegica
Mullein, Dark = Verbascum nigrum
  Great = Verbascum thapsus
  Hoary = Verbascum pulverulentum
  Moth = Verbascum blattaria
  Twiggy = Verbascum virgatum
  White = Verbascum lychnitis
Musk-mallow = Malva moschata
Mustard, Black = Brassica nigra
  Garlic = Alliaria petiolata
  Hedge = Sisymbrium officinale
  Tower = Turritis glabra
  Treacle = Erysimum cheiranthoides
  White = Sinapis alba

Naiad, Holly-leaved = Najas marina
  Slender = Najas flexilis
Navelwort = Umbilicus rupestris
Nettle, Common = Urtica dioica
  Small = Urtica urens
Nightshade, Black = Solanum nigrum
  Deadly = Atropa belladonna
Nipplewort = Lapsana communis
Nit-grass = Gastridium ventricosum

Oak, Pedunculate = Quercus robur
  Sessile = Quercus petraea
Oat-grass, Downy = Avenula pubescens
  False = Arrhenatherum elatius
  French = Gaudinia fragilis
  Meadow = Avenula pratensis
  Yellow = Trisetum flavescens
Onion, Wild = Allium vineale
Orache, Babington's = Atriplex glabriuscula
  Common = Atriplex patula
  Early = Atriplex praecox
  Frosted = Atriplex laciniata
  Grass-leaved = Atriplex littoralis
  Long-stalked = Atriplex longipes
  Spear-leaved = Atriplex prostrata
Orchid, Bee = Ophrys apifera
  Bird's-nest = Neottia nidus-avis

  Bog = Hammarbya paludosa
  Burnt = Neotinea ustulata
  Coralroot = Corallorhiza trifida
  Dense-flowered = Neotinea maculata
  Early-purple = Orchis mascula
  Fen = Liparis loeselii
  Fly = Ophrys insectifera
  Frog = Coeloglossum viride
  Ghost = Epipogium aphyllum
  Green-winged = Anacamptis morio
  Lady = Orchis purpurea
  Lizard = Himantoglossum hircinum
  Loose-flowered = Anacamptis laxiflora
  Man = Orchis anthropophora
  Military = Orchis militaris
  Monkey = Orchis simia
  Musk = Herminium monorchis
  Pyramidal = Anacamptis pyramidalis
  Small-white = Pseudorchis albida
Orpine = Sedum telephium
Osier = Salix viminalis
Oxlip = Primula elatior
Oxtongue, Bristly = Helminthotheca echioides
  Hawkweed = Picris hieracioides
Oxytropis, Purple = Oxytropis halleri
  Yellow = Oxytropis campestris
Oysterplant = Mertensia maritima

Pansy, Dwarf = Viola kitaibeliana
  Field = Viola arvensis
  Mountain = Viola lutea
  Wild = Viola tricolor s.s.
Parsley, Corn = Petroselinum segetum
  Cow = Anthriscus sylvestris
  Fool's = Aethusa cynapium
  Garden = Petroselinum crispum
  Stone = Sison amomum
Parsley-piert = Aphanes arvensis s.s.
  Slender = Aphanes australis
Parsnip, Wild = Pastinaca sativa
Pasqueflower = Pulsatilla vulgaris
Pea, Marsh = Lathyrus palustris
  Sea = Lathyrus japonicus
Pear = Pyrus communis s.s.
  Plymouth = Pyrus cordata
  Wild = Pyrus pyraster
Pearlwort, Alpine = Sagina saginoides
  Annual = Sagina apetala s.s.
  Heath = Sagina subulata
  Knotted = Sagina nodosa

    Procumbent = Sagina procumbens
    Sea = Sagina maritima
    Slender = Sagina filicaulis
    Snow = Sagina nivalis
Pellitory-of-the-wall = Parietaria judaica
Penny-cress, Alpine = Noccaea caerulescens
    Field = Thlaspi arvense
    Perfoliate = Microthlaspi perfoliatum
Pennyroyal = Mentha pulegium
Pennywort, Marsh = Hydrocotyle vulgaris
Pepper-saxifrage = Silaum silaus
Pepperwort, Field = Lepidium campestre
    Narrow-leaved = Lepidium ruderale
    Smith's = Lepidium heterophyllum
Periwinkle, Lesser = Vinca minor
Persicaria, Pale = Persicaria lapathifolia
Pheasant's-eye = Adonis annua
Pigmyweed = Crassula aquatica
Pignut = Conopodium majus
    Great = Bunium bulbocastanum
Pillwort = Pilularia globulifera
Pimpernel, Bog = Anagallis tenella
    Scarlet = Anagallis arvensis
    Yellow = Lysimachia nemorum
Pine, Scots = Pinus sylvestris
Pink, Cheddar = Dianthus gratianopolitanus
    Childing = Petrorhagia nanteuilii
    Deptford = Dianthus armeria
    Maiden = Dianthus deltoides
    Proliferous = Petrorhagia prolifera
Pipewort = Eriocaulon aquaticum
Plantain, Buck's-horn = Plantago coronopus
    Greater = Plantago major
    Hoary = Plantago media
    Ribwort = Plantago lanceolata
    Sea = Plantago maritima
Ploughman's-spikenard = Inula conyzae
Plum, Wild = Prunus domestica
Polypody = Polypodium vulgare s.s.
    Intermediate = Polypodium interjectum
    Southern = Polypodium cambricum
Pond-sedge, Greater = Carex riparia
    Lesser = Carex acutiformis
Pondweed, American = Potamogeton epihydrus
    Blunt-leaved = Potamogeton obtusifolius
    Bog = Potamogeton polygonifolius
    Broad-leaved = Potamogeton natans
    Curled = Potamogeton crispus
    Fen = Potamogeton coloratus
    Fennel = Potamogeton pectinatus

    Flat-stalked = Potamogeton friesii
    Grass-wrack = Potamogeton compressus
    Hairlike = Potamogeton trichoides
    Horned = Zannichellia palustris
    Lesser = Potamogeton pusillus
    Loddon = Potamogeton nodosus
    Long-stalked = Potamogeton praelongus
    Opposite-leaved = Groenlandia densa
    Perfoliate = Potamogeton perfoliatus
    Red = Potamogeton alpinus
    Sharp-leaved = Potamogeton acutifolius
    Shetland = Potamogeton rutilus
    Shining = Potamogeton lucens
    Slender-leaved = Potamogeton filiformis
    Small = Potamogeton berchtoldii
    Various-leaved = Potamogeton gramineus
Poppy, Common = Papaver rhoeas
    Long-headed = Papaver dubium s.s.
    Opium = Papaver somniferum
    Prickly = Papaver argemone
    Rough = Papaver hybridum
    Welsh = Meconopsis cambrica
    Yellow-juiced = Papaver lecoqii
Primrose = Primula vulgaris
    Bird's-eye = Primula farinosa
    Scottish = Primula scotica
Privet, Wild = Ligustrum vulgare
Purple-loosestrife = Lythrum salicaria

Quaking-grass = Briza media
    Lesser = Briza minor
Quillwort = Isoetes lacustris
    Land = Isoetes histrix
    Spring = Isoetes echinospora

Radish, Sea = Raphanus maritimus
    Wild = Raphanus raphanistrum
Ragged-Robin = Silene flos-cuculi
Ragwort, Common = Senecio jacobaea
    Fen = Senecio paludosus
    Hoary = Senecio erucifolius
    Marsh = Senecio aquaticus
    York = Senecio eboracensis
Ramping-fumitory, Common = Fumaria muralis
    Martin's = Fumaria reuteri
    Purple = Fumaria purpurea
    Tall = Fumaria bastardii
    Western = Fumaria occidentalis
    White = Fumaria capreolata

Rampion, Round-headed = Phyteuma orbiculare
    Spiked = Phyteuma spicatum
Ramsons = Allium ursinum
Rannoch-rush = Scheuchzeria palustris
Raspberry = Rubus idaeus
Redshank = Persicaria maculosa
Reed, Common = Phragmites australis
Restharrow, Common = Ononis repens
    Small = Ononis reclinata
    Spiny = Ononis spinosa
Rock-cress, Alpine = Arabis alpina
    Bristol = Arabis scabra
    Hairy = Arabis hirsuta
    Northern = Arabidopsis petraea
Rocket, Sea = Cakile maritima
Rock-rose, Common
      = Helianthemum nummularium
    Hoary = Helianthemum oelandicum
    Spotted = Tuberaria guttata
    White = Helianthemum apenninum
Rose, Burnet = Rosa spinosissima
Roseroot = Sedum rosea
Rowan = Sorbus aucuparia
Rupturewort, Fringed = Herniaria ciliolata
    Smooth = Herniaria glabra
Rush, Alpine = Juncus alpinoarticulatus
    Baltic = Juncus balticus
    Blunt-flowered = Juncus subnodulosus
    Bulbous = Juncus bulbosus
    Chestnut = Juncus castaneus
    Compact = Juncus conglomeratus
    Dwarf = Juncus capitatus
    Frog = Juncus ranarius
    Hard = Juncus inflexus
    Heath = Juncus squarrosus
    Jointed = Juncus articulatus
    Leafy = Juncus foliosus
    Pigmy = Juncus pygmaeus
    Round-fruited = Juncus compressus
    Saltmarsh = Juncus gerardii
    Sea = Juncus maritimus
    Sharp = Juncus acutus
    Sharp-flowered = Juncus acutiflorus
    Thread = Juncus filiformis
    Three-flowered = Juncus triglumis
    Three-leaved = Juncus trifidus
    Toad = Juncus bufonius s.s.
    Two-flowered = Juncus biglumis
Rustyback = Asplenium ceterach
Rye-grass, Perennial = Lolium perenne

Saffron, Meadow = Colchicum autumnale
Sage, Wood = Teucrium scorodonia
Sainfoin = Onobrychis viciifolia
Saltmarsh-grass, Borrer's = Puccinellia fasciculata
    Common = Puccinellia maritima
    Reflexed = Puccinellia distans
    Stiff = Puccinellia rupestris
Saltwort, Prickly = Salsola kali
Samphire, Rock = Crithmum maritimum
Sand-grass, Early = Mibora minima
Sandwort, Arctic = Arenaria norvegica
    Fine-leaved = Minuartia hybrida
    Fringed = Arenaria ciliata
    Mountain = Minuartia rubella
    Recurved = Minuartia recurva
    Sea = Honckenya peploides
    Slender = Arenaria leptoclados
    Spring = Minuartia verna
    Teesdale = Minuartia stricta
    Three-nerved = Moehringia trinervia
    Thyme-leaved = Arenaria serpyllifolia s.s.
Sanicle = Sanicula europaea
Saw-wort = Serratula tinctoria
    Alpine = Saussurea alpina
Saxifrage, Alpine = Saxifraga nivalis
    Drooping = Saxifraga cernua
    Highland = Saxifraga rivularis
    Irish = Saxifraga rosacea
    Kidney = Saxifraga hirsuta
    Marsh = Saxifraga hirculus
    Meadow = Saxifraga granulata
    Mossy = Saxifraga hypnoides
    Purple = Saxifraga oppositifolia
    Rue-leaved = Saxifraga tridactylites
    Starry = Saxifraga stellaris
    Tufted = Saxifraga cespitosa
    Yellow = Saxifraga aizoides
Scabious, Devil's-bit = Succisa pratensis
    Field = Knautia arvensis
    Small = Scabiosa columbaria
Scurvygrass, Common = Cochlearia officinalis
    Danish = Cochlearia danica
    English = Cochlearia anglica
    Mountain = Cochlearia micacea
    Pyrenean = Cochlearia pyrenaica
Sea-blite, Annual = Suaeda maritima
    Shrubby = Suaeda vera
Sea-buckthorn = Hippophae rhamnoides
Sea-heath = Frankenia laevis
Sea-holly = Eryngium maritimum

Sea-kale = Crambe maritima
Sea-lavender, Broad-leaved
    = Limonium auriculae-ursifolium
    Common = Limonium vulgare
    Lax-flowered = Limonium humile
    Matted = Limonium bellidifolium
    Rock = Limonium binervosum agg.
Sea-milkwort = Glaux maritima
Sea-purslane = Atriplex portulacoides
    Pedunculate = Atriplex pedunculata
Sea-spurrey, Greater = Spergularia media
    Lesser = Spergularia marina
    Rock = Spergularia rupicola
Sedge, Bird's-foot = Carex ornithopoda
    Bottle = Carex rostrata
    Bristle = Carex microglochin
    Brown = Carex disticha
    Carnation = Carex panicea
    Club = Carex buxbaumii
    Common = Carex nigra
    Curved = Carex maritima
    Cyperus = Carex pseudocyperus
    Davall's = Carex davalliana
    Dioecious = Carex dioica
    Distant = Carex distans
    Divided = Carex divisa
    Dotted = Carex punctata
    Downy-fruited = Carex filiformis
    Dwarf = Carex humilis
    Elongated = Carex elongata
    Estuarine = Carex recta
    False = Kobresia simpliciuscula
    Few-flowered = Carex pauciflora
    Fingered = Carex digitata
    Flea = Carex pulicaris
    Glaucous = Carex flacca
    Green-ribbed = Carex binervis
    Grey = Carex divulsa
    Hair = Carex capillaris
    Hairy = Carex hirta
    Hare's-foot = Carex lachenalii
    Long-bracted = Carex extensa
    Oval = Carex leporina
    Pale = Carex pallescens
    Pendulous = Carex pendula
    Pill = Carex pilulifera
    Prickly = Carex muricata
    Remote = Carex remota
    Rock = Carex rupestris
    Russet = Carex saxatilis
    Saltmarsh = Carex salina
    Sand = Carex arenaria
    Sheathed = Carex vaginata
    Slender = Carex lasiocarpa
    Smooth-stalked = Carex laevigata
    Soft-leaved = Carex montana
    Spiked = Carex spicata
    Star = Carex echinata
    Stiff = Carex bigelowii
    String = Carex chordorrhiza
    Tawny = Carex hostiana
    Three-nerved = Carex trinervis
    Water = Carex aquatilis
    White = Carex canescens
Selfheal = Prunella vulgaris
    Cut-leaved = Prunella laciniata
Service-tree = Sorbus domestica
    Wild = Sorbus torminalis
Sheep's-bit = Jasione montana
Sheep's-fescue = Festuca ovina s.s.
    Fine-leaved = Festuca filiformis
    Viviparous = Festuca vivipara
Shepherd's-needle = Scandix pecten-veneris
Shepherd's-purse = Capsella bursa-pastoris
Shield-fern, Hard = Polystichum aculeatum
    Soft = Polystichum setiferum
Shoreweed = Littorella uniflora
Sibbaldia = Sibbaldia procumbens
Silky-bent, Loose = Apera spica-venti
Silverweed = Potentilla anserina
Skullcap = Scutellaria galericulata
    Lesser = Scutellaria minor
Small-reed, Narrow = Calamagrostis stricta
    Purple = Calamagrostis canescens
    Scandinavian = Calamagrostis purpurea
    Scottish = Calamagrostis scotica
    Wood = Calamagrostis epigejos
Snapdragon, Lesser = Misopates orontium
Sneezewort = Achillea ptarmica
Snowflake, Summer = Leucojum aestivum
Soapwort = Saponaria officinalis
Soft-brome = Bromus hordeaceus
Soft-grass, Creeping = Holcus mollis
Soft-rush = Juncus effusus
Solomon's-seal = Polygonatum multiflorum
    Angular = Polygonatum odoratum
    Whorled = Polygonatum verticillatum
Sorrel, Common = Rumex acetosa
    Mountain = Oxyria digyna
    Sheep's = Rumex acetosella

Sow-thistle, Marsh = Sonchus palustris
    Perennial = Sonchus arvensis
    Prickly = Sonchus asper
    Smooth = Sonchus oleraceus s.s.
Spearwort, Adder's-tongue
        = Ranunculus ophioglossifolius
    Creeping = Ranunculus reptans
    Greater = Ranunculus lingua
    Lesser = Ranunculus flammula
Speedwell, Alpine = Veronica alpina
    Fingered = Veronica triphyllos
    Germander = Veronica chamaedrys
    Heath = Veronica officinalis
    Ivy-leaved = Veronica hederifolia
    Marsh = Veronica scutellata
    Rock = Veronica fruticans
    Spiked = Veronica spicata
    Spring = Veronica verna
    Thyme-leaved = Veronica serpyllifolia
    Wall = Veronica arvensis
    Wood = Veronica montana
Spider-orchid, Early = Ophrys sphegodes
    Late = Ophrys fuciflora
Spignel = Meum athamanticum
Spike-rush, Common = Eleocharis palustris
    Dwarf = Eleocharis parvula
    Few-flowered = Eleocharis quinqueflora
    Many-stalked = Eleocharis multicaulis
    Needle = Eleocharis acicularis
    Northern = Eleocharis mamillata
    Slender = Eleocharis uniglumis
Spindle = Euonymus europaeus
Spleenwort, Black = Asplenium adiantum-nigrum
    Forked = Asplenium septentrionale
    Green = Asplenium viride
    Irish = Asplenium onopteris
    Lanceolate = Asplenium obovatum
    Maidenhair = Asplenium trichomanes
    Sea = Asplenium marinum
Spotted-orchid, Common = Dactylorhiza fuchsii
    Heath = Dactylorhiza maculata
Spring-sedge = Carex caryophyllea
    Rare = Carex ericetorum
Spurge, Broad-leaved = Euphorbia platyphyllos
    Caper = Euphorbia lathyris
    Dwarf = Euphorbia exigua
    Hairy = Euphorbia villosa
    Irish = Euphorbia hyberna
    Petty = Euphorbia peplus
    Portland = Euphorbia portlandica

    Purple = Euphorbia peplis
    Sea = Euphorbia paralias
    Sun = Euphorbia helioscopia
    Upright = Euphorbia stricta
    Wood = Euphorbia amygdaloides
Spurge-laurel = Daphne laureola
Spurrey, Corn = Spergula arvensis
    Sand = Spergularia rubra
Squill, Autumn = Scilla autumnalis
    Spring = Scilla verna
Squinancywort = Asperula cynanchica
St John's-wort, Hairy = Hypericum hirsutum
    Imperforate = Hypericum maculatum
    Marsh = Hypericum elodes
    Pale = Hypericum montanum
    Perforate = Hypericum perforatum
    Slender = Hypericum pulchrum
    Square-stalked = Hypericum tetrapterum
    Toadflax-leaved = Hypericum linariifolium
    Trailing = Hypericum humifusum
    Wavy = Hypericum undulatum
St Patrick's-cabbage = Saxifraga spathularis
Starfruit = Damasonium alisma
Star-of-Bethlehem = Ornithogalum umbellatum
    Early = Gagea bohemica
    Spiked = Ornithogalum pyrenaicum
    Yellow = Gagea lutea
Star-thistle, Red = Centaurea calcitrapa
Stitchwort, Bog = Stellaria alsine
    Greater = Stellaria holostea
    Lesser = Stellaria graminea
    Marsh = Stellaria palustris
    Wood = Stellaria nemorum
Stock, Sea = Matthiola sinuata
Stonecrop, Biting = Sedum acre
    English = Sedum anglicum
    Hairy = Sedum villosum
    Mossy = Crassula tillaea
    Rock = Sedum forsterianum
    White = Sedum album
Stork's-bill, Common = Erodium cicutarium
    Musk = Erodium moschatum
    Sea = Erodium maritimum
    Sticky = Erodium lebelii
Strapwort = Corrigiola litoralis
Strawberry, Barren = Potentilla sterilis
    Wild = Fragaria vesca
Strawberry-tree = Arbutus unedo
Succory, Lamb's = Arnoseris minima

Sundew, Great = Drosera anglica
    Oblong-leaved = Drosera intermedia
    Round-leaved = Drosera rotundifolia
Sweet-briar = Rosa rubiginosa
    Small-flowered = Rosa micrantha
    Small-leaved = Rosa agrestis
Sweet-grass, Floating = Glyceria fluitans
    Plicate = Glyceria notata
    Reed = Glyceria maxima
    Small = Glyceria declinata
Swine-cress = Lepidium coronopus

Tansy = Tanacetum vulgare
Tare, Hairy = Vicia hirsuta
    Slender = Vicia parviflora
    Smooth = Vicia tetrasperma
Tasselweed, Beaked = Ruppia maritima
    Spiral = Ruppia cirrhosa
Teasel, Small = Dipsacus pilosus
    Wild = Dipsacus fullonum
Thistle, Carline = Carlina vulgaris
    Cotton = Onopordum acanthium
    Creeping = Cirsium arvense
    Dwarf = Cirsium acaule
    Marsh = Cirsium palustre
    Meadow = Cirsium dissectum
    Melancholy = Cirsium heterophyllum
    Milk = Silybum marianum
    Musk = Carduus nutans
    Slender = Carduus tenuiflorus
    Spear = Cirsium vulgare
    Tuberous = Cirsium tuberosum
    Welted = Carduus crispus
    Woolly = Cirsium eriophorum
Thorow-wax = Bupleurum rotundifolium
Thrift = Armeria maritima
    Jersey = Armeria alliacea
Thyme, Basil = Clinopodium acinos
    Breckland = Thymus serpyllum
    Large = Thymus pulegioides
    Wild = Thymus polytrichus
Timothy = Phleum pratense
Toadflax, Common = Linaria vulgaris
    Pale = Linaria repens
    Small = Chaenorhinum minus
Tongue-orchid, Small-flowered
        = Serapias parviflora
Toothwort = Lathraea squamaria
Tor-grass = Brachypodium rupestre

Tormentil = Potentilla erecta
    Trailing = Potentilla anglica
Traveller's-joy = Clematis vitalba
Tree-mallow = Malva arborea
    Smaller = Malva pseudolavatera
Trefoil, Hop = Trifolium campestre
    Lesser = Trifolium dubium
    Slender = Trifolium micranthum
Tufted-sedge = Carex elata
    Scarce = Carex cespitosa
    Slender = Carex acuta
Turnip = Brassica rapa
Tussock-sedge, Fibrous = Carex appropinquata
    Greater = Carex paniculata
    Lesser = Carex diandra
Tutsan = Hypericum androsaemum
Twayblade, Common = Neottia ovata
    Lesser = Neottia cordata
Twinflower = Linnaea borealis

Valerian, Common = Valeriana officinalis
    Marsh = Valeriana dioica
Venus's-looking-glass = Legousia hybrida
Vernal-grass, Sweet = Anthoxanthum odoratum
Vervain = Verbena officinalis
Vetch, Bithynian = Vicia bithynica
    Bush = Vicia sepium
    Common = Vicia sativa
    Horseshoe = Hippocrepis comosa
    Kidney = Anthyllis vulneraria
    Spring = Vicia lathyroides
    Tufted = Vicia cracca
    Wood = Vicia sylvatica
Vetchling, Grass = Lathyrus nissolia
    Meadow = Lathyrus pratensis
    Yellow = Lathyrus aphaca
Violet, Fen = Viola persicifolia
    Hairy = Viola hirta
    Marsh = Viola palustris
    Sweet = Viola odorata
    Teesdale = Viola rupestris
Viper's-bugloss = Echium vulgare
    Purple = Echium plantagineum
Viper's-grass = Scorzonera humilis

Wallflower = Erysimum cheiri
    Perennial = Diplotaxis tenuifolia
Wall-rue = Asplenium ruta-muraria
Water-cress = Nasturtium officinale s.s.
    Narrow-fruited = Nasturtium microphyllum

Water-crowfoot, Brackish = Ranunculus baudotii †
    Common = Ranunculus aquatilis s.s. †
    Fan-leaved = Ranunculus circinatus †
    Pond = Ranunculus peltatus †
    River = Ranunculus fluitans †
    Stream = Ranunculus penicillatus †
    Thread-leaved = Ranunculus trichophyllus †
Water-dropwort, Corky-fruited
       = Oenanthe pimpinelloides
    Fine-leaved = Oenanthe aquatica s.s.
    Hemlock = Oenanthe crocata
    Narrow-leaved = Oenanthe silaifolia
    Parsley = Oenanthe lachenalii
    River = Oenanthe fluviatilis
    Tubular = Oenanthe fistulosa
Water-lily, Fringed = Nymphoides peltata
    Least = Nuphar pumila
    White = Nymphaea alba
    Yellow = Nuphar lutea
Water-milfoil, Alternate
       = Myriophyllum alterniflorum
    Spiked = Myriophyllum spicatum
    Whorled = Myriophyllum verticillatum
Water-parsnip, Greater = Sium latifolium
    Lesser = Berula erecta
Water-pepper = Persicaria hydropiper
    Small = Persicaria minor
    Tasteless = Persicaria mitis
Water-plantain = Alisma plantago-aquatica
    Floating = Luronium natans
    Lesser = Baldellia ranunculoides
    Narrow-leaved = Alisma lanceolatum
    Ribbon-leaved = Alisma gramineum
Water-purslane = Lythrum portula
Water-soldier = Stratiotes aloides
Water-speedwell, Blue
       = Veronica anagallis-aquatica s.s.
    Pink = Veronica catenata
Water-starwort, Autumnal
       = Callitriche hermaphroditica
    Blunt-fruited = Callitriche obtusangula
    Common = Callitriche stagnalis s.l.
    Narrow-fruited = Callitriche palustris
    Pedunculate = Callitriche brutia s.s.
    Short-leaved = Callitriche truncata
    Various-leaved = Callitriche platycarpa
Water-violet = Hottonia palustris
Waterweed, Esthwaite = Hydrilla verticillata
Waterwort, Eight-stamened = Elatine hydropiper
    Six-stamened = Elatine hexandra

Wayfaring-tree = Viburnum lantana
Weld = Reseda luteola
Whin, Petty = Genista anglica
Whitebeam, Common = Sorbus aria agg.
    Rock = Sorbus rupicola
Whitlowgrass, Common = Erophila verna s.s.
    Glabrous = Erophila glabrescens
    Hairy = Erophila majuscula
    Hoary = Draba incana
    Rock = Draba norvegica
    Wall = Draba muralis
    Yellow = Draba aizoides
Whorl-grass = Catabrosa aquatica
Wild-oat = Avena fatua
Willow, Almond = Salix triandra
    Bay = Salix pentandra
    Creeping = Salix repens
    Dark-leaved = Salix myrsinifolia
    Downy = Salix lapponum
    Dwarf = Salix herbacea
    Eared = Salix aurita
    Goat = Salix caprea
    Grey = Salix cinerea
    Mountain = Salix arbuscula
    Net-leaved = Salix reticulata
    Purple = Salix purpurea
    Tea-leaved = Salix phylicifolia
    White = Salix alba
    Whortle-leaved = Salix myrsinites
    Woolly = Salix lanata
Willowherb, Alpine = Epilobium anagallidifolium
    Broad-leaved = Epilobium montanum
    Chickweed = Epilobium alsinifolium
    Great = Epilobium hirsutum
    Hoary = Epilobium parviflorum
    Marsh = Epilobium palustre
    Pale = Epilobium roseum
    Rosebay = Chamerion angustifolium
    Short-fruited = Epilobium obscurum
    Spear-leaved = Epilobium lanceolatum
    Square-stalked = Epilobium tetragonum
Winter-cress = Barbarea vulgaris
Wintergreen, Common = Pyrola minor
    Intermediate = Pyrola media
    One-flowered = Moneses uniflora
    Round-leaved = Pyrola rotundifolia
    Serrated = Orthilia secunda
Woad = Isatis tinctoria
Woodruff = Galium odoratum

Wood-rush, Curved = Luzula arcuata
    Fen = Luzula pallescens
    Field = Luzula campestris
    Great = Luzula sylvatica
    Hairy = Luzula pilosa
    Heath = Luzula multiflora
    Southern = Luzula forsteri
    Spiked = Luzula spicata
Wood-sedge = Carex sylvatica
    Starved = Carex depauperata
    Thin-spiked = Carex strigosa
Woodsia, Alpine = Woodsia alpina
    Oblong = Woodsia ilvensis
Wood-sorrel = Oxalis acetosella
Wormwood = Artemisia absinthium
    Field = Artemisia campestris
    Sea = Artemisia maritima
Woundwort, Downy = Stachys germanica
    Field = Stachys arvensis
    Hedge = Stachys sylvatica
    Limestone = Stachys alpina
    Marsh = Stachys palustris

Yarrow = Achillea millefolium
Yellow-cress, Creeping = Rorippa sylvestris
    Great = Rorippa amphibia
    Marsh = Rorippa palustris
    Northern = Rorippa islandica s.s.
Yellow-rattle = Rhinanthus minor
    Greater = Rhinanthus angustifolius
Yellow-sedge, Common = Carex demissa
    Large = Carex flava s.s.
    Long-stalked = Carex lepidocarpa
    Small-fruited = Carex oederi
Yellow-vetch = Vicia lutea
Yellow-wort = Blackstonia perfoliata
Yew = Taxus baccata
Yorkshire-fog = Holcus lanatus

Other BSBI publications

The Botanical Society of Britain and Ireland

The BSBI is for everyone interested in the flora of Great Britain and Ireland. Details of membership may be obtained from the Membership Secretary, Mr Gwynn Ellis, 41 Marlborough Road, Roath, Cardiff, CF23 5BU (E-mail gwynn.ellis@bsbi.org). The following books are available from the official agents for BSBI publications, Summerfield Books, Unit L, Skirsgill Business Park, Penrith, CA11 0FA (Tel. 01768 210793; E-mail info@summerfieldbooks.com). Fuller details can be found at http://www.summerfieldbooks.com under "BSBI Publications".

BSBI handbooks

Each handbook deals in depth with one or more difficult groups of British and Irish plants.

- No. 1 **Sedges of the British Isles** – A. C. Jermy, D. A. Simpson, M. J. Y. Foley & M. S. Porter. Third edition, 2007.
- No. 2 **Umbellifers of the British Isles** – T. G. Tutin. 1980.
- No. 3 **Docks and Knotweeds of Britain and Ireland** – J. R. Akeroyd. 2014. Replaces *Docks and Knotweeds of the British Isles* by J. E. Lousley & D. H. Kent (1981).
- No. 4 **Willows and Poplars of Great Britain and Ireland** – R. D. Meikle, 1984.
- No. 5 **Charophytes of Great Britain and Ireland** – Jenny A. Moore. 1986, reprinted 2005 and 2014 with a new preface and corrections by C. D. Preston.
- No. 6 **Crucifers of Great Britain and Ireland** – T C. G. Rich. 1991.
- No. 7 **Roses of Great Britain and Ireland** – G. G. Graham & A. L. Primavesi. 1993, reprinted with corrections 2005.
- No. 8 **Pondweeds of Great Britain and Ireland** – C. D. Preston. 1995, reprinted with minor alterations 2003 and 2015.
- No. 9 **Dandelions of Great Britain and Ireland** – A. A. Dudman & A. J. Richards. 1997, reprinted with minor alterations 2000 and 2014.
- No. 10 **Sea Beans and Nickar Nuts** – E. Charles Nelson. 2000.
- No. 11 **Water-starworts (*Callitriche*) of Europe** – R. V. Lansdown. 2008,
- No. 12 **Fumitories of Britain and Ireland** – R. J. Murphy. 2009.
- No. 13 **Grasses of the British Isles** – Tom Cope & Alan Gray, 2009.
- No. 14 **Whitebeams, Rowans and Service Trees of Britain and Ireland** – Tim Rich, Libby Houston, Ashley Robertson & Michael Proctor, 2010.
- No. 15 **British Northern Hawkweeds** – Tim C. G. Rich & Walter Scott, 2011.
- No. 16 **Evening Primroses (Oenothera) of Britain and Ireland** – R.J. Murphy, 2016.

No17 **Violas of Britain and Ireland** – Michael Porter & Michael Foley, 2017.

Other BSBI publications

Hybrid Flora of the British Isles – C.A. Stace, C.D. Preston & D.A. Pearman, 2015. 510pp., with distribution maps and text for 909 hybrids recorded in the Wild, with notes on another 156 further hybrids, which are either erroneous or which might possibly occur here. Some colour photographs.

Alien Plants of the British Isles – E. J. Clement & M. C. Foster. 1994.

Alien Grasses of the British Isles –

T. B. Ryves, E. J. Clement & M. C. Foster. 1996, reprinted with addenda 2008.

Illustrations of Alien Plants of the British Isles – E. J. Clement, D. P. J. Smith & I. R. Thirlwell. 2005.

Plant Crib – T. C. G. Rich & A. C. Jermy. 1998, reprinted 2012 (with corrigenda included).

Change in the British Flora 1987–2004 (A report on the BSBI Local Change survey) – M. E. Braithwaite, R. W. Ellis & C. D. Preston, 2006.

50 Years of Mapping the British and Irish Flora 1962–2012 – Michael Braithwaite & Kevin Walker. 2012.

A Vascular Plant Red List for England – P. A. Stroh *et al.* 2014.

Threatened plants in Britain and Ireland – K.J. Walker, P.A. Stroh & R.W. Ellis. 2017

Atlas of British and Irish Brambles – A. Newton & R. D. Randall. 2004.

Atlas of British and Irish Hawkweeds – David McCosh & Tim Rich. 2011.

British Alpine Hawkweeds – David Tennant & Tim Rich. 2007.

List of Vascular Plants of the British Isles – D. H. Kent. 1992. Supplied with five errata lists. Three supplements (published 1996, 2000 and 2006) are also available.

Vice-county Census Catalogue of Vascular Plants of Great Britain, the Isle of Man and the Channel Islands – C. A. Stace, R. G. Ellis, D. H. Kent & D. J. McCosh (eds), 2003.

Botanical Links in the Atlantic Arc – S. J. Leach, C. N. Page, Y. Peytoureau & M. N. Sandford (eds), 2006.

Current Taxonomic Research on the British and European Flora – J. P. Bailey & R. G. Ellis (eds), 2006.

Other publishers' books

Scarce Plants in Britain – A. Stewart, D. A. Pearman & C. D. Preston (eds), 1994.

Aquatic Plants in Britain and Ireland – C. D. Preston & J. M. Croft. 1997.

New Atlas of the British & Irish Flora – C. D. Preston, D. A. Pearman & T. D. Dines (eds), 2002. Out of print, but *Online Atlas of the British & Irish Flora* (a joint project of the BSBI, Biological Records Centre and JNCC) can be accessed at http://www.brc.ac.uk/plantatlas/.

The Vegetative Key to the British Flora – John Poland & Eric Clement. 2009.